WAVELENGTH DIVISION MULTIPLEXING

WILEY SERIES IN PURE AND APPLIED OPTICS

Founded by Stanley S. Ballard, University of Florida

EDITOR: Glenn Boreman, University of North Carolina at Charlotte

A complete list of the titles in this series appears at the end of this volume.

WAVELENGTH DIVISION MULTIPLEXING

A Practical Engineering Guide

KLAUS GROBE
MICHAEL EISELT

WILEY

1

INTRODUCTION TO WDM

1.1 WDM THEORY

Wavelength division multiplexing (WDM) refers to a multiplexing and transmission scheme in optical telecommunications fibers where different wavelengths, typically emitted by several lasers, are modulated independently (i.e., they carry independent information from the transmitters to the receivers). These wavelengths are then multiplexed in the transmitter by means of passive WDM filters, and likewise they are separated or demultiplexed in the receiver by means of the same filters or coherent detection that usually involves a tunable local oscillator (laser).

WDM is an efficient means for increasing the transport capacity, or usable bandwidth, particularly of optical single-mode fibers. It also allows the separation of different customers' traffic in the wavelength (or optical frequency) domain and as such can be used as a multiple-access mechanism. The respective scheme is called wavelength-division multiple access (WDMA).

Modulated and multiplexed signals must be separated from each other or demultiplexed in order to be demodulated (otherwise, cross talk may appear). For separation, each pair of the respective signals must support orthogonality. For any two signals to be orthogonal, their scalar product must be zero:

$$(\underline{f},\ \underline{g}^*) = \underline{f}^{\mathrm{T}} \cdot \underline{g}^* = \sum_i f_i \cdot g_i^* \overset{!}{=} 0, \quad \text{with} \quad f_i \cdot g_i^* = \int_a^b f_i(x) \cdot g_i^*(x)\mathrm{d}x, \ i = 1, \cdots, N. \tag{1.1}$$

Wavelength Division Multiplexing: A Practical Engineering Guide, First Edition.
Klaus Grobe and Michael Eiselt.

(f, g^*) is the scalar product of complex functions, where * denotes complex conjugation. Equation (1.1) is also written for vector functions in order to be able to consider effects of orthogonally polarized signals.

The vanishing scalar product of the two signals is equivalent to a vanishing cross-correlation product or cross-correlation function (CCF). For the CCF, meaningful integration bounds must be considered, for example, integration over one symbol period. For optical WDM, the requirement (1.1) is easily fulfilled. Given that the different wavelength channels, including the Fourier transform-induced broadening due to the modulation, are properly spaced in the wavelength domain, any two different passbands of the WDM multiplexing (MUX) and demultiplexing (DMX) filters are orthogonal with respect to each other. In reality, Eq. (1.1) may not be achieved exactly, but only approximately due to linear or nonlinear cross talk.

WDM is the generalization of frequency-domain multiplexing that is long known from radio and coaxial transmissions. With a WDM channel, it can be combined with any other of the known electrical multiplexing or multiple-access schemes. These include electrical frequency-domain multiplexing, which is then referred to as subcarrier multiplexing (SCM), time-domain multiplexing (TDM), and code-domain multiplexing. One scheme of particular interest for both the multiplexing and multiple access is orthogonal frequency-domain multiplexing (OFDM), which can be applied within a wavelength channel or covering the optical frequencies of several wavelength channels. The respective multiple access schemes are time-domain multiple access (TDMA), subcarrier multiple access (SCMA), frequency-domain multiple access (FDMA), and code-domain multiple access (CDMA).

1.2 HISTORY OF WDM

The development toward commercial WDM transport systems as the common basis of all metropolitan area, regional, national, and international telecommunications networks was enabled by a number of relevant milestones:

- 1960: first laser developed [1]
- 1966: first description of dielectric waveguides as a potential means for data transmission by Kao and Hockham [2]
- 1970: first low-loss optical fiber produced (~20 dB/km) [3,4]
- 1976: first InGaAsP diode laser for 1300 nm window produced [5]
- 1978: first low-loss single-mode fiber produced (~0.2 dB/km) [6]
- 1978: first experimental WDM systems developed [7]
- 1987: first Erbium-doped fiber amplifier (EDFA) developed [8,9]
- 1995: first commercial WDM systems available

These milestones were accompanied by the development of ever-improved components (e.g., diode lasers for the 1550 nm window) and various types of single-mode fibers.

2

OPTICAL FIBER EFFECTS

2.1 LINEAR EFFECTS

Wavelength division multiplexing (WDM) transmission heavily depends on the fiber type that is used, and the related transmission effects and characteristics. Single-mode fibers are transmission systems that show frequency dependence (i.e., the pulse response is not the Dirac delta function), time variance, and weak nonlinear behavior. The resulting transmission impairments are as follows:

- *Linear Effects*
  - Attenuation (intrinsic loss, bending loss)
  - Polarization-dependent loss (PDL)
  - Chromatic dispersion (CD) due to material waveguide, and profile dispersion. These dispersion mechanisms lead to
    - Group-velocity dispersion (GVD)
    - Dispersion slope or higher order CD
  - Polarization-mode dispersion (PMD) including higher order PMD
- *Nonlinear Effects*
  - Self-phase modulation (SPM)
  - Cross-phase modulation (XPM, IXPM)
  - Four-wave mixing (FWM, IFWM)

Wavelength Division Multiplexing: A Practical Engineering Guide, First Edition.
Klaus Grobe and Michael Eiselt.

- Modulation instability
- Nonlinear phase noise (NLPN)
- Cross-polarization modulation (XPolM)
- Stimulated Raman scattering (SRS)
- Stimulated Brillouin scattering (SBS)

Detailed discussions of these effects can be found in the literature [1,2]. For WDM long-haul transmission, all these effects and their interactions have to be considered.

2.1.1 Attenuation

Attenuation in silica optical single-mode fibers is caused by different effects:

- Intrinsic loss of silica glass
  - Rayleigh scattering
  - Infrared absorption
- Extrinsic loss due to impurity of the silica glass
- Bending loss
  - Macrobending loss
  - Microbending loss
- Polarization-dependent loss

Attenuation in an optical transmission fiber can be described with a linear, frequency-dependent and (with the exception of PDL) time-invariant transfer function.

2.1.1.1 Intrinsic Loss Intrinsic loss in silica fibers leads, in lateral or z-direction, to an exponential power loss. This power loss can be described by a transfer function $H_a(z, \omega)$:

$$H_a(z, \omega) = e^{-\alpha_p(\omega) z}. \tag{2.1}$$

Here, ω is the angular frequency, and $\alpha_p(\omega)$ is the attenuation constant with dimension (1/m).

Rayleigh scattering is caused by density fluctuations that occur in microscopic regions with dimensions smaller than the optical wavelengths. They are unavoidably caused during manufacturing of the glass rod when a certain stochastic density distribution of the glass is fixed. These density fluctuations cause scattering of electromagnetic waves. Light being scattered out of the z-direction has a randomly distributed phase. Due to interference, the resulting light propagating perpendicular to the z-direction cancels almost exactly. Only a small fraction interferes constructively and is scattered out of the original (z) direction. This is the Rayleigh scattering loss. The light scattered in forward direction adds coherently without any further attenuation. This mechanism is also called *elastic scattering*.

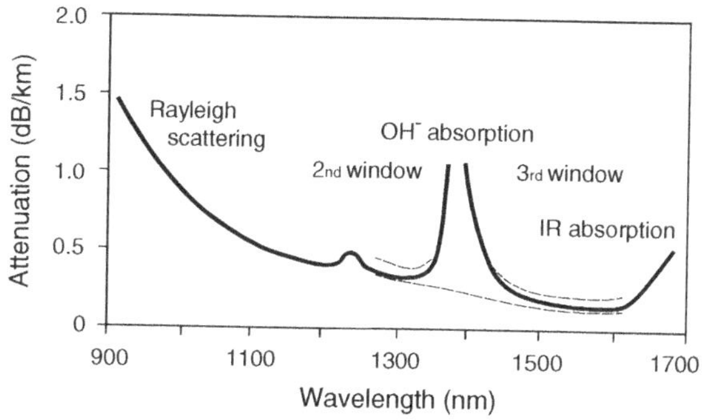

FIGURE 2.1 Spectral loss in single-mode fibers.

Rayleigh scattering is proportional to λ^{-4}, where λ is the optical wavelength. At 1550 nm, the attenuation caused by Rayleigh scattering is in the range of 0.12–0.16 dB/km. In this wavelength region, Rayleigh scattering is the dominant source of intrinsic loss [1].

Infrared absorption is caused by absorption through vibrational resonances in the infrared regime with wavelengths larger than 7 μm. In silica glass, there are also electronic resonances in the UV regime below 400 nm that do not contribute to loss significantly. Due to the amorphous structure of glass, the vibrational resonances form absorption bands that are spaced as far as into the regime of visible optical wavelengths. In the range of 800–1550 nm, the loss caused by infrared absorption is below 0.1 dB/km [1].

The spectral loss caused by intrinsic effects and impurity is shown in Fig. 2.1. The attenuation peak in the wavelength region around 1400 nm is caused by OH^- ions. In old fibers (vintages up into the 1980s), this peak was very pronounced, while in new fibers, it has been eliminated by means of better production methods (better clean-room environment). The dashed lines in Fig. 2.1 indicate the minimum and maximum loss tolerance as specified in the ITU-T Recommendation G.695 for standard single-mode fibers.

2.1.1.2 Fiber Bending Loss Fiber-optic waveguides not only exhibit loss due to intrinsic effects and impurity but also exhibit loss caused by perturbations of the ideal waveguide geometry. Such perturbations can result from bending the fiber, where bending radii with $R \gg \lambda$ and $R \approx \lambda$ with λ the optical wavelength have to be considered separately. Loss caused by bending radii $R \gg \lambda$ is called macrobending loss, for $R \approx \lambda$, microbending loss occurs. Both kinds of bending loss can result from incorrect handling of fibers like bad cabling with tight bends.

Macrobending loss occurs because part of the electromagnetic field that is normally guided in and near the core of the fiber is radiated out of the fiber in a bend. This happens because *phase planes* propagate in the fiber as long as the medium is homogenous and isotropic. The phase planes are orthogonal to the wave vector k. In a fiber bend, propagation is still based on phase planes that are orthogonal

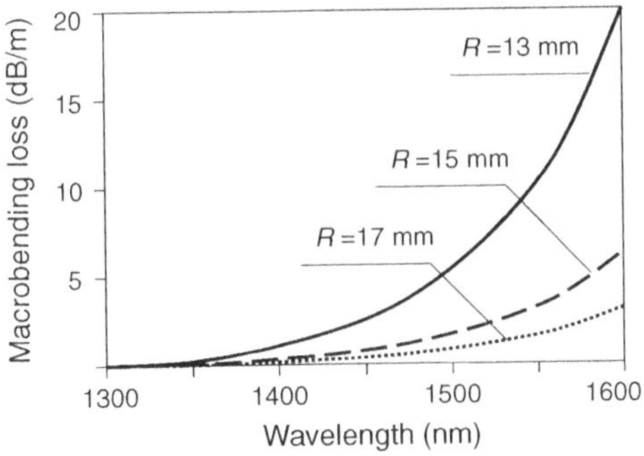

FIGURE 2.2 Macrobending loss in standard single-mode fibers (G.652A).

to the direction of propagation and hence follow the geometry of the bend. In order to maintain phase planes orthogonal to k, the planes now have to propagate at constant *angular* velocity. Those parts of the plane that are *outside* the middle axis of the fiber have to propagate at higher velocity than the ones *inside*. Along the middle axis of the fiber, the plane propagates with a velocity of $c_{\text{Co}} = c_0/n_{\text{Co}} = R \cdot \mathrm{d}\phi/\mathrm{d}t$, with c_{Co} the velocity of light in the core of the fiber, c_0 the velocity of light in vacuum, n_{Co} the refractive index of the core, R the bending radius, and $\mathrm{d}\phi/\mathrm{d}t$ the angular velocity. At a critical radius $R_{\text{C}} = n_{\text{Co}}/n_{\text{Cl}} \cdot R$, the maximum allowed velocity of $c_{\text{Cl}} = c_0/n_{\text{Cl}}$ is exceeded in the optical cladding of the fiber. Here, n_{Cl} is the refractive index of the optical cladding. Beyond R_{C}, no phase planes orthogonal to k are possible anymore, hence the respective parts of the field must be radiated out of the fiber and contribute to bending-induced loss.

Macrobending loss can be calculated using mode coupling or antenna theory [3–6]. It increases exponentially with increasing wavelength and decreasing bending radii. Typical macrobending loss for a single-mode fiber (according to ITU-T Recommendation G.652A) is shown in Fig. 2.2.

For wavelengths <1300 nm and bending radii >20 mm, macrobending loss can be neglected in most cases. The DWDM wavelength region around 1550 nm can already be affected significantly, for example, by sharp bends of patch cables. The region >1600 nm is very sensitive to macrobending. For these reasons, low-bending loss fibers have been developed more recently (they are specified in the ITU-T Recommendation G.657).

Microbending loss is caused when the fiber, including its buffer/coating, is subject to radial pressure. This can happen when a fiber is brought into a cable that is then put under radial mechanical pressure [7,8]. The pressure forces deformations of the boundary between core and cladding of the fiber that have similar dimension as the wavelength. This causes interference between light that is scattered out of its original direction and consequently loss. Microbending loss can be described as a stochastic process. If fibers are cabled correctly, it can usually be neglected. The effect, however, can be useful in certain fiber sensors (e.g., for detecting mechanical pressure).

2.1.1.3 Polarization-Dependent Loss WDM transmission systems with optical amplifiers and other passive components in the optical path are vulnerable to performance degradation due to PDL. A similar effect in active components is polarization-dependent gain (PDG) [9]. PDL can be caused by optical components such as directional couplers and isolators, while PDG is caused by polarization hole burning in optical amplifiers. Due to PDL and PDG, a polarized signal may be attenuated or amplified differently than unpolarized noise, and the signal-to-noise ratio (SNR) is modified accordingly. These SNR modifications lead to performance degradation. Since the state of polarization (SOP) of the transmitted light is randomly changed in time, PDL and PDG may cause the system performance to vary in a random manner as a function of time.

Distributed PDL and PDG in a high-PMD link also changes the characteristics of the PMD. The principal states of polarization (PSP) will not be orthogonal anymore. PMD-induced pulse spreading can become larger. Also, the distribution of the differential group delay (DGD) may not be Maxwellian anymore. This in turn can lead to severe under- or overestimation of DGD.

With a growing number of PDL elements, the probability density function (PDF) of PDL (in dB) approaches a Maxwellian distribution. The mean PDL then grows with increasing number of elements N like [10]

$$\langle \mathrm{PDL} \rangle = \mathrm{PDL}_i \cdot \sqrt{\frac{8}{3\pi} \cdot N}. \tag{2.2}$$

Effects of PDL and PDG are difficult to be quantized due to their interaction with PMD. In long transmission links with a large number of PDL-prone components, a PDL-related penalty must be taken into consideration. An example for the PDL penalty for an ultralong-haul link is given in Fig. 2.3 [9].

For this example, a transmission link of almost 9000 km total length with 270 EDFAs has been considered.

2.1.2 Chromatic Dispersion

CD causes signal distortions through the dependence of the velocity of propagation (the group velocity) on the frequency of the respective spectral components. Since every information transmission requires at least Fourier-limited spectral broadening

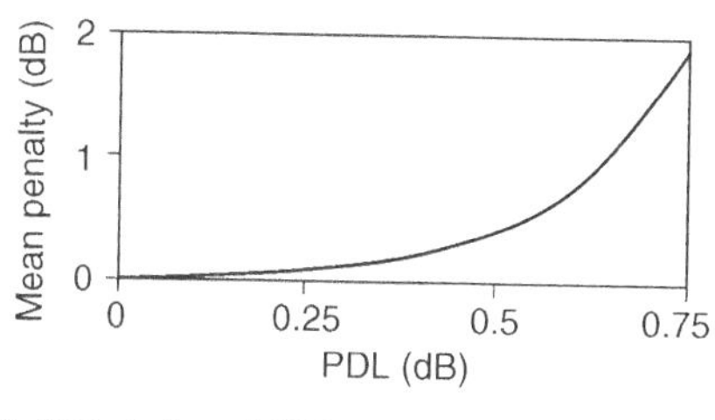

FIGURE 2.3 PDL-induced SNR penalty in long-haul transmission.

due to the respective modulation of the carrier wave, every information transmission in optical fibers is subject to CD.

CD can be described by a linear, (mostly) time-invariant system with memory. Since different spectral components of a signal, that is, a pulse carrying digital information, propagate at different velocities, the resulting effect is a temporal spread of the signal. In digital transmission, this spread leads to intersymbol interference (ISI). The effect is also known as intramode dispersion [11]. It can further be split into

- material dispersion,
- waveguide dispersion,
- profile dispersion.

Material dispersion results from interactions of a propagating electromagnetic wave with the bonding electrons of the surrounding matter. It depends on the frequency ω of the wave and can be described as the frequency dependence of the refractive index $n(\omega)$. $n(\omega)$ can be derived from the Sellmeier series that considers the relevant resonances that are responsible for the wave-matter interactions [2]:

$$n^2 = 1 + \sum_{i=1}^{3} B_i \cdot \frac{\omega_i^2}{\omega_i^2 - \omega^2}. \tag{2.3}$$

The resonance frequencies ω_i are weighted with weights B_i. The relevant values for silica fibers are summarized in Table 2.1. Here, $\lambda_i = 2\pi \cdot c/\omega_i$ and c are the respective wavelength and the speed of light in silica glass, respectively.

Material dispersion can also be assigned a material dispersion parameter D_{mat}:

$$D_{\mathrm{mat}} = -\frac{\lambda}{c}\frac{\mathrm{d}^2 n}{\mathrm{d}\lambda^2}. \tag{2.4}$$

Here, n is the refractive index of the bulk material. The negative sign indicates that low-wavelength components arrive before higher wavelength components

Waveguide dispersion appears because the light waves are guided in a waveguide with certain geometrical shape and dimensions. The effect is caused by the dependence of the eigenvalues of the wave propagation equation on the relation between waveguide cross-sectional dimensions and wavelengths of the propagating signal. Compared to material dispersion, waveguide dispersion is a weak effect but as compared to bulk silica, it shifts the wavelength region of lowest CD toward longer

TABLE 2.1 Sellmeier Coefficients

$B_1 = 0.6961663$	$\lambda_1 = 0.0684043\ \mu\mathrm{m}$
$B_2 = 0.4079426$	$\lambda_2 = 0.1162414\ \mu\mathrm{m}$
$B_3 = 0.8974794$	$\lambda_3 = 9.8961610\ \mu\mathrm{m}$

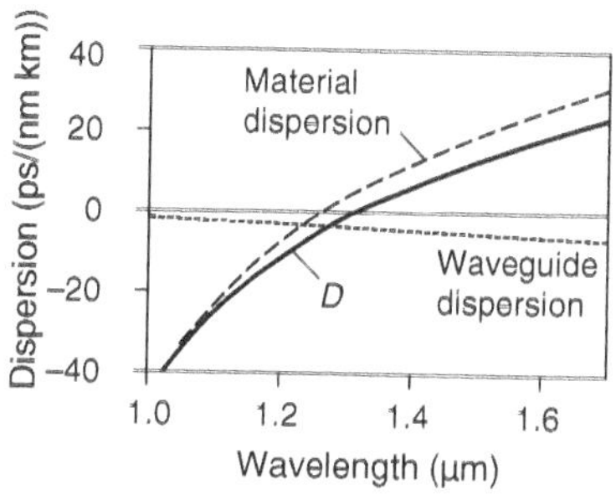

FIGURE 2.4 Waveguide and material dispersion.

wavelengths. Lowest-dispersion wavelengths are $\lambda_0 \approx 1270\,\text{nm}$ for bulk silica and $\lambda_0 \approx 1310\,\text{nm}$ for standard single-mode fibers [2]. This effect is shown in Fig. 2.4.

Profile dispersion describes the dependence of the radial refractive index profile $n(r)$ on the wavelength. $n(r)$ is determined by the refractive indices of the core and its cladding(s), and the core diameter. It depends on the wavelength because core and cladding(s) consist of differently doped glass with different dependencies of the refractive indices on wavelength [12]. It is the weakest of the chromatic dispersion effects.

In the absence of nonlinearity and birefringence, propagation of a light wave with envelope E is given by

$$E(z,t) = E_0\,\mathrm{e}^{j(\omega t - kz)} = E_0\,\mathrm{e}^{-\alpha z}\,\mathrm{e}^{j(\omega t - \beta z)}. \tag{2.5}$$

The effect of chromatic dispersion is described by the complex wave number $k(\omega)$ that splits into attenuation constant $\alpha(\omega)$ and phase constant $\beta(\omega)$:

$$k(\omega) = \beta(\omega) - j\alpha(\omega). \tag{2.6}$$

For practical calculations, $\alpha(\omega)$ is often considered a constant. Alternatively, it can be calculated via the Sellmeier series [1], or via Hilbert transform of $\beta(\omega)$. The phase constant is usually developed into a Taylor series around a mean carrier frequency ω_0, for example, 1550 nm:

$$\beta(\omega) = \beta_0 + \beta_1(\Delta\omega) + \frac{\beta_2}{2!}(\Delta\omega)^2 + \frac{\beta_3}{3!}(\Delta\omega)^3 + \cdots, \tag{2.7}$$

$\Delta\omega$ is given by $\omega - \omega_0$. The Taylor coefficients β_i can be derived as

$$\beta_i = \left.\frac{\mathrm{d}^i\beta(\omega)}{\mathrm{d}\omega^i}\right|_{\omega_0}. \tag{2.8}$$

From the Taylor series, some velocities and dispersion parameters can be derived.

The *phase velocity* is given by $v_\mathrm{P} = \omega_0/\beta_0 = c_0/n$. It is irrelevant for information transmission since it relates to an unmodulated (monochromatic) wave. Under certain conditions, the phase velocity is also allowed to exceed the vacuum speed of light c_0 (which is forbidden for any information–transmission velocity).

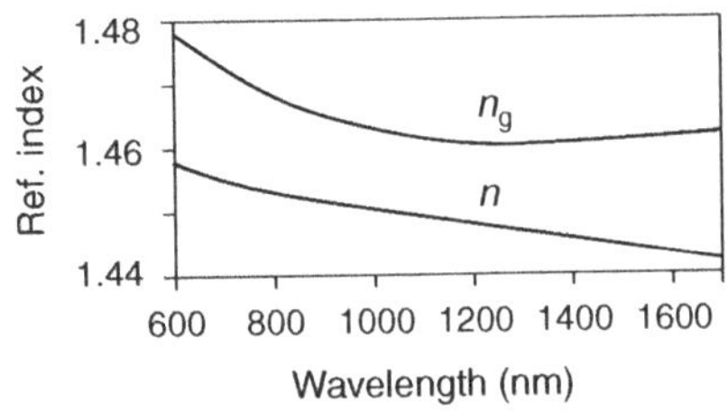

FIGURE 2.5 Fiber group index.

The coefficient β_1 leads to the *group velocity* v_g and the group index n_g. The group index is inverse proportional to v_g that is the relevant parameter for the velocity of the information transfer in media with *normal* dispersion (i.e., $dn/d\omega > 0$, and $0 < v_g < c_0$). However, under certain conditions of *anomalous* dispersion, the group velocity may be greater than the speed of light or even negative and then does not represent the velocity of energy flow or information transmission. In this case, information–transmission velocity is given by the so-called group-front velocity that is smaller than c_0 [13,14].

The group velocity is given through the Taylor coefficient β_1:

$$v_g = 1/\beta_1 = c_0/n_g. \tag{2.9}$$

The group velocity is also given through the fiber group index that is shown in Fig. 2.5.

The next coefficient, β_2, describes chromatic group-velocity dispersion. For the description of transmission systems with up to ~10 Gb/s, this coefficient is sufficient. If transmission at very high bandwidths—ultrahigh-speed TDM and/or WDM—must be calculated, the next coefficient, β_3, has to be considered as well. β_3 describes the dispersion slope of the respective fibers, or higher order dispersion.

β_2 is often described via the dispersion parameter D [1,2]:

$$D = -\omega^2\beta_2/2\pi c_0. \tag{2.10}$$

This parameter has dimension [ps/(nm km)]. It describes the temporal spread (in ps) of signal pulses with a certain optical bandwidth (in nm) over a certain transmission distance (in km).

Figure 2.6 lists D parameters for relevant single-mode fiber types (according to ITU-T Recommendations G.652, G.653, G.655), as well as brands of fibers (G.655 TW-RS®, G.655 LEAF®, SMF-LS™ NZ-DSF).

For almost all single-mode fibers, the D parameter crosses zero somewhere between 1300 and 1550 nm, that is, in the wavelength region of interest. Around that particular wavelength, fibers have low chromatic dispersion. The zero-crossing also splits the spectrum into two distinct domains: smaller wavelengths exhibit normal dispersion (negative D, positive β_2), and wavelengths above the zero-crossing have

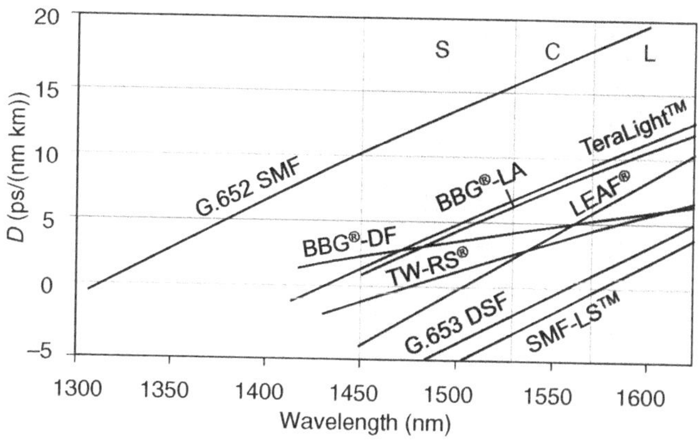

FIGURE 2.6 *D* parameters of various single-mode fibers.

anomalous dispersion (positive *D*). For normal dispersion, spectral components with *higher* frequencies (blue-shifted) are propagating *slower* than those at lower frequencies (red-shifted). For anomalous dispersion, blue-shifted components are traveling faster than red-shifted components.

The *D* parameter can be used to define a linear transfer function that considers CD:

$$H_{\mathrm{CD}}(z, j\omega) = \exp\left[j\left(\beta_0 + \beta_1(\omega - \omega_0) - \frac{1}{2}\frac{\lambda_0^2}{2\pi c_0}D(\omega - \omega_0)^2\right)z\right]. \tag{2.11}$$

Propagation of a slowly varying pulse envelope $A(\omega, l)$, is then given by

$$A(\omega, L) = \exp\left(\frac{j}{2}\cdot\beta_2\omega^2 L\right)A(\omega, 0). \tag{2.12}$$

The transfer function $H_{\mathrm{CD}}(z, j\omega)$, together with a suitable consideration of the attenuation (which may be a constant attenuation coefficient), can be used for calculations or simulations of a linear fiber model that neglects the nonlinear effects.

The *D* parameter can be used for a simplified approximation of the bandwidth that can be utilized for a transmission fiber link with given length. The 3 dB bandwidth $f_{3\,\mathrm{dB}}(z)$ is

$$f_{3\,\mathrm{dB}}(z) \approx \frac{3}{4|D|z\,\Delta\lambda}. \tag{2.13}$$

Here, $\Delta\lambda$ is the optical bandwidth. Equation (2.13) allows an approximation of the maximum transparent field length (i.e., the maximum distance between two terminations or 3R-regenerators, and without CD compensation) for transmission at a given bit rate and using a given modulation scheme, for example, NRZ-OOK with direct detection. The corresponding GVD-limited maximum link length depending on the bit rate is shown for NRZ-OOK in Fig. 2.7.

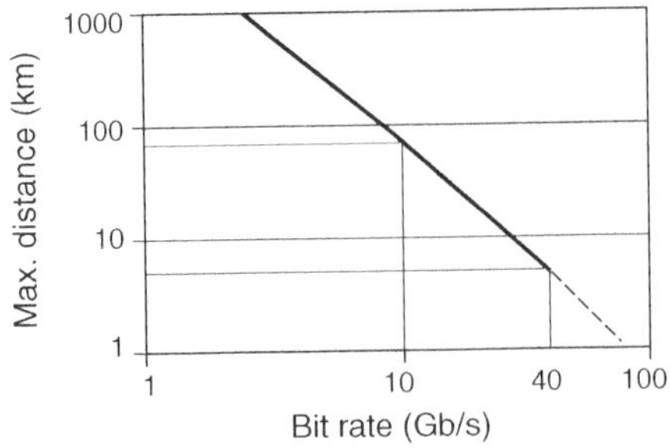

FIGURE 2.7 Maximum distance limitation caused by uncompensated GVD on G.652 fibers.

It is obvious from Fig. 2.7 that simple on–off keying (OOK) is not sufficient for very high bit rates in excess of 10 Gb/s. However, OOK can still be used for serial short-reach interfaces at 40 Gb/s that aim at a maximum reach of 2 km, and similar applications.

2.1.3 Polarization-Mode Dispersion

2.1.3.1 PMD and DGD Polarization-mode dispersion is caused by birefringence of the transmission fibers, that is, the effective refractive index is different between two orthogonal polarizations. This birefringence, in turn, is caused by fiber geometry imperfections or by lateral stress on the fiber, leading to two orthogonally polarized (OP) modes propagating at slightly different velocities [15]. The geometry imperfections occur during fiber manufacturing and cabling of the fibers (i.e., bringing the fibers into a cable), and also during deployment of the fiber-optic cables. They cannot be fully suppressed during manufacturing and deployment.

While in a perfectly circular fiber no particular pair of orthogonal polarization modes is distinguished, birefringence leads to two particular modes standing out. These modes are referred to as the principal states of polarization. They usually depend on the frequency (or wavelength) of the optical signal. The PSPs are defined at the input to the fiber as those states of polarization, for which, when slightly varying the signal frequency, the output polarization remains constant. Even if a single polarization mode is excited at the transmitter—like in a typical laser—this mode not necessarily coincides with a fiber PSP such that both PSPs are excited within the fiber.

Over a given transmission length, the time delay between both PSPs is given by the DGD. The PSPs as well as the DGD are wavelength dependent and also vary over time with changing environmental impact. As the statistics over time and wavelength theoretically follow the same Maxwellian distribution, the average value of the DGD over time and over wavelength are equal. This number is commonly referred to as the PMD of a fiber span. The effect of PMD is visualized in Fig. 2.8.

Polarization and its related effects are represented by the Stokes parameters and can be visualized on the Poincaré sphere. The relationship of the Stokes parameters (as components of the Stokes vector $\vec{S}$) to intensity and polarization ellipse

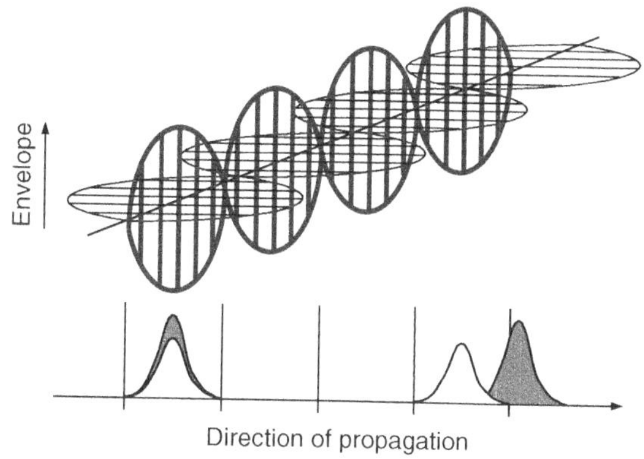

FIGURE 2.8 Intersymbol interference caused by PMD.

parameters is given by

$$\begin{aligned} S_0 &= I = |E_x|^2 + |E_y|^2, \\ S_1 &= Q = |E_x|^2 - |E_y|^2, \\ S_2 &= U = 2\text{Re}\ (E_x E_7^*), \\ S_3 &= V = 2\text{Im}\ (E_x E_7^*). \end{aligned} \tag{2.14}$$

Here, I is the total intensity. The four Stokes parameters S_0, S_1, S_2, and S_3 are sometimes also denoted as I, Q, U, and V, respectively.

The degree of polarization (DOP) is the portion of the optical signal that is polarized. Using the Stokes parameters, the DOP is given by

$$\text{DOP} = \sqrt{\frac{S_1^2 + S_2^2 + S_3^2}{S_0^2}} = \sqrt{s_1^2 + s_2^2 + s_3^2}. \tag{2.15}$$

A graphical representation of the Poincaré sphere together with the S_1-, S_2-, S_3-axes and the related SOP is shown in Fig. 2.9. The Poincaré sphere describes orientation and rotation of polarized light.

A long optical fiber can be modeled as a cascade of many pieces of birefringent sections with a random polarization orientation. For a particular frequency, for the cascade of these birefringent sections, two PSPs and a DGD between these PSPs can be calculated. The two PSPs are randomly oriented and the DGD is a random variable, depending on the alignment of individual polarized fiber pieces. Time varying external stress changes both, the PSP and DGD.

The PMD vector can now be stated:

$$\vec{\Omega} = \Delta\tau \cdot \vec{q}, \tag{2.16}$$

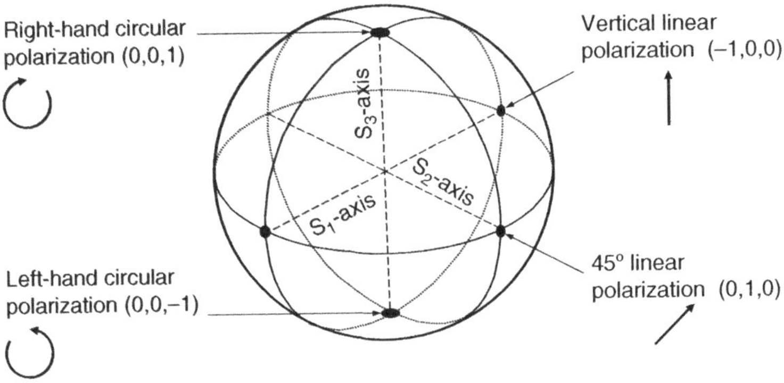

FIGURE 2.9 Graphical representation of states of polarization on Poincaré sphere.

$\Delta\tau$ is the DGD (first-order PMD) and $\vec{q}$ is a unit vector, which points in the direction of the fast principal axis.

The effect of PMD can now be described as a rotation of the Stokes vector $\vec{S}$ around the PSP axis by an angle dependent on the optical frequency and proportional to the DGD.

$$\Delta\tau = \frac{\Delta\phi}{\Delta\omega}. \tag{2.17}$$

This rotation is visualized in Fig. 2.10.

PMD results in pulse broadening ΔT^{PMD}, which can be explained either by the different propagation delays between the PSPs as shown in Fig. 2.8, or by the depolarization of the signal's spectral components, leading to an effectively narrower signal spectrum.

The PMD of a long fiber is a function of the fiber length L and the so-called PMD parameter D_{PMD}:

$$\Delta T^{\mathrm{PMD}} = D_{\mathrm{PMD}}\sqrt{L}. \tag{2.18}$$

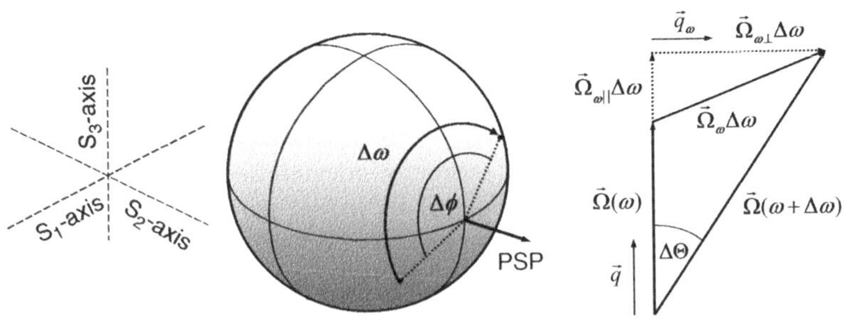

FIGURE 2.10 Rotation of Stokes vector around PSP.

For optical paths with M fiber spans, the resulting total PMD yields

$$\Delta T_{\text{Total}}^{\text{PMD}} = \sqrt{\sum_{k=1}^{M} \left(\Delta T_k^{\text{PMD}}\right)^2} = \sqrt{\sum_{k=1}^{M} D_{\text{PMD}}^2(k) L(k)}. \qquad (2.19)$$

The PMD parameter D_{PMD} has the dimension (ps/km$^{1/2}$). The ITU-T Recommendations G.652A/C, G.653, and G.655A/B state a maximum of $D_{\text{PMD}} < 0.5$ ps/km$^{1/2}$, the Recommendations for newer fiber types G.652B/D, G.655C, G.656, and G.657 allow for 0.2 ps/km$^{1/2}$. Latest fiber brands, for example, OFS FITEL AllWave®, are specified for ≤ 0.1 ps/km$^{1/2}$, and standardization of further fiber types (G.652E) is likely to go the same direction.

While the fiber parameter PMD is an average over time or wavelength, the actual pulse broadening in an optical signal is related to the instantaneous DGD and also depends on the actual SOP of the signal relative to the current PSPs. The pulse broadening is therefore a stochastic value, changing in time. The probability of particular DGD values follows a Maxwellian distribution function, as shown in Fig. 2.11.

The *average* DGD is referred to as PMD. As most transmission systems are impacted by the instantaneous DGD, for those systems a DGD tolerance is specified. Since values of DGD > 3 PMD only occur with a probability of 4×10^{-5} (equaling ~20 min per year), the specified tolerance of most transmission systems with respect to PMD is usually given as a factor of 3 smaller than the DGD tolerance, to yield sufficiently low outage probability due to PMD.

In order to avoid severe PMD penalties on the receive-end SNR of a transmission system, PMD-induced broadening should be less than a fraction of the bit (or symbol) period:

$$\Delta T_{\text{Total}}^{\text{PMD}} < \varepsilon T_{\text{B}}. \qquad (2.20)$$

Here, T_{B} is the bit duration. It has to be replaced by the symbol duration for higher order modulation schemes. The fraction ε also depends on the pulse shape, that is, NRZ versus RZ. Table 2.2 lists ε values for both NRZ and RZ and also for different penalties.

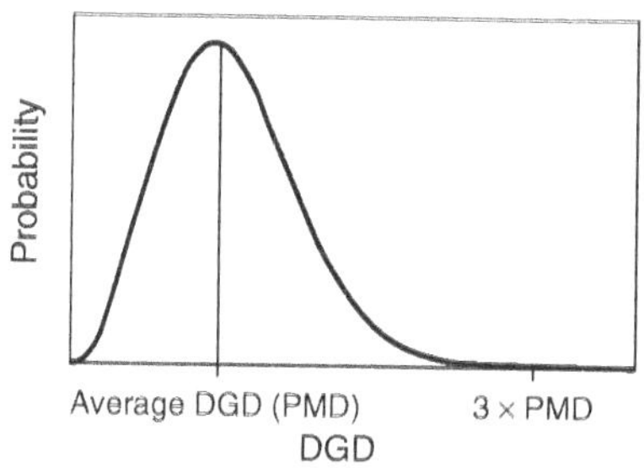

FIGURE 2.11 Distribution function for DGD.

TABLE 2.2 Tolerable Bit-Period Fraction and Induced System Penalty

System Penalty Versus Worst-Case DGD	1 dB	2 dB	3 dB
NRZ	$0.30T_B$	$0.45T_B$	$0.50T_B$
RZ	$0.45T_B$	$0.55T_B$	$0.65T_B$

In order to derive the PMD transfer function, one needs the field representation with two orthogonal unit vectors for the input field $\hat{\varepsilon}_{a+}$ and $\hat{\varepsilon}_{a-}$ according to Fig. 2.12.

Input field $\vec{E}_{\text{in}}$ and output field $\vec{E}_{\text{out}}$ can now be written as

$$\begin{aligned}\vec{E}_{\text{in}} &= E_0(t)\left(\sqrt{\gamma}\hat{\varepsilon}_{a+} + \sqrt{1-\gamma}\hat{\varepsilon}_{a-}\right),\\ \vec{E}_{\text{out}} &= E_0(t-\tau_0)\sqrt{\gamma}\hat{\varepsilon}_{b+} + E_0(t-\tau_0-\Delta\tau)\sqrt{1-\gamma}\hat{\varepsilon}_{b-}.\end{aligned} \tag{2.21}$$

Here, γ is the power splitting ratio between the two orthogonal PSPs, and $\Delta\tau$ is the DGD, respectively. From this, the PMD transfer function can be derived:

$$H(jf) = \sqrt{\gamma}\hat{\varepsilon}_{b+} + \sqrt{1-\gamma}\,\exp\left[-j2\pi f\,\Delta\tau\right]\hat{\varepsilon}_{b-}. \tag{2.22}$$

If the duration of the bits or symbols (for higher order modulation) becomes comparable to DGD, higher order PMD must be taken into account. This considers the frequency dependence of both, the PSPs and DGD.

Second-order PMD (SOPMD) is defined as the frequency dependence of the PMD vector:

$$\frac{\mathrm{d}\vec{\Omega}}{\mathrm{d}\omega} = \frac{\mathrm{d}\,\Delta\tau}{\mathrm{d}\omega}\cdot\vec{q} + \Delta\tau\cdot\frac{\mathrm{d}\vec{q}}{\mathrm{d}\omega}. \tag{2.23}$$

The first term, $(\mathrm{d}\Delta\tau/\mathrm{d}\omega)\vec{q}$, is referred to as Polarization-dependent Chromatic Dispersion, PCD [10]. Since $\vec{q}$ is a unit vector, its magnitude is given by $|\Delta\tau_\omega|$ where the subscript ω denotes derivation with respect to frequency. The second term is called depolarization because it is orthogonal to the PSP $\vec{q}$, refer to Fig. 2.10. Note that in some references [16], PCD is defined as

$$\text{PCD (s/m)} = \left(-\frac{\pi c_0}{\lambda^2}\right)\cdot\frac{\mathrm{d}(\Delta\tau)}{\mathrm{d}\omega}. \tag{2.24}$$

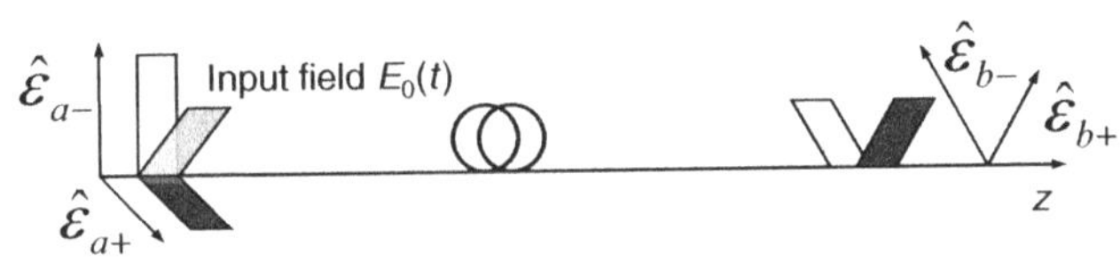

FIGURE 2.12 PMD transfer function.

TABLE 2.3 Relative RMS Magnitudes of Higher Order PMD Taylor Series Terms

	DGD	Second Order	Third Order	Fourth Order	Fifth Order
$\Delta\omega = 0.5\Delta\omega_{PSP}$	1	0.25	0.06	0.014	0.003
$\Delta\omega = \Delta\omega_{PSP}$	1	0.49	0.24	0.11	0.045
$\Delta\omega = 2\Delta\omega_{PSP}$	1	0.98	0.97	0.87	0.72

It then contributes to an effective dispersion $(\mathrm{DL})_{\mathrm{eff}}$ over fiber length L by

$$(\mathrm{DL})_{\mathrm{eff}} = \mathrm{DL} \pm \mathrm{PCD}. \tag{2.25}$$

Here, the plus and minus signs correspond to the two PSPs, and D is the well-known chromatic dispersion parameter.

The relation between symbol duration and DGD has severe impact on the relevance of the higher order PMD terms. According to Ref. [17], the relative RMS magnitudes of the higher order terms of the Taylor series expansion of the PMD vector listed in Table 2.3 result.

Generally, the distortion of a signal degraded by first- and second-order PMD and the corresponding outage probability OP can be quantified by numerical integration over a set of five parameters as reported in Ref. [18]:

$$\mathrm{OP} = \underset{\mathrm{Penalty}>\mathrm{Limit}}{\int\int\int\int\int} p_j^* \, \mathrm{d}|\vec{\Omega}|\mathrm{d}|\vec{\Omega}'|\mathrm{d}\theta \, \mathrm{d}\phi \, \mathrm{d}\xi. \tag{2.26}$$

The measurement and calculation of the OP can be accelerated by a reduction of the PMD emulation to only two parameters. First, scrambling of the SOP is performed during BER acquisition to eliminate the need to set the two angles θ and ϕ describing the orientation relative to the PSP vector. Second, the mutual orientation ξ between the first- and second-order PMD vectors Ω and Ω_ω is set to 90° in the PMD emulator. This corresponds to a worst-case distortion of the second-order PMD (maximum depolarization and no PMD-induced chromatic dispersion). This reduction leads to the two-dimensional PDF p_j by the integration of p_j^* over the angles θ, ϕ, and ξ. Thus, the outage probability can be expressed by

$$\mathrm{OP} = \underset{\mathrm{Penalty}>\mathrm{Limit}}{\iint} p_j \mathrm{d}|\vec{\Omega}|\mathrm{d}|\vec{\Omega}'| \equiv \underset{\mathrm{Penalty}>\mathrm{Limit}}{\iint} p_j \, \mathrm{dDGD} \, \mathrm{d}|\mathrm{SOPMD}|. \tag{2.27}$$

Finally, the outage probability is the numerical integration of the PDF p_j over the area with the OSNR penalty exceeding the system limit. The two integration parameters are the DGD in ps and the SOPMD in ps/GHz or ps^2, respectively.

Figure 2.13 shows the calculated OP derived from measurement of the two-dimensional distortion plane for 43 Gb/s NRZ-OOK. The X-axis indicates the level of both first- and second-order PMD. An OSNR penalty limit of 2 dB has been

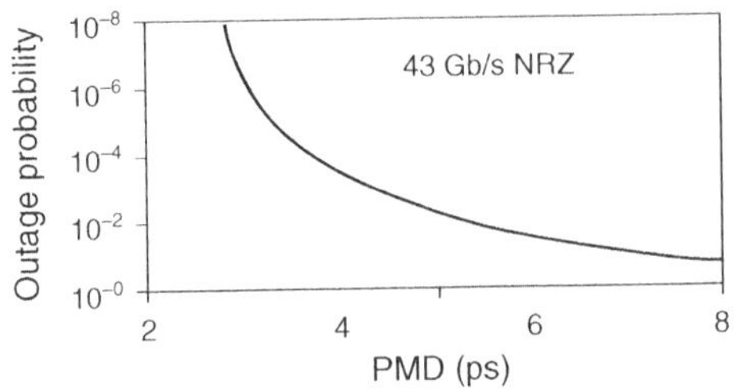

FIGURE 2.13 Outage probability versus PMD at serial 43 Gb/s NRZ-OOK.

applied. If the outage probability that is considered acceptable is 10^{-5}, then the maximum PMD tolerance at 2 dB of OSNR penalty is 2.7 ps.

Depending on the fiber characteristics, PMD can lead to severe limitations of the maximum transmission distance. Since analog optical PMD compensation is difficult at best, PMD can pose the ultimate limit in incoherent, uncompensated high-speed transmission. In incoherent 10 Gb/s NRZ-OOK transmission, the PMD tolerance is ~10 ps for <1 dB OSNR penalty. This limits PMD (in ps) to $0.1T_B$ as stated in Table 2.2. With other symbol rates in uncompensated systems using the same modulation, the PMD tolerance develops as shown in Fig. 2.14.

Transmission systems can tolerate large DGD values (up to 0.65 × symbol duration, see Table 2.2); however, the averaged PMD has to be lower. For incoherent (nondigitally compensated) high-speed transmission beyond 10 Gb/s per channel, PMD and its compensation must be considered closely. With modulation format and pulse shaping kept constant, the resistance against PMD decreases by a factor of 4 when quadrupling the symbol rate.

PMD compensation techniques are discussed, for example, in Refs. [19,20].

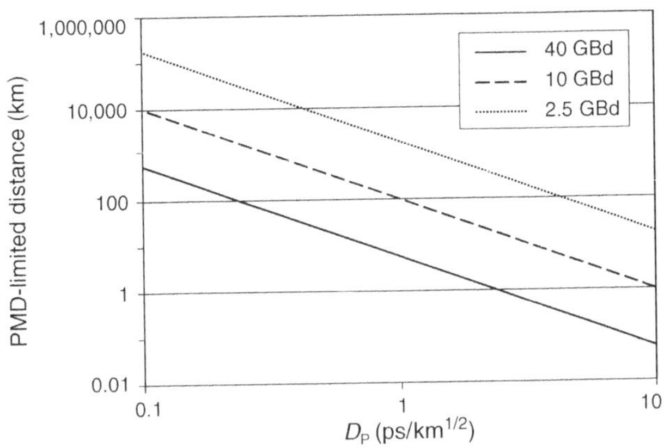

FIGURE 2.14 Distance limitation for NRZ through uncompensated PMD.

2.1.3.2 PMD in Recirculating Loops Due to its statistical nature, PMD is of particular interest in recirculating loops. Recirculating loops are periodic transmission links where signals propagate through a closed loop a certain number of circulations. They are coupled into the loop via a directional coupler, and after the requested circulation number has been achieved, the loop is opened via a fast switch and the signal is directed toward the receiver. Recirculating loops can hence be used for ultralong-haul transmission experiments in laboratories without providing hundreds of amplifiers and thousands of kilometers of transmission fiber. Since PMD is relevant for ultrahigh-speed and long-haul transmission, it must be considered in recirculating loops. The problem, as compared to an installed real-world link, is the periodic characteristic of recirculating loops. Since the circulations are not statistically independent of each other with regard to PMD, recirculating loops do not necessarily lead to correct results with regard to PMD. The basic recirculating loop setup is shown in Fig. 2.15.

Using PMD concatenation [10], the development of the mean DGD with the number of circulations is given by

$$\langle \mathrm{DGD} \rangle = \mathrm{DGD_{Loop}} \cdot \sqrt{\frac{8}{3\pi} \cdot \mathrm{No_{Circulations}}}. \qquad (2.28)$$

After 40 circulations, mean DGD of $\langle \mathrm{DGD} \rangle = \mathrm{DGD_{Loop}} \cdot 5.83$ results. In order to emulate DGD in a recirculating loop correctly, it is necessary to implement an interloop polarization decorrelator [21]. Without such a decorrelator, no Maxwellian DGD distribution results even if multisection PMD emulators are used. With polarization decorrelation, simple (even single stage) PMD emulation in the loop can be used.

2.1.3.3 Real-World Fiber Plant PMD Audits It has been shown in several extensive audits (measurements) that even fibers that were installed after Y2000 can produce PMD in excess of 1 ps/km$^{1/2}$. The first of these audits [22], issued in 1997 by Bellcore and conducted on more than 1000 installed fibers, showed that G.655 fibers installed in the mid-1990s were prone to high PMD (see Fig. 2.16). More than 20% of all fibers showed PMD with $D_P > 1$ ps/km$^{1/2}$.

The vintages of the different fiber types and brands are listed in Table 2.4 for relevant fibers.

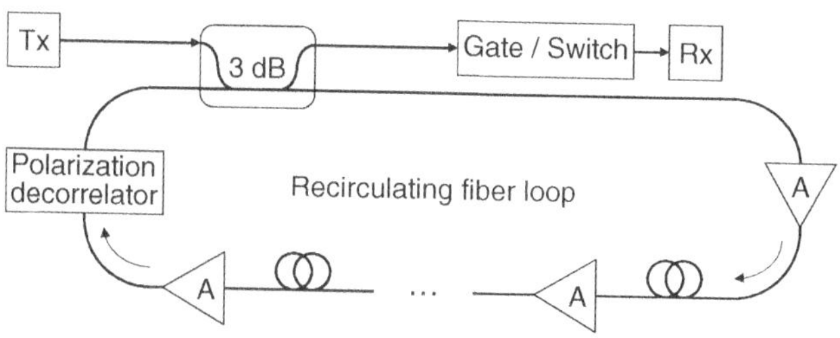

FIGURE 2.15 Simplified representation of a recirculating loop.

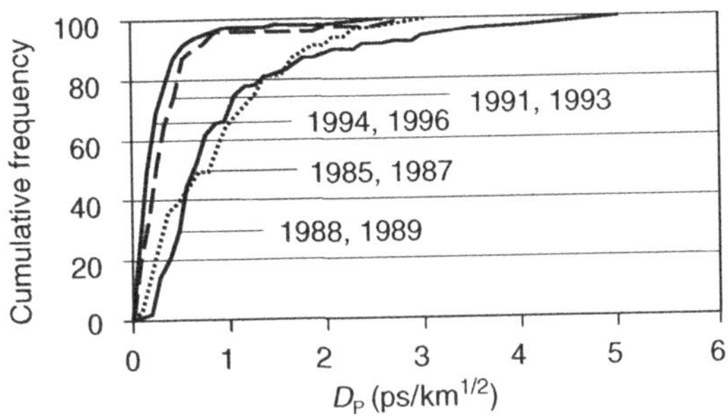

FIGURE 2.16 PMD characteristics as a function of the year of installation of the fibers [22].

TABLE 2.4 Vintages of SMF, DSF, and NZ-DSF

Fiber Type	Vintage	Fiber Type	Vintage
SMF	1980–	TrueWave® Classic	1992–1996
LEAF®	1996–	DSF	1990–1992
TrueWave® RS	1996–	SMF-LS™	1992–1995

The audit also showed that severe differences exist between fibers of different manufacturers (i.e., different manufacturing techniques). On the other hand, it also showed that only minor differences exist between buried fibers, aerials, and optical ground wires (OPGW).

In addition to the Bellcore study, a newer Brazilian audit issued in early 2005 showed that even fibers installed in the decade after Y2000 can have very high PMD [23]. This audit was conducted on more than 2000 installed fibers, covering different fiber types, manufacturers, and cabling techniques (buried vs. OPGW). Results are shown in Fig. 2.17 that summarizes the cumulative frequency of the PMD parameter D_P as a function of the year of installation of the respective fibers. Here, more than 20% of all fibers were exceeding 0.5 ps/km$^{1/2}$.

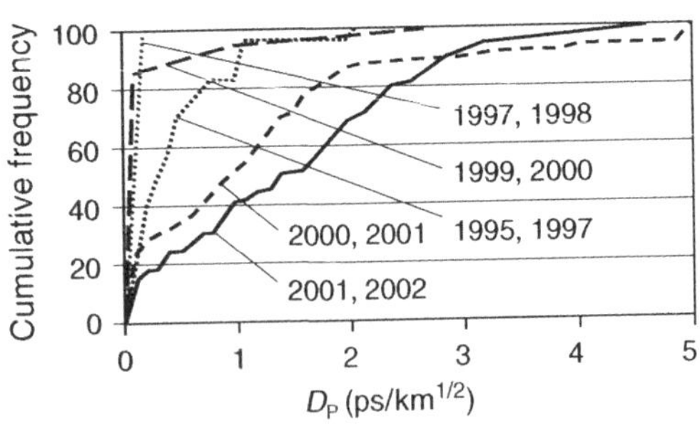

FIGURE 2.17 PMD characteristics as a function of the year of installation of the fibers [23].

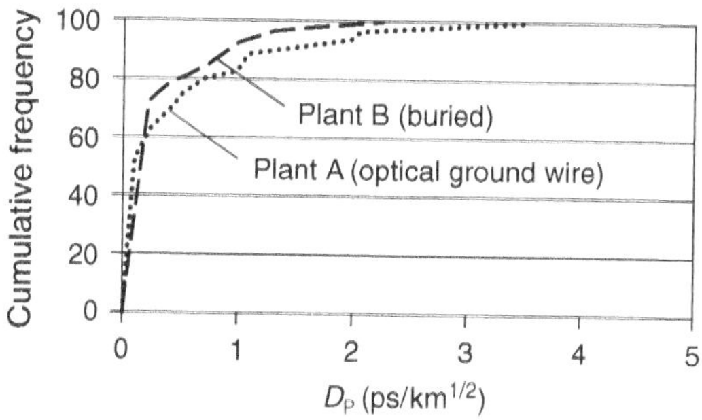

FIGURE 2.18 PMD characteristics as a function of type of installation of the fibers [23].

The Brazilian audit confirmed that the type of cabling (buried vs. OPGW) is of minor importance. This result is shown in Fig. 2.18. The figure contradicts the assumption that PMD can be caused by wind loads working on OPGWs. The audit also confirmed that there is a strong correlation between high PMD and certain manufacturing techniques.

For ultralong-haul, high-speed transmission, PMD should be measured on those fibers that are to be used, not other fibers within the same cable. For such transmission, either low-PMD fibers (e.g., G.652D) or highly tolerant transmission schemes (i.e., coherent intradyne modulation with digital equalization) have to be used.

2.2 NONLINEAR FIBER EFFECTS

Silica fibers exhibit weak, cubic nonlinearity. Typically, this nonlinearity only becomes apparent at high (per-channel and accumulated) power levels and for signal propagation over very long transparent links (without 3R regeneration, that is, only optically amplified). The latter is achieved in all long-haul WDM transport, and it has been shown that the fiber capacity is ultimately limited by this nonlinearity [24].

The cubic fiber nonlinearity fundamentally splits into two classes of effects, Kerr effects that are caused by a nonlinear refractive index n_2, and (stimulated, that is, laser-like) scattering effects that are caused by interactions with molecular or acoustic phonons [25]. The Kerr coefficient n_2 is described as a memory-less system, leading to instantaneous nonlinear reaction covering the whole WDM spectrum of interest. The scattering effects exhibit, via the related phonons, memory (or pulse response), consequently they also have dedicated spectral dependencies.

Overviews of fiber nonlinearity are given, for example, in Refs [2,26–31].

2.2.1 Kerr Effects

The Kerr effects can be separated into inter (WDM)-channel and intrachannel effects, and also into effects caused by signal–signal or signal–noise interaction, respectively. The resulting Kerr effects are classified in Fig. 2.19 [32].

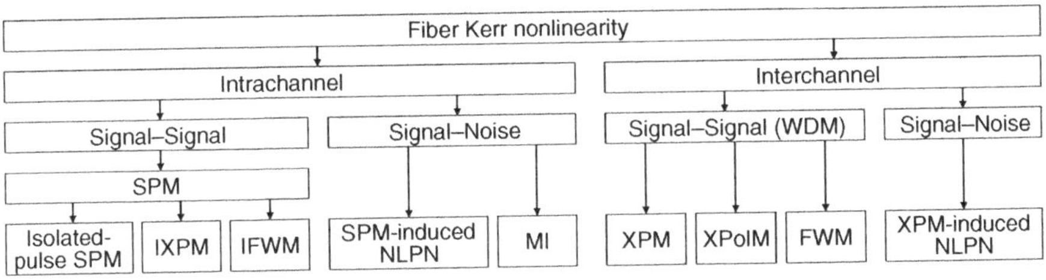

FIGURE 2.19 Kerr effect classification. SPM: self-phase modulation, MI: modulation instability, NLPN: nonlinear phase noise, XPolM: cross-polarization modulation, and (I)XPM and (I)FWM: (intrachannel) cross-phase modulation and four-wave mixing, respectively.

The excitation of the individual effects, that is, the question which of the Kerr effects is the most relevant (deteriorating) one in a given situation, depends on power levels, fiber type (in particular CD and effective area), and also on the per-channel bit rate. An assignment to these parameters according to Fig. 2.20 has been given in Ref. [32].

The different Kerr effects can be described by a common model, together with a correct representation of the participating signals.

The propagation of an optical signal in the fiber is governed by the propagation equation (2.5). The propagation constant β is proportional to the (effective) refractive index n_{eff} seen by the electrical field.

$$\beta = \frac{2\pi \cdot n_{\mathrm{eff}}}{\lambda}. \tag{2.29}$$

The effective refractive index is wavelength dependent, as explained in Section 2.1.2. In addition, the refractive index is dependent on the instantaneous power in the fiber [33–35]. This effect is known as the *Kerr effect*:

$$n_{\mathrm{eff}} = n_0(\lambda) + n_2 \frac{P}{A_{\mathrm{eff}}}. \tag{2.30}$$

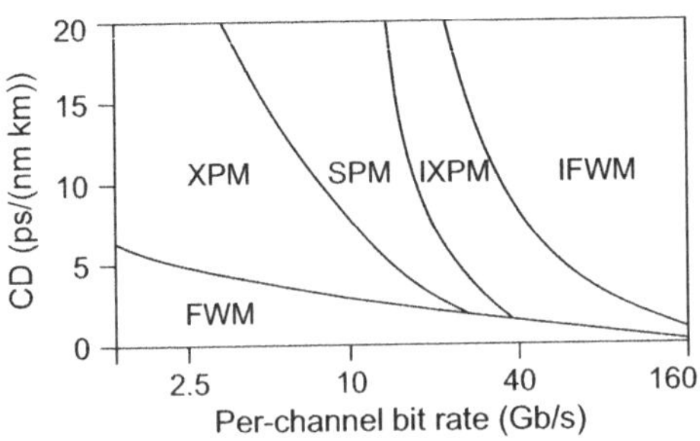

FIGURE 2.20 Dominance of nonlinear effects.

TABLE 2.5 Values of Nonlinearity for Various Types of Transmission Fiber

Fiber Type	A_{eff} Type	γ Type at 1550 nm
SSMF G.652	80 μm^2	1.3 W^{-1} km^{-1}
TrueWave® G.655	55 μm^2	1.8 W^{-1} km^{-1}
LEAF® G.655	72 μm^2	1.4 W^{-1} km^{-1}
DSF G.653	55 μm^2	1.8 W^{-1} km^{-1}

Here, n_2 is the nonlinear fiber coefficient, A_{eff} is the effective area of the fiber, which is the average area occupied by the field in the fiber, and P is the instantaneous signal power. A nonlinearity coefficient γ is often used to describe the nonlinear properties of the fiber, which includes the fiber parameter A_{eff} in addition to the nonlinear fiber coefficient n_2. The nonlinearity coefficient is defined as

$$\gamma = \frac{2\pi n_2}{\lambda A_{eff}}. \tag{2.31}$$

A typical value for γ is about 1.3 W^{-1} km^{-1} in standard single-mode fiber at a wavelength of 1550 nm. Typical values for other fiber types are shown in Table 2.5.

The Kerr effect can be included in the propagation equation as an additional imaginary term, yielding the *nonlinear Schrödinger equation* (NLSE) [2,26–28]:

$$\frac{\partial A}{\partial z} = -\frac{\alpha}{2}A - \frac{j}{2}\beta_2\frac{\partial^2 A}{\partial T^2} + j\gamma|A|^2 A. \tag{2.32}$$

Neglecting the dispersive term (proportional to β_2) in Eq. (2.32), the equation can be solved analytically, resulting in

$$A(z) = A(0)\cdot \exp\left(-\frac{\alpha}{2}z\right)\cdot \exp\left(j\gamma\int_{x=0}^{z}|A(x)|^2\mathrm{d}x\right). \tag{2.33}$$

This shows that the nonlinearity impacts the phase of the optical signal, proportional to the integrated power over the length of a span. The first exponential term in Eq. (2.33) represents the fiber loss. The equation can be solved using

$$|A(x)|^2 = |A(0)|^2\cdot \exp(-\alpha z). \tag{2.34}$$

This results in

$$A(z) = A(0)\cdot \exp\left(-\frac{\alpha}{2}z\right)\cdot \exp\left(j\gamma|A(0)|^2 L_{eff}\right). \tag{2.35}$$

L_{eff} is called the effective length of the fiber and defined as

$$L_{eff} = \frac{\int_{x=0}^{z}|A(x)|^2\mathrm{d}x}{|A(0)|^2} = \frac{1-\exp(-\alpha z)}{\alpha}. \tag{2.36}$$

In a WDM system, multiple waves propagate in a fiber such that the overall field is comprised of signals at different frequencies

$$A = \sum_n A_n \cdot \exp\left[-j(\omega_n - \omega_0)t - j(\phi_n - \phi_0)\right]. \tag{2.37}$$

The NLSE for this multiplexed signal is then given by

$$\frac{\partial A}{\partial z} = -\frac{\alpha}{2}A - \frac{j}{2}\beta_2 \sum_n \frac{\partial^2 A_n}{\partial T^2} + j\gamma |A|^2 A. \tag{2.38}$$

While each slowly varying complex signal pulse amplitude A_n can be considered separately for each channel in a WDM system, the expression $|A|^2$ in the third term on the right-hand side of Eq. (2.38) is the aggregate instantaneous power of all signal channels at a position z in the fiber. This nonlinear term couples the propagation of all waves present in the fiber. The propagation of a single channel A_n is then described by

$$\begin{aligned}\frac{\partial A_n}{\partial z} &= -\frac{\alpha}{2}A_n - \frac{j}{2}\beta_2 \frac{\partial^2 A_n}{\partial T^2} \\ &\quad + j\gamma\left(|A_n|^2 + 2\sum_{i \neq n}|A_i|^2 + \sum_i \sum_{k \neq i}|A_i A_k| \cos\left[(\omega_i - \omega_k)t + \phi_i - \phi_k\right]\right)A_n.\end{aligned} \tag{2.39}$$

The last summand represents the nonlinear propagation effect. The phase is impacted by three distinct power terms (given in parentheses in Eq. (2.39)). The power related to the considered channel itself (A_n^2) leads to SPM as discussed in Section 2.2.1.1. Signal power components from other channels of the WDM system (A_n^2) lead to XPM, discussed in Section 2.2.1.2. Please note that the phase impact of other channels due to XPM is twice as large as SPM. Finally, varying power terms from beating between different signal channels lead to FWM, as discussed in Section 2.2.1.4. We will consider these effects separately in the following sections, but keep in mind that SPM, XPM, and FWM all arise from the power-dependent refractive index of the transmission fiber (Kerr effect). Modulation instability, as discussed in Section 2.2.1.5, combines the effects of SPM, XPM, and FWM and can lead to an enhancement of noise power.

The NLSE can be extended to a pair of coupled equations to account for the propagation of two orthogonally polarized modes (e.g., to include effects of birefringence, polarization multiplexing, and XPolM, or to massively extend the optical bandwidth that is covered) [2]:

$$\begin{aligned}\frac{\partial U}{\partial z} + \beta_{1U}\frac{\partial U}{\partial t} + \beta_{2U}\frac{j}{2}\frac{\partial^2 U}{\partial t^2} + \frac{\alpha_U}{2}U &= j\gamma_U\left(|U|^2 + 2|V|^2\right)U, \\ \frac{\partial V}{\partial z} + \beta_{1V}\frac{\partial V}{\partial t} + \beta_{2V}\frac{j}{2}\frac{\partial^2 V}{\partial t^2} + \frac{\alpha_V}{2}V &= j\gamma_V\left(|V|^2 + 2|U|^2\right)V.\end{aligned} \tag{2.40}$$

Now, U and V are the slowly varying envelopes of the two orthogonally polarized signals, and β_{iU}, β_{iV}, and γ_U, γ_V are the respective phase and nonlinear coefficients. Similar coupled NLSEs can also be used to calculate effects between two successive pulses of on WDM channel, in order to cover intrachannel interactions caused by IXPM.

The NLSE (2.32) does not consider Raman and Brillouin scattering effects, and it is limited in its accuracy with respect to ultrabroadband signals due to a number of simplifications that were made. A more general (numerical) description has to be based on Wiener–Volterra series; see, for example, Refs. [36–38]. It is, however, possible to extend Eq. (2.32) by both, higher order CD and a Raman shock term [2,39,40]. The resulting equation is referred to as generalized nonlinear Schrödinger equation (GNLSE) (here, again without attenuation term):

$$\frac{\partial A}{\partial z}=\sum_{i=1}^{3}\beta_i\frac{(-j)^{i+1}}{i!}\frac{\partial^i A}{\partial t^i}-\frac{j\beta_0 n_2}{n_0}\left(1+\frac{j}{\omega_0}\frac{\partial}{\partial t}\right)\left(A\int_0^\infty R(\tau)|A(z,t-\tau)|^2\mathrm{d}\tau\right). \quad (2.41)$$

Equation (2.32) can efficiently be solved numerically by means of the (symmetrical) split-step Fourier algorithm [41]. The GNLSE (2.41) has to be solved with more complex (and more CPU-demanding) pseudospectral methods, involving fast convolutions [2,42].

2.2.1.1 Self-Phase Modulation As the refractive index depends on the instantaneous signal power, the phase that a signal acquires during propagation in the fiber depends on the time-dependent power and the shape of the signal. At the signal transitions between high and low power, the acquired phase will change, resulting in a frequency shift of the signal. The leading edge of a pulse will be shifted toward lower frequencies (red shift), while the trailing edge will experience a blue shift, resulting in a spectral broadening of the signal spectrum. This frequency shift alone does not constitute a signal distortion as long as all optical filters in the transmission path have sufficient bandwidth to let pass the widened signal spectrum and as long as the signal spectrum broadening does not impact neighboring channels. However, when chromatic dispersion is taken into consideration, the wider spectrum increases the temporal broadening and deteriorates the system performance. On the other hand, self-phase modulation modifies the phase of the signal, just as chromatic dispersion does. Under the right conditions, especially $\beta_2 < 0$, SPM partly cancels the effect of chromatic dispersion. This can be used to design transmission systems with appropriate pulse conditions and dispersion management.

Excitation of the different aspects of SPM mainly depends on the Baud rate, as already shown in Fig. 2.20. For WDM channels with OOK (see Section 5.3.1.1), the dominating effects with increasing Baud rate are FWM, XPM, and SPM, with its variants single-pulse SPM, IXPM, IFWM [31]. The main effects of IXPM and IFWM are added chirp and amplitude and phase jitter (IXPM), and the generation of so-called ghost pulses in OOK signals at blank ("0") positions (IFWM) [43]. These effects can be influenced by the design of dispersion maps (in cases where optical dispersion compensation is used, refer to Sections 3.2.5 and 6.5.3.2). Work in Refs [44,45] indicates that intrachannel effects can be mitigated by so-called constrained coding techniques.

Solitons. One special case including chromatic dispersion and self-phase modulation can be derived from Eq. (2.32). If the second and third terms on the RHS of Eq. (2.32) cancel each other except for a phase shift from propagation, the signal only experiences a relative phase shift and fiber attenuation. This regime, which is known as the soliton regime, requires the signal pulse to have a specific shape [2,46–51]:

$$A_{\text{Soliton}}(z,T) = \sqrt{P_0}\,\text{sech}\left(\frac{T}{T_0}\right)\cdot\exp\left(j\frac{\gamma P_0}{2}z\right), \quad \text{with} \quad T_0 = \sqrt{\frac{-\beta_2}{\gamma P_0}}. \tag{2.42}$$

The soliton pulse width, T_0, depends on the ratio of chromatic dispersion and pulse peak power. Therefore, to maintain a constant pulse shape, frequent signal amplification is required to maintain a constant signal amplitude despite the fiber attenuation. Alternatively, the dispersion of the fiber could be designed to decrease with propagation distance such that the reduced self-phase modulation with decreasing signal power is met by decreased fiber dispersion [52].

Dispersion-Managed Solitons. An alternative to the "true" soliton is the so-called dispersion-managed soliton (DMS). To achieve this pulse propagation behavior, the chromatic dispersion of a fiber span is considered together with the dispersion compensation following the fiber span and the loss in the transmission fiber. While the signal shape and spectrum is changing during propagation along the fiber, both are restored after compensation of the chromatic dispersion. Therefore, at the input into each fiber span, the same signal launch conditions exist.

The DMS can be approximated by a Gaussian shape. A transmission link as shown in Fig. 2.21 is assumed: at the transmitter, chirp is added to the signal, for example, by some amount of dispersion compensation. Each transmission section or span consists of a length of dispersive fiber, followed by a dispersion compensation module (DCM) and by an amplifier to maintain a constant signal power into each span. The DCM can be located at the mid-stage of the amplifier to reduce the noise generated in the amplifier (see Section 3.2.5). A number N of these spans are cascaded before the receiver, where an additional amount of dispersion compensation is located.

The propagation of a signal pulse $A(z,t) = \sqrt{a(z)}\cdot q(z,t)$ over such a line with dispersion $\beta_2(z)$ can be described by the NLSE (Eq. (2.32)). The energy of the pulse

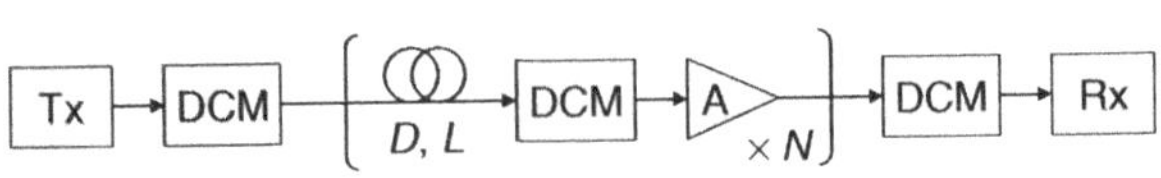

FIGURE 2.21 Transmission link with N repeated transmission sections.

evolves with $a(z)$, determined by the fiber attenuation such that $a(z) = a(0) \cdot \exp(-\alpha z)$, where α is the fiber attenuation coefficient (see Section 2.1.1).

The pulse shape $q(z, t)$ can be approximated by a Gaussian function [53]

$$q(z,t) = \sqrt{B}\,\exp\left[-(1+jb)\frac{t^2}{2W^2} + j\phi - j\omega t\right]. \tag{2.43}$$

The energy of this pulse is defined as

$$E = \int_{t=-\infty}^{\infty} |q(z,t)|^2 \mathrm{d}t = B \cdot W \cdot \sqrt{\pi}. \tag{2.44}$$

This energy has to remain constant during propagation over the distance z.

The pulse width W and the chirp parameter b both depend on the distance z, and they are related to the two parameters g and C by

$$\begin{aligned} gW^2 &= 1 + g^2C^2, \\ b &= -gC. \end{aligned} \tag{2.45}$$

The spectral width of the Gaussian signal pulses is determined by [2]

$$\Delta\omega = \frac{\sqrt{(1+b^2)}}{W} = \sqrt{g}. \tag{2.46}$$

Therefore, the parameter g is a direct measure for the spectral width of the signal and remains constant for linear propagation. The parameter C corresponds to the accumulated chromatic dispersion for linear propagation. With nonlinear propagation, the following equations govern the evolution of g and C over the distance z:

$$\frac{\mathrm{d}C}{\mathrm{d}z} = \left(C^2 - \frac{1}{g^2}\right)\frac{a(z)\gamma E}{W^3\sqrt{2\pi}} - \beta_2, \tag{2.47}$$

$$\frac{\mathrm{d}g}{\mathrm{d}z} = -2gC\frac{a(z)\gamma E}{W^3\sqrt{2\pi}}. \tag{2.48}$$

The evolution of the parameters g and C over 100 km of standard single-mode fiber (G.652 fiber) is shown in Fig. 2.22 for different initial chirp ("precompensation") and power values. At the fiber input, where nonlinear effects are present, the spectral width of the signal, represented by the parameter g, can change strongly, while deeper into the fiber, the spectral width remains constant and only the accumulated dispersion C is increased. The corresponding evolution of the pulse width (W) and the spectral width of the Gaussian signal are shown in Fig. 2.23. In general, the higher the signal power, the smaller is the resulting pulse width after

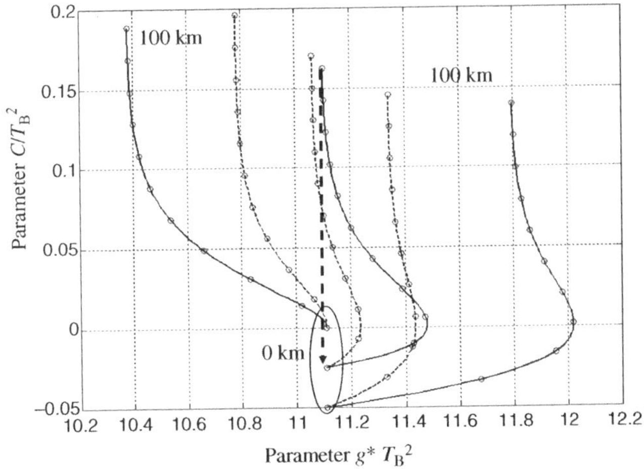

FIGURE 2.22 Evolution of Gaussian pulses over G.652 fiber. Pulse FWHM is 50 ps, data rate 10 Gb/s. Precompensation is 0, −250, and −500 ps^2, average signal power is 0 dBm (dashed lines) and 4 dBm (solid lines). The dashed arrow shows linear CD compensation equivalent to 93.6 km of SSMF.

propagation. Initial precompensation of the chromatic dispersion (prechirp) leads to a reduced pulse width after transmission.

Dispersion compensation at the end of the fiber span only reduces the parameter C, as shown by the arrow in Fig. 2.22. As linear dispersion compensation is assumed, the spectral width of the signal is not impacted by dispersion compensation.

Dispersion-managed soliton propagation is achieved if, after linear dispersion compensation of a fiber span, the same pulse shape and spectrum as described by the parameter set (g, C) is obtained as at the input to the fiber span. In this case, the input to each fiber span is the same. However, as can be seen in Fig. 2.22, this condition can only be achieved for specific values of pulse power, pulse width, and

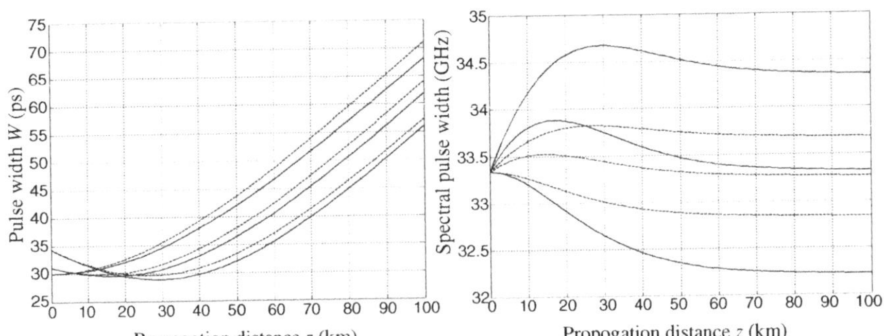

FIGURE 2.23 Evolution of temporal (left) and spectral (right) pulse widths of the Gaussian pulses according to Fig. 2.22. Dashed lines for 0 dBm, solid lines for 4 dBm average signal power.

precompensation. Especially, the spectral width after propagation must be the same as at the span input. For example, in Fig. 2.22, this is obtained for a precompensation of $-250\,\mathrm{ps}^2$ and an average power of 4 dBm. Chromatic dispersion compensation equivalent to 93.6 km of SSMF brings the pulse parameters back to the initial values. This is equivalent to a residual dispersion per span of 100 ps/nm (see Section "Dispersion Map").

If the initial pulse parameters are not chosen to achieve soliton pulse evolution, the evolution of the (g, C) pulse parameters can still be obtained using Eqs. (2.47) and (2.48). The (g, C) parameters evolve from span to span along the transmission system. For some range of initial values, the evolution is approximately periodic. The periodicity depends on the parameters. Figure 2.24 shows, as an example, the evolution of g and C (at the span input points) along a system for an initially unchirped pulse ($C=0$). If the initial pulse parameters and the per-span dispersion compensation are chosen appropriately, the parameter curve collapses to a single point, describing the DMS condition.

Dispersion Map. As has already been shown, higher power signal propagation in the transmission fiber leads to spectral broadening due to the nonlinear effect of self-phase modulation. Chromatic dispersion (see Section 2.1.2), on the other hand, broadens the signal in time. Chromatic dispersion can be compensated using components with inverse sign of the chromatic dispersion than the fiber dispersion, as discussed in Section 3.2.5. As self-phase modulation partly compensates the effect of chromatic dispersion (and can do so completely in the case of solitons, see Section "Solitons"), the required dispersion compensation is usually less than the accumulated fiber dispersion. Part of the span dispersion remains uncompensated. A typical undercompensation value for G.652 fiber is 100 ps/nm per span. This is very similar

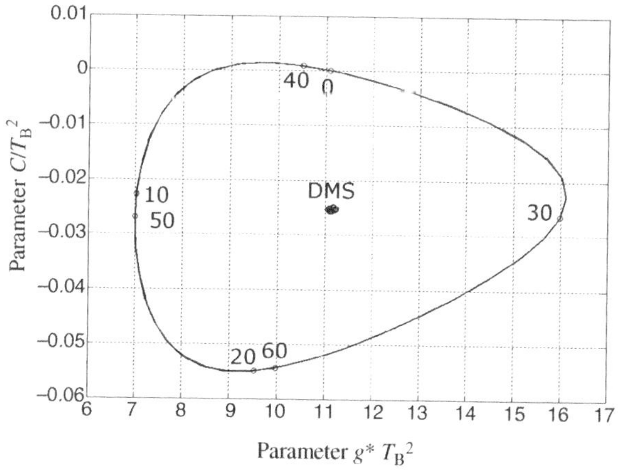

FIGURE 2.24 Evolution of the parameters g, C at the span input points along a 60-span system on G.652 fiber. The precompensation is 0, the average power is 4 dBm. The evolution is periodic with approximately 39 spans.

to the dispersion-managed soliton condition discussed in Section "Dispersion-Managed Solitons."

2.2.1.2 *Cross-Phase Modulation* As seen in Eq. (2.39), the phase of the signal wave is impacted by the power copropagating in all other channels of the WDM system. This part of the Kerr effect is called XPM. XPM has the most severe impact in a WDM system, when the impacted channel is phase modulated and the neighboring channels are amplitude modulated. But even if all WDM channels are amplitude modulated, XPM can lead to amplitude distortions and to timing jitter of the optical pulses. XPM is partly mitigated by chromatic dispersion of the transmission fiber. CD results in a walk-off between the signals in different channels, such that the phase modulation experienced by a signal channel is a time average of the amplitude in the neighboring channels. All channels propagate with different velocities (see Section 2.1.2):

$$v_n = \frac{1}{\beta_1(\omega_n)}.$$

For the analytical description of the effect, we introduce retarded coordinate systems that are separate for each channel such that

$$T_n = t - \beta_1(\omega_n)z. \tag{2.49}$$

The signals in channels 1 and 2 propagate with different velocities v_1 and v_2, respectively. During propagation over a distance z, the signals in both channels walk off by time difference of

$$\Delta T = [\beta_1(\omega_1) - \beta_1(\omega_2)]z = \Delta\beta_1 z. \tag{2.50}$$

Using the retarded coordinates, the nonlinear phase shift in channel 1 due to power in channel 2 is proportional to the accumulated power in channel 2, which passes the time slot T_1 in channel 1:

$$\phi_{\mathrm{NL}}(z, T_1) = 2\gamma \int_{x=0}^{z} P_2(x, T_2 = T_1 - \Delta\beta_1 x)\mathrm{d}x. \tag{2.51}$$

If we assume no major pulse shape distortion and negligible attenuation in the region of the pulse interference, such that

$$P_n(z, T_n) = P_n(0, T_n), \tag{2.52}$$

Equation (2.51) can be further evaluated using Eq. (2.52) and changing the integration constants:

$$\phi_{\mathrm{NL}}(z, T_1) = \frac{2\gamma}{\Delta\beta_1} \int_{T_2=T_1}^{T_1-\Delta\beta_1 z} P_2(0, T_2)\mathrm{d}T_2. \tag{2.53}$$

The nonlinear phase shift induced by cross-phase modulation is essentially proportional to the energy in the interfering channel passing the considered time slot of the impacted channel. The velocity difference between both channels can be expressed in terms of the chromatic dispersion parameter β_2:

$$\Delta\beta_1 = \beta_1(\omega_1) - \beta_1(\omega_2) = \beta_2 \cdot (\omega_1 - \omega_2). \tag{2.54}$$

The XPM impact of a full crossing of a particular pulse in channel 2 (with a given energy) is inversely proportional to the velocity difference. This is due to the fact that the fiber length over which the interference occurs is smaller for a faster walk-off between the channels. Higher chromatic dispersion, therefore, leads to lower cross-phase modulation impact.

The phase modulation directly results in a signal distortion of phase-modulated signals. For amplitude-modulated signals, the spectral broadening from XPM together with the fiber chromatic dispersion also leads to amplitude distortions, reducing the quality of the received signal.

In addition to amplitude distortions, XPM also leads to timing jitter. The nonlinear phase, as a function of distance and time, as expressed in Eq. (2.53) corresponds to a frequency modulation of the impacted signal, which can be evaluated by taking the derivative of the nonlinear phase with respect to time.

$$\Delta\omega(z, T_1) = \frac{\mathrm{d}\phi_{NL}(z, T_1)}{\mathrm{d}T_1} = \frac{2\gamma}{\Delta\beta_1}[P_2(0, T_1 - \Delta\beta_1 z) - P_2(0, T_1)]. \tag{2.55}$$

The nonlinearity-induced frequency offset at a particular position in the fiber depends on the difference between the momentary interfering power at the input to the fiber and the momentary interfering power at the particular position (terms in brackets in Eq. (2.55)). Therefore, if a full pulse crosses the impacted time slot, the frequency offset will first change and then change back. The section of the signal that experiences this temporary frequency shift, will travel with a different velocity during the pulse crossing due to chromatic dispersion. This will result in timing jitter, when the signal is detected. An estimate of the amount of time shift during a pulse walk-through is

$$|\Delta T| = \frac{2\gamma E}{|\beta_2|(\omega_1 - \omega_2)^2}. \tag{2.56}$$

Here, E is the energy in the interfering pulse. For example, in a 10 Gb/s system on nonzero dispersion-shifted fiber (NZ-DSF), with a channel separation of 50 GHz and RZ pulses with 50 ps pulse width and 8 mW peak power (average signal power: 3 dBm), each pulse crossing with a neighboring channel pulse induces a timing shift of approximately 2 ps.

On the other hand, if a pulse in the interfering channel overlaps with the signal channel at the fiber input, and the interfering pulse walks off, a permanent frequency offset will occur, leading to an amplitude modulation in the signal. Both effects, amplitude distortions and timing jitter, lead to a deteriorating signal quality [54].

As chromatic dispersion is a main factor in the signal deterioration due to XPM, part of the XPM impact can be mitigated by appropriate dispersion compensation. However, optimum compensation for XPM effects usually means a full compensation of a fiber span dispersion. This is in conflict to the optimum chromatic dispersion compensation for SPM effects, which usually requires undercompensation of the chromatic dispersion. A trade-off for the optimum dispersion map needs to be found for each system, taking into account both, SPM and XPM effects.

2.2.1.3 Cross-Polarization Modulation XPolM is closely related to XPM. XPM is caused by nonlinear variations of the refractive index by other WDM channels, XPolM describes nonlinear variations of the refractive index difference between the polarizations. Both result in (mostly unwanted) signal modulations. In XPM, the resulting phase of the probe signal (the WDM channel under consideration) is proportional to the sum of the powers of the interfering channels. For XPolM, the resulting polarization (i.e., the phase difference) depends on the sum of the Stokes vectors of the interferers [55]. The XPM intensity varies with walk-off between the interfering channels, that is, with group-velocity differences. Likewise, the XPolM-related probe SOP varies with PMD, that is, when the group velocity between interferer and probe differs due to birefringence. XPolM has first been described in 1995 by Mollenauer in the context of soliton collisions [56].

XPolM is an interchannel effect that leads to fast polarization modulation between signals and to depolarization of the probe signal. This can result in fading and cross talk for dual-polarization (DP) signals [30]. The depolarization depends on the variance of the resulting Stokes vector of the aggregated copropagating channels. The variation of the probe-signal SOP can be described by its Stokes vector $\vec{S}$ [57,58]:

$$\frac{\partial \vec{S}}{\partial z} = \frac{8}{9}\gamma \vec{S}_{\mathrm{Tot}} \times \vec{S}, \tag{2.57}$$

$\vec{S}_{\mathrm{Tot}}$ is the total Stokes vector of the probe and the aggregated interferers, and γ is the nonlinear coefficient. XPolM leads to a rotation of the probe-signal Stokes vector around the one of the interferers. This is shown in Fig. 2.25 for two probes (WDM channels) with Stokes vectors $\vec{S}_x$ and $\vec{S}_y$, respectively.

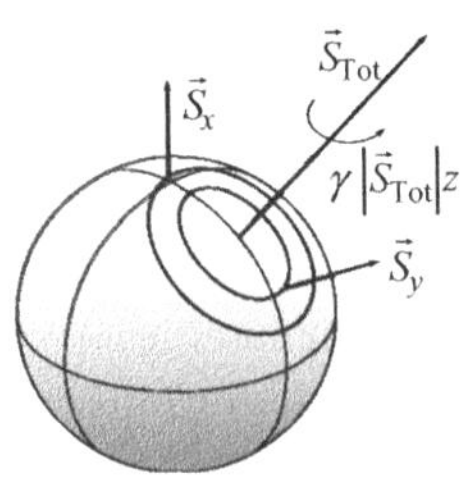

FIGURE 2.25 XPolM-induced polarization rotations.

XPolM causes symbol-to-symbol fluctuations around the mean SOP that cannot be fully compensated [55]. The resulting depolarization creates detrimental cross talk between the polarization tributaries in dual-polarization modulation schemes (refer to Sections 5.1 and 5.4). XPolM is, therefore, harmful for polarization-sensitive receivers, it is of minor importance to 10 Gb/s OOK systems. For DP-QPSK systems, however, it is relevant and has influence on the window length of the digital polarization demultiplexer (which, for QPSK, typically is the constant-modulus algorithm, CMA). In DP-QPSK, the CMA tracks average polarization. The fast XPolM misalignments (with speeds of the order of the symbol duration) are not tracked by the CMA and produce fading of the desired component (X or Y polarization tributary), as well as cross talk from the orthogonal polarization tributary [30].

Depending on the walk-off between the WDM channels (i.e., presence of dispersion and other means leading to differential delay), the depolarization can lead to substantial penalties over the typically long measurement time of the CMA receiver. Even with an infinite walk-off, the depolarization does not fully disappear. This is a fundamental difference between XPolM and XPM.

2.2.1.4 *Four-Wave Mixing* A nonlinear effect involving more than two copropagating waves is FWM. Based on the Kerr effect, two or more waves generate additional optical waves at different frequencies. FWM can be considered as the generation of modulation sidebands due to phase modulation of an optical carrier wave.

The power of two copropagating waves E_1 and E_2 with

$$E_k = |E_k(z)| \cdot \exp\left[j(\omega_k t - \beta(\omega_k)z + \phi_k)\right], \quad k = 1, 2 \tag{2.58}$$

can be considered as the (constant) power in each individual wave plus the beat between both waves at the intermediate frequency:

$$P(z,t) = P_1(z,t) + P_2(z,t) + 2\sqrt{P_1 P_2}\cos\left[(\omega_1 - \omega_2)t - (\beta(\omega_1) - \beta(\omega_2))z + \phi_1 - \phi_2\right]. \tag{2.59}$$

Here, $P_k = |E_k|^2$ was substituted.

The power variations modulate the refractive index according to Eq. (2.30) such that

$$n(z,t) = n_0 + n_2 \cdot \frac{P(z,t)}{A_{\text{eff}}}. \tag{2.60}$$

We now consider a third wave, E_3 according to Eq. (2.58) with $k = 3$, which is copropagating with E_1 and E_2. E_3 experiences the time varying refractive index and therefore undergoes a phase modulation. Propagating over the infinitesimal section dz, E_3 experiences a varying phase shift of

$$\tilde{\phi}(z,t) = 2\gamma\sqrt{P_1 P_2}\cos\left[(\omega_1 - \omega_2)t - (\beta(\omega_1) - \beta(\omega_2))z + \phi_1 - \phi_2\right]\mathrm{d}z. \tag{2.61}$$

The nonlinear coefficient γ is defined in Eq. (2.31). It is well known that this phase modulation with frequency $\Delta\omega = (\omega_1 - \omega_2)$ and modulation amplitude,

$$\widehat{\Phi} = 2\gamma\sqrt{P_1 P_2}\,\mathrm{d}z, \tag{2.62}$$

generates modulation sidebands at $\pm\Delta\omega$ around the carrier. The amplitudes and phases of these first-order modulation sidebands are $J_{\pm 1}(\widehat{\Phi})$ and $\pm\pi/2$, respectively. $J_\nu(z)$ is the Bessel function of the first kind and νth order. For small values of z,

$$J_1(z) \approx \frac{z}{2}, \quad J_{-1}(z) \approx -\frac{z}{2}.$$

The sidebands generated by phase modulation in the section dz are therefore

$$\mathrm{d}E_{+1} = j\gamma\sqrt{P_1 P_2 P_3}\exp\{j[(\omega_3 + \omega_1 - \omega_2)t - (\beta(\omega_3) + \beta(\omega_1) - \beta(\omega_2))z + \phi_3 + \phi_1 - \phi_2]\}\mathrm{d}z \tag{2.63}$$

and

$$\mathrm{d}E_{-1} = j\gamma\sqrt{P_1 P_2 P_3}\exp\{j[(\omega_3 - \omega_1 + \omega_2)t - (\beta(\omega_3) - \beta(\omega_1) + \beta(\omega_2))z + \phi_3 - \phi_1 + \phi_2]\}\mathrm{d}z. \tag{2.64}$$

The sidebands propagate along the fiber with their respective propagation constants $\beta(\omega_3 + \omega_1 - \omega_2)$ and $\beta(\omega_3 - \omega_1 + \omega_2)$ and their attenuation coefficient α, while in each consecutive fiber section dz other sideband components are generated with phases depending on the propagation constants of the three contributing waves, $\beta(\omega_1)$, $\beta(\omega_2)$, and $\beta(\omega_3)$. At location L in the fiber, the sidebands have therefore accumulated, taking into account coherent addition of the generated sideband components.

$$E_{+1}(L,t) = j\gamma\exp\{j[(\omega_3 + \omega_1 - \omega_2)t + \phi_3 + \phi_1 - \phi_2]\}\cdot\exp\left(-\frac{\alpha L}{2}\right) \times \sqrt{P_1(0)P_2(0)P_3(0)} \times \int_{z=0}^{L}\exp(-\alpha z)\exp\{-j[\beta(\omega_3) + \beta(\omega_1) - \beta(\omega_2)]z\}\cdot\exp\{-j[\beta(\omega_3 + \omega_1 - \omega_2)][L - z]\}\mathrm{d}z. \tag{2.65}$$

The generated wave at $\omega_3 + \omega_1 - \omega_2$ grows maximally, if the phase term in the integral of Eq. (2.65) is constant over z. This condition is met, when

$$\beta(\omega_3 + \omega_1 - \omega_2) = \beta(\omega_3) + \beta(\omega_1) - \beta(\omega_2). \tag{2.66}$$

Equation (2.66) is called the phase matching condition of four-wave mixing [2,25,28,59]. With the phase mismatch term

$$\Delta\beta = \beta(\omega_3) + \beta(\omega_1) - \beta(\omega_2) - \beta(\omega_3 + \omega_1 - \omega_2), \tag{2.67}$$

equation (2.65) can be further evaluated and the power of the generated wave at location L, based on this phase modulation approach, can be calculated. For the power of the contributing waves, an evolution due to fiber attenuation is assumed, $P_k(z) = P_k(0) \cdot \exp(-\alpha z)$:

$$P_{+1} = \gamma^2 P_1(0) P_2(0) P_3(0) \exp(-\alpha L) \frac{1 - 2\exp(-\alpha L)\cos(\Delta\beta L) + \exp(-2\alpha L)}{\alpha^2 + \Delta\beta^2}. \tag{2.68}$$

As previously discussed, Eq. (2.68) shows the evolution of one modulation sideband at the frequency $\omega_3 + \omega_1 - \omega_2$ due to the beat between E_1 and E_2 and modulation of E_3. At the same frequency, however, sidebands will occur from the beat between E_2 and E_3 resulting in a modulation of E_1. As this second sideband adds with the same phase to the first sideband, the power is four times the power given in Eq. (2.68). Using the definition of the effective length in Eq. (2.36), the power in the mixing product at $\omega_3 + \omega_1 - \omega_2$ is

$$P_4 = (D_{123}\gamma L_{\text{eff}})^2 P_1 P_2 P_3 \cdot \exp(-\alpha L) \frac{\alpha^2}{\alpha^2 + \Delta\beta^2} \left[1 + \frac{4\exp(-\alpha L)\sin^2(\Delta\beta L/2)}{(1 - \exp(-\alpha L))^2} \right]. \tag{2.69}$$

For the mixing product stemming from three different waves ("three-tone product"), the multiplier D_{123} equals 2. Mixing products are, however, also generated for only two waves propagating. In this case, the beat between the two waves modulates one of the waves, resulting in only one contributor to the generated sideband. In this case of a two-tone product, $D_{123} = 1$.

The phase mismatch term $\Delta\beta$ strongly determines the growth of the mixing products. Its value depends on the chromatic fiber dispersion. As shown in Section 2.1.2, the propagation constant $\beta(\omega)$ can be developed in a Taylor series around ω_0:

$$\beta(\omega) = \beta_0 + \beta_1(\omega - \omega_0) + \frac{\beta_2}{2}(\omega - \omega_0)^2 + \frac{\beta_3}{6}(\omega - \omega_0)^3 + \cdots. \tag{2.70}$$

Using the Taylor series for all frequencies in Eq. (2.67), the resulting phase mismatch term depends on the frequency separation of the waves and the chromatic dispersion at a frequency between ω_3 and ω_1:

$$\Delta\beta = -(\omega_1 - \omega_2) \cdot (\omega_3 - \omega_2) \cdot \left[\beta_2(\omega_0) + \beta_3 \cdot \left(\frac{\omega_1 + \omega_3}{2} - \omega_0 \right) \right]. \tag{2.71}$$

More practically, the fiber dispersion value D (in ps/nm) is used rather than the Taylor series coefficients β_2 and β_3. The phase mismatch can then be found to be a function of the frequency spacing between the involved waves and the chromatic dispersion at the frequency centered between f_1 and f_2:

$$\Delta\beta = \frac{2\pi\lambda^2}{c}(f_1 - f_2)\cdot(f_3 - f_2)\cdot D\left(\frac{f_1 + f_3}{2}\right). \qquad (2.72)$$

In an optical transmission system, a large number of FWM products are generated from the combinations of all optical channels. At each frequency, several mixing products are superimposed with phases depending on the optical phases of the involved channels. For amplitude-modulated signals, the amplitude of the mixing components also depends on the instantaneous amplitudes of the involved channels. In Ref. [60], the statistics of FWM was analyzed, yielding an exponential distribution of the FWM amplitude. The overall FWM power in one channel also depends on the position in the spectrum. Central channels experience a larger number of mixing products than edge channels. The analysis reported was carried out under the assumption of OOK signals, which are all copolarized. In a practical transmission system this would be the upper limit, as the polarization of channels in a WDM spectrum usually evolve differently due to birefringence and polarization-mode dispersion (see Section 2.1.3). An approximate formula for the system parameter requirements to keep the penalty due to FWM in DWDM system below 1 dB was given in Ref. [60]:

$$\frac{\gamma\cdot P_{\text{avg}}\sqrt{N_{\text{Spans}}}}{|D|\cdot\Delta f^2} \leq 7\times10^{-4}. \qquad (2.73)$$

In this quantity equation, γ is given in $1\,\text{W}^{-1}\,\text{km}^{-1}$, P_{avg} is the average signal power per channel in mW, N_{Spans} is the number of amplified spans, D is given in ps/(nm km), and Δf is the channel spacing in GHz. As an example, a 10 Gb/s system on G.655 fiber [$\gamma = 1.4\,\text{W}^{-1}\,\text{km}^{-1}$, $D = 4$ ps/(nm km)] using NRZ signals with an average power of 2 mW and 50 GHz channel spacing would start experiencing problems from FWM after about six spans.

As has been discussed above, the interference between two optical waves and the resulting modulation of the nonlinear refractive index of the fiber lead to a phase modulation of all involved waves and to the generation of modulation sidebands. If two waves are injected into the fiber, two new sidebands are generated, while the injection of three waves leads to the generation of nine mixing products. Figure 2.26 shows the special case that the spacing between the three injected waves is uniform, as is the case in standard DWDM systems. Here, some of the mixing products have the same frequency and coincide with the injected waves. This leads to a distortion of the data signals in a DWDM system.

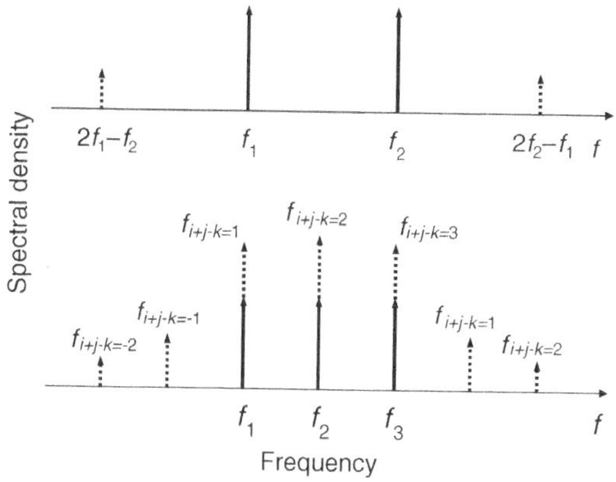

FIGURE 2.26 Generation of new optical waves by four-wave mixing between two (top) or three (bottom) injected optical waves.

For N incident WDM (pump) channels, the number M of created sidebands is given by

$$M = \frac{1}{2}\left(N^3 - N^2\right). \tag{2.74}$$

Hence, for $N = 3$, nine additional sidebands are generated, and so on. The power levels of these new FWM-generated spectral components at frequencies $f_{ijk} = f_i + f_j - f_k$ can be expressed as a function of the FWM efficiency η [61]:

$$P_{ijk}(L) = \left(\frac{1024\pi^6}{n^4\lambda^2c^2}\right)\left(D\chi^{(3)}\right)^2 \frac{P_i(0)P_j(0)P_k(0)}{A_{\text{eff}}}\left|\frac{\{e^{(j\,\Delta\beta-\alpha)L}-1\}}{j\,\Delta\beta-\alpha}\right|^2. \tag{2.75}$$

The phase mismatch $\Delta\beta$ is given by $\Delta\beta = \beta_i + \beta_j - \beta_k - \beta_{ijk} \propto \Delta f^2\, D$. A_{eff} is the effective cross-sectional area that is given by [2,35]

$$A_{\text{eff}} = \left(\int_0^{2\pi}\int_0^{\infty} |F(\mathbf{r},t)|^2 \rho\, \mathrm{d}\rho\, \mathrm{d}\phi\right)^2 \Bigg/ \int_0^{2\pi}\int_0^{\infty} |F(\mathbf{r},t)|^4 \rho\, \mathrm{d}\rho\, \mathrm{d}\phi. \tag{2.76}$$

$F(\mathbf{r}, t)$ is the modal distribution (of the fundamental mode), which can be approximated as a Gaussian distribution across the fiber radius. Practical single-mode transmission fibers have effective areas in the range of 75–150 μm^2, whereas negative-dispersion NZ-DSF (used for dispersion-managed subsea links) can have effective areas as small as 25–50 μm^2. For positive-dispersion transmission fibers (G.652, G.654, G.655), A_{eff} values $>100\ \mu m^2$ are obtained by newer large effective area fibers.

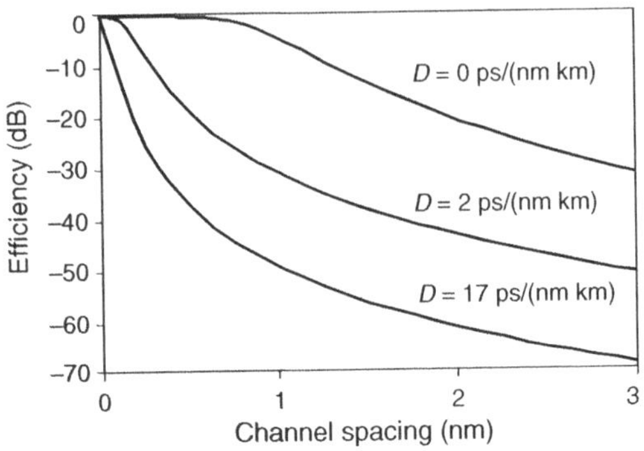

FIGURE 2.27 Relative efficiency of the four-wave mixing effect for different values of the fiber chromatic dispersion.

$\chi^{(3)}$ is the cubic nonlinear susceptibility that connects to the nonlinear refractive index via

$$n_2 = 3\chi^{(3)}/8n_0. \tag{2.77}$$

Finally, D in Eq. (2.75) is the *degeneracy factor*, which can select $D = 1, 3$, and 6, depending on whether three, two, or none of the frequencies f_i, f_j, and f_k are the same.

One way to reduce the signal distortion by four-wave mixing products is to choose the channel frequencies off-grid, that is, to vary the channel separation over the spectrum [62]. With this unequal channel spacing, the mixing products fall in the spectral range between the data carrying waves. However, the deviation from the standard frequency spacing must be larger than the spectral width of the modulated data signals to avoid a partial overlap of the mixing products with the data signals. This reduces the usable optical spectrum and therefore the spectral efficiency. Also, with a large number of data carrying waves, the choice of channel frequencies becomes very cumbersome.

A more practical way to reduce the effect of four-wave mixing is the use of higher dispersion transmission fiber. As was shown above, the growth of the four-wave mixing product is limited by the phase mismatch factor $\Delta\beta$, which in turn is proportional to the chromatic fiber dispersion. Figure 2.27 shows the relative efficiency of the four-wave-mixing effect for various values of the fiber chromatic dispersion [63].

It can be seen that while quite large FWM products are generated even for large channel spacing beyond 1 nm for zero dispersion fiber, even moderate values of chromatic dispersion reduce the effect of four-wave mixing. This is the reason why WDM does not work efficiently on G.653 fibers and why WDM systems with high channel count perform best on G.652 and G.654 fibers. It is also the driver behind latest G.654 fibers (where CD has been increased as compared to G.652 fibers).

2.2.1.5 Modulation Instability In the previous sections, we have considered the contributions to the nonlinear Schrödinger equation (Eq. (2.32)) separately. It was shown that phase matching between the optical waves contributing to four-wave mixing is critical for the generation of FWM products and we have shown the impact of chromatic dispersion on the generation of FWM products. Essentially, due to a large phase mismatch $\Delta\beta$ in highly dispersive fiber, the generation of FWM products is kept in bay. There is a special case, however, when the interplay between self-phase modulation, cross-phase modulation, and chromatic dispersion can lead to the generation of FWM products even in the presence of high chromatic dispersion. It has been shown in Eq. (2.30) that the refractive index of the fiber is not only wavelength dependent but also has a nonlinear component. This nonlinear component can be included in the Taylor series of the propagation constant in Eq. (2.70) as an extra term

$$\beta_{NL} = \gamma(P_{SPM} + 2P_{XPM}). \tag{2.78}$$

The power responsible for self-phase modulation, that is, the power in the signal whose propagation constant is derived, contributes less to the nonlinear phase shift than the power in all other channels, as shown in Eq. (2.39). Considering the nonlinear term for the phase mismatch $\Delta\beta$, it can be seen that the phase matching condition can be met even for a high chromatic dispersion value, if the power levels at the frequencies are different:

$$\begin{aligned}\Delta\beta = &-(\omega_1 - \omega_2)\cdot(\omega_3 - \omega_2)\cdot\beta_2\left(\frac{\omega_1 + \omega_3}{2}\right) - \gamma[P(\omega_1) + P(\omega_3)\\ &- P(\omega_2) - P(\omega_1 + \omega_3 - \omega_2)].\end{aligned} \tag{2.79}$$

A typical example is the amplification of optical noise in the vicinity of a strong optical signal via two-tone FWM. Here, a strong signal at $\omega_1 = \omega_3$ leads to the phase matching condition being met ($\Delta\beta = 0$) at a frequency separation

$$(\omega_1 - \omega_2)_{PM} = \sqrt{\frac{2\gamma P(\omega_1)}{-\beta_2(\omega_1)}}. \tag{2.80}$$

For typical values of G.652 fiber parameters $\beta_2 = -20.4\,\text{ps}^2/\text{km}$ and $\gamma = 1.3\,\text{W}^{-1}\,\text{km}^{-1}$ and a signal power level of 4 dBm (2.5 mW), the phase matching condition is met at a frequency separation of 2.8 GHz. For typical G.655 fiber parameters ($\beta_2 = -6\,\text{ps}^2/\text{km}$ and $\gamma = 1.8\,\text{W}^{-1}\,\text{km}^{-1}$), the phase matching condition is met at a frequency separation of 6.2 GHz. In these cases, noise is being amplified within the spectrum of a data signal and the data signal is additionally distorted.

2.2.1.6 Nonlinear Phase Noise WDM transmission using coherent or differentially coherent detection is subject to impairment by phase noise. Amplified spontaneous emission (ASE) from in-line amplifiers is a major source of phase noise, and is referred to as linear phase noise. In Ref. [64], it was shown that signals and ASE can also interact via the fiber Kerr nonlinearity to produce NLPN. NLPN is also known under the name Gordon–Mollenauer effect.

According to Fig. 2.19, NLPN splits into single-channel (SPM-induced NLPN) and WDM multichannel (XPM-induced NLPN, sometimes also referred to as X-NLPN) effects. Depending on CD compensation schemes (i.e., effectiveness of XPM suppression, or the allowance of walk-off), SPM- or XPM-induced NLPN can become the dominant effect [65].

Nonlinear phase noise degrades the performance of both PSK and DPSK systems (refer to Section 5.3.2). Because NLPN is correlated with the received intensity, this intensity can also be used to correct NLPN [66,67].

Added to the signal phase, SPM-induced NLPN is non-Gaussian distributed, but XPM-induced NLPN is Gaussian distributed. This Gaussian convergence comes from the central-limit theorem since X-NLPN is induced by many WDM channels. The PDFs are relevant for system performance evaluation [65].

For an N-span transmission system, the total SPM-induced NLPN is given by [66]

$$\varphi_{\mathrm{NL}} = \gamma L_{\mathrm{eff}}\left(|E_0 + n_1|^2 + |E_0 + n_1 + n_2|^2 + \cdots + |E_0 + n_1 + \cdots + n_N|^2\right). \tag{2.81}$$

Here, E_0 is the transmitted signal under consideration, n_k with $k = 1, \ldots, N$ are the optical amplifier noise contributions per span (where the n_k are independent identically distributed complex zero-mean circular Gaussian random variables), γ is the fiber nonlinear coefficient, and L_{eff} is the effective length per span. In Eq. (2.81), all fiber spans are assumed to be identical with same launched power, same fiber length and span loss, and using optical amplifiers with identical noise figures. Amplifier noise from the polarization orthogonal to the considered signal tributary is ignored. Equation (2.81) considers NLPN that is induced within a bandwidth matched (limited) to the transmitted signal. X-NLPN is not included.

The received electric field is then given by

$$E_{\mathrm{Rx}} = (E_0 + n_1 + \cdots + n_N)\exp\left(-j\varphi_{\mathrm{NL}}\right). \tag{2.82}$$

ϕ_{NL} is the nonlinear phase rotation caused by NLPN. It does not change the received amplitude. However, it can be compensated by minimizing the variance of the residual NLPN in (2.82). This can be done using the received intensity.

X-NLPN can be described, for two interacting WDM channels, with the so-called pump–probe model. This model can be extended to multichannel and multispan scenarios. The power spectral density (PSD) of the XPM-induced NLPN from channel 2 (pump) to channel 1 (probe) is given by [65,68]

$$\Phi_{\varphi_1}(f) = \Phi_{P_2}(f)|H_{12}(f)|^2. \tag{2.83}$$

$\Phi_{P_2}(f)$ is the PSD of the pump intensity, and $H_{12}(f)$ is the transfer function. For a single fiber span, the transfer function is given by

$$H_{12}(f) = 2\gamma \frac{1 - e^{-\alpha L + j2\pi f d_{12} L}}{\alpha - j2\pi f d_{12}}. \tag{2.84}$$

γ is the fiber Kerr coefficient, α is fiber attenuation coefficient, L is the span length, $d_{12} \approx D\,\Delta\lambda$ is the relative walk-off between the two WDM channels with wavelength separation of $\Delta\lambda$, and D is the fiber CD coefficient. The transfer function (2.84) ignores distortions of the pump caused by fiber transmission. For pump channels running at 10 Gb/s and WDM channel spacing of 50 GHz, the channel walk-off has a larger effect than pump distortion. For (10 Gb/s) WDM channels with on–off keying (see Section 5.3.1.1), the PSD is given by $\Phi_{P_2}(f) = P_0\left(\delta(f) + T_B(\sin(\pi f T_B)/\pi f T_B)^2\right)$, where T_B is the bit duration.

For K fiber spans, the transfer function becomes [68]

$$H_{12}(f) = 2\gamma \frac{1 - e^{-\alpha L + j2\pi f d_{12} L}}{\alpha - j2\pi f d_{12}} \times \frac{1 - e^{-j2\pi f(1-\kappa) d_{12} K L}}{\alpha - j2\pi f(1-\kappa) d_{12} L}. \tag{2.85}$$

Here, κ is the fraction of optical per-span CD compensation, that is, $\kappa = 1$ for perfect per-span CD compensation, and $\kappa = 0$ without optical CD compensation, respectively. The transfer function (2.85) is valid for K identical spans.

In Ref. [69], X-NLPN-related penalties on a reference SNR for directly detected DQPSK and coherent DP-QPSK (see Sections 5.3.2.3 and 5.4, respectively) have been derived. For DQPSK, the penalty is given by

$$\mathrm{PEN}_{\mathrm{Sens}} \cong -8.5 \log_{10}(1 - \mathrm{SNR}_{\mathrm{ref}} \mathrm{Var}[\varphi_{\mathrm{NL}}(t)]). \tag{2.86}$$

For DP-QPSK, the penalty yields

$$\mathrm{PEN}_{\mathrm{Sens}} \cong -7.3 \log_{10}(1 - 1.75 \cdot \mathrm{SNR}_{\mathrm{ref}} \mathrm{Var}[\varphi_{\mathrm{NL}}(t)]), \tag{2.87}$$

Again, ϕ_{NL} is the XPM-induced Gaussian phase-noise process.

For the strength of X-NLPN and consequently the ratio between SPM-induced NLPN and X-NLPN, the dispersion map of possible optical CD compensation as well as the CD of the transmission fiber (G.652 SSMF vs. G.653 DSF or G.655 NZ-DSF) are relevant. If strong channel walk-off (differential delay between WDM channels) is given, X-NLPN becomes less important. This also explains the significant difference in penalties that are caused by Dispersion Compensating Fibers (DCFs) versus channelized fiber Bragg gratings (FBGs) as CD compensators on coherent 100 Gb/s systems [70]. The FBGs allow channel walk-off, whereas the DCFs periodically (per span) compensate the walk-off.

2.2.2 Scattering Effects

2.2.2.1 Stimulated Raman Scattering Raman scattering is a nonlinear effect that depends on the material properties of silica glass [2,25,39,40]. The effect has memory, and is time-invariant to good approximation. It leads to scattering of photons at glass molecules with certain nuclear and vibrational states. These states become dependent on the incoming intensity that leads to nonlinear behavior. Since the molecular states cover a certain frequency spectrum, the scattered photons are shifted in their frequency by this spectrum. Due to the amorphous structure of silica glass, harmonic broadening occurs, leading to continuous scattering spectra. The Raman shift only depends on the properties of the silica material, it is not depending on the incoming frequencies of the light waves. Raman scattering is also called *inelastic scattering*, unlike Rayleigh scattering where frequencies are conserved.

During scattering, the system molecule-plus-photon is in a virtual, that is, forbidden, energy state. Hence, scattering takes place very quickly, within $\sim$1 ps [2]. Regarding the resulting frequency shift, two cases must be separated:

- Generation of a Stokes wave
- Generation of an anti-Stokes wave

Raman scattering primarily occurs in the Stokes band. Then, photons (of the incident pump wave) loose part of their energy that is converted into molecular vibrational energy (i.e., phonons). Hence, scattered photons with lower energy and lower frequency—Stokes band—must be emitted. If photons are scattered by excited molecules, emission in the anti-Stokes band at higher frequencies is possible by converting part of the molecular energy onto the photon. This is also called superelastic scattering [71].

In optical fibers, (stimulated) Raman scattering always leads to emissions in the Stokes band, that is, with lower frequencies as compared to the pump waves. Due to the necessary matching of the photon and phonon wave vectors, anti Stokes emission is possible in gases, plasmas, or bulk silica only. It is used for example in coherent anti-Stokes Raman spectroscopy (CARS) [71].

The energy states during Stokes and anti-Stokes Raman scattering are depicted in Fig. 2.28. For comparison, the figure also shows elastic scattering, absorption, and fluorescence.

Spontaneous Raman scattering is of minor importance in fiber-optic transmission due to its statistical, noise-like nature. Above certain threshold intensity however, the effect can be stimulated, leading to coherent emission. This SRS is relevant in fiber optics. The change of the molecular state and the emission of (Stokes) photons is stimulated by already emitted photons. This requires suitably high intensity of the incident (pump) wave and is similar in principle to light amplification in a traveling wave amplifier. This is particularly true for the phase of Stokes photons that are automatically phase-matched to the stimulating Stokes photons. Unlike FWM, which requires phase matching of the optical waves, SRS only depends on intensity. The phase is automatically matched by the phonons.

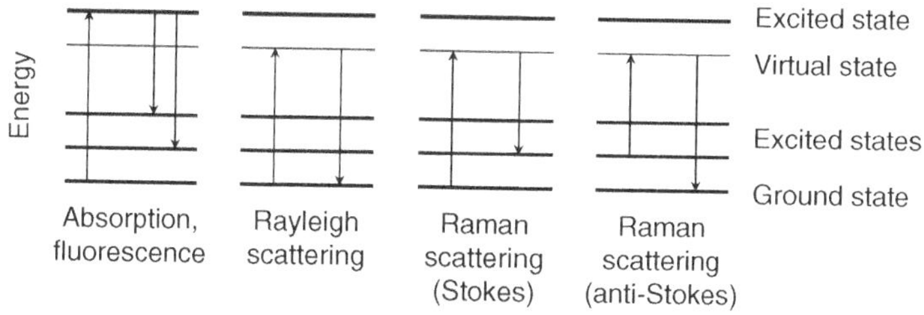

FIGURE 2.28 Elastic and inelastic scattering effects.

The gain of a Stokes band signal S and the corresponding attenuation of the pump signal P are given by a pair of coupled propagation equations [2]:

$$\begin{aligned} \frac{\mathrm{d}P_S}{\mathrm{d}z} &= -\alpha_S P_S + \frac{g_R}{A_{\mathrm{eff}}} P_P P_S, \\ \frac{\mathrm{d}P_P}{\mathrm{d}z} &= -\alpha_P P_P - \frac{\lambda_S}{\lambda_P}\frac{g_R}{A_{\mathrm{eff}}} P_S P_P. \end{aligned} \tag{2.88}$$

α_S and α_P are the attenuation coefficients of the amplified Stokes signal and the pump, respectively, and A_{eff} is the effective fiber cross-sectional area. Neglecting fiber attenuation (i.e., assuming strong Raman amplification), amplification of Stokes waves of incident power $P_S(0)$ along a fiber of length L through SRS is then given by

$$P_S(L) = P_S(0) \cdot \exp\left[G_R(\Delta f) I_0 L\right]. \tag{2.89}$$

Here, I_0 is the intensity of the optical pump, and $G_R(\Delta f)$ is the frequency-dependent Raman gain. It has a maximum of $G_R(\Delta f_R) \approx 10^{-13}$ m/W at a Stokes shift of $\Delta f_R \approx 13$ THz [2,26–28]. $G_R(\Delta f)$ is shown in Fig. 2.29.

Figure 2.29 also shows triangle (dashed) and single Lorentzian (dotted) approximations that were sometimes used for numerical descriptions [40,42,72].

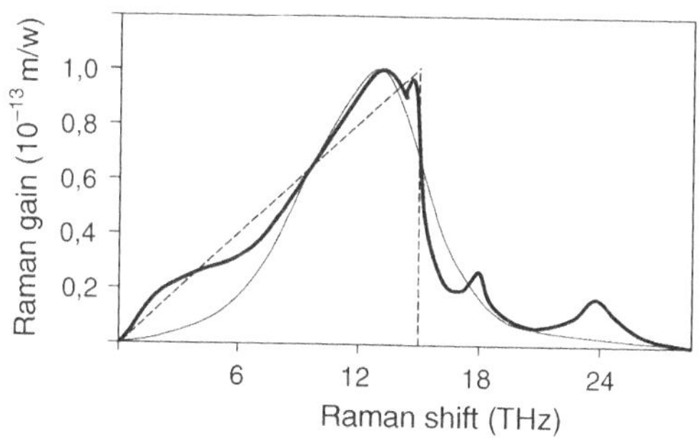

FIGURE 2.29 Raman gain spectrum.

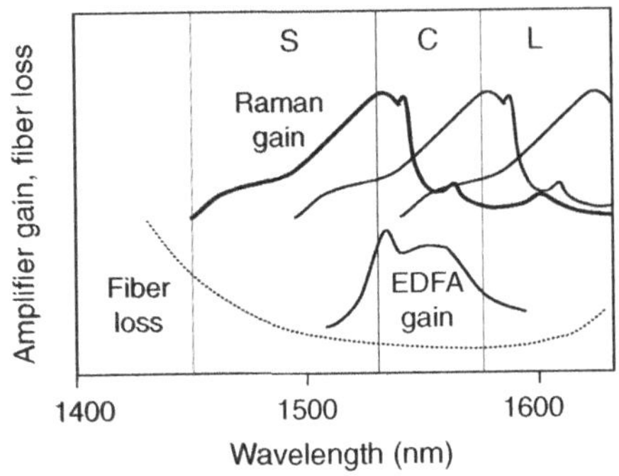

FIGURE 2.30 Multiple pump Raman gain versus EDFA gain.

Raman gain from a single optical pump does not provide equal gain for broadband multichannel systems. It is, however, possible to combine several pump lasers at different frequencies, together with gain flattening filters. This way, the gain spectrum can be broadened, equalized, and shifted into the wavelength region of interest. This is schematically shown in Fig. 2.30 that also shows the gain spectrum of an EDFA for comparison.

Raman amplifiers can be used as counterdirectionally pumped preamplifiers for (standard) Erbium-doped fiber amplifiers (EDFAs), or in exclusively Raman-amplified systems. The latter can make use of co- and counterdirectionally pumped Raman amplifiers for maximum span lengths and optimum noise figure performance [73].

SRS is not only a means for broadband (and low-noise) amplification but also leads to several detrimental effects in WDM multichannel transmission. SRS can lead to cross talk between WDM channels (where higher-frequency channels pump lower-frequency channels that fall into their Stokes bands). In WDM systems, this leads to power loss of higher frequency channels and consequently a gain tilt, which is shown in Fig. 2.31.

The SRS-induced mean tilt can be approximated for multichannel WDM systems:

$$\bar{a}_{\mathrm{SRS,tilt}}\,(\mathrm{dB}) = 10\log{}_{e}\Delta\lambda \cdot N_{\mathrm{ch}}(N_{\mathrm{ch}}-1)\frac{L_{\mathrm{eff}}}{2\cdot A_{\mathrm{eff}}}\frac{\partial g_{\mathrm{R}}}{\partial\lambda}P_{\mathrm{SRS}}\,(\mathrm{mW}). \tag{2.90}$$

Here, N_{ch} is the WDM channel count, A_{eff} is the effective fiber cross-sectional area [2,35], and L_{eff} is the effective interaction length given by $L_{\mathrm{eff}} = [1 - \exp(-\alpha L)]/\alpha$,

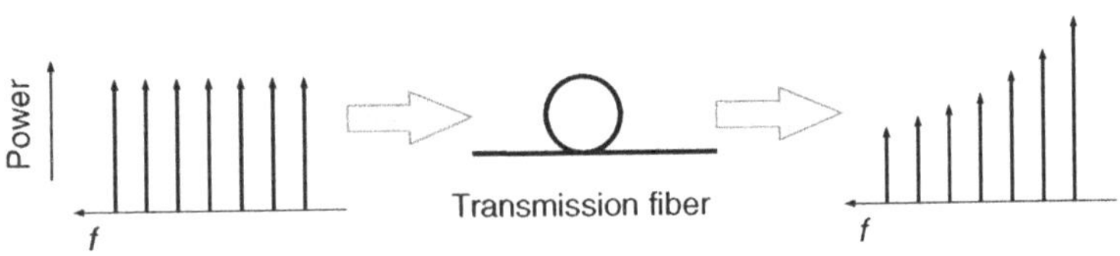

FIGURE 2.31 Gain tilt in WDM channels due to SRS.

with α the fiber absorption coefficient at the respective pump wavelength. $\partial G_R/\partial\lambda$ can be approximated as $\partial G_R/\partial\lambda = G_{R\,peak}/\delta\lambda_{peak}$, where G_R is the Raman gain as per Fig. 2.29.

In a multichannel WDM system, each (anti-Stokes) channel contributes equally to SRS-induced cross talk. Then, gain is $\propto \Delta\lambda$, and walk-off between the channels is $\propto 1/\Delta\lambda$. The channel walk-off leads to fast decorrelation of all channels (for small walk-off, that is, channel spacing, also the Raman gain coefficient is small). This leads to statistical (noise-like) accumulation of SRS with increasing span number N_{sp}:

$$P_{SRS} = P_1\sqrt{N_{sp}}. \tag{2.91}$$

If the Raman-induced loss for the anti-Stokes channels has to be <1 dB, the product of total bandwidth and total power of all channels has to be <50 GHz W [28]. In addition, SRS leads to cross talk in the Stokes channels according to the (amplitude-modulated) bit patterns in the anti-Stokes channels.

2.2.2.2 Brillouin Scattering Brillouin scattering can be described as inelastic scattering caused by spatial fluctuations of the refractive index that in turn is caused by density fluctuations induced by hypersonic sound waves [25]. The fluctuations can also be caused by applying electrical fields, which is known as electrostriction. Brillouin scattering only leads to emissions in the Stokes band. Due to matching of the wave vectors of the pump and Stokes wave and the hypersonic phonons, the scattered Stokes wave must propagate in the counter direction of the pump. This follows from the conservation law of energy. Since the phase constants β_S and β_P of the Stokes-shifted (amplified) optical wave and the optical pump wave are orders of magnitude larger than the phase constant β_A of the acoustic phonon wave, $\beta_S \approx \beta_P \gg \beta_A$, it follows immediately for the wave vectors k_S, k_P, and k_A that

$$k_A = k_P + k_S, \quad k_P \approx -k_S. \tag{2.92}$$

Stimulated Brillouin scattering occurs when the effect is self-amplifying. The optical pump and the counterpropagating Brillouin Stokes wave amplify the hypersonic sound wave that can be triggered by thermal oscillations. Similar to SRS, SBS does not require phase matching between the optical waves. This is automatically provided by the acoustic wave.

SBS has the lowest power threshold of all nonlinear effects and can hence limit multichannel WDM systems in their maximum per-channel power. The maximum Brillouin gain can be as high as $G_B(\Delta f_B) \approx 4 \times 10^{-11}$ m/W and is achieved at a Stokes shift of ~11 GHz. Depending on fiber geometry, the SRS gain bandwidth for optical pumps in the 1550 nm window is limited to 20–100 MHz.

Amplification of Stokes signals of incident power $P_S(0)$ is described in analogy to SRS:

$$P_S(L) = P_S(0) \cdot \exp\left[G_B(\Delta f)I_0L\right]. \tag{2.93}$$

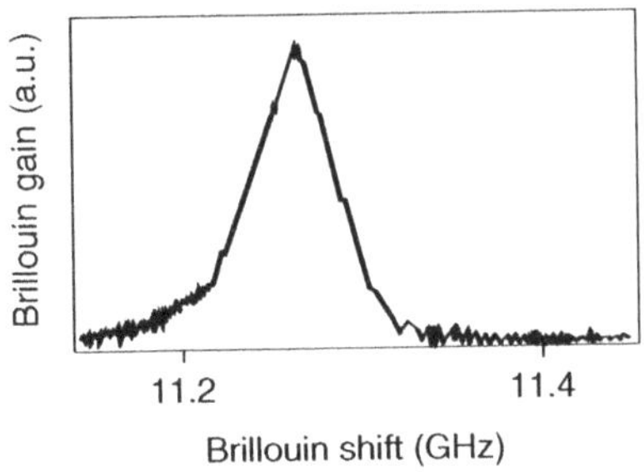

FIGURE 2.32 Brillouin gain of a G.653 fiber [2].

Again, I_0 is the intensity of the optical pump, L is the length of the fiber, and $G_B(\Delta f)$ is the frequency-dependent Brillouin gain. An example of Brillouin gain of a dispersion-shifted SMF is shown in Fig. 2.32 [2].

SRS causes backscattering and can hence be blocked in dual-fiber transmission (i.e., unidirectional transmission per fiber) by means of optical isolators. However, due to the low SBS threshold, per-channel power must be <10 dBm in order to avoid reflection (and hence loss) of significant amounts of power [26–28]. Due to the narrowband gain characteristics, SBS is also reduced by widening the laser linewidths within a WDM channel. This can be achieved through additional slow amplitude modulation (in the MHz range, which can be used for modulating pilot-tone-based in-band communication channels), or by using directly modulated lasers where the linewidth is broadened by parasitic effects. Figure 2.33 shows the efficiency of slow (FM) pilot tones for SBS suppression.

SBS can become a problem in coherent ultradense WDM passive optical (access) networks (UDWDM-PONs). These PONs can provide dedicated per-wavelength broadband access with a potentially very large number of ultradensely (3–6 GHz) spaced wavelengths. As PON access systems, they must support single fiber working. Then, upstream and downstream wavelengths propagate in the same fiber

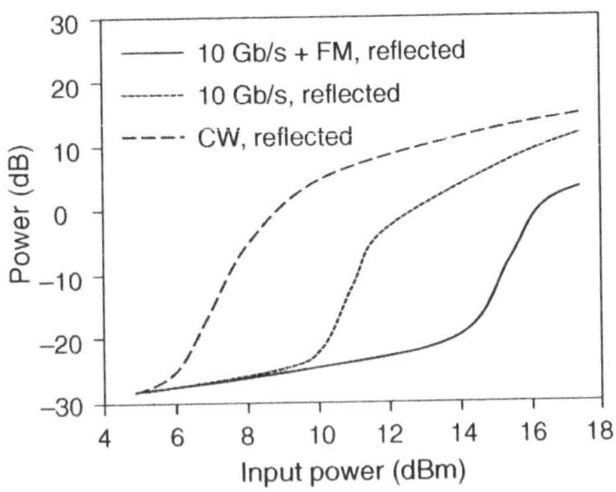

FIGURE 2.33 SBS thresholds with and without pilot tones.

(i.e., isolators must not be used in the fiber links), and pairs of upstream and downstream wavelengths can fall within the Brillouin Stokes shift. Since these PONs also aim at coherent detection (otherwise, the wavelengths cannot be separated), the use of lasers with large linewidths is also prohibited.

REFERENCES

1. G.P. Agrawal, Fiber-Optic Communication Systems, John Wiley & Sons, New York, 1992.
2. G.P. Agrawal, Nonlinear Fiber Optics, 2nd ed., Academic Press, San Diego, 1995.
3. D. Marcuse, Curvature loss formula for optical fibers, J. Opt. Soc. Am., Vol. 66, No. 3, March 1976, pp. 216 ff.
4. D. Marcuse, Influence of curvature on losses of doubly clad fibers, Appl. Opt., Vol. 21, No. 23, December 1982, pp. 4208 ff.
5. S.B. Andreasen, New bending loss formula explaining bends on loss curve, Electron. Lett., Vol. 23, No. 21, October 1987, pp. 1138 ff.
6. I.A. White, Radiation from bends in optical waveguides: the volume current method, J. Microwave, Opt. Acoust., Vol. 3, No. 5, September 1979, pp. 186 ff.
7. W.B. Gardner, Microbending loss in optical fibers, Bell Syst. Tech. J., Vol. 54, No. 2, 1975, pp. 457–465.
8. J.-I. Sakai and T. Kimura, Practical microbending loss formula for single-mode optical fibers, IEEE J. Quantum Electron., Vol. QE-15, No. 6, 1979, pp. 497–500.
9. E. Lichtman, Limitations imposed by polarization-dependent gain and loss on all-optical ultralong communication systems, IEEE J. Lightwave Technol., Vol. 13, No. 5, May 1995.
10. H. Kogelnik, R.M. Jopson, and L.E. Nelson, Polarization-mode dispersion, in I. Kaminow and T. Li (eds), Optical Fiber Telecommunications IV B, Elsevier Science, San Diego, 2002, pp. 725–861.
11. G. Grau and W. Freude, Optische Nachrichtentechnik, 3rd ed., Springer, Berlin, 1991.
12. S. Geckeler, Lichtwellenleiter für die optische Nachrichtenübertragung, Springer, Berlin, 1986.
13. Z.M. Zhang and Keunhan Park, On the group front and group velocity in a dispersive medium upon refraction from a nondispersive medium, J. Heat Transfer, Vol. 126, April 2004, pp. 244–249.
14. J.C. Garrison et al., Superluminal signals: causal loop paradoxes revisited, Phys. Lett. A, Vol. 245, 1998, pp. 19–25.
15. J.P. Gordon and H. Kogelnik, PMD fundamentals: polarization mode dispersion in optical fibers, PNAS, Vol. 97, No. 9, April 2000, pp. 4541–4550.
16. G.J. Foschini et al., Probability densities of second-order polarization mode dispersion including polarization dependent chromatic fiber dispersion, IEEE Photonics Technol. Lett., Vol. 12, No. 3, March 2000.
17. H. Kogelnik et al., Emulation and inversion of polarization-mode dispersion, IEEE J. Lightwave Technol., Vol. 21, No. 2, February 2003.
18. H. Bülow, System outage probability due to first- and second-order PMD, IEEE Photonics Technol. Lett., Vol. 10, No. 5, May 1998.

19. Recommendation ITU-T G.666, Characteristics of PMD Compensators and PMD Compensating Receivers, March 2008.
20. F. Buchali and H. Bülow, Adaptive PMD compensation by electrical and optical techniques. IEEE J. Lightwave Technol., Vol. LT-22, No. 4, April 2004.
21. S. Lee et al., A short recirculating fiber loop testbed with accurate reproduction of Maxwellian PMD statistics, Proceedings of the OFC 2001, March 2001, Paper WT2.
22. J. Peters et al., Bellcore's fiber measurement audit of existing cable plant for use with high bandwidth systems, NFOEC1997, San Diego, CA, September 1997.
23. S. Barcelos et al., Polarization mode dispersion (pmd) field measurements: an audit of Brazilian newly installed fiber networks, OFC2005, Anaheim, CA, March 2005.
24. R.-J. Essiambre et al., Capacity limits of optical fiber networks, IEEE J. Lightwave Technol., Vol. 28, No. 4, 2010, pp. 662–701.
25. D.L. Mills, Nonlinear Optics, Springer, Berlin, 1991.
26. R.H. Stolen, Nonlinear properties in optical fibers, in S.E. Miller and A.G. Chynoweth (eds), Optical Fiber Telecommunications, Academic Press, San Diego, 1979.
27. R.H. Stolen, Nonlinearity in fiber transmission, Proc. IEEE Vol. 68, No. 10, 1980, pp. 1232 ff.
28. A.R. Chraplyvy, Limitations on lightwave communications imposed by optical-fiber nonlinearities, J. Opt. Commun., Vol. 8, No. 10, 1990, pp. 1548 ff.
29. D. Marcuse et al., Effect of fiber nonlinearity on long-distance transmission, IEEE J. Lightwave Technol., Vol. 9, No. 1, 1991, pp. 121 ff.
30. A. Bononi et al., Transmission limitations due to fiber nonlinearity, OFC2007, Los Angeles, March 2007, Paper OWO7.
31. A. Bononi et al., Modeling of signal-noise interactions in nonlinear fiber transmission with different modulation formats, Invited Paper, ECOC2009, Vienna, September 2009, Paper 10.4.6.
32. P.J. Winzer and R.-J. Essiambre, Advanced modulation formats for high-capacity optical transport networks, invited paper, IEEE J. Lightwave Technol., Vol. 24, No. 12, 2006, pp. 4711–4728.
33. R.H. Stolen and C. Lin, Self-phase-modulation in silica optical fibers, Phys. Rev. A, Vol. 17, No. 4, 1978, pp. 1448 ff.
34. A. Boskovic et al., Direct continuous-wave measurement of n_2 in various types of telecommunication fiber at 1.55 μm, Opt. Lett., Vol. 21, No. 24, 1996, pp. 1966 ff.
35. K.S. Kim et al., Measurement of the nonlinear index of silica-core and dispersion-shifted fibers, Opt. Lett., Vol. 19, No. 4, 1994, pp. 257 ff.
36. M. Schetzen, The Volterra and Wiener Theories of Nonlinear Systems, John Wiley & Sons, Inc., New York, 1980.
37. D. Stoll, Nichtlineare Wellenausbreitung in Single-Mode-Fasern, VDI Verlag, Düsseldorf, 1994.
38. K. Grobe and H. Braunisch, A broadband model for single-mode fibers including nonlinear dispersion, Prog. Electromagn. Res., Vol. PIER 22, 1999, pp. 131–148.
39. Y.S. Kivshar and B.A. Malomed, Raman-induced optical shocks in nonlinear fibers, Opt. Lett., Vol. 18, No. 7, 1993, pp. 485 ff.
40. R.H. Stolen et al., Raman response function of silica-core fibers, J. Opt. Soc. Am. B, Vol. 6, No. 6, 1989, pp. 1159 ff.

41. T.R. Taha and M.J. Ablowitz, Analytical and numerical aspects of certain nonlinear evolution equations. II. Numerical, nonlinear Schrödinger equation, J. Comput. Phys., Vol. 55, No. 2, 1984, pp. 203–230.
42. P.L. Francois, Nonlinear propagation of ultrashort pulses in optical fibers: total field formulation in the frequency domain, J. Opt. Soc. Am. B, Vol. 8, No. 2, 1991, pp. 276 ff.
43. M.F. Ferreira, Nonlinear Effects in Optical Fibers, John Wiley & Sons, Inc., Hoboken, 2011.
44. I.B. Djordjevic et al., Constrained coding techniques for the suppression of intrachannel nonlinear effects in high-speed optical transmission, IEEE J. Lightwave Technol., Vol. 24, No. 1, 2006, pp. 411–419.
45. I.B. Djordjevic et al., Suppression of intrachannel nonlinear effects using pseudoternary constrained codes, IEEE J. Lightwave Technol., Vol. 24, No. 2, 2006, pp. 769–774.
46. L.F. Mollenauer et al., Experimental observation of picosecond pulse narrowing and solitons in optical fibers, Phys. Rev. Lett., Vol. 45, No. 13, 1980, pp. 1095 ff.
47. R.K. Dodd et al., Solitons and Nonlinear Wave Equations, Academic Press, New York, 1982.
48. F.M. Mitschke and L.F. Mollenauer, Discovery of the soliton self-frequency shift, Opt. Lett., Vol. 11, No. 10, 1986, pp. 659 ff.
49. L.F. Mollenauer et al., Wavelength division multiplexing with solitons in ultra-long distance transmission using lumped amplifiers, IEEE J. Lightwave Technol., Vol. 9, No. 3, 1991, pp. 362 ff.
50. S.G. Evangelides et al., Polarization multiplexing with solitons, IEEE J. Lightwave Technol., Vol. 10, No. 1, 1992, pp. 28 ff.
51. A. Hasegawa and Y. Kodama, Solitons in Optical Communications, Clarendon Press, Oxford, 1995.
52. L.F. Mollenauer et al., Demonstration of soliton WDM transmission at up to 8×10 Gb/s, error-free over transoceanic distances, Optical Fiber Communication Conference, Vol. 2 of 1996 OSA Technical Digest Series, OSA, 1996, Paper PD22.
53. F. Favre et al., Experimental evidence of pseudoperiodical soliton propagation in dispersion-managed links, IEEE J. Lightwave Technol., Vol. 17, No. 6, 1999, pp. 1032 ff.
54. M. Eiselt et al., Contribution of timing jitter and amplitude distortion to XPM system penalty in WDM systems, IEEE Photonics Technol. Lett., Vol. 11, No. 6, 1999, pp. 748ff.
55. M. Winter et al., Polarization-multiplexed system outage due to nonlinearity-induced depolarization, ECOC MMX, Turino, September 2010, Paper Th.10.E.3.
56. L.F. Mollenauer et al., Polarization scattering by soliton–soliton collisions, Opt. Lett., Vol. 20, 1995, pp. 2060–2062.
57. H. Louchet et al., Comparison of XPM and XpolM-induced impairments in mixed 10G–100G transmission, ICTON 2011, Stockholm, June 2011, Paper We.B1.1.
58. M. Karlsson and H. Sunnerud, Effects of nonlinearities on PMD-induced system impairments, IEEE J. Lightwave Technol., Vol. 24, No. 11, 2006, pp. 4127–4137.
59. K.-H. Löcherer and C.-D. Brandt, Parametric Electronics, Springer, Berlin, 1982.
60. M. Eiselt, Limits on WDM systems due to four-wave mixing: a statistical approach, IEEE J. Lightwave Technol., Vol. 17, No. 11, 1999, pp. 2261ff.
61. N. Shibata et al., Phase-mismatch dependence of efficiency of wave generation through four-wave mixing in a single-mode optical fiber, IEEE J. Quantum Electron., Vol. QE-23, No. 7, 1987, pp. 1205–1210.

62. F. Forghieri et al., Reduction of four-wave mixing crosstalk in WDM systems using unequally spaced channels, IEEE Photonics Technol. Lett., Vol. 6, No. 6, 1994, pp. 754ff.
63. I. Kaminov and T. Koch (eds), Optical Fiber Telecommunications IIIa, Academic Press, San Diego, 1997.
64. J.P. Gordon and L.F. Mollenauer, Phase noise in photonic communications systems using linear amplifiers, Opt. Lett., Vol. 15, No. 23, 1990, pp. 1351–1353.
65. K.-P. Ho, Error probability of DPSK signals with cross-phase modulation induced nonlinear phase noise, IEEE J. Sel. Top. Quantum Electron., Vol. 10, No. 2, 2004, pp. 421–427.
66. K.-P. Ho, The optimal compensator for nonlinear phase noise, Opt. Commun., Vol. 221, 2003, pp. 419–425.
67. K.-P. Ho, Probability density of nonlinear phase noise, J. Opt. Soc. Am. B, Vol. 20, No. 9, 2003, pp. 1875–1879.
68. K.-P. Ho, Can 100-Gb/s QPSK Signal Locate Adjacent to Legacy 10-Gb/s OOK Signal without Guard-Band? Available at arXiv:1004.1832v1 [physics.optics] 11 April, 2010.
69. A. Bononi et al., Cross-phase modulation induced by OOK channels on higher-rate DQPSK and coherent QPSK channels, IEEE J. Lightwave Technol., Vol. 27, No. 18, 2009, pp. 3974–3983.
70. C. Xie, Suppression of inter-channel nonlinearities in WDM coherent PDM-QPSK systems using periodic-group-delay dispersion compensators, ECOC2009, Vienna, September 2009, Paper P4.08.
71. W. Demtröder, Laser Spectroscopy: Basic Concepts and Instrumentation, 3rd ed., Springer, Berlin, 2002.
72. K.J. Blow and D. Wood, Theoretical description of transient stimulated Raman scattering in optical fibers, IEEE J. Quantum Electron., Vol. 23, No. 12, 1989, pp. 2665 ff.
73. M. Premaratne, Analytical characterization of optical power and noise figure of forward pumped Raman amplifiers. Opt. Express, Vol. 12, No. 18, September 2004, pp. 4235–4245.

3

COMPONENTS AND SUBSYSTEMS

End-to-end wavelength division multiplexing (WDM) transmission consists of a transmitter, a transmission channel, and a receiver. The transmission channel in all relevant WDM applications consists of a silica single-mode fiber (SMF). Different SMF types are described in Section 3.2. WDM transmitters and receivers consist of several functional stages, as shown in Fig. 3.1.

Necessary functions in the transmit/receive end include channel coding/decoding (i.e., the application of forward error correction (FEC)), modulation/demodulation, and WDM and optional polarization multiplexing/demultiplexing. Optional functions in WDM transmission systems include encryption/decryption, time-domain multiplexing/demultiplexing, and optical booster and/or preamplification. Source coding, as a means for redundancy reduction, is typically performed in the client layers. However, in certain cases interframe gaps are removed, or re- or transcoding is applied (e.g., from 8B/10B to 64B/66B) in order to reduce source redundancy and to increase transport efficiency.

3.1 TRANSMITTERS

Here, we consider semiconductor laser diodes as means for WDM transmitters only.

3.1.1 Laser Diodes

3.1.1.1 Principle of Diode Lasers Laser diodes are based on semiconductor p–n junctions [1,2]. As compared to edge-emitting LEDs, they require an optical cavity

Wavelength Division Multiplexing: A Practical Engineering Guide, First Edition.
Klaus Grobe and Michael Eiselt.

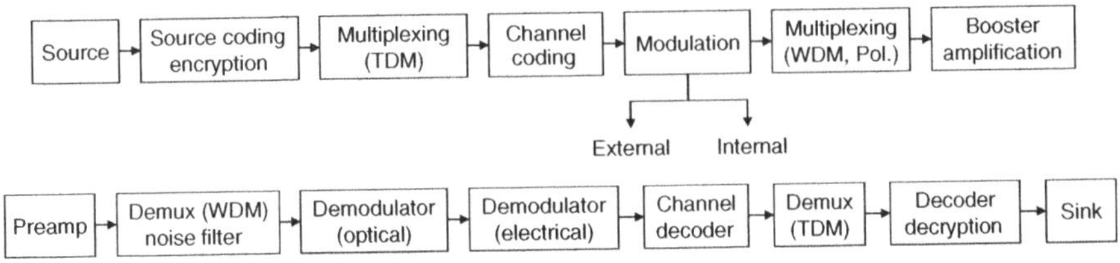

FIGURE 3.1 Transmit end (top) and receive end (bottom) of WDM transmission system.

(i.e., a resonator) that provides feedback in order to elevate the intensity of stimulated emission, as shown in the right-hand part of Fig. 3.2. This enables an output of continuous coherent radiation. A simple cavity is provided by cleaving the ends of the crystal, followed by polishing. Then, photons reflected from the cleaved surfaces stimulate more photons of the same frequency.

The wavelength λ of the radiation that escalates in the cavity is dependent on the length L of the cavity (the resonant length). Only multiples of $\lambda/2$ can be excited.

$$m\frac{\lambda}{2n} = L. \tag{3.1}$$

Here, n is an integer. Separation between the potential modes that can develop, or allowed wavelengths, is determined by

$$\Delta\lambda = \frac{\lambda^2}{2nL}. \tag{3.2}$$

The output spectrum of the laser diode depends on the geometry of the optical cavity and wavelength-dependence of the optical gain.

A *homojunction* laser diode is one where the p–n junction uses the same direct-bandgap semiconductor material throughout the entire component. A drawback of homojunction structures is the high threshold current density that requires operation at very low temperatures. This is improved in *heterostructure* laser diodes. Reduction of threshold current to usable level is achieved by an improvement of the rate of stimulated emission as well as the efficiency of the optical cavity [2].

- *Carrier Confinement.* Injected electrons and holes are confined to a narrow region around the junction. This allows less current to establish the required concentration of electrons for population inversion.

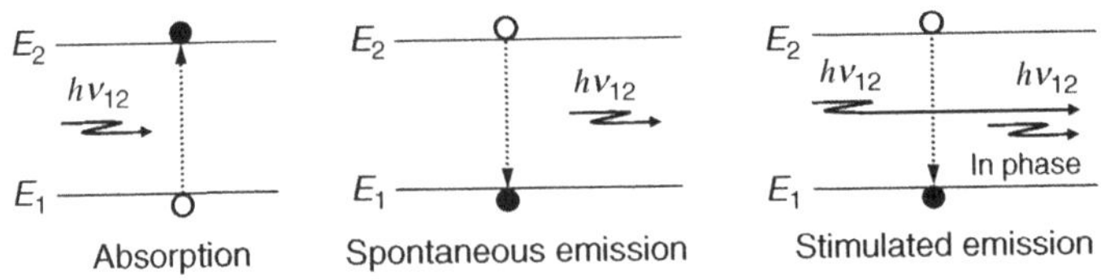

FIGURE 3.2 Absorption and emission in diode lasers.

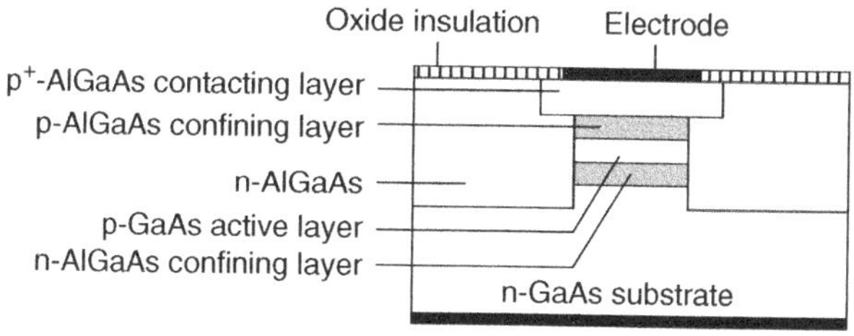

FIGURE 3.3 Cross section of a buried double-heterostructure diode laser.

- *Optical Confinement.* A dielectric waveguide is constructed around the optical gain region to increase the photon concentration and elevate the probability of stimulated emission.

The waveguide is typically constructed in a *double-heterostructure*, where the active region consists of one material (e.g., GaAs), and the outer p- and n-layers of a second material (e.g., AlGaAs). A semiconductor with wider bandgap (here: AlGaAs) also has lower refractive index (than GaAs in this case). This refractive index difference establishes an optical dielectric waveguide that confines photons to the active region. In addition, the active layer is also confined to a stripe geometry by cladding it laterally with lower index material. An example of cross section of this so-called buried heterostructure is shown in Fig. 3.3.

In the example shown in Fig. 3.3, the substrate is n-GaAs. The confining layers are n-AlGaAs and p-AlGaAs, and the active layer again is p-GaAs (emission at 870–900 nm). The additional p-GaAs contacting layer allows better electrode contact. The p-AlGaAs and n-AlGaAs layers provide carrier and optical confinement by forming heterojunctions with the active p-GaAs layer. The stripe geometry (lateral confinement by n-AlGaAs) forces the current to be maximum along the central path, and to diminish on either side. The reduced contact of the stripe geometry reduces the threshold current. It also reduced the emission area at the end facet that makes light coupling to fibers easier.

Semiconductor materials for the wavelength range of 300 nm to 10 μm are summarized in Fig. 3.4 [3,4]. Various options for the relevant WDM range exist.

Lasing occurs when the supply of free electrons exceeds the losses in the cavity. The current through the junction and the electron supply are directly proportional. With an applied voltage greater than the (heterojunction) bandgap energy, the barrier potential is reduced to 0 allowing electrons to flow into the space-charge layer (SCL) and over to the p+-side to establish a diode current. A similar reduction in barrier potential for holes flowing from the p+-side to the n+-side occurs. As a result, the SCL is no longer depleted, and more electrons can be generated in the conduction band at energies near E_C than electrons in the valence band near E_V. This results in population inversion in energies near E_C and E_V (the energy levels in the conduction and valence bands, respectively). The region where population inversion occurs develops a narrow layer along the junction called inversion layer or active region.

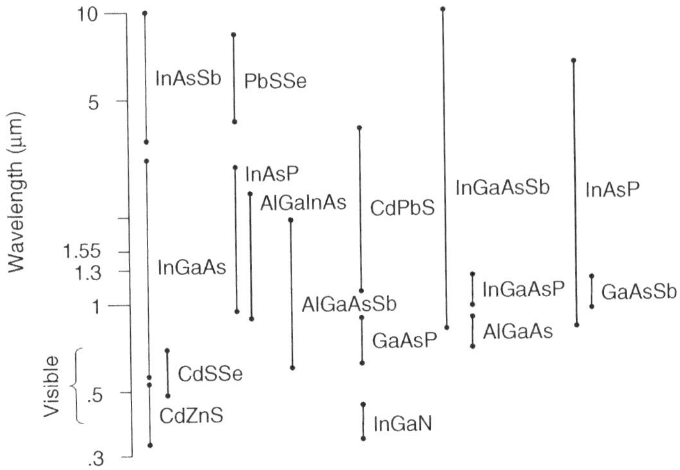

FIGURE 3.4 Semiconductor materials for laser diodes.

Population inversion through a forward diode current is also called *injection pumping*.

Lasing oscillations occur when the optical gain G in the cavity exceeds photon losses. This point where laser modes or resonant frequencies resonate within the cavity is determined by

$$G \geq (1/L)\ln(1/R) + \alpha_{\mathrm{I}} = a_{\mathrm{M}} + a_{\mathrm{I}}, \tag{3.3}$$

α_{I} and a_{M} are the average internal loss and the mirror loss from both facets, respectively, and the reflectivity R is given by

$$R = \left(\frac{n-1}{n+1}\right)^2. \tag{3.4}$$

Above threshold, the optical output power P_0 from both facets approximately is a linear function of the drive current I and can be described as $P_0 = \eta(I - I_{\mathrm{th}})$, where $\eta = \mathrm{d}P_0/\mathrm{d}I$ is the slope efficiency and I_{th} is the threshold current. The *external quantum efficiency* η_{E}, which is the ratio between photons that are coupled out of the cavity and electrons that are injected into the cavity and contribute to stimulated emission, is related to η by

$$\eta_{\mathrm{E}} = \frac{\eta q}{h\nu} = \frac{\eta_{\mathrm{I}}(G_{\mathrm{th}} - \alpha_{\mathrm{I}})}{G_{\mathrm{th}}}. \tag{3.5}$$

Here, q is the electron charge and $h\nu$ is the photon energy. The *internal quantum efficiency* η_{I} is the number of photons emitted per recombination event, it has been determined empirically to be 0.65 ± 0.05 for diode lasers. G_{th} is the threshold gain for lasing. The *total quantum efficiency* (TQE) is then given by the emitted optical power divided by the provided electrical power, $TQE = h\nu\eta_E/qV$. For GaAs laser diodes, $TQE \approx 50\%$, and for InGaAsP laser diodes, $TQE \approx 20\%$, respectively.

Direct current modulation of laser diodes causes both intensity and frequency modulation. The latter is also referred to as (frequency) chirp. As the electron density changes through current modulation, the gain (imaginary part of refractive index n_i) and the real part of the refractive index (n_r) both change. The susceptibility of a laser diode to chirp is characterized by the *transient chirp parameter* or *linewidth enhancement factor* α:

$$\alpha = \frac{\partial n_r}{\partial N} \bigg/ \frac{\partial n_i}{\partial N}. \tag{3.6}$$

N is the electron density. An approximated model of chirp is given by [5,6]

$$\Delta\nu(t) = -\frac{\alpha}{4\pi}\left(\frac{\partial P(t)/\partial t}{P(t)} + \kappa P(t)\right). \tag{3.7}$$

κ is the *adiabatic chirp parameter*. It is proportional to the nonlinear gain compression and inverse proportional to efficiency, photon energy, and cavity volume, respectively. $P(t)$ is the time-dependent laser output power per facet, before coupling into the fiber. If κ is relatively large and/or the bit rate is low, adiabatic chirp dominates, and the laser FM approximately follows the intensity modulation. If κ is relatively small and/or the bit rate is high, transient chirp (the derivative term in the equation) prevails that causes FM spikes at the pulse edges [6]. For harmonic $P(t)$, the frequency chirp can be derived as

$$\frac{\Delta\nu}{\Delta P} = -j\frac{\alpha f}{2P_0} + \kappa \quad \text{for } P(t) = P_0 + \Delta P\, e^{j\omega t}. \tag{3.8}$$

For good laser diodes, $\alpha = 1$–3 is achieved. Larger α implies higher chirp. Further, chirp gets worse with higher frequencies. Hence, relaxation oscillations that produce large $dP(t)/dt$ also lead to large chip. Consequently, damping of relaxation oscillations reduces chirp. Adjustment of material composition and laser mode volume can also reduce α.

In standard single-mode fibers (SSMF), chromatic dispersion (CD) is anomalous at 1550 nm. The refractive index of the fiber increases with wavelength. Therefore, longer wavelength signal components travel at slower speed. If a signal pulse has higher frequency components at the beginning of the pulse and lower frequency at the tail, it will experience broadening during propagation. If the order of higher frequency and lower frequency content at the leading and trailing pulse edges is reversed through intentional laser chirp, partial CD compensation can be accomplished.

If the laser transmitter does not turn off completely during the transmission of a "0," then the extinction ratio r_{EX} (i.e., the ratio of power P_0 transmitted during a "0" to power P_1 during a "1") will cause a bit error rate (BER) penalty and a reduction in

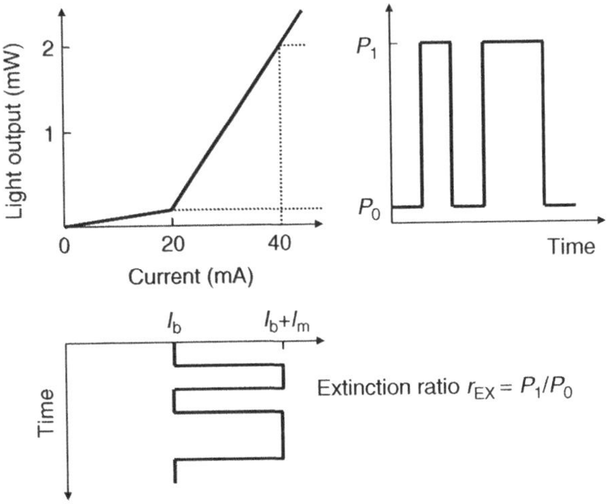

FIGURE 3.5 Extinction ratio.

sensitivity. The extinction-ratio penalty corrects for these emission during "0" bits. For $r_{EX} = P_0/P_1$, it is given by

$$\delta_{EX} = 10 \log \left(\frac{1 + r_{EX}}{1 - r_{EX}} \right). \tag{3.9}$$

If the extinction ratio is 0.1, the penalty is 0.87 dB. The dependence of the extinction ratio on the laser bias current is shown in Fig. 3.5.

Temperature drift and aging can shift the threshold current. If only constant-average-power control is used, the extinction ratio will be decreased. If fixed-bias control is used, average power and extinction ratio will be decreased, and chirp will be generated. Good stability of the output power is therefore achieved by automatic current control (ACC) together with temperature control (by means of a thermo-electrical cooler, TEC). Without cooling, ACC should not be used since output power can severely deviate [7].

Constant-current sources for CW laser diodes require tackling of robustness, stability, noise, and other issues such as protection against ESD (electrostatic discharge) damage, start-up spikes, overshoot, and possible fluctuation arising from external optical feedback [8]. A basic ACC is shown in Fig. 3.6 [9].

The basic laser-diode ACC supplies up to 250 mA via Q1. The amplifier controls laser current by maintaining the 1 Ω shunt voltage at a potential that the programming input dictates. Local compensation at the amplifier stabilizes the loop, and the 0.1 μF capacitor filters input commands. This prevents overshoot due to input dynamics. The enable input turns off the current source by simultaneously grounding the Q1 base and starving the amplifier's "+" input while biasing the "−" input high. This combination also ensures that the amplifier smoothly ramps to the desired output current when the enable switches low.

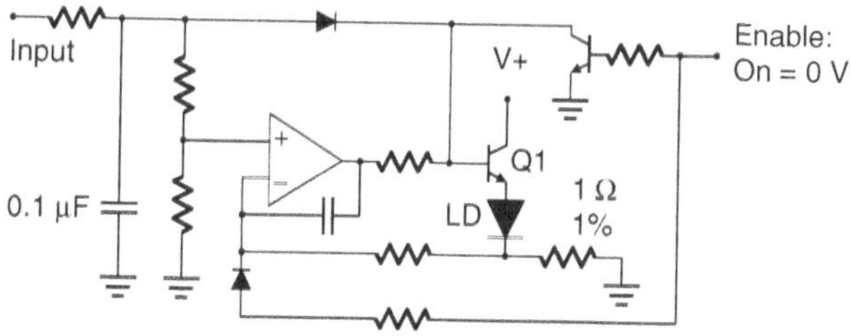

FIGURE 3.6 Basic ACC for laser diodes.

Automatic power control (APC) requires a feedback signal that is generated by a monitoring photodiode (MD). Three different connections of LD and MD are possible, N-type, P-type, and M-type. The common LD–MD port is connected to the enclosure of the laser diode, which typically is electrically and thermally connected to the chassis [7]. Figure 3.7 shows basic APC for N-type and P-type LD–MD combinations.

The N-type requires an output driver with negative supply voltage, the P-type requires positive voltage. The M-type (not shown here) requires dual supply voltage.

Many variants of LD drivers (LDDs) exist. An example of an APC driver for directly modulated LDs that also considers the modulation including duty cycle (mark density) is shown in Fig. 3.8.

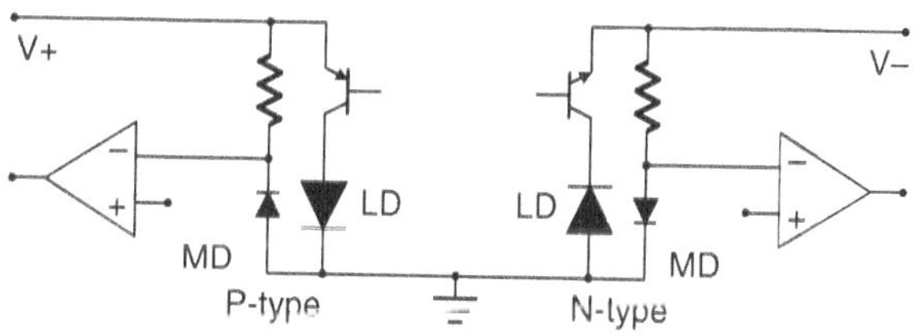

FIGURE 3.7 Basic APC for laser diodes.

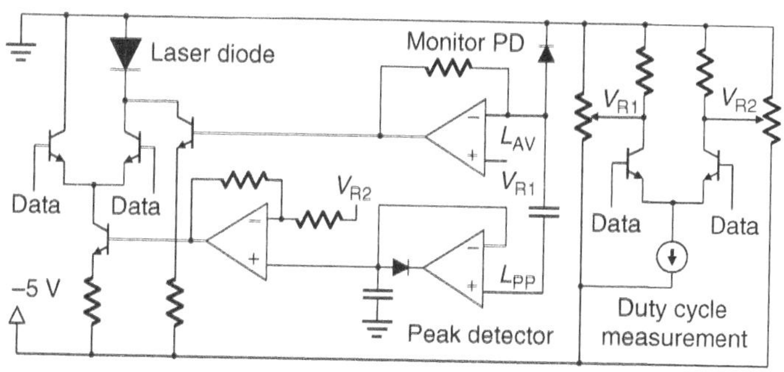

FIGURE 3.8 Laser diode control for average power, duty cycle, and modulation.

For low-speed, low-cost applications (like optical access), temperature control by means of a TEC can be omitted. Ideally, high-temperature substrates are used then (possibly together with a small heater to again keep temperature nearly constant).

In addition to the extinction-ratio penalty, an intensity-noise penalty can be defined that accounts for power fluctuations during "1" bits. Let r_I be the inverse of the SNR of the transmitter and amplifier output. If the SNR of the amplified signal is 13 dB = 20, then $r_I = 0.05$ and the power at the receiver has to be increased by the intensity-noise penalty δ_I to maintain the target SNR.

$$\delta_I = -10 \ \log\left(1 - r_I^2 Q^2\right). \tag{3.10}$$

Q is the Q-factor (see Section 5.3.1.2). If target $Q = 6$, a penalty of $\delta_I = 0.41$ dB follows.

The intensity fluctuations are characterized by the relative intensity noise (RIN) that is defined as the ratio of the mean value of the optical intensity fluctuation squared to the mean optical power squared at a specified frequency in a 1 Hz bandwidth.

Assuming 100% modulation depth and a ½ decision level, minimum BER as a function of total RIN can be estimated as BER = 0.5 erfc[1/(2√(2RIN))].

3.1.1.2 Fabry–Pérot Laser Diodes Fabry–Pérot laser diodes (FP-LD) are the simplest form of double-heterojunction laser diodes according to Fig. 3.3. The active layer forms a Fabry–Pérot cavity. Several lasing modes are allowed according to Eq. (3.2) and will consequently be excited. This leads to multilongitudinal mode (MLM) lasers, as opposed to single-longitudinal mode (SLM) lasers that require additional effort for mode selection. The number of excited modes, and hence the bandwidth, depends on the gain spectrum of the semiconductor material, as shown in Fig. 3.9. Though individual modes can be spectrally very narrow, it is the spectral width of their envelope that determines in particular the tolerance with regard to chromatic dispersion.

The simplified longitudinal structure of an FP-LD is shown in Fig. 3.10. This structure still applies to the buried double-heterostructure shown in Fig. 3.3. In the

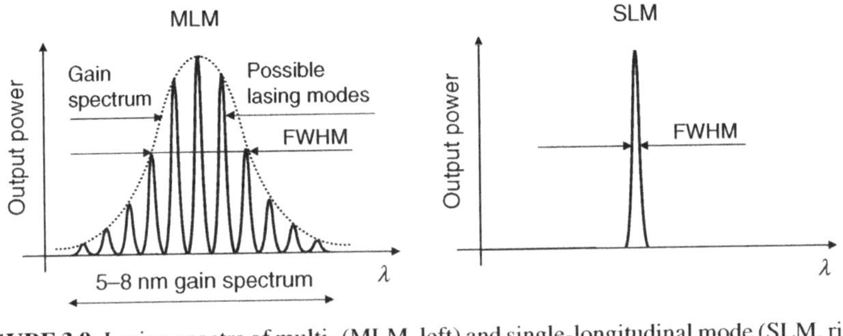

FIGURE 3.9 Lasing spectra of multi- (MLM, left) and single-longitudinal mode (SLM, right) lasers.

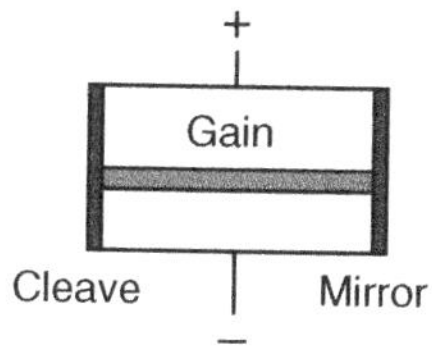

FIGURE 3.10 Fabry–Pérot laser diode.

longitudinal direction, the active layer consists of a Fabry–Pérot cavity, allowing several lasing modes.

The end facets are cleaved and polished (end facet, highly reflective mirror) or roughened (front facet, to decrease reflectivity).

Due to the broad lasing spectrum, FP-LDs are suitable for bit rates up to 2.5 Gb/s and distances up to 100 km (i.e., bit rate × length products of <250 Gb km/s) only. Within these limits, they are a common choice for low-cost (WDM) transport.

3.1.1.3 Distributed Feedback and Distributed Bragg Reflector Laser Diodes Distributed feedback (DFB) and distributed Bragg reflector (DBR) lasers are the most commonly used SLM lasers for high bit rate, long-distance WDM transmission. As compared to FP-LDs, a mode-selecting periodic grating is added to the longitudinal structure of Fig. 3.10. This grating is etched into the semiconductor material, either underneath the active layer (DFB LD) or adjacent to it (DBR LD). The two structures are shown in Fig. 3.11.

The main difference between the two structures is tunability. The DFB structure can inherently be tuned only within a narrow bandwidth via thermal tuning (with all implications like temperature-dependent gain). The DBR structure separates the gain section from the mode-selecting section (the grating section), which leads to broadband-tunable concepts as described in Section 3.1.1.5.

The periodic structures in DFB and DBR lasers constitute of *Bragg* gratings. Consequently, the wavelength of DFB and DBR laser diode is given by

$$\lambda = \lambda_B \pm \frac{\lambda_B^2 (m + 1/2)}{2nL}, \quad \lambda_B = \frac{2n\Lambda}{k}, \tag{3.11}$$

λ_B is the respective Bragg wavelength, and m is an integer.

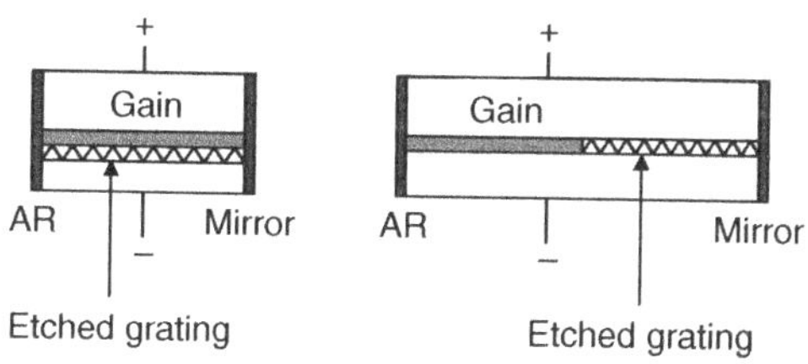

FIGURE 3.11 DFB (left) and DBR (right) laser diodes.

DFB and DBR LDs can achieve narrow linewidth suitable for (coherent) high-speed WDM transmission, high output power and low noise. They suppress side modes efficiently. In addition, DBR laser can be made broadband tunable. The relevant parameters are given in Section 3.1.1.5.

3.1.1.4 Vertical-Cavity Surface-Emitting Lasers Vertical-cavity surface-emitting lasers (VCSEL) use vertical current injection and provide vertical light emission. The first VCSEL was presented in 1979, but devices for CW operation at room temperature were not reported until 1988 [10,11]. Today, VCSELs have replaced edge-emitting lasers in short-range applications such as GbE LAN and Fibre Channel SAN (Storage Area Networking). They are currently being considered as low-cost source for WDM-PON. In contrast to edge-emitting semiconductor lasers, VCSELs have a radially symmetric Gaussian near-field that simplifies coupling to optics or fibers. In addition, VCSEL technology allows the fabrication of two-dimensional laser arrays.

VCSELs use a thin active region that leads to comparatively low gain. Hence, high reflectivity at the cavity boundaries is required. Therefore, a stack of alternating low- and high-index dielectrics is used below and above the active layer. The simplified structure of a VCSEL is shown in Fig. 3.12. In addition, VCSELs suffer from large Ohmic resistance. Hence, heat dissipation can become a problem that further limits output power.

The frequency (i.e., longitudinal mode) spacing of VCSELs is given by $\Delta f = c/2nl$, with l the cavity length. If l is made sufficiently small, mode spacing increases beyond the cutoff of the gain-region bandwidth and single-mode operation is achieved.

3.1.1.5 Tunable Laser Diodes Tunable lasers are necessary in flexible WDM transport networks. They are required for network-capacity optimization, working together with colorless and directionless reconfigurable add–drop multiplexers (ROADMs). In addition, protection mechanisms based on wavelength protection become simpler, or more powerful (or even enabled at all). Tunable lasers are also required to reduce the number of different transmitter variants. This allows reducing the CapEx of the devices, and it also helps reducing inventory cost. In WDM-PON, tunable lasers are one way to achieve colorless ONUs. Colorless ONUs are not wavelength-specific (which will not be allowed in mass deployments). The

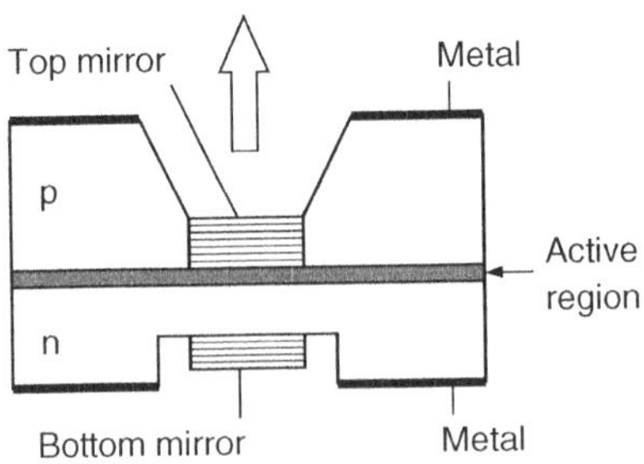

FIGURE 3.12 VCSEL diode.

alternative to tunable lasers for ONUs are seeded reflective ONUs, based on reflective components such as reflective semiconductor optical amplifiers (RSOAs), REAMs, or injection-locked (IL)-FP lasers.

In principle, tunability of a laser can be achieved by [12–16]

- electronic tuning,
- thermal tuning,
- mechanical tuning.

In addition, arrays of DFB lasers can be considered as a quasi-tunable transmitter [17]. As a further potential laser source, quantum-dot lasers can be considered. So far, these lasers have only supported the wavelength range of 1050–1320 nm [18,19].

Electronic tuning refers to the principle of applying an electric field that directly tunes the laser frequency. Several possibilities exist, with derivatives of the DBR laser being the most important ones:

- Three-section DBR
- Sampled grating DBR laser (SG-DBR)
- Superstructure grating DBR (SSG-DBR) laser
- Digital supermode DBR (DS-DBR) laser
- Grating-assisted codirectional coupler laser with sampled grating reflector (GCSR)
- Y-Junction laser (Y3 laser, transverse interferometric type)
- Vertical Mach–Zehnder Laser (VMZ, vertical interferometric type)

One relevant type of thermally tuned lasers exists, the matrix-grating strongly gain-coupled (MG-SGC) DFB laser [15]. For mechanical tuning, two relevant classes can be identified:

- Microelectromechanical vertical-cavity surface-emitting laser (MEM-VCSEL)
- External cavity laser (ECL)

From this list, multisection DBR (DS-DBR, SG-DBR) lasers have gained commercial relevance as tunable WDM transmitters.

Mode number m and wavelength λ of a laser diode depend on effective refractive index and cavity dimension (effective length L):

$$m\lambda/2 = \bar{n}L. \tag{3.12}$$

Consequently, tunability of a laser diode can be achieved by tuning any of these parameters

$$\frac{\Delta\lambda}{\lambda} = \frac{\Delta\bar{n}}{\bar{n}} + \frac{\Delta L}{L} - \frac{\Delta m}{m}. \tag{3.13}$$

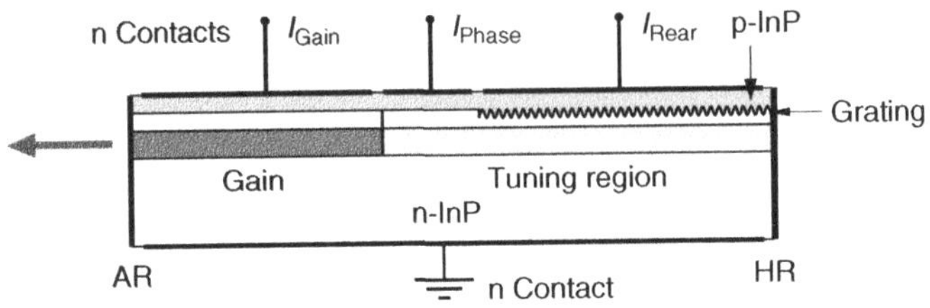

FIGURE 3.13 Three-section DBR.

With regard to the first two terms on the right-hand side, electronic, thermal, or mechanical tuning may apply. During tuning, the third term may lead to mode jumps and consequent jumps in the wavelength.

Tunable *multisection DBR lasers* have been described, for example, in Refs [12–14,20,21]. They are in mass production today. The advantage of these lasers is the fact that they are monolithically integrated, which allows low cost.

The basis of any tunable monolithically integrated multisection DBR laser is the three-section DBR laser. Its schematic cross-sectional diagram is shown in Fig. 3.13. Not shown here are the thermoelectrical cooler, also referred to as Peltier cooler, and a wavelength locker for wavelength tuning or stabilization.

The relevant three sections are the gain, phase, and rear sections. The gain section provides the lasing gain. It is not wavelength selective and provides gain in a relatively broad spectrum. The rear section is able to select a single laser line by means of the Bragg grating that is etched into the substrate. The phase section can provide some degree of fine tuning. Therefore, the three-section DBR laser cannot provide full-band (e.g., across full C-band) tuning capability. It is, however, the basis for improved tunable DBR lasers.

One relevant derivative of the three-section DBR laser is the DS-DBR laser [20]. Its cross-sectional schematic diagram is shown in Fig. 3.14. The DS-DBR laser has the potential for uncooled operation (which is relevant for WDM-PON for reduced cost and power consumption), and uncooled operation has been demonstrated, for example, in Refs [21,22].

The DS-DBR laser shown in Fig. 3.14 is a five-section DBR laser. As compared to the three-section DBR laser, one grating section has been added for improved tuning range. The fifth section is an integrated semiconductor optical amplifier (SOA). This section provides both broadband modulation capability (up to 10 Gb/s) and high

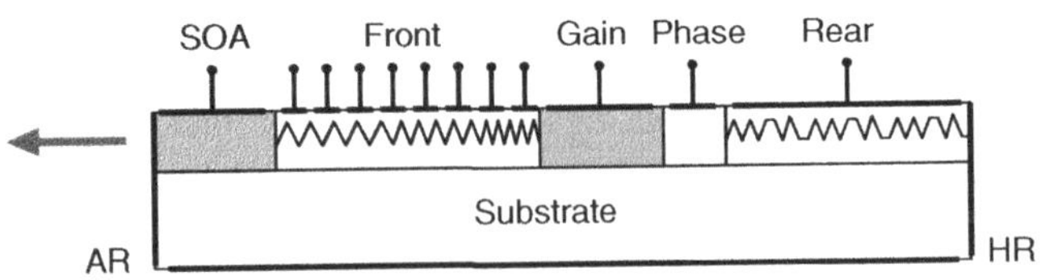

FIGURE 3.14 Five-section DS-DBR laser.

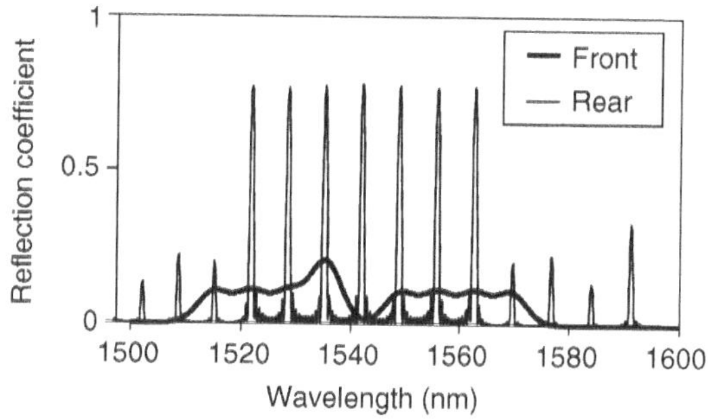

FIGURE 3.15 Reflection coefficients of front and rear grating of a DS-DBR.

per-channel output power (up to 15 dBm CW). During tuning, it can also be used as a fast shutter. Similar five-section designs exist, for example, for the SG-DBR laser.

Wider tuning range comes from a grating—in the rear section—with several reflection peaks. This is combined with a second grating in the front section for selecting between these peaks. This is done by activating the respective zone in the front grating as shown in Fig. 3.15.

Each peak of the rear section corresponds to a so-called *supermode*. These modes have a spacing of ~8 nm (~1 THz). Each supermode in a DS-DBR laser behaves like a standard three-section DBR laser. Additionally, the DS-DBR has longitudinal modes that are spaced ~37 GHz. Within one longitudinal mode, the frequency of the DS-DBR laser can be fine-tuned via the phase section. By setting the operation point to the middle of a longitudinal mode, a fine-tuning range of approximately +/−20 GHz can be achieved.

Tuning is relatively insensitive to actual front grating currents, which allows a broad tuning range per supermode. If this range is to be exceeded, a supermode jump becomes necessary.

Superside-mode suppression ratio (SMSR) is in the range of 45 dB. Further relevant optical characteristics are RIN of < −150 dB/Hz, and the laser linewidth in the range of 500 kHz for all channels.

The DS-DBR laser has the potential to work without TEC. However, temperature drift will not only cause wavelength drift (which can be counteracted through grating current tuning) but also variations of the maximum output power. At 25 °C, the output power stays above 14 dBm for all channels. However, at 45 °C, output power of some channels drops to slightly below 12 dBm. Hence, avoiding the TEC may lead to somewhat reduced output power. Other parameters such as SMSR are less affected by temperature drift.

In order to tackle the temperature drift problem, high-temperature materials, for example, AlGaInAs/InP are considered [23]. This approach can support uncooled operation in the 70 °C region, possibly supported by a small low-power heater in order to suppress mode jumps. Target power consumption is in the 1 W range.

More detailed comparisons between multisection DBR lasers can be found in Refs [12,13].

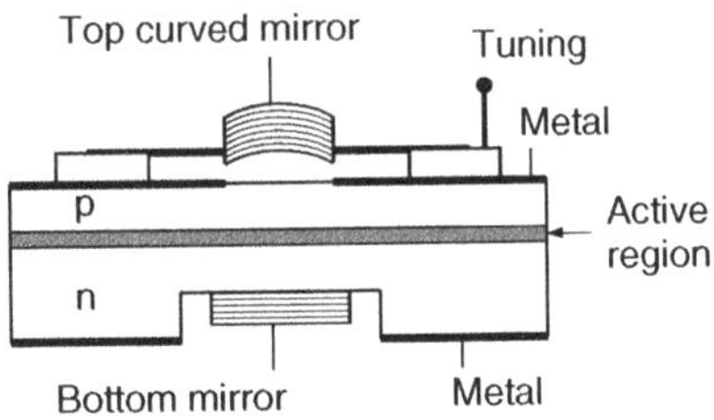

FIGURE 3.16 Cross-sectional diagram of a MEM-VCSEL.

Microelectromechanically tuned VCSELs (MEM-VCSEL) are an alternative to multisection DBR lasers as full-band tunable transmitters. VCSELs have low-cost potential, and by adding MEM means, they can be tuned. Tunable MEM-VCSELs have been described for metro-area applications [24].

The basic concept of the MEM-VCSEL is based on a DBR-VCSEL with an attached micromirror that is thermally tunable. (In Ref. [24], application of an electrostatic force to the mirror led to tuning.) A schematic diagram of a MEM-VCSEL is shown in Fig. 3.16 [25].

MEM-tunable VCSELs are a hybrid two-chip assembly of a microelectromechanical system (MEMS) with concave AlGaAs-GaAs mirror membrane, and an InP-based semiconductor cavity with tunnel junction aperture. Using electrothermal MEMS actuation, the included air-gap can be expanded and the cavity resonance be tuned to longer wavelengths. The active part (according to Fig. 3.16) is formed by compressively strained InGaAlAs quantum wells embedded between an n-type and a thin p-type InGaAlAs confinement layer grown by molecular beam epitaxy (MBE). A centrally limited buried tunnel junction (BTJ) defines the active diameter and reduces the inefficiently conducting p-type region. The BTJ is overgrown with a binary n-InP current- and heat-spreading layer in a subsequent MBE process, followed by an evaporated dielectric backside DBR composed of pairs of calcium fluoride and zinc sulfide with gold coating and an electroplated substrate. The back mirror size is slightly larger than the BTJ. The top DBR consists of MBE-grown GaAs-AlGaAs quarter wavelength layers with refractive index contrast. The inclusion of graded indium induces compressive strain, resulting in a curved mirror membrane with radius of curvature of 2–3 mm, suspended on four arms. The upward movement is achieved by Joule heating of the upper membrane.

Slightly different designs for MEM-VCSELs exist, namely the semiconductor-coupled cavity (SCC) and the extended-cavity (EC) designs. From the basic SCC design (bottom Bragg mirror, cavity, air-gap and curved mirror), the EC scheme can be obtained by placing an antireflection (AR) coating at the semiconductor–air interface of the lower half VCSEL [25]. The extended tuning range of EC devices has to be paid at the expense of lower delivered power and higher threshold. The laser can be continuously tuned over almost the whole free spectral range (FSR) and it stays always single mode with high SMSR. The FSR is given by the distance between two adjacent longitudinal modes, which is related to the length of the air-gap. The FSR sets the upper limit for the tuning range. The latter can also be limited by the material

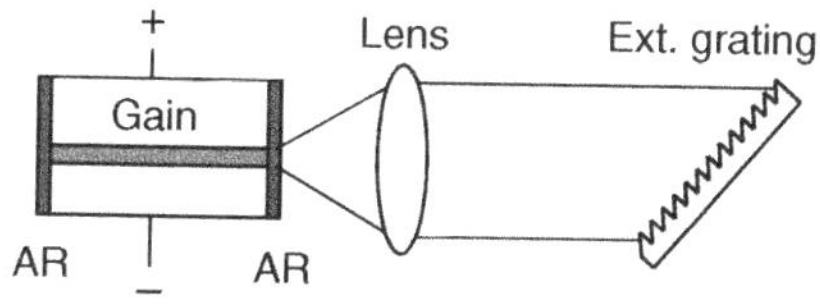

FIGURE 3.17 Schematic diagram of an ECL.

gain spectrum, which has to compensate for the wavelength-dependent threshold conditions. The limitations due to the DBR bandwidth are negligible in both designs.

Tuning ranges up to 60 nm and single-mode output power up to 2 mW with high side-mode suppression and stable polarization have been demonstrated.

External-cavity lasers are an obvious choice for tunable transmitters because of their inherent capability to be tuned via the external cavity. They are described in Refs [12–14]. The basic ECL consists of a broadband gain section to which an external cavity is coupled via an AR coating and a collimating lens. This setup is shown in Fig. 3.17. The gain section can be a simple Fabry–Pérot laser diode. Wavelength tuning is possible by tuning the external grating. This can be done mechanically or thermally. One contributor to cost is the optical coupling between the components that typically has to be done via a lens. This is a disadvantage over monolithically integrated solutions. It also leads to lower output power due to the loss of the free-space region and coupling inefficiency.

Mechanical tuning often is implemented using MEM components. Thermal tuning is often applied to etalons. It can also be applied to fiber Bragg grating (FBGs) that in addition offers the chance for better and more cost-effective integration. The corresponding devices are also referred to as FBG-ECLs [26], a principle diagram is shown in Fig. 3.18.

The DFB laser is not broadband tunable because gain and grating sections are not decoupled. However, *integrated arrays of DFB lasers* can be made quasi-tunable and hence are an alternative to DBR, ECL, and VCSEL. DFB arrays with 12 DFB lasers spaced 10 μm apart in position and 450 GHz apart in frequency are commercially available [27]. The chip size is 500 μm × 800 μm, which is comparable to high-performance single-element devices.

To maintain high yield, the lasers are phase-shifted with AR coatings on both facets, which implies lower efficiency since some power is wasted from the backside. The grating and phase shift are designed such that twice as much power exits the front facet than the rear.

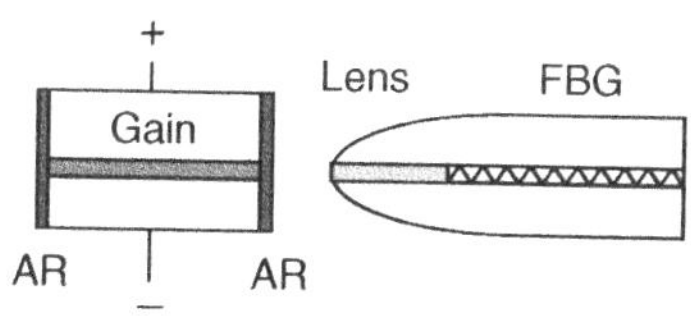

FIGURE 3.18 Principle of an FBG-ECL.

TABLE 3.1 Comparison of Relevant Parameters of Tunable Lasers

	DS-, SG-DBR	FBG-ECL	MEM-ECL	MEM-VCSEL	DFB-Array
Launch power (dBm)	>13 CW	>13 CW	>13 CW	<5 CW	>13 CW
Power dissipation (W)	<2.5 (<1 w/o TEC)	<3.5	<3.5	<1	<4.5
Linewidth (MHz)	<0.5	<25	<0.5	<10	<0.5
Tuning range (THz)	>4	>4	>4	>4	>4
RIN (dB/Hz)	<−150	<−140	<−150	<−120	<−150
SMSR (dB)	>45	>30	>50	>30	>50
Tuning speed	Fast	Medium	Fast	Fast	Medium
Frequency stability w/o wavelocker	Good	Poor	Poor	Poor	Good

An advantage of a laser array is redundancy: any error of a specific laser means that a subset of the entire tuning range can still be used. This way, the array yield can be improved.

The optical performance of the DFB array is within the range that can be expected from a DFB laser. Linewidth is below 500 kHz for all channels, and RIN is better than −153 dB/Hz. Fiber-coupled CW output power in the range of +15 dBm, and SMSR of >55 dB has been demonstrated. A disadvantage is the relatively high power consumption that is in the range of 5 W for a 12-channel device. Here, future improvements are expected to be relatively low [12].

Relevant parameters of the lasers discussed so far are listed in Table 3.1 [28]. Output power refers to unmodulated output. Modulated launch power can be lower by 7–8 dB.

From viewpoint of lowest power consumption, multisection DBR lasers without TEC (uncooled operation of both the DS-DBR and the SG-DBR laser has been demonstrated [21,22]) and MEM-VCSELs are the most promising devices. From viewpoint of RIN, SMSR, linewidth and also tuning range, all lasers can be considered applicable for WDM. Regarding laser output power, multisection DBR, ECL, and DFB arrays have advantages over VCSELs. The last relevant parameter is the future potential for low-cost mass production in the WDM-PON context. This parameter correlates with the components that are necessary within the transceiver package. Here, advantages can be identified for the monolithically integrated multisection DBR lasers with their potential for TEC-less design and VCSELs.

3.1.2 External Modulators

External modulators are used for high-speed WDM transmission. They allow broad modulation bandwidth, (partly) avoid the chirp associated with directly modulated lasers, and they can provide means for complex (I + Q) modulation schemes. Basically, electro-optic modulation and electroabsorption modulation can be

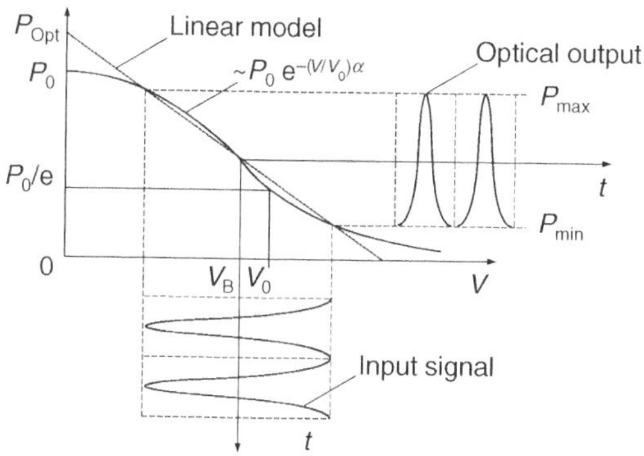

FIGURE 3.19 Modulation of an EAM.

used. In addition, semiconductor optical amplifiers can be used as modulators, especially in reflective structures.

3.1.2.1 Electroabsorption Modulators The electroabsorption modulator (EAM) is a reverse-biased p-i-n diode with a bulk active region or multiple quantum wells (MQWs) as absorption layer [29,30]. Compared to bulk absorption layers, the absorption spectrum edge in MQWs is sharper and moves faster with reverse-biased electric field. MQW EAMs have low driving voltage, broad bandwidth, compact size, and monolithic integration capability with other devices (i.e., laser diodes). They have potential for low power consumption and as low-cost modulators for high-speed transmission. On the other hand, they are mainly restricted to intensity modulation.

The output intensity of an EAM is a nonlinear function of the input (modulating) voltage, as shown in Fig. 3.19. For intensity modulation at high extinction ratio and low insertion loss, small bias voltage has to be used. For very high bias voltage, extinction ratio decreases and insertion loss increases (i.e., output power decreases).

The insertion loss IL_{EAM} of an EAM depends on the length L of the absorption region and the internal loss α at $V=0$ and is given by

$$IL_{EAM} = \frac{P_{in} - P_{out}(V=0)}{P_{in}} = 1 - e^{-\alpha(0)L}. \tag{3.14}$$

This suggests using short device lengths. On the other hand, the contrast ratio $R_{on/off}$ that relates to the extinction ratio is given by

$$\begin{aligned} R_{on/off} &= \frac{P_{out}(V_{on}=0)}{P_{out}(V_{off}=V)} = \frac{e^{-\alpha(0)L}}{e^{-\alpha(V)L}}, \\ R_{on/off}(\mathrm{dB}) &= 10 \log (R_{on/off}) = 4.343 \cdot [\alpha(V) - \alpha(0)]L. \end{aligned} \tag{3.15}$$

The contrast ratio should be large. This can be achieved by increasing the length of the modulator. Hence, a careful trade-off is required.

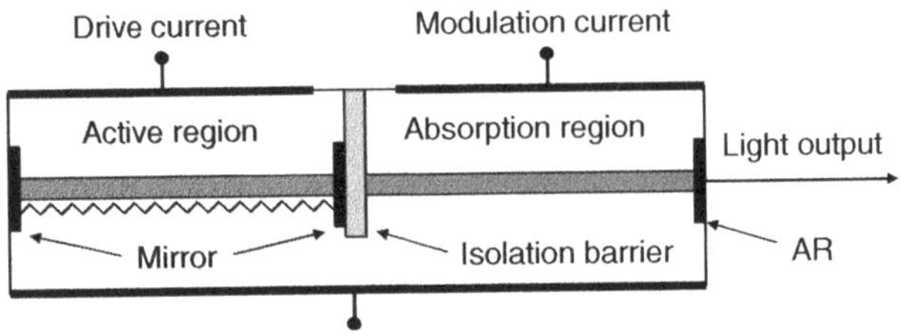

FIGURE 3.20 Integrated DFB-EAM.

EAMs can be monolithically integrated with laser diodes, for example, DFB lasers [31,32]. A simplified lateral structure of an integrated DFB-EAM is shown in Fig. 3.20. This device consists of two regions, an active and an absorption region, which are separated by an insulation barrier.

One advantage of an integrated laser-EAM device is reduced chirp because adiabatic laser chirp is avoided. However, due to the similar semiconductor structure, the EAM suffers similar transient chirp that is again described by the linewidth enhancement factor α:

$$\frac{\mathrm{d}\phi}{\mathrm{d}t} = \frac{\alpha}{2\pi S}\left(\frac{\mathrm{d}S}{\mathrm{d}t}\right), \quad S = |E_{\mathrm{out}}|^2. \tag{3.16}$$

The amount of chirp generated depends on α and the signal waveform. S is the output photon number (it equals the EAM cavity photon number since the output-facet reflectivity $R = 0$). For large-signal modulation, blueshift at the beginning of a pulse and redshift at the tail result.

In an EAM, unlike a laser diode, there is loss instead of gain. With the increase of carrier density due to signal absorption, one can expect the linewidth enhancement factor to become negative if the generated carriers can be removed from the cavity quickly enough (i.e., no saturation of the absorption occurs). This will lead to positive chirp when the signal has negative slope and vice versa.

EAMs are mainly used for WDM transmission at 10 Gb/s. Performance on G.652 SSMF is a function of modulator chirp, optical extinction ratio and average transmitted optical power [33]. Chirp and extinction ratio depend on the change in the refractive index of the absorption layer in the waveguide with the applied voltage that is described by the α-parameter and the device absorption. A negative α-parameter, which is desirable for transmission on fibers with positive D-parameter (e.g., SSMF) typically occurs in a region of high absorption [34]. This leads to difficulties regarding insertion loss and extinction ratio. In Ref. [35], negative α-parameters was obtained for near-zero bias voltages (i.e., at a bias for proper insertion loss and extinction ratio). Using negative low-chirp integrated EAM-DFB lasers, 10 Gb/s NRZ-OOK penalty-free transmissions have been reported over more than 100 km of G.652 standard fiber [34,36].

System optimization specific to EAMs can enhance the maximum reach between regenerators. Electronic dispersion compensator (EDC) at the receiver or electrical

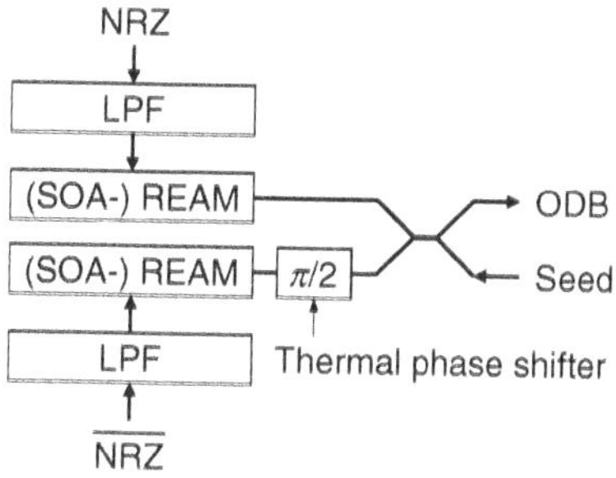

FIGURE 3.21 ODB modulator based on two reflective EAMs.

predistortion can be applied. This holds for intensity modulation (which is mostly used for 10 Gb/s WDM transmission).

Recently, a modulator structure suitable for optical duobinary (ODB) modulation has also been demonstrated [37]. This structure is based on two *reflective* EAMs (REAMs) and is shown in Fig. 3.21.

The structure in Fig. 3.21 consists of two REAM arms that are modulated with the original and the inverse of the ODB-filtered NRZ signal and shifted against each other by π. This phase shift can be achieved by traversing a dedicated $\pi/2$ phase shifter twice (forward and reverse, as indicated in Fig. 3.21). If a 3 dB coupler is used, a dedicated phase shifter can be omitted. The π shift is then produced in the coupler (see Section 3.2.6.1).

REAMs are used to allow seeding with an arbitrarily colored laser, that is, to achieve a colorless transmitter design. For improved launch power, the REAM can be complemented by a SOA.

The REAM structure according to Fig. 3.21 performs similarly to MZM-based duobinary modulation (and hence slightly worse than chirp-managed laser (CML), see Section 3.1.3.3), as can be seen from Fig. 3.22 [37]. It also shows similar behavior with regard to ODB low-pass filtering (LPF).

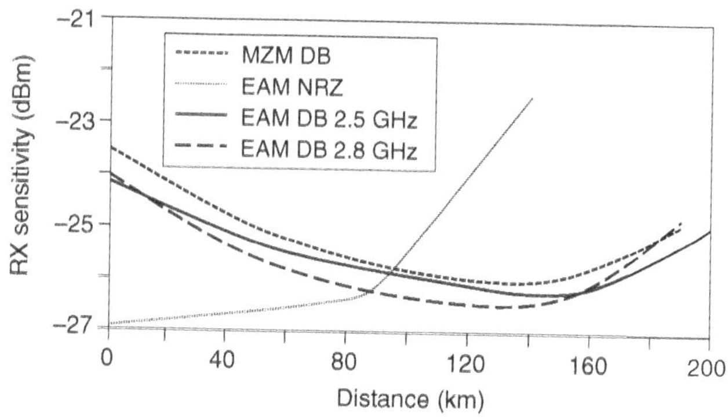

FIGURE 3.22 Performance of EAM-based Duobinary.

Cost of EAM-DFBs has decreased significantly over time. As a result, the EAM is emerging as the winner over DML for many metro WDM applications. Commercial EAM modulators for 10 Gb/s are characterized by low drive voltages (around 2 V) and low average α (-0.5). They can achieve low dispersion penalty up to 1600 ps/nm. EAMs for 40 Gb/s are also commercially available, and in Ref. [38], transmission at 40 Gb/s was demonstrated. On the other hand, EAM-based higher order QAM modulation will become difficult due to their internal physics. These modulation formats require precise control of the signal AM and PM.

3.1.2.2 Electro-Optic Modulators In an electro-optic modulator, the distribution of electrons within the medium is distorted if an electrical field is applied so that the refractive index of the medium changes anisotropically. Along the medium, this leads to a phase change of traversing optical signals and can be used for modulation. The basic setup is shown in Fig. 3.23.

Suitable materials for electro-optic modulators should have high electro-optic coefficients, high optical transparency in the wavelength range used for WDM transmission, and they should be mechanically and chemically stable and well suited for manufacturing. $LiNbO_3$ (lithium niobate) is one such material. Other crystals include $LiTaO_3$, KDP (KH_2PO_4), and ADP ($NH_4H_2PO_4$) [39,40].

Similar to the nonlinear effects in transmission fibers, the electro-optic effects can be described by the induced polarization P of the medium (here in a simplified scalar way):

$$P = \varepsilon_0\chi^{(1)}E + \varepsilon_0\chi^{(2)}E^2 + \varepsilon_0\chi^{(3)}E^3 + \cdots. \tag{3.17}$$

E is the applied electric field, ε_0 is the dielectric constant of vacuum, and the susceptibilities $\chi^{(i)}$ are constant. In linear optics, the refractive index is then given by $n^2 = 1 + \chi^{(1)}$. In general, the electric displacement field follows from E and P as $D = \varepsilon_0 E + P$.

The second-order term, $\chi^{(2)}EE$, is responsible for frequency doubling and sum/difference generation, given the nonlinear susceptibility $\chi^{(2)}$ is nonzero ($\chi^{(2)} = 0$, for example, in transmission fibers). The third-order term, $\chi^{(3)}EEE$, produces third harmonics, Raman and Brillouin scattering, and the Kerr effects.

If we consider an optical signal with frequency ω and a DC field $E = E_0 + E_\omega \cos(\omega t)$, several polarization components can be generated. $\chi^{(2)}E_0E_\omega\cos(\omega t)$ refers to the *linear electro-optic effect* discovered by Pockels in 1893 [41]. It modifies the refractive index with a DC field and can be used for optical switching and for phase

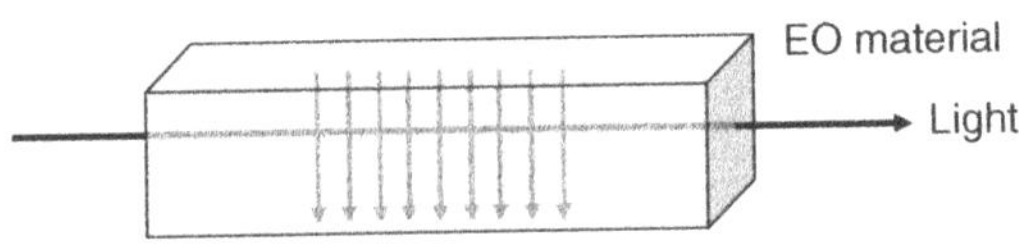

FIGURE 3.23 Electro-optic modulator.

TABLE 3.2 Parameters for Electro-Optic Effect

	Pockels Effect	Kerr Effect
Types of coefficient	$r = 10^{-12}$ to 10^{-10} m/V	$s = 10^{-18}$ to 10^{-14} m^2/V^2
Δn for $E = 10^6$ V/m (crystals)	10^{-6} to 10^{-4}	10^{-6} to 10^{-2}

modulation of light. The polarization component $\chi^{(3)} E_0^2 E_\omega \cos(\omega t)$ relates to the *quadratic electro-optic effect* or *DC Kerr effect* (discovered 1875 [42]). This effect is not to be confused with the *optical* or *AC Kerr effect* that results from the polarization component $\chi^{(3)} E_\omega^3 \cos(\omega t) = \chi^{(3)} E_\omega^2 (E_\omega \cos(\omega t))$. This effect looks like a refractive index that depends on the optical intensity $n = n_0 + n_2 I$ as described in Section 2.2.

Similar to the fiber's nonlinear refractive index, the change in refractive index n in an electro-optic medium can be written in terms of a linear electro-optic coefficient r (Pockels effect) and a quadratic electro-optic coefficient s (Kerr effect).

$$n(E) = n + a_1 E + \frac{1}{2} a_2 E^2 + \cdots. \tag{3.18}$$

The coefficients a_1, a_2 relate to the linear electro-optic (Pockels) effect and the quadratic electro-optic (AC Kerr) effect and their coefficients r, s via

$$n(E) = n - \frac{1}{2} r \cdot n^3 E, \quad r = -2\frac{a_1}{n^3}, \quad n(E) = n - \frac{1}{2} s \cdot n^3 E^2, \quad s = -\frac{a_2}{n^3}. \tag{3.19}$$

Parameters for electro-optic crystals are summarized in Table 3.2.

With the change of refractive index Δn, the resulting phase change of an optical signal propagating through an electro-optic medium (e.g., an optical phase modulator) is given by

$$\Delta\phi = \Delta\beta_0 L = k_0 \, \Delta n L. \tag{3.20}$$

Electro-optic modulators can be used as phase modulators or as nearly chirp-free amplitude modulators. For amplitude modulation, the electro-optic medium is used in a Mach–Zehnder interferometer (MZI) structure according to Fig. 3.24. This structure is also referred to as Mach–Zehnder modulator (MZM).

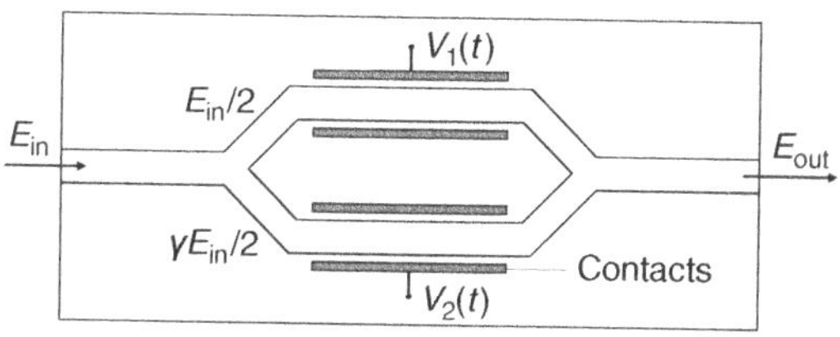

FIGURE 3.24 MZM structure.

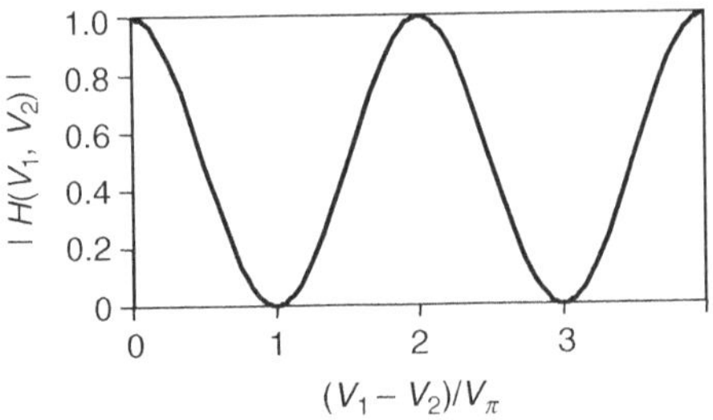

FIGURE 3.25 Absolute value of MZM transfer function.

The output field E_{out} according to Fig. 3.24 can be expressed as (where $\gamma < 1$ accounts for an arbitrary asymmetric split in the MZI)

$$E_{out} = \frac{E_{in}}{2}\left[e^{j\phi_1(t)} + \gamma\, e^{j\phi_2(t)}\right] = \frac{E_{in}}{2}\left[e^{j\pi V_1(t)/V_\pi} + \gamma\, e^{j\pi V_2(t)/V_\pi}\right]. \qquad (3.21)$$

Consequently, the transfer function of the MZM, for $\gamma = 1$, can be written as a function of the modulator voltages V_1 and V_2:

$$H(V_1, V_2) = E_0 \cos\left[\pi(V_1 - V_2)/2V_\pi\right] \cdot \exp\left[j\pi(V_1 + V_2)/2V_\pi\right]. \qquad (3.22)$$

The cosine term accounts for amplitude modulation, and the exponential term for phase modulation or chirp. Equation (3.22) also shows a fundamental behavior of the MZM with regard to the absolute value of the transfer function: its cosine form. This is also shown in Fig. 3.25.

A chirp parameter $P_C = (V_1 + V_2)/(V_2 - V_1)$ can be defined. Typically, $V_1(t)$ and $-V_2(t)$ are not independent of each other. They are related to each other and the constant bias voltage V_{DC} via $V_2(t) = V_{DC} - V_1(t)$. Two standard cases of MZM biasing are often used, normal bias and bias at extinction. Normal bias is defined by $V_{DC} = V_\pi/2$. The transfer function results as

$$H(V_1, V_\pi/2) = \cos\left[\pi\left(\frac{V_1}{V_\pi} - \frac{1}{4}\right)\right] \approx \frac{1}{\sqrt{2}}\left[1 + \frac{\pi}{V_\pi}V_1\right]. \qquad (3.23)$$

Bias at extinction is given for $V_{DC} = V_\pi$. It leads to the following transfer function:

$$H(V_1, V_\pi) = \cos\left[\pi\left(\frac{V_1}{V_\pi} - \frac{1}{2}\right)\right] \approx -\frac{\pi}{V_\pi}V_1. \qquad (3.24)$$

Similar to other modulators, the parameters insertion loss and extinction ratio are relevant for MZM. They are given by IL (dB) $= 10\log_{10}(I_{max}/I_0)$ and ER (dB) $= -10\log_{10}(I_{min}/I_{max})$, respectively. The intensity values I_0, I_{min}, and I_{max} are defined in Fig. 3.26. In addition, a modulator response M can be defined that caters for the

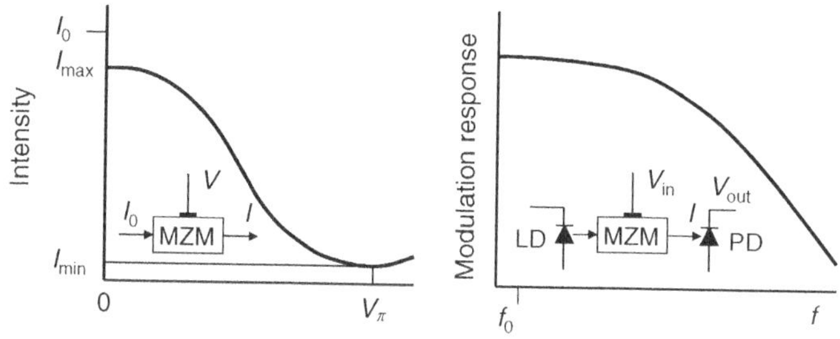

FIGURE 3.26 MZM switching curve (left) and modulation response (right).

frequency response of the MZM:

$$M(\text{dB}) = 20 \log_{10}\left(\frac{V_{out}(f)}{V_{out}(f_0)}\right). \tag{3.25}$$

The frequency response is also shown in Fig. 3.26.

MZM can achieve bandwidths in excess of 100 GHz, and they can provide very high extinction ratio in the range of 15–20 dB.

3.1.2.3 Reflective Semiconductor Optical Amplifiers Reflective semiconductor optical amplifiers are used as modulators mainly in WDM-PON [43,44]. Typically, SOAs are used as amplifiers or fast switches (shutters), but not as modulators. Reasons include bandwidth limitations and chirp. Both are less critical in WDM-PON running at limited bit rates. Reflective SOAs are *seeded* with an external light source and hence allow the construction of *colorless* transmitters/modulators (i.e., a unified broadband device that can be used for many different WDM channels). This principle is similar to the REAM concept described at the end of Section 3.1.2.1. It is shown in Fig. 3.27.

RSOAs can be operated in the whole wavelength range where they produce gain. Hence a single device can cover large bandwidths in excess of 40 nm, and RSOAs can in general be built for a very broad wavelength range (see Section 3.2.4.3 about SOAs).

Unlike (R)EAM and MZM, RSOAs can provide gain rather than insertion loss. They can be used as transmitter/modulator combinations. Output power and gain depend on the seed input that also specifies the transmit wavelength of the RSOA. This is shown in Fig. 3.28.

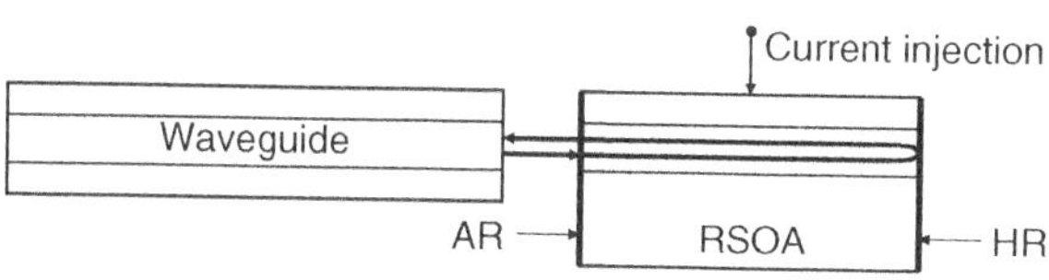

FIGURE 3.27 RSOA principle.

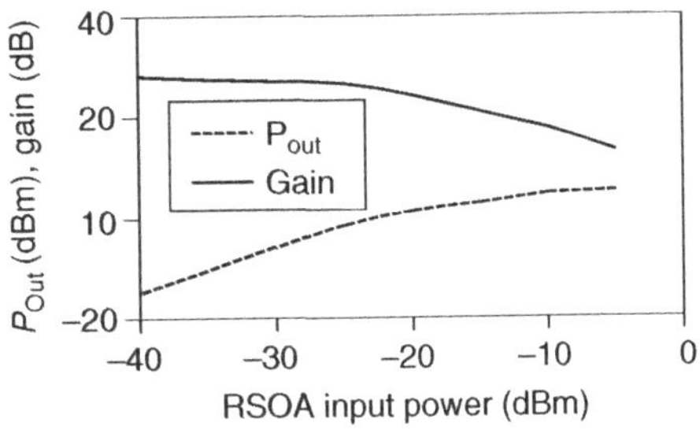

FIGURE 3.28 Output power and gain versus RSOA input power.

Like EAMs, RSOAs can be based on bulk active regions or (multiple) quantum wells. The two variants have different relevant characteristics that are listed in Table 3.3. Both types are suitable for uncooled operation over a temperature range up to 60 °C.

Bulk-type RSOAs are less prone to polarization (PDG), but have less gain. In addition, they have lower yield. Saturation output power of both types is in the same range.

RSOAs can be seeded with broadband light sources (BLS) or lasers. LEDs or amplifiers (amplified spontaneous emission, ASE) can be used as BLSs. Since they must be filtered (spectrally sliced, for example, with an arrayed waveguide gratings (AWG)), they provide only small injection power to the RSOA. In addition, performance of ASE-injected RSOAs is noise limited. With laser-injected RSOAs, a penalty due to reflection occurs. For the same reason, laser-injected RSOAs should be used with angled connectors or waveguide (fiber) adaptation, respectively.

The reflection problem is shown in Fig. 3.29. Obviously, any reflection points in the transmission link can lead to multiple reflections that will be detected at the

TABLE 3.3 RSOA Comparison

	Bulk Type	QW Type
PDG	<2 dB	~10 dB
Gain	~20 dB	~28 dB
Saturation output power	~3 dBm	~2 dBm

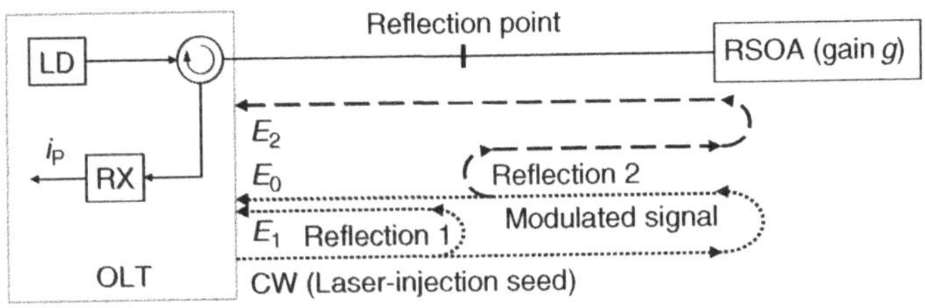

FIGURE 3.29 Reflections in RSOA-based WDM-PON.

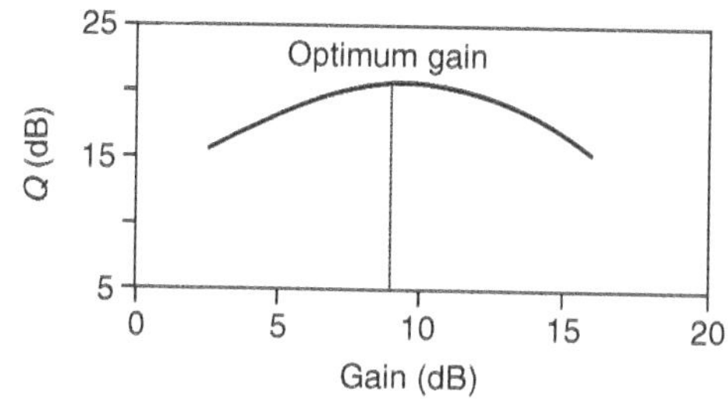

FIGURE 3.30 Q limitation due to reflection noise.

receiver at the far end (in this case, an optical line termination (OLT) in a WDM-PON system using a single transmission fiber).

The detector (RX) in the OLT receives the modulated signal (E_0) together with the reflections. If higher order ("Ping-Pong") reflections are neglected, the RX photodiode current becomes

$$i_p \propto E_0^2 + 2\sum_{k=1}^{2} E_0 E_k \cos \Delta\varphi_{0k} + E_1^2 + E_2^2 + 2E_1E_2 \cos \Delta\varphi_{12}. \qquad (3.26)$$

The relative intensities of the terms containing E_1, E_2 depend on the location of the reflection point (and hence fiber attenuation), and of the strength of the respective reflection. In a setup according to Fig. 3.29, the RSOA gain can be optimized with regard to reflection. Since $E_0^2 \propto g$, $E_1^2 \propto r$, and $E_2^2 \propto g^2 r$, the gain g can be traded off against reflections to achieve maximum resulting Q. Here, r is the reflectivity of the reflection points. The result is shown in Fig. 3.30 for a bulk-type RSOA. Laser-injected schemes can have reflection penalties in the range of 3 dB.

Further methods for reduction of back reflection in single fiber WDM-PON are discussed in Refs [45–47].

A further problem of RSOAs is their limited modulation bandwidth. Since gain recovery is in order of a bit period of a 10 Gb/s signal, RSOAs cannot be modulated at higher Baud rates.

A comparison of the different modulator options is given in Table 3.4.

TABLE 3.4 Comparison of Modulators

Modulator	Chirp Parameter	Voltage for ER = 10 (V)	BW (GHz)	IL (dB)
Laser diode (DFB)	$\alpha > -3$	1.5	30	
$LiNbO_3$	$-1 < \alpha < 1$	5–8	110	>5
(Reflective) EAM	$\alpha < -2$	2	>40	8–10
RSOA	$\alpha > -3$?	2	<10	
DFB-MZ	$-1 < \alpha < 1$	5	55	
DFB-EAM	$\alpha < -2$	2	20	

From Table 3.4, $LiNbO_3$ (Mach–Zehnder) modulators can be identified as components for the high-performance (ultrahigh bit rates, low chirp) domain. For less demanding WDM applications (metro access, WDM-PON), direct modulation of lasers (see next section), EAM or RSOAs can be used.

3.1.3 Direct-Modulation Techniques

3.1.3.1 Directly Modulated Laser Directly modulated lasers are attractive because they are the simplest single-frequency sources that can be modulated. Ten gigabits per second devices are commercially available, and modulation speed up to 40 Gb/s has been experimentally demonstrated.

DMLs are affected by frequency chirp. An approximated model of chirp is [5,6]

$$\Delta f(t) = \frac{\alpha}{4\pi}\left[\frac{1}{P(t)} \cdot \frac{\mathrm{d}}{\mathrm{d}t}P(t) + \kappa P(t)\right], \tag{3.27}$$

which requires two constants, α and κ, to be fully characterized. $P(t)$ is the laser output power per facet (before coupling loss into the fiber). Depending on the relative values of α and κ and on the modulation rate, different lasers can have widely different chirp behavior. Specifically, if κ is relatively large and/or the bit rate is low, it is the adiabatic chirp that dominates, and the laser FM approximately follows the same waveform as the intensity modulation. If κ is relatively small and/or the bit rate is high then it is transient chirp (the derivative term in the equation) that prevails causing FM spikes at the pulse edges [6].

Chirp can be beneficial, depending on its qualitative and quantitative features and on dispersion. The large positive dispersion of standard G.652 fibers interacts with typical laser chirp (especially the transient type) in a way that pulses get quickly degraded [48]. At 10 Gb/s, the typical 2 dB penalty distance is 5–10 km, as opposed to the theoretical 60–70 km of a chirpless source. Countermeasures to chirp include

- transmission at 1300 nm (near-zero dispersion in G.652 fibers), a technique from the early (non-WDM) days that has been revamped for Metro Ethernet [49],
- optical filtering techniques to reduce spectral broadening [50],
- keeping the laser well above threshold on "0"s and reducing extinction ratio [51].

Though effective in specific contexts, all of the above have major limitations that prevent a generalized adoption. Recently, the use of adaptive electronic equalizers or EDCs in reception has been proposed [52]. It has been shown that a properly fitted finite impulse response (FIR) equalizer can lead to a doubling of the 2 dB penalty distance, approximately, reaching about 15 km at 10 Gb/s on G.652.

In conclusion, DMLs will likely remain the components of choice for WDM transport with up to 2.5 Gb/s only. At that rate, the 2 dB limit is about 100 km on G.652 fiber, which is sufficient, for example, for WDM-PON applications. The use of DMLs at 10 Gb/s or higher, and above 10–15 km remains unlikely.

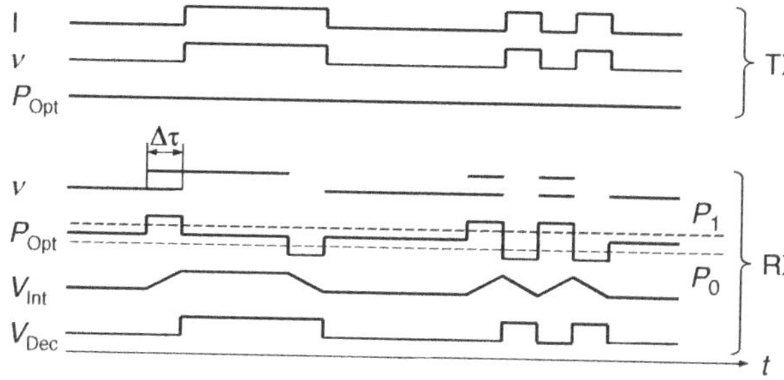

FIGURE 3.31 Principle of pure-FSK dispersion-supported transmission [55].

3.1.3.2 Dispersion-Supported Transmission Dispersion-supported transmission (DST) is one example for constructively using the frequency modulation that occurs in DML. It was developed in the early 1990s for 10 Gb/s transmission.

In DST, a (highly biased, DFB) laser generates a frequency-modulated output signal [53–55]. The chromatic dispersion of the transmission fiber then converts the frequency modulation into amplitude modulation (FM-AM conversion). This AM is detected in the receiver, the principle is shown in Fig. 3.31. The upper part of the figure shows the pulse shapes of the driving signal I, the optical frequency ν, and the optical power P_{Opt} of the transmitter (TX). For simplification and clarification, no intensity modulation is assumed here ($P_{\text{Opt}}(t) = \text{const.}$). The optical frequency ν is switched by the incoming NRZ binary data signal between two frequencies, which leads to a wavelength shift of $\Delta\lambda = -\Delta\nu\lambda^2/c$. Due to CD, the signal components with different wavelengths arrive at different times at the receiver (RX) after traversing fiber of length L. The time difference $\Delta\tau$ is given by $\Delta\tau = \Delta\lambda DL$, where D is the dispersion parameter of the fiber. The resulting AM of the received optical power P_{Opt} is also shown in Fig. 3.31. The received power is no longer constant because of overlap of the time-shifted components with different frequencies.

For constructive overlap, a positive peak of $P_{\text{Opt}}(t)$ results. In regions where no signal components are present, $P_{\text{Opt}}(t)$ shows negative peaks. If the optical transmitter is modulated with bit rate B and $\Delta\tau$ is adjusted to $|\Delta\tau| = 1/B$, a three-level signal $P_{\text{Opt}}(t)$ is expected at the receiver input. After direct detection, the original NRZ data signal can be derived by using an electrical integrator in conjunction with a decision circuit (lower part of Fig. 3.31).

If the spectral width of the transmitted signal is given solely by the modulation rate (i.e., by Fourier spread) by using chirp-free external modulation, the CD limit for an RX optical power penalty of 1 dB is estimated by $B^2DL = c/(2\lambda^2)$. For $\lambda = 1532$ nm, $D = 16.2$ ps/(nm km), and $B = 9.95$ Gb/s, this results in a link length of $L \approx 40$ km. Using DML with uncompensated chirp, the maximum link length is even smaller.

The maximum reach L that is achievable with DST can be estimated by assuming *minimum-shift keying* (MSK). Then, the frequency shift $|\Delta\nu|$ is adjusted to half the bit rate B. With $|\Delta\tau| = 1/B$, $\Delta\lambda = -\Delta\nu\lambda^2/c$, and $\Delta\tau = \Delta\lambda DL$, assuming $D > 0$, this results in $B^2DL = 2c/\lambda^2$. The link length L now is four times the conventional dispersion

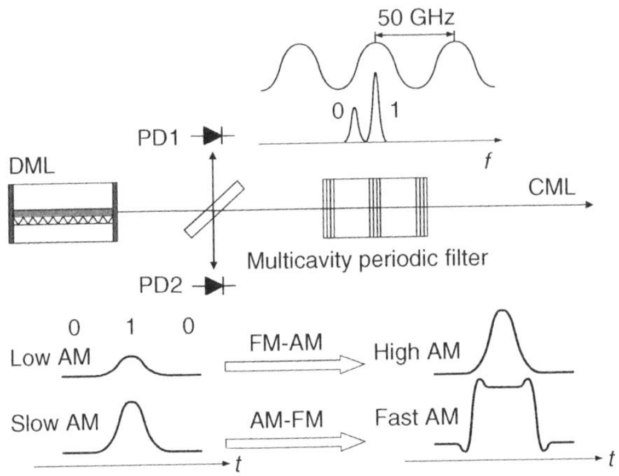

FIGURE 3.32 CML principle [56].

limit for external modulation, hence with the same parameters used above, $L \approx 160$ km. This reach can be extended by allowing $|\Delta\nu|$ to be slightly smaller than $B/2$, $|\Delta\tau|$ to be slightly larger than $1/B$, and also accepting a certain increase of the related RX power penalty into the range of 200 km at 10 Gb/s [55].

3.1.3.3 Chirp-Managed Laser A chirp-managed laser essentially consists of a DML and a so-called optical spectrum reshaper (OSR) filter [56]. A standard DFB laser can be used that is biased high above threshold (e.g., at ~5 × threshold) and directly modulated. This results in a low ER (~2 dB) intensity modulation and also generates FSK. Further benefits of high bias include high output power, wide modulation bandwidth, low timing jitter, and *suppression of transient chirp*.

FSK results from gain compression in the laser, which generates *adiabatic chirp in proportion to the output intensity*: "1" bits are blueshifted relative to the "0" bits. The laser wavelength is aligned on the transmission *edge* of the OSR filter in order to pass the blueshifted "1" bits and attenuate the redshifted "0" bits (see top of Fig. 3.32). Hence, the OSR creates FM-AM conversion that increases the ER at the output of the OSR to >10 dB.

Thermal chirp of the laser, caused by excess heating of the active region through injection has the opposite sign to adiabatic chirp and can result in amplitude variation at the output of the OSR for modulation patterns with low-frequency content at <20 MHz. This is mitigated by a low-frequency circuit that slowly modulates the bias of the laser with the low-pass-filtered copy of the incoming data. When the pattern contains long strings of "1"s, which tend to redshift the laser output, the compensation circuit increases bias slightly, causing an adiabatic blueshift to cancel the thermal chirp. For long strings of "0"s, compensation works similarly. This compensation works down to the so-called thermal-dip frequency (~few 100 kHz) where thermal chirp efficiency equals adiabatic chirp efficiency. Therefore, DFB lasers with low thermal-dip frequency are required.

The DFB wavelength is locked to the transmission edge of the OSR using two photodiodes (PD1, PD2) that monitor the optical power from the DFB and reflection from the OSR, respectively. A locking circuit adjusts the laser wavelength to keep the ratio PD2/PD1 constant. The CML principle is shown in Fig. 3.32 [56].

The limited BW of the OSR filter (7.1 GHz at 3 dB for transmission of 10.6 Gb/s) removes ringing in the intensity profile, and increases the rise and fall times. After the OSR, the ER is increased to >10 dB owing to the high slope of the OSR (2.2 dB/GHz). Such FM-AM conversion has been previously applied to DFB lasers either using filters [57] or by using the transmission fiber itself in DST (see Section 3.1.3.2). The function of the OSR filter in CML, however, is fundamentally different.

The peak-to-peak adiabatic chirp of the DFB is set to be half the bit rate frequency, that is, the MSK condition. Hence, the phase of a 10 Gb/s signal varies slowly with a total shift of $2\pi \times 5\,\mathrm{GHz} \times 100\,\mathrm{ps} = \pi$ during "0" bits. The OSR changes this frequency profile: while the rise and fall times for intensity become longer, those for the chirp are significantly reduced and the frequency profile becomes flat-topped over the entire "1" pulses. This creates nearly chirp-free pulses except for the tails. The chirp waveform shows sharp spikes with excursions as large as -30 GHz at the intensity minima of the isolated "0" bits over a short duration of ~25 ps. The constant-phase envelope, together with the abrupt π phase shifts at intensity minima during "0" bits, is the reason for the dispersion tolerance of CML. After propagation through fiber, the "1" bits overlap and interfere destructively in the "0" bits separating them, producing an open eye and a low BER. This phase correlation between the bits is similar to that for ODB (see Chapter 5.3.5.1.). The CML format also suppresses the optical carrier and occupies half the bandwidth of standard NRZ. Output intensity of CML remains NRZ and operates with a standard square-law receiver.

The reshaping of the frequency profile is induced by the OSR filter slope. Consider passing an optical field with amplitude envelope $A(\mathrm{t})$ and adiabatic chirp through an OSR with a linear transmission edge function $T(\omega) = a + b\ (\omega - \omega_0)$, where b is the slope and a is transmission at center frequency. According to Fourier theory, the linear filter adds the time derivative of the input signal to itself with a $\pi/2$ phase shift. The frequency profile of the output field $\Delta\omega_{\mathrm{Out}}$ is altered because of the $\pi/2$ phase shift and can be approximated by [56]

$$\Delta\omega_{\mathrm{Out}}(t) = \Delta\omega_{\mathrm{AC}}(t) - \frac{\mathrm{d}}{\mathrm{d}t}\tan^{-1}\left[\frac{bA'(t)}{A(t)(a + b\,\Delta\omega_{\mathrm{AC}}(t))}\right], \tag{3.28}$$

where $\Delta\omega_{\mathrm{AC}}$ is input adiabatic chirp. The OSR takes weighted derivatives of the input amplitude and symmetrically adds the resulting blueshifted peaks to the adiabatic chirp at both the 1-to-0 and 0-to-1 bit transitions. This added chirp, which scales with the slope of the OSR b, significantly reduces the rise and fall times, and creates the flat-top (FT) chirp profile across each pulse. This chirp entirely contains the intensity profile whose rise and fall times are increased due to the BW limitation by the OSR. Equation (3.28) also explains the abrupt phase shifts at the local minima in intensity after the OSR. Spike chirp occurs when the transmitted field given by the denominator is small, while the rate of change of the input field given by the numerator is

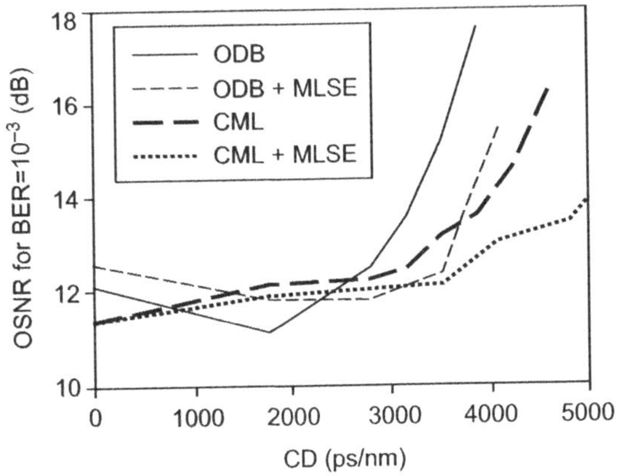

FIGURE 3.33 Comparison of CML with ODB (with and without MLSE) [58].

nonzero. This results in very rapid change in chirp near the intensity minima due to the inverse tangent function.

Reach of CML at 10 Gb/s on uncompensated SSMF can exceed 200 km (or 3500–4000 ps/nm). It can further be extended by using electronic postcompensation (maximum-likelihood sequence estimation, MLSE) [58] or digital pulse preshaping that is applied through a digital signal processor (DSP)-enabled laser driver [59]. Due to the loss of phase information in intensity demodulation, reach increase through RX-based MLSE is limited. As seen from Fig. 3.33, CML can achieve somewhat *better* reach performance on uncompensated fiber as compared to optical duobinary generated with external modulators. This holds for both standard intensity demodulation and MLSE.

Using DSP-based pulse preshaping, reach can be increased to beyond 6000 ps/nm (360 km on SSMF) [59]. These reach capabilities make CML one of the choices for 10 Gb/s WDM transmission over distances up to ~300 km. Tunability can be achieved by applying the CML principles to multisection DBR lasers (e.g., DS-DBR, SG-DBR) instead of DFB lasers. Applicability of CML to higher bit rates, for example, 28 Gb/s, is an ongoing research topic.

3.2 TRANSMISSION LINE

3.2.1 Single-Mode Fiber Types

The characteristics of the transmission fibers have major impact on the transmission and accordingly on cost-optimized solutions in the metro and regional area. Neglecting specialty and multimode fibers (the latter are not suitable for long-haul, high-capacity transmission), the most relevant single-mode fibers are the following:

- Standard single-mode fibers, according to ITU-T Recommendation G.652A–D.

- Dispersion-shifted (single-mode) fibers (DSF), according to ITU-T G.653.
- Cutoff-shifted (single-mode) fibers (CSF), according to ITU-T G.654A–C.
- Nonzero dispersion-shifted (single-mode) fibers (NZ-DSF), ITU-T G.655A-E, including large effective-area fibers (LEAF) and reduced-slope (RS) fibers.
- Ultrabroadband, nonzero dispersion-shifted single-mode fibers, ITU-T G.656.
- Low bending loss single-mode fibers, ITU-T G.657A/B.

Obviously, all relevant single-mode fibers (and also some of the multimode fibers) have been specified in ITU-T Recommendations (G Series), for further details refer to these. Apart from these Recommendations, newer developments include polymer, few-mode, multicore, and photonic-crystal fibers. These have not yet been fully standardized.

SSMF are the oldest single-mode fibers, with first low-loss samples going back to 1978. They were designed for lowest CD in the second transmission window around 1310 nm, where laser diodes were available. Later, it turned out that their high CD around 1550 nm is beneficial for suppressing nonlinear effects in long-haul transmission, in particular four-wave mixing (FWM). Over time, several Recommendations (G.652A to G.652D) have been finalized that mainly address loss in the wavelength region around 1400 nm (where OH^- ion absorption may occur) and the PMD dispersion parameter D_{PMD}. New variants (G.652E) will likely combine low bending loss (similar to G.657A fibers) with low PMD.

DSF were an early attempt to construct fibers that combine lowest loss and lowest CD around 1550 nm. This development had single-channel transmission (i.e., non-WDM) in mind. For reasons that are now well-understood, it turned out that these fibers are critical at best for DWDM transmission because they allow relatively easy phase matching between WDM channels that in turn can lead to disastrous FWM. DSF can still be found in certain subsea systems and in terrestrial deployments in Japan and few other countries.

CSF are single-mode fibers with cutoff wavelength shifted into the region of 1530 nm, and with ultralow loss. For wavelengths higher than cutoff, a fiber becomes a single-mode fiber. Cutoff shifting can be achieved by increasing the fiber core area. Hence, these fibers can have large core areas, which decrease nonlinear effects. Fibers according to G.654B and G.654C have low PMD, and the G.654B type has CD *increased* as compared to SSMF. Therefore, these fibers are suitable for (ultra)long-haul and submarine applications.

NZ-DSF are an improvement of DSF with regard to higher CD in the third transmission window. The attempt was to find a compromise between suitably low CD for high CD-limited reach and sufficiently high CD for efficient suppression of nonlinear effects. Hence, over time and with developing Recommendations, CD was increased. In addition, the latest standard defines lowest PMD.

G.656 fibers are the extension of G.655 fibers toward even higher CD. In addition, the Recommendation defines higher CD for an increased bandwidth (making the fibers suitable for the complete CWDM wavelength range), low PMD, and also low loss due to decreased water absorption.

TABLE 3.5 Macrobending Loss in G.657A/B Fibers

	G.657A		G.657B		
Radius (mm)	15	10	15	10	7.5
Number of turns	10	1	10	1	1
Max. loss at 1550 nm (dB)	0.25	0.75	0.03	0.1	0.5
Max. loss at 1625 nm (dB)	1.0	1.5	0.1	0.2	1.0

G.657 low bending-loss fibers were developed to cope with the increasing demands for better control of macrobending loss in particular in shelf and rack cabling, and for in-house cabling (where sharp bends around corners, and so on, cannot be avoided). It is relevant to note that the Recommendations refer to single-mode fibers that are splice-compatible (G.657A) with SSMF at reasonably increased tolerance against macrobending loss, and incompatible with SSMF at massively improved bending loss, respectively. The respective characteristics are summarized in Table 3.5.

The D parameters for chromatic dispersion have already been shown in Fig. 2.6. Where all G.652 SSMF have very similar CD, Fig. 2.6 also shows that G.655 fibers can have severely different dispersion slope S. This parameter has been reduced for some G.655 fibers in order to achieve similar dispersion characteristics for all channels in ultrabroadband WDM transmission systems. The dispersion parameters (D, dispersion slope S, zero dispersion wavelength λ_0, PMD parameter D_{PMD}) for relevant fibers are listed in Table 3.6.

Many G.655 fibers have smaller effective core areas than G.652 fibers. This results from the necessary manipulation of the waveguide dispersion contribution to the resulting effective CD that often leads to multiple claddings, and also decreased core diameter (see Fig. 3.34). Decreased core diameters lead to higher power densities as compared to G.652 fibers that have ~9 μm effective core diameter. Hence, G.655 NZ-DSF can show stronger nonlinear behavior as compared to standard SMF. In addition, old G.655A fibers can have very small CD—down to 0.1 ps/(nm·km)—which again contributes to high nonlinear effects. Newer G.655B/C and in particular G.655E fibers

TABLE 3.6 Dispersion Characteristics of Relevant Single-Mode Fibers

	G.652A/C	G.652B/D	G.654B	G.654C	G.655A
D @ 1550 nm (ps/(nm·km))	16–21	16–21	22	20	0.1–6.0
λ_0 (nm) (zero dispersion)	1311.5 ± 10	1311.5 ± 10	(~1300)	(~1300)	1480 ± 30
S @ 1550 nm (ps/(nm^2 km))	≤0.092	≤0.092	0.07	0.07	≤0.05
D_{PMD} (ps/km$^{1/2}$)	0.5	0.2	0.2	0.2	0.5
	G.655B	G.655C	G.655E	G.656	G.657A
D @ 1550 nm (ps/(nm km))	1.0–6.0	1.0–6.0	5.5–10	2.0–14	16–21
λ_0 (nm) (zero dispersion)	1480 ± 30	1480 ± 30	1440	1480 ± 30	1312 ± 12
S @ 1550 nm (ps/(nm^2 km))	≤0.05	≤0.05	≤0.052	≤0.05	≤0.092
D_{PMD} (ps/km$^{1/2}$)	0.5	0.2	0.2	0.2	0.2

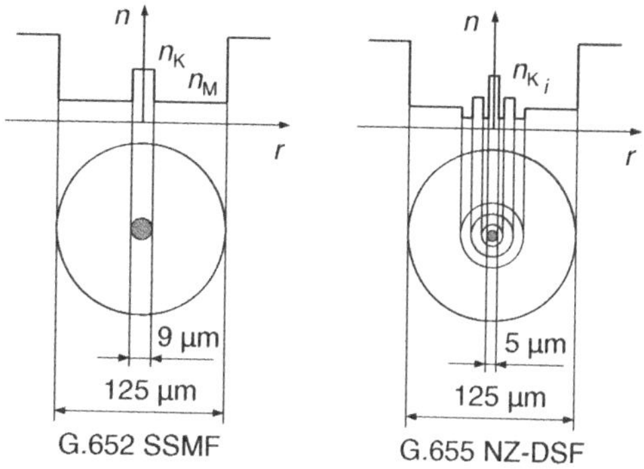

FIGURE 3.34 Refractive index profile $n(r)$ for G.652 SSMF (left) and typical G.655 NZ-DSF (right).

increase the lower CD limit to >1.0 ps/(nm km), thus suppressing FWM more efficiently. They also can have increased effective areas—so-called large area (LA) or LEAF—which further reduces nonlinear effects. Most recently, the use of few-mode fibers (FMFs) has been proposed in order to allow increased effective areas. Few-mode fibers have cross-sectional dimensions that shift the cutoff wavelength to >1530 nm. In the WDM regime around 1550 nm (especially in the C-band), they may support the next higher modes (i.e., the two LP_{11} modes).

Due to the efficient suppression of nonlinear effects (in particular FWM), new fiber deployments that have to support long-haul, coherent ultrahigh capacity WDM transport should be based on latest G.652(B/D) SSMF or G.654(B/C) CSF. In metropolitan areas with their typical service mixture (which for quite some time to come will not allow purely coherent systems) and reduced reach requirements, new G.655 may also be deployed.

3.2.2 Novel and Specialty Fibers

Today, almost all transmission fibers for (WDM) transport ranging from long haul via data center interconnections down to Fiber-to-the-Home installations are solid silica glass single-mode fibers. Only in LANs, intradata center and related in-house installations, silica graded-index (GI) multimode fibers are used as well.

There are four areas where new fiber types or materials are being investigated. The respective fibers may become relevant in WDM networking. The fiber types are FMFs, multicore fibers (MCF), polymer optical fibers (POF), and photonic crystal fibers (PCF), respectively.

3.2.2.1 Few-Mode Fibers The technically achievable maximum transport capacity of SSMF is quickly approaching [60]. Since disruptively new fiber design like PBG fibers (see Section 3.2.2.4) are still years away in long-haul transmission, one of the few ways to further improve fiber capacity is to use few-mode fibers, where the respective guided modes are modulated and demodulated separately. So far, experiments have been done on fibers supporting three modes (LP_{01} and the two degenerated LP_{11} modes) mainly. One challenge of this approach is the need for low-loss mode multiplexers and demultiplexers. So far, only early experimental devices have been used that did not fully solve the low-loss requirement. Further disadvantages include strong differential mode delay. This effect can be counteracted by strong mode coupling in the transmission fiber between the supported modes. Then, a high-performance MIMO (multiple inputs multiple outputs) receiver is required. The MIMO receiver, however, will be required anyway since mode de-/multiplexers and other lumped devices will also cause mode coupling. For a FMF supporting the LP_{01} and the two degenerated LP_{11} modes, a 6×6 MIMO is required (each mode has two polarization modes). Work on such receivers is ongoing.

A FMF is similar to an SSMF and hence relatively simple in manufacturing. As compared to G.652 SSMF, it has increased effective mode area such that the next (two) modes can be excited. FMFs can be regarded an extension of CSF as described in ITU-T Recommendation G.654. For CSFs, the cutoff wavelength, that is, the wavelength above which the fiber is a SMF, was already shifted into the 1530 nm region. This was done to allow larger effective mode areas for ultralong-haul fibers. Hence, FMFs can be derived from CSFs by properly adjusting the cutoff wavelength such that in the desired wavelength region (probably the C-band) the first three modes can be used. This will lead to fiber designs with mode-filed diameters somewhere in the region of 10–12 μm.

FMFs have the advantage of being easily splicable (similar to SMF, CSF). Given proper Erbium-doped fiber amplifier (EDFA) design (with EDF core diameters not stripping off the higher modes), all modes in an FMF can be amplified simultaneously by means of a single EDFA. Problems include the (cost) effective production of mode de-/multiplexers, and the necessity of powerful MIMO equalizers for cross talk cancellation at the receiver.

3.2.2.2 Multicore Fibers Multicore fibers have recently been considered as an approach to increasing the fiber capacity [61,62]. Multiple single-mode cores are integrated in one fiber. Ideally, the multiple cores can be amplified by single fiber amplifiers (which are also based on multicore fibers), however problems may occur with proper splicing. In general, MCFs are being investigated since the late 1990s [63,64].

So far, different MCF designs have been proposed and tested. In most cases, these are based on 3, 7, and 19 cores (see Fig. 3.35). Other designs, for example, with six

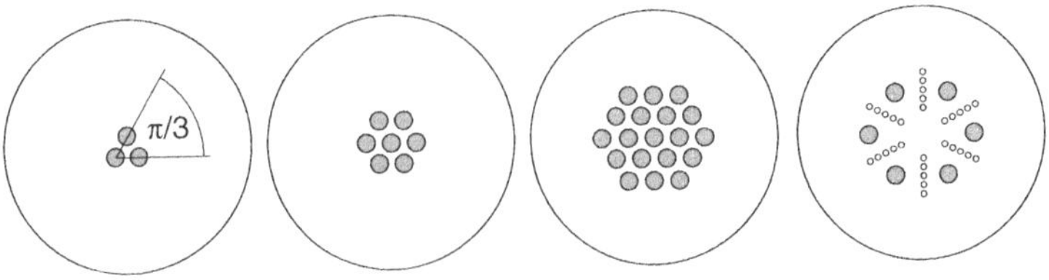

FIGURE 3.35 Multicore fibers with 3, 7, and 19 cores, and 6 cores with holey barriers (FLTR).

large-area cores (in a configuration similar to the 7-core arrangement, but without central core) and additional holey barriers between the cores for cross talk suppression have been reported [65].

MCF can show strong cross talk between the cores. Then, so-called supermodes are excited, and the fiber can be regarded as sort of a multimode fiber. If cross talk between the cores exists, MIMO receivers are required, similar to FMF. However, cross talk can also be suppressed by proper fiber design (e.g., by holey barriers between the cores). A MIMO may nonetheless be required because of cross talk caused in fiber bends and lumped components.

In addition to tackling cross talk, new devices for separating the cores are required. Such devices can be based on extensions of fiber tapers. Further problems include the necessary splicing to amplifiers.

3.2.2.3 Polymer Optical Fibers Polymer optical fibers have been investigates for decades [66,67]. Main objective is the production of short-reach, thick-core, multimode fibers that tolerate sharp bends and very inexpensive connector technology. Main application areas are in-house LANs and cabling in airplanes and large vehicles (because copper wiring of large vehicles acts as very efficient antennas and can add severe weight). These applications typically do not include the use of WDM, but multiple wavelengths are supported by these fibers. In addition, POF have also been developed as SMF, with very similar characteristics (regarding dimensions, spectral loss, dispersion) as compared to silica SMF. Due to advantages of the silica material system (availability of silica, nonpoisonous fabrication), it is doubtful that single-mode POF will play an important role against silica SMF.

POF for in-house cabling (Fiber-in-the-Home, FITH) are mostly based on step-index or graded-index multimode fibers with core diameters as thick as 0.5–1.0 mm. Due to strong mode dispersion, these fibers are typically restricted to bandwidth × length products in the range of 2 Gb/s × 100 m. Leading-edge experiments report $B \times L > 1$ Gb/s km [68].

The main advantage of POF with thick-core diameters is the very simple connector technology. A typical POF connector consists of two inexpensive plastic parts only which are crimped onto the fiber. At the end of the ferrule, the fiber is then cut with a sharp blade, which usually gives better results than attempts to further polish the fiber surface. Hence, POF are very easy to install, even by unskilled staff. In addition, the large core diameter makes them insensitive to bending loss.

3.2.2.4 PCF and PBG Fibers Photonic crystal fibers have a microstructured cross section, in most cases axial holes running along the fiber. The rationale is to increase the fiber's confinement, or to allow air-guiding fibers. This allows improving certain fiber parameters including bending loss, PMD, nonlinear effects, and multicore cross talk (which can be either decreased or increased according to the needs).

PCF split into two categories: index-guided and photonic bandgap (PBG)-guided fibers [69–71]. Index-guided PCF typically have a solid core and surrounding holes, the aim of PBG-guided fibers is to allow air-guiding, hollow-core fibers. Cross sections of these fiber types are shown in Fig. 3.36.

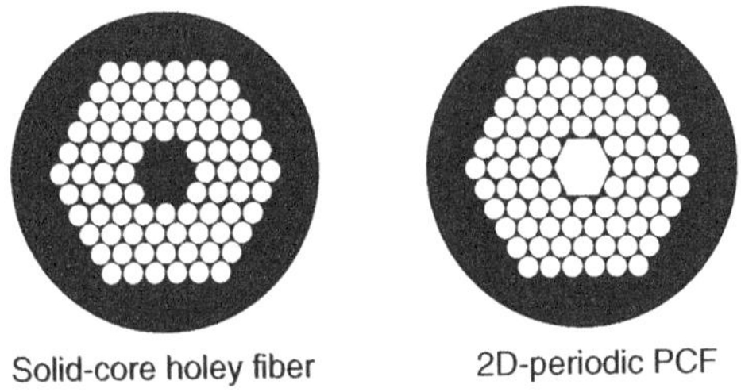

FIGURE 3.36 Holey photonic crystal fibers. Left: index-guided, right: bandgap-guided.

Solid-core, index-guided fibers with holey cladding have lower *effective* refractive index in the cladding and consequently can have higher contrast than doped silica fibers. They produce stronger confinement that allows enhanced (stronger!) nonlinearities, birefringence, or weaker bending sensitivity. The best attenuation reported so far is almost in the region of standard silica SMFs, it is 0.28 dB/m at 1550 nm. The effective index of the core region is also very similar to SSMF. Hence, from today's perspective, it is not clear if these fibers will play a more important role in WDM transmissions. One relevant application may be the cross talk suppression provided by holey barriers in MCF.

PBG-guided fibers aim at air guidance that would allow very low effective index (i.e., increased propagation velocity), and almost zero nonlinearity and PMD. Therefore, it is likely that they will play an increasingly important role in WDM transmissions. PBG-guided fibers again split into two categories, omnidirectional Bragg mirrors (Bragg fibers), and hollow-core PCFs. Their cross sections are shown in Fig. 3.37. Bragg fibers utilize the fact that 1D crystals can reflect light from all angles and polarizations. They can form hollow-core waveguides with modes similar to hollow metal waveguides. They support the TE_{01} as the lowest-loss mode, which is immune to PMD.

Hollow-core PCFs consist of 2D-periodic photonic crystals and a hollow core. Here, the PBG—which prohibits photons to be guided—is distorted by the hollow

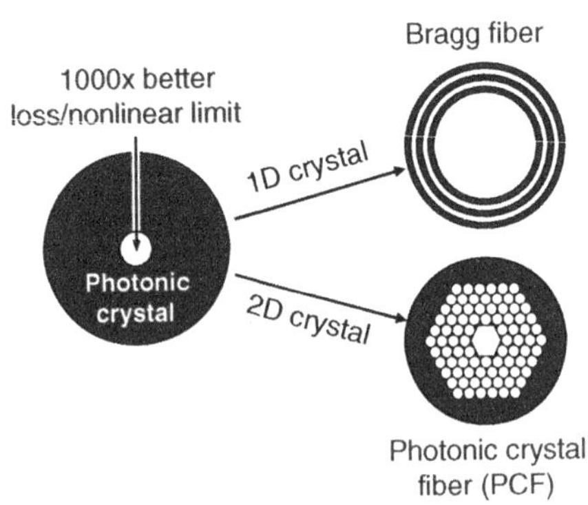

FIGURE 3.37 Air-guiding photonic bandgap fibers.

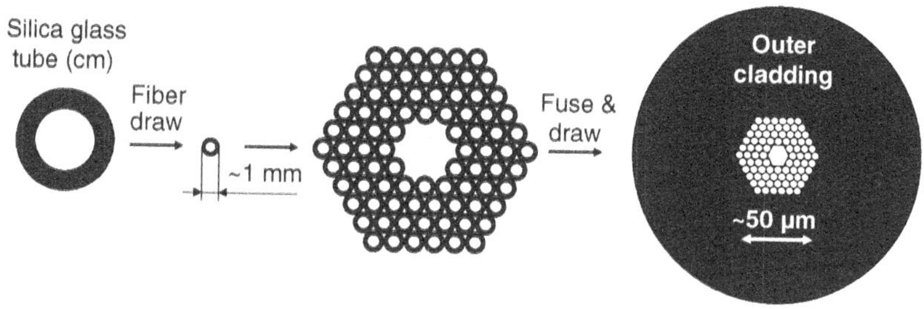

FIGURE 3.38 Fabrication of air-guiding PBG fibers.

core. Hence, the field is confined into the air-guiding core region. In theory, this allows ultralow nonlinearity, low PMD (2D-PCFs do not support the TE_{01} and hence are subject to PMD), and also low effective refractive index. The latter can make them suitable for "fast" fibers where the group velocity exceeds the one known from silica SMF. However, best results so far only report 1.7 dB/km at 1570 nm over 800 m fiber length [72,73]. Hence, wider spread use may take some further years for optimizing the fabrication of these fibers.

Fabrication of PCFs is similar to drawing capillary tubes. This is shown in Fig. 3.38.

First, silica glass tubes are produced. These are drawn, assembled in 2D-symmetric clusters, fused, and drawn again. Currently, the production process is more complex as compared to standard silica fibers, limiting both fiber parameters and maximum lengths.

3.2.3 Fiber-Optic Cables

Optical fibers need to be protected against mechanical loads and sharp bends. Hence, they need to be brought into cables that provide (at least) strain relief and that avoid low bending radii by their design. Three cable types, with various derivatives of each, exist:

- Indoor cables
- Outdoor cables
- Submarine cables

Indoor cables are metal-free in almost all cases. They typically contain smaller numbers of fibers, starting with single fibers and ranging up to 144 fibers. Single-fiber cables often are built with connectors, for example, patch cables. For all single-mode fibers, connectors with physical end-face contact must be used. Various connector brands exist, for example, FC-PC, LC, or E2000. These connectors come in nonangled versions with typical return loss <-40 dB, or with end face angled at 8° in order to achieve return loss of <-60 dB. They are antitwist protected to prevent scratches on the end faces when connecting them. Cables with all combinations of (single-mode) fiber types can be produced. For further information, see [74].

Outdoor cables can have up to 768–864 fibers. They can be divided into three categories:

- Underground cables
- Directly buried cables
- Aerial cables

Underground cables are generally installed in conduits. These cables are designed for high pulling tension, and often lubricants are used to reduce friction when pulling them into longer conduits. Alternatively, special lightweight cables are blown into the ducts with compressed air.

Directly buried cables are placed underground without a conduit. These cables are more rugged, they are often armored to sustain pressure from the soil in which they are buried.

Aerial cables are installed at poles. Some aerial cables are lashed to other cables, others are self-supporting. Optical ground wires (OPGW) are used by utilities for high-voltage distribution lines. These are electrical cables with fibers that are protected in a hermetically sealed metal tube. They are installed like standard electrical conductors.

Fibers can be protected in a cable in different ways:

- Tight buffer coating
- Loose tubes

Tight buffer coating uses a soft protective coating applied directly to the 250 μm primary fiber coating. This provides additional protection for the fiber that is sufficient to allow direct connector termination. In a similar way, further strength members, including high-strength stranded steel wires, can be wrapped directly around the buffer.

Loose-tube cables contain thin plastic tubes (submarine designs and certain outdoor cables like OPGW: metal tubes) containing up to 12,250 μm-buffered fibers. This isolates the fibers from pulling tensions. The tubes are often jelly filled to prevent moisture entry.

Strength members usually consist of aramid yarn, also known by the DuPont trade name Kevlar®. Aramid fibers are very strong, relative to their weight. If pulled hard, they will not stretch but eventually break at a point beyond their strength limit. This ensures that these fibers always break first when the cable is being pulled too hard (i.e., the optical fibers only break after the strength members are already mechanically destroyed).

The outer jacket is the outermost cable layer. Outdoor cables use black polyethylene that resists moisture and sunlight exposure. Indoor cables use flame-retardant jackets that can be color coded to identify the fibers inside the cable. Some outdoor cables may have double jackets with armor between them.

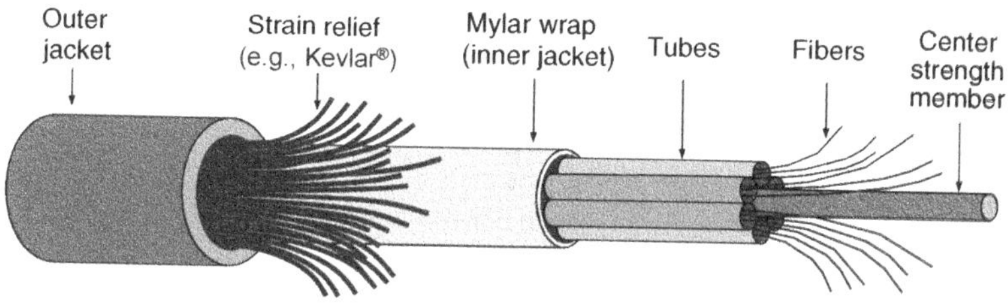

FIGURE 3.39 Fiber-optic outdoor cable.

An example of a fiber-optic outdoor cable in loose-tube design is shown in Fig. 3.39.

Submarine cables require an even more ruggedized design. They must resist very high strain during deployment, in particular when deep-sea deployment is considered. They must also withstand pressure of up to 800 bar in various transoceanic deployments.

Where possible, these cables are carved ~1 m deep into the ocean floor with a remotely operated plough. In these areas, with typical depth of >2500 m, a lightweight cable without heavy outer armor can be used. In other areas, for example, where steep gradients have to be traversed, the cables are additionally armored with double-layer stranded steel wire. A typical submarine cable design is shown in Fig. 3.40 [75,76]. It uses the loose-tube design.

The core of the cable consists of a tube that contains up to 48 fibers. This tube is strain-relieved by two layers of high-strength stranded steel wires. These are wrapped with a copper sheath that is required for high-voltage power supply. The power supply works against seawater ground, with voltages at the land terminals exceeding 3 kV (see Section 6.5.8). The inner part of the cable, including the polyethylene jacket, can be used as the lightweight version.

3.2.4 Optical Amplifiers

In optical amplifiers, the WDM signals are amplified in the optical domain, without conversion into the (electrical) baseband. Optical amplifiers are transparent with regard to protocols carried on the individual WDM channels, and most optical

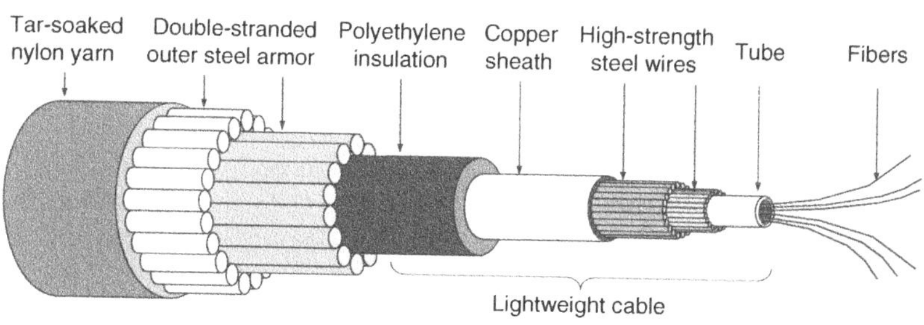

FIGURE 3.40 Submarine fiber-optic cable.

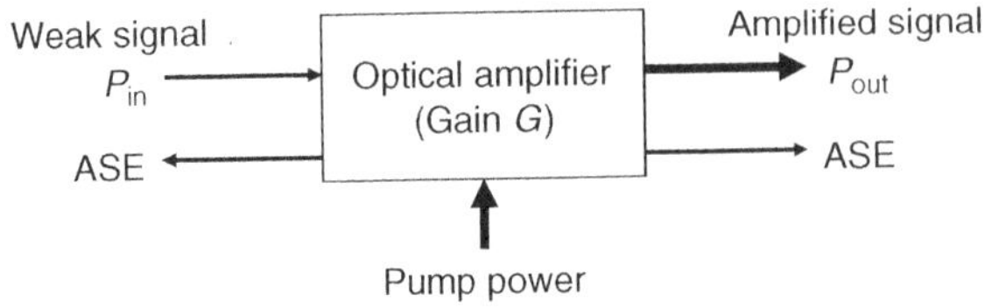

FIGURE 3.41 Generic optical amplifier.

amplifiers are also transparent with regard to the per-channel bit rates. In some optical amplifiers, the entire WDM optical multiplex section (OMS) can be amplified simultaneously. Alternatively, (sub) groups of WDM channels or even individual channels can be amplified.

Unlike 3R (reamplification, reshaping, retiming) regenerators, optical amplifiers are purely analog intensity amplifiers (the sole exception are parametric optical amplifiers that amplify the propagating fields rather than intensity [77–80]). Without added means, they do not provide any equalization or pulse reshaping. They also cannot counteract accumulated jitter. As analog amplifiers, they *add* noise instead. In optical amplifiers, this noise is referred to as ASE. The generic diagram of an optical amplifier with gain G is shown in Fig. 3.41.

Weak incoming signals P_{in} are amplified, and ASE is added. Generally, the ASE can propagate in both fiber directions, that is, co- and counterpropagating to the amplified signal P_{out}. The counterpropagating part can easily be blocked by means of optical isolators, whereas the copropagating part can accumulate in several amplifiers toward the receiver. The power that is required to amplify the incoming signals must be provided by a suitable *pump* source. Similar to lasers, pumps can be optical or electrical (where optical pumps can be delivered through the same transmission fiber).

The optical amplifiers most relevant to WDM transmission are rare-earth-ion-doped fiber amplifiers (in the form of the EDFA), Raman fiber amplifiers, and SOAs. Parametric optical amplifiers (based on the fiber's Kerr nonlinearity [77–80]) so far have not proven relevance, and Brillouin fiber amplifiers (which are, like Raman amplifiers, based on stimulated scattering and as such are part of the fiber's cubic nonlinearity) cannot be used for WDM signals due to lack of amplification bandwidth (refer to Fig. 2.32) [81].

3.2.4.1 EDFAs and Other Rare-Earth-Doped Fiber Amplifiers Optical amplifiers based on EDFAs, are the most relevant amplifiers for WDM transmission [82–85]. Originally developed for WDM C-band transmission [86,87], similar rare-earth-ion-doped fiber amplifiers exist for other wavelength bands as well. Examples include (codoped) L-band EDFAs, TDFAs (Thulium-doped) for S-band amplification [88,89], and PDFAs (Praseodymium-doped) for O-band amplification around 1310 nm [90–92]. An assignment of amplifier technologies suitable for WDM systems is shown in Fig. 3.42 [93].

Erbium as the dopant for C-band (and later also L-band) amplifiers was chosen because a material was required that produces gain in the wavelength region 1530–

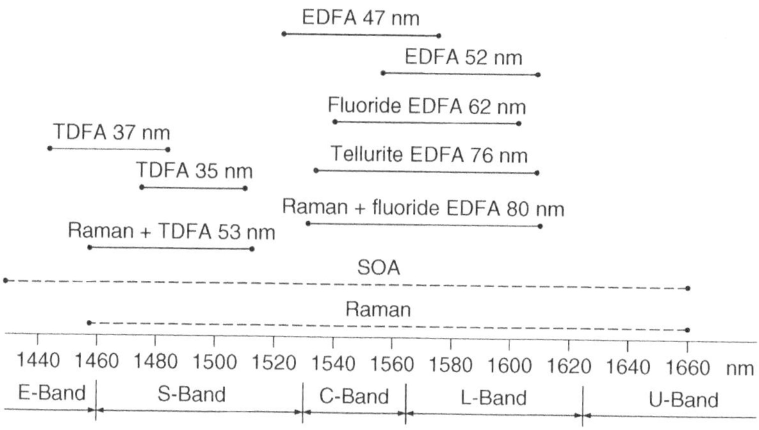

FIGURE 3.42 WDM amplifier technologies.

1610 nm, has absorption bands preferably elsewhere that are accessible with commercially producible pump lasers, and has very long lifetime in the excited state. The latter allows almost complete population inversion at reasonable pump power levels. The absorption and emission bands of an EDFA are shown in Fig. 3.43. Absorption and emission are given in decibel per meter of Erbium-doped fiber. The emission band covers the C-band. Two absorption bands can be identified, one centered around 980 nm, the other overlapping with the emission band. Efficient pumping is possible with wavelengths around 1480 nm (which is not considered a relevant WDM payload wavelength).

The spectral shape of the absorption and emission bands is determined (broadened) by the amorphous nature of silica and the codopants. Both effects can be described in an energy model with a ground state and several excited states. For most considerations, this model can be simplified to ground state and two excited states, thus neglecting higher excited states. The two excited states correspond with 1480 and 980 nm pumping, respectively, the higher excited states can be pumped via further absorption bands around 810, 670, and 540 nm, respectively (which correspond with increasing pump frequencies and hence increasing energy levels). An energy model that includes the 810 nm state is shown in Fig. 3.44. Between the

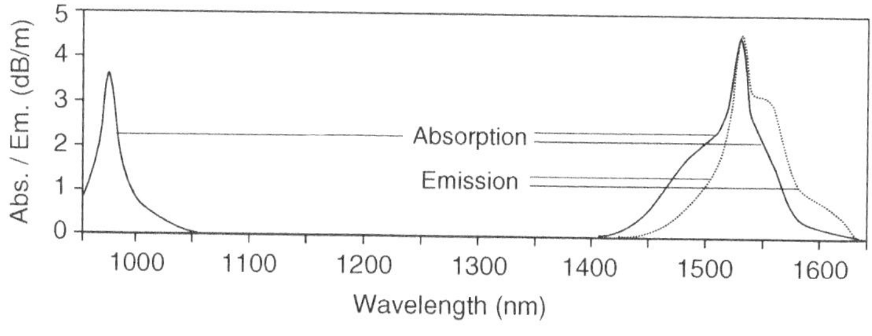

FIGURE 3.43 Absorption and emission bands of an EDFA.

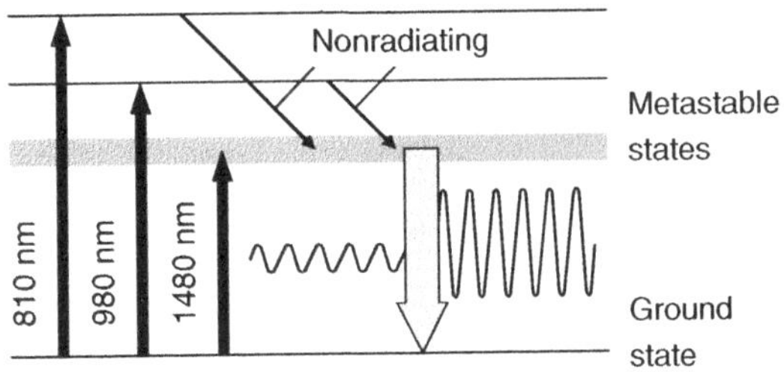

FIGURE 3.44 Simplified EDFA energy band model.

higher exited state and the 1480 nm level, nonradiating transitions occur, and electrons falling back from the 1480 nm level emit around 1550 nm.

The energy model for the (relevant) case of 1480 nm pumping can further be simplified to a two-stage model. Then, a simplified ordinary differential equation that describes the EDFA can be derived [94]:

$$\left(\tau_0 \frac{\mathrm{d}}{\mathrm{d}t} + 1\right)\overline{N}_2(t) = -\frac{1}{\ell\varsigma}\sum_{i=1}^{N} P_i^{\mathrm{in}}(t)\left(\exp\left[\left((\gamma_i + \alpha_i)\overline{N}_2(t) - \alpha_i\right)\ell\right] - 1\right). \quad (3.29)$$

Here, $\overline{N}_2(t) = \frac{1}{\ell}\int_0^{\ell} N_2(z,t)\mathrm{d}z$, N_2 is the fraction of atoms in the excited state, τ_0 is the spontaneous lifetime in the upper level, $\varsigma = \rho S/\tau_0$, S is the Erbium-fiber's cross section, ρ is the density of active Er^+ ions, l is the length of the Erbium fiber, and N is the number of WDM wavelength propagating through the EDFA. Further, $\gamma_n = \rho\Gamma_n\sigma_n^{\mathrm{e}}$ and $\alpha_n = \rho\Gamma_n\sigma_n^{\mathrm{a}}$, with emission cross section σ_n^{e}, absorption cross section σ_n^{a}, and confinement factor Γ_{n}, respectively. From the differential equation (3.29), the gain $\overline{g}_n(t)$ and output power $P_n^{\mathrm{out}}(t)$ can be derived:

$$\overline{g}_n(t) = (\gamma_n + \alpha_n)\overline{N}_2(t) - \alpha_n \quad \text{and} \quad P_n^{\mathrm{out}}(t) = P_n^{\mathrm{in}}(t)\exp\left[\overline{g}_n(t)\ell\right]. \quad (3.30)$$

The assumptions for the simplified differential equation and gain and output power expressions are a two-level model with homogeneously broadened absorption and emission bands, neglecting ASE and excited-state absorption, slow power variation along the Erbium fiber, and strong confinement of the Erbium ions. Then, the physical meaning of the differential equation is given by

$$\underbrace{\frac{\mathrm{d}}{\mathrm{d}t}\left[\rho S\ell\overline{N}_2(t)\right]}_{\substack{\text{Rate of change}\\ \text{in upper level}}} + \underbrace{\frac{\rho S\ell}{\tau_0}\overline{N}_2(t)}_{\substack{\text{Spontaneous}\\ \text{decay rate}}} = \underbrace{\sum_{i=\mathrm{Pump}}\left[P_i^{\mathrm{in}}(t) - P_i^{\mathrm{out}}(t)\right]}_{\substack{\text{Absorption rate of}\\ \text{pump photons}}} - \underbrace{\sum_{i=\mathrm{Signal}}\left[P_i^{\mathrm{out}}(t) - P_i^{\mathrm{in}}(t)\right]}_{\substack{\text{Stimulated emission rate}\\ \text{of signal photons}}}. \quad (3.31)$$

Since an EDFA produces additional ASE (refer to Fig. 3.41), optical isolators are necessary to prevent back propagation of noise. A simple resulting block diagram of

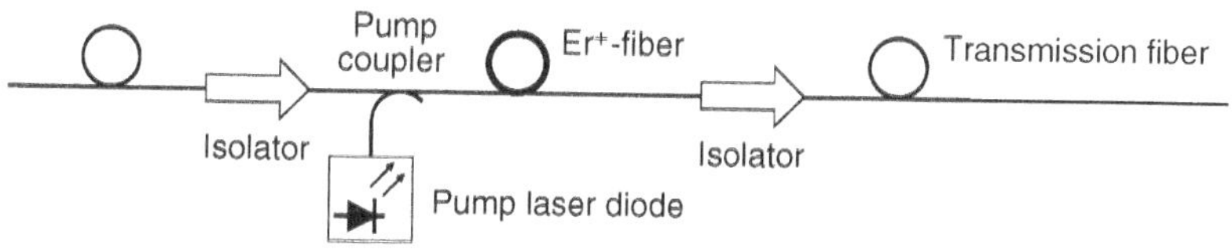

FIGURE 3.45 Principle EDFA configuration with isolators.

an EDFA is shown in Fig. 3.45. Further components of a fully functional EDFA are discussed later.

Both the ASE spectrum and the gain characteristics of an EDFA strongly depend on the (WDM) input signals. As seen from Fig. 3.46, a free-running EDFA (without WDM payload) produces much stronger ASE as an EDFA that actually amplifies WDM channels. Hence, EDFAs are also frequently used as broadband noise sources.

Strong input signals can saturate an EDFA and lead to a certain gain depression around this saturating signal by reducing the average ion population. This effect is referred to as spectral hole burning, or SHB. The width of the hole (the spectral area with depressed gain) can be 3–10 nm, and gain depression can be in the range of 0.1–0.4 dB. Generally, the 1530 nm region is more sensitive to SHB than the 1550 nm region. SHB usually is not an issue in DWDM transport, but it can affect the accuracy of some lightwave measurements.

As already indicated in Fig. 3.46, the level and spectral shape of an EDFA depend on input power. This behavior can more clearly be derived from Fig. 3.47 where spectral gain is shown as a function of input power level.

Obviously, an EDFA saturates with high input power, an effect known from all amplifiers. Saturation output power is achieved for input power levels where gain has decreased by 3 dB as compared to very small input power (the small-signal-gain region) (see Fig. 3.48).

The gain dependence on input power leads to the requirement of power-level control specially in a network context with cascaded amplifiers and frequent add–drop of individual WDM channels. Power levels can be controlled at the transmit side (by controlling launch power) and in in-line sites that can perform equalization through variable optical gain/attenuation. The latter can be achieved with variable optical attenuators (VOAs) that work on a per-WDM channel basis. This function is

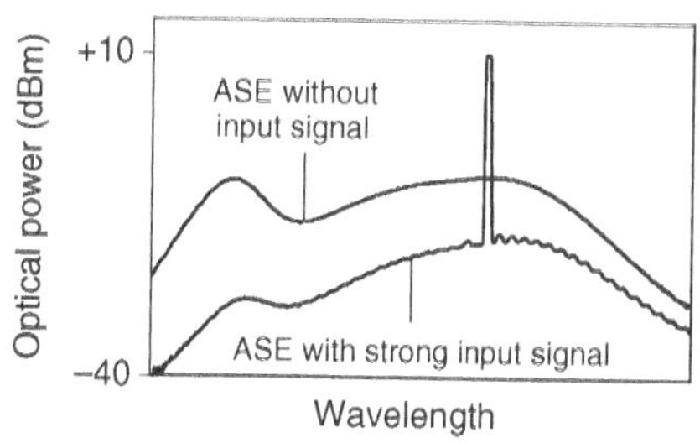

FIGURE 3.46 EDFA ASE spectrum.

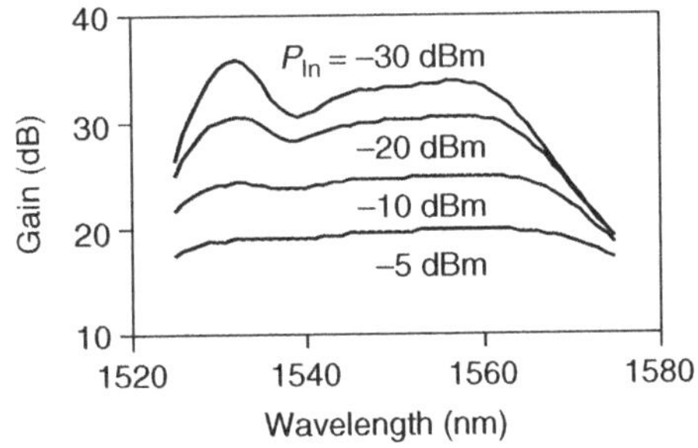

FIGURE 3.47 EDFA gain as a function of input power level.

typically implemented in ROADM sites. In the network context, power-level control also requires signaling that can be implemented, for example, in a suitable control plane (CP).

When cascading EDFAs, the spectral gain shape must also be taken into consideration. Similar to filter cascades (where the filter function becomes steeper), the nonflatness of EDFA gain increases with concatenation of more amplifiers in the absence of countermeasures. This effect is shown in Fig. 3.49.

The spectral gain shape of concatenated amplifiers leads to the requirement of equalization, as already discussed. In addition to CP-controlled equalization, EDFAs can also be complemented by a passive optical filter with approximated inverse gain characteristic, as shown in Fig. 3.50. Due to the shape-dependence of gain on input power (see Fig. 3.47), WDM channel power control is still required.

An EDFA that is fully functional in the network context requires further additions. Power levels of individual signals and the multiplex sections must by measured at several points, and various parameters (of the pump laser(s), of power supply units (PSUs), and also environmental parameters such as temperature) must be captured and processed such that they can bc transferred by a suitable DCN (Data Communications Network) channel (e.g., an Optical Supervisory Channel, OSC) to both centralized Network Management Systems (NMS) entities and CP connection

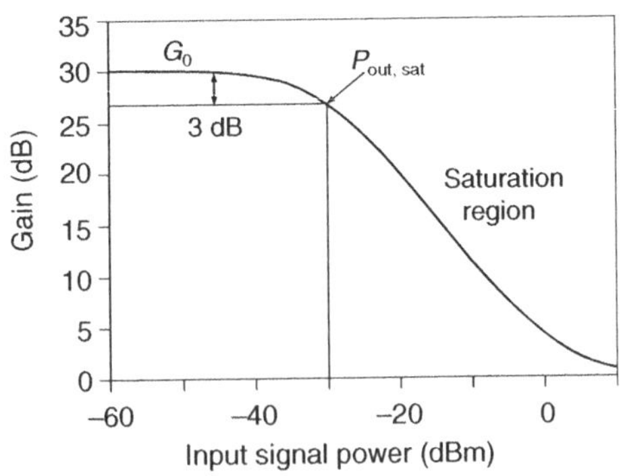

FIGURE 3.48 Input power-dependent EDFA gain and saturation power.

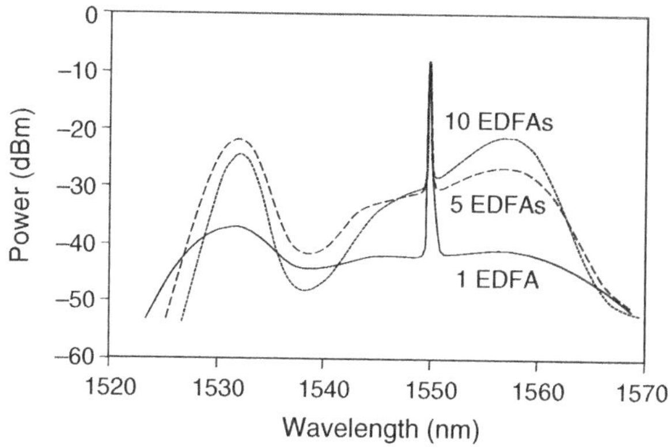

FIGURE 3.49 Spectra of concatenated EDFAs.

controllers (refer to Section 7.6 for further details). A block diagram of an EDFA including these monitoring, control, and signaling means is shown in Fig. 3.51.

In Fig. 3.51, PL is the pump laser, MPs are monitoring ports (e.g., fused asymmetric fiber couplers), OSC is an Optical Supervisory Channel (an instantiation of the DCN on a dedicated wavelength outside the EDFA gain band that carries the management information), WDM1 are the filters to add and drop the OSC, and CDR is the clock and data recovery for the OSC assembler. WDM2 is the wavelength filter for adding the pump signal. The environmental parameters T and RH are temperature and relative humidity, respectively.

As mentioned at the beginning of this section, different rare-earth-ion-doped fiber amplifiers exist for the different wavelength bands. A comparison of the most relevant parameters of these amplifiers is given in Table 3.7.

As shown in Fig. 3.42, EDFAs can be codoped for larger wavelength bands. From the three amplifiers listed in Table 3.7, only the EDFA is widely used. TDFAs may become more important in the attempts to increase total fiber transmission capacity, whereas it is unlikely from today's perspective that PDFAs will play an important role in WDM transmission. Given a massive cost decrease, they may, however, become more important in future (WDM) PONs (passive optical networks).

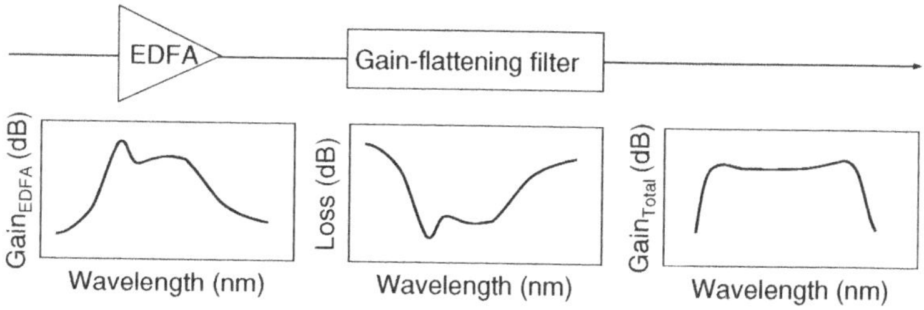

FIGURE 3.50 Principle of EDFA with gain-flattening filter.

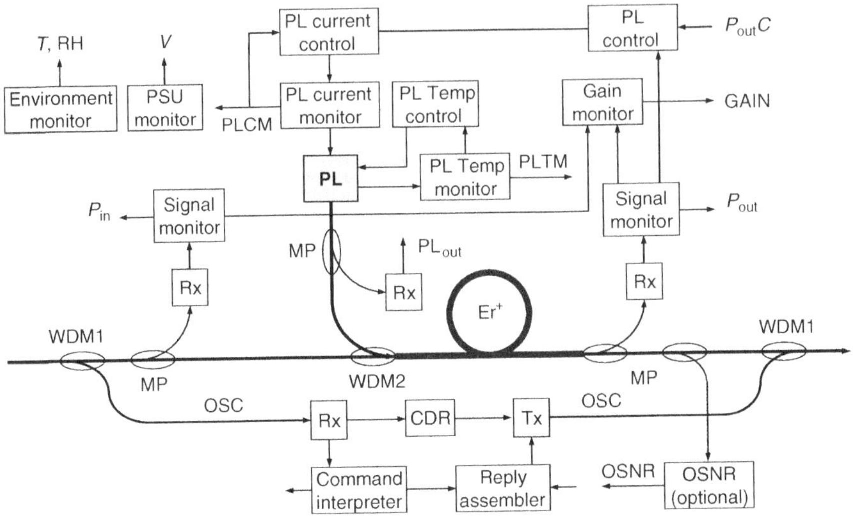

FIGURE 3.51 In-line EDFA with supervision and control means. PL: pump laser.

3.2.4.2 Raman Amplifiers The effect of Stimulated Raman Scattering has been described in Section 2.2.2.1. In WDM systems, SRS acts as a detrimental effect, leading to Raman gain tilt among WDM channels and the related cross talk, but it can also be used as an efficient, broadband, low-noise amplifier. The major characteristic of SRS as a means for amplification is its independence of the wavelength band since the gain spectrum only depends on the spectral location of the Raman pumps. Given the availability of cost-effective high-power pumps, Raman gain can be achieved throughout the entire low-loss region of the optical fiber (this also implies that Raman amplification is difficult at best in the water absorption region around 1400 nm).

The SRS gain spectrum, relative to its pump wavelength, is shown in Fig. 2.29. With two pumps, gain bandwidths exceeding 90 nm can be achieved [95], as is indicated in the top insert of Fig. 3.52. Although lumped Raman amplifiers (with spools containing >10 km of fiber) can be built, most Raman amplifiers use the transmission fiber itself as gain medium. Two modes of operation are possible (and can be combined), with pumps propagating co- and counterdirectionally with respect

TABLE 3.7 Relevant Rare-Earth Ion-Doped Fiber Amplifiers' Parameters

	EDFA	TDFA	PDFA
Wavelength band	C- and L-bands	S-band	O-band
3 dB bandwidth (nm)	30–60	35	35
Noise figure, type (dB)	~5	~5	~6
Saturation power (dBm)	22	20	20
Excitation-state lifetime (ms)	10	~1	~0.4

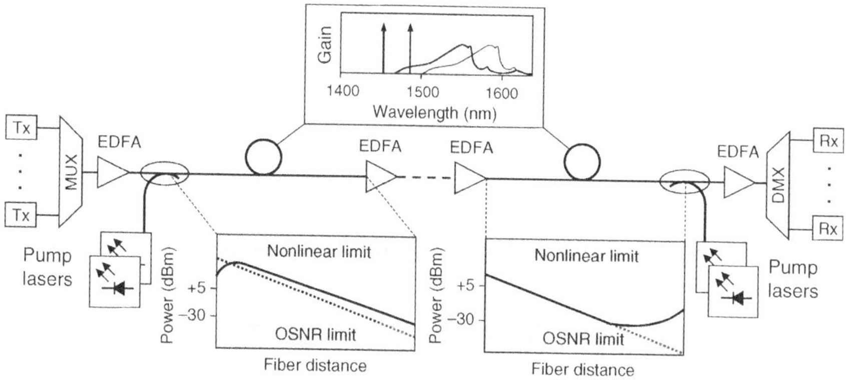

FIGURE 3.52 Raman amplification with co- and counterpropagating pumps.

to the WDM signals to be amplified [96–98]. Most Raman amplifiers make use of counterpropagating pumps because the forward-pumped scheme requires very high pump power levels (which is difficult to be achieved cost-efficiently). In many cases, SRS is then used as a preamplifier that is followed by an EDFA [99,100], also refer to Section 4.2.3. Both operation modes are shown in Fig. 3.52. Mid-link EDFAs can also be complemented by Raman amplification.

Distributed Raman amplification depends on the type of transmission fiber that is used [101]. For S-band amplification, the use of SRS may become impossible on fibers with strong water absorption peak due to strong pump attenuation. On various G.655 NZ-DSF, C-band amplification becomes less effective because the pumps fall into the region of zero CD. In general, Raman amplification will most efficiently work on newer G.652D fibers.

Raman amplifiers require high pump power levels. This narrows down the availability of cost-effective pump lasers for arbitrary gain bands. Commercially, Raman amplifiers are available for C- and L-band. High pump power also leads to the requirement of laser-safety mechanisms. Fast pump-laser shutdown must be implemented for cases of open fiber ends. These shutdown mechanisms require power monitoring (at the far end) and fast signaling.

Due to the ultrashort time constant of SRS (which is in the femtosecond range), more sophisticated transient and power-level control is required as compared to EDFAs. Transient control can be based, for example, on gain clamping [102–104].

3.2.4.3 Semiconductor Optical Amplifiers Semiconductor optical amplifiers are the third amplifier type with certain relevance for WDM systems. Basically, there are two types of SOAs, simple SOAs (sometimes referred to as Fabry–Pérot amplifiers) and traveling-wave semiconductor laser amplifiers (TWSLA) [105].

Simple SOAs are almost the same as regular index-guided Fabry–Pérot LDs. The back facet is pigtailed to allow the throughput of signal light. The main problem is that it is difficult to produce SOAs longer than about 500 μm. Over this short distance, there is not sufficient gain available on a single pass-through for many

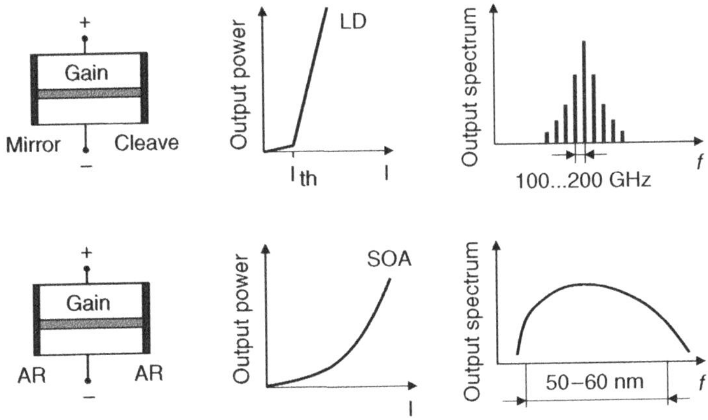

FIGURE 3.53 Comparison of a FP-LD and a SOA (TWSLA).

amplification applications. Gain can be increased by retaining the reflective facets (mirrors) characteristic of FP-LF operation. With typical mirror reflectivity of around 30%, the signals can reflect a few times within the cavity and obtain higher amplification. Simple SOAs are suitable for single-channel or few-channel (CWDM) applications.

TWSLAs are different from simple SOAs in that the cavity is lengthened (approximately tripled) to allow for sufficient gain in a single pass-through. Devices with cavities as long as 2 mm are available. The back facet is AR coated and pigtailed. The front facet of the amplifier is just the same as for a laser except that it is also AR coated. Because of the absence of feedback the TWSLA can be operated above lasing threshold. This gives higher gain per unit of length than the simple SOA. Gain of up to 25 dB over a bandwidth of 40 nm is possible. The absence of feedback leads to a continuous gain spectrum, as compared to the typical FP-LD output (see Fig. 3.53).

SOAs have broad gain bandwidth, as indicated in Fig. 3.53. A wavelength region of 1280–1650 nm can be covered by different devices. For several CWDM applications, SOAs are the only cost-efficient amplification technology (with Raman amplifiers potentially being too expensive for single or few CWDM channels).

SOAs and TWSLAs have several limitations. Fiber-chip coupling is lossy. The amplifier must have additional gain to overcome these losses. This is also the major reason for the typical noise-figure range of 6–8 dB.

Unlike fiber amplifiers (rare-earth-doped, Raman), SOAs exhibit a certain degree of polarization dependence due to the cross-section shape of the gain region. The resulting polarization dependence of the gain is typically <1 dB, and <0.5 dB in most cases. Figure 3.54 shows the polarization dependence of the gain as a function of the pump current.

Due to their fast dynamics, SOAs can produce severe pulse distortions and cross talk when multiple optical channels are amplified. Without countermeasures, this makes them unusable as amplifiers in WDM systems. For the same reason, SOAs can

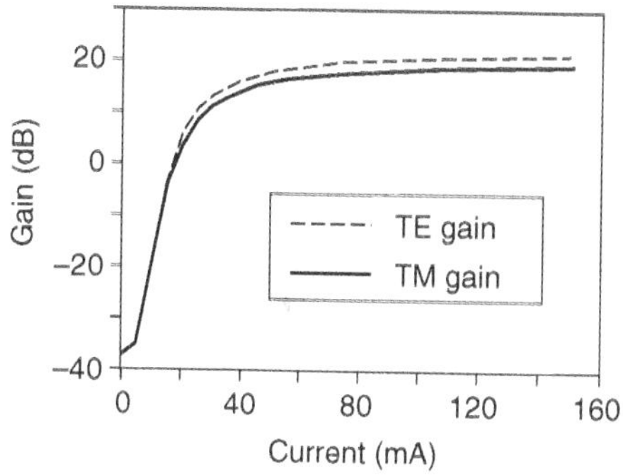

FIGURE 3.54 SOA gain versus current, injection optical power −20 dBm.

be used as wavelength converters and as simple logic gates in optical network systems. The effect of gain compression and the related time constant are shown in Fig. 3.55.

For multichannel WDM operation, control mechanisms at either the device or system level must be implemented. At system level, a saturating reservoir channel or small-signal operation (linear regime) can be implemented [106,107]. At device level, laser gain clamping can be implemented. In a gain-clamped SOA, the gain medium is shared between the SOA and a laser. Hence, the device is operated above threshold and produces constant gain. Lasing occurs at wavelengths different from the amplification band. In order to determine these wavelengths, DBR gratings are integrated into the device (see Fig. 3.56). The laser signals propagate into both fiber directions.

An advantage of SOAs is that they can be integrated with other components on a single planar substrate. For example, a multichannel WDM transmitter may be constructed including, for example, 10 lasers and a coupler all on the same substrate. An SOA can be integrated to overcome coupling losses [23]. Such devices can be used, for example, in WDM-PON.

Table 3.8 lists relevant SOA parameters in comparison with those of fiber amplifiers.

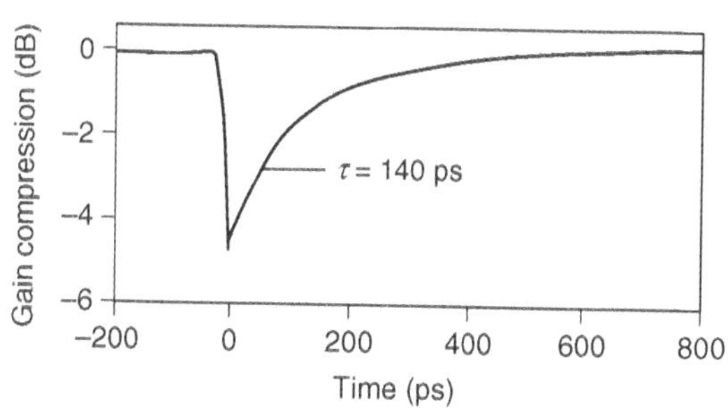

FIGURE 3.55 Gain compression in SOA.

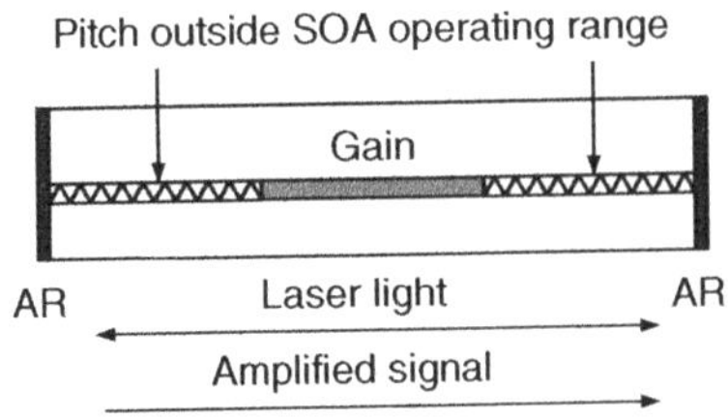

FIGURE 3.56 Gain-clamped SOA.

Similar to other semiconductor-based fiber-optic devices (e.g., EAMs), SOAs can be based on bulk or quantum-dot techniques. SOAs having nanosized semiconductor particles (quantum dots) can have ultrawide operating wavelength range, suppressed waveform distortion in high power output, and the capability of noise reduction (signal regeneration) by limiting amplification [108]. Quantum-dot SOAs are presently in the stage of commercialization. Applications are not limited to optical amplifiers, but also include light sources for sensors and gyroscopes, and gain elements integrated into wavelength-tunable or mode-locked lasers.

3.2.5 Dispersion Compensation

Compensation of chromatic dispersion became necessary in the 1990s with WDM channel bit rates increasing from 2.5 to 10 Gb/s per channel [109]. Due to its linear, time-invariant (neglecting weak temperature dependence) and deterministic nature, CD can be compensated with reasonable effort. In its beginnings, compensation was based on quasi-static techniques in the analog optical domain. Examples include the use of:

- dispersion-compensating fibers (DCF) [110–112],
- dispersion-compensating gratings, for example, fiber Bragg grating [113–115],

TABLE 3.8 Comparison of Relevant SOA and Fiber-Amplifier Parameters

	EDFA	Raman	SOA
Max. gain (dB)	>40	>25	>30
Wavelength region (nm)	1530–1610	1280–1650	1280–1650
3 dB bandwidth (nm)	30–60	Pump dependent	60
P_{Sat} (dBm)	22	0.75 × pump	18
Polarization dependence (dB)	0	0	<0.5
Noise figure, type (dB)	4–6	3–5	6–8
Pump power	25 dBm	>30 dBm	<400 mA
Time constant	10 ms	fs	0.2 ns
Size, form factor	Shelf module	Transmission fiber	Very compact, can be integrated
Switchable	No	No	Yes

- electronic prechirp of the laser transmitter [116–118],
- supportive FM-AM conversion, for example, DST [53–55] (see Section 3.1.3.2),
- mid-span spectral inversion, spectral mirroring through FWM [119–122],
- constructive use of self-phase modulation (SPM), also known as soliton effect [123–125] (see Chapter 2, Section "Solitons").

Though many different proposals for CD compensation have been made, only few of them have been relevant for installed WDM systems. These include the use of DCF or FBG, and the optional additional use of SPM for RZ-coded transmission schemes. Prechirp as a means for CD compensation is limited in the achievable link lengths. In addition, it suffers from added spectral broadening (also see Section 3.1.3.1). Similar restrictions regarding achievable reach hold for DST. Mid-span spectral inversion has been discussed, and lab-tested, for more than 25 years, without having seen significant deployments.

In addition to the quasi-static link-compensation schemes, various tunable optical dispersion compensators (TODC) have been developed, especially for the directly detected and self-coherent 40 Gb/s WDM systems. TODCs became necessary to compensate the remaining receive-end residual CD. Various optical compensators have been developed, two of which are discussed in more detail hereinafter.

After Y2000, digital electronics were powerful enough to catch up with the requirements of real-time dispersion equalization in high-speed fiber-optic transmission. The respective techniques are mostly applied to the receive end. Alternatively, adaptive digital predistortion at the transmitter is possible. The most commonly used digital equalizers are FIR filters (also known as feed-forward equalizers, FFEs) and MLSE. Digital equalization is described in Sections 3.4.3 and 5.3.5.1.

More details on dispersion compensation techniques can be found in Refs [126,127].

3.2.5.1 Dispersion-Compensating Fibers In the first decade after Y2000, most transmission links were fitted with DCFs. One problem is that the compensating fibers have to be adapted to both the type of transmission fiber used for the respective link and the span lengths. This leads to the requirement that respective compensating fibers have to be available for all relevant transmission fibers, and also with certain granularity, that is, DCF lengths. In particular, various compensating fibers for the different brands of G.655 fibers have to be available (refer to Fig. 2.6).

The relevant parameter for successful compensation is the ratio R_{DS} of dispersion parameter D and slope S for both fibers:

$$R_{DS} = D/S. \tag{3.32}$$

Misadaptation of the DCF type to the transmission-fiber type means that CD is compensated exactly only in the direct vicinity of a center frequency, say 1550 nm,

TABLE 3.9 Dispersion Characteristics of Transmission Fibers and DCF

	D (ps/(nm km))	S (ps/(nm^2 km))	D/S (nm)	A_{eff} (μm^2)
Standard G.652	16.7	0.056	298	80
Hi-slope DCF	−105	−0.35	300	20
Low-slope G.655 (TWRS®)	6.6	0.045	147	50
Ultraslope DCF	−115	−0.78	147	18
Low-slope NZ-DSF	4.5	0.045	100	(50)
SC-DCF	−70	−0.70	100	17

and that mismatch increases toward the edges of the WDM transmission spectrum. This is the reason why the dispersion slope of the transmission fibers needs to be considered. Full compensation, including slope compensation, is achieved if the slope-compensation rate R_{SC} approaches unity (where the subscript TMF denotes the transmission fiber):

$$R_{\text{SC}} = \frac{R_{\text{DS,DCF}}}{R_{\text{DS,TMF}}} \overset{!}{\approx} 1. \tag{3.33}$$

Table 3.9 lists pairs of fibers (standard G.652 SMF and two types of reduced-slope G.655 NZ-DSF) together with well-adapted compensation fibers [128,129].

DCF can have attenuation in the range of 0.23–0.6 dB/km [129,130]. DCF for longer spans can thus have accumulated insertion loss of 10 dB and more, they must hence be connected to the mid-stage access of dual-stage in-line amplifiers. Due to the reduced effective area, input power levels must be reduced as compared to the transmission fiber in order to limit nonlinear effects.

In long-haul transmission, DCFs should be placed frequently along the transmission line, for example, per span. This is possible in long-haul transmission systems with close-to equidistant amplifier spacing. In (ultralong-haul) subsea links, the DCF is directly brought into the cable, as part of the transmission fiber (dispersion management, see Chapter 2, Section "Dispersion Map" and Section 6.5.9). In metro and regional systems with their potentially highly irregular span lengths, per-span compensation sometimes is not possible. The main reason for per-span compensation is the provisioning of the power budget that is required to accommodate the DCFs. This is usually done in the mid-stage access of dual-stage amplifiers.

Various schemes of per-span exact compensation, under- and overcompensation have been investigated extensively. These must be combined with respective pre- and postcompensation at the transmit and receive ends, respectively, which can compensate for mismatch of the DCFs, for example, with respect to the span or link lengths. For transmission systems optimized for transport of WDM channels operating at 10 Gb/s, the two most relevant schemes are distributed undercompensation (DUCS), and full-in-line compensation and optimized postcompensation (FOCS). It has been show that DUCS is more advantageous for 10G systems [131]. This scheme is not compatible with the self-coherent 40G systems that were deployed from 2005. These systems require slight overcompensation together with negative precompensation for optimum results. If these systems have to operate on 10G-optimized links, they will not achieve their optimum performance, and they require (tunable optical or electronic) postcompensation at the receivers. For more details on dispersion maps and design rules, refer to Chapter 2, Section "Dispersion Map" and Section 6.5.3.2.

When using DCFs, mismatch to both fiber type and span/link length needs to be taken into account. It is not possible for transmission systems with irregular span lengths to exactly adopt DCF modules to the respective span lengths. Together with fiber-type mismatch, this is the second contribution to suboptimum compensation and the related system performance. This contribution can be kept small in static (ultra)long-haul point-to-point links. Here, the per-span mismatches can be designed to approximately cancel each other (irrespective of the compensation scheme (FOCS, DUCS)). This can be achieved by successively over- and undercompensating the spans, with regard to the target compensation (i.e., considering FOCS, DUCS). In large reconfigurable (meshed, ROADM) networks, the accumulated compensation mismatch can cause performance penalties.

For information on DCF design, refer to Ref. [132].

3.2.5.2 Chirped Fiber Gratings Chirped fiber Bragg gratings are lumped components that were frequently used for in-line CD compensation. As compared to DCF, chirped FBG have the advantages of lower insertion loss, lower cost, and smaller form factor. Unlike DCF, they produce a certain amount of phase ripple that can lead to performance penalties.

Two different FBG types exist, *channelized* (periodic) FBGs with a free spectral range according to the WDM grid (50 or 100 GHz), and *broadband* FBGs that offer linear group delay over the whole transmission band (C-band). Typically, for CD compensators, channelized FBGs are used. They offer periodic passbands with piecewise linear group delay for each individual WDM channel. They are also referred to as periodic group delay (PGD) compensators [133]. This is depicted in Fig. 3.57. They have been used for 10.7 and 43 Gb/s WDM transmission [134–137].

Cascaded FBGs lead to signal distortions that are caused by imperfections in the group delay, known as group delay ripples (GDR). This leads to distortions of the phase response (phase ripple) [138]. The influence of the phase ripple depends on the modulation format, the data rate, the shape of the ripple, and the FBG type [139,140]. The GDR can take values in the range of 6 ps, leading to OSNR penalties for long-haul links in the range of 0.5 dB [141].

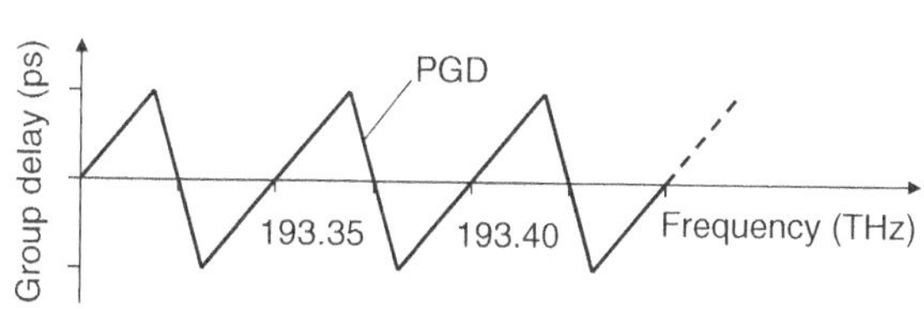

FIGURE 3.57 Group delay of a periodic group delay compensator.

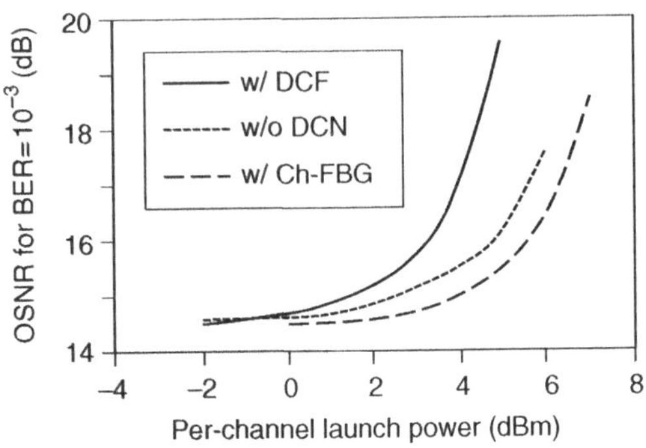

FIGURE 3.58 Effect of FBG-CD compensation on QPSK.

In Ref. [142], it has been found that channelized FBGs, when used for CD compensation, can have a positive effect on coherent DP-QPSK WDM signals. The resulting negative penalty (see Fig. 3.58) is related to suppression of XPM in the PGD compensators. As can be seen, DCF do not have this advantage. They lead to a penalty for coherent intradyne systems over uncompensated links (or links with PGD compensators). This penalty is related to walk-off compensation in the DCF.

The compensation effect of a chirped FBG results from reflections of specific spectral components that are distributed along the fiber grating (also see Section 3.2.6.4). Hence, the reflected components have to traverse different distances inside the FBG that counteracts the distortions caused by CD. Since FBGs are reflective devices, they have to be used in configurations incorporating optical circulators (see Fig. 3.59).

Circulators are nonreciprocal three-port devices that use Faraday rotators and waveplates to achieve the nonreciprocal behavior [143]. A block diagram of a circulator is shown in Fig. 3.60. In order to achieve polarization-independent characteristics, the device uses polarization beam splitters (PBS) and a dual-path configuration. Circulators can have insertion loss below 1 dB, and isolation in the range of 20–40 dB.

According to Ref. [133], suppression of interchannel nonlinear cross talk can be achieved with PGD compensators in coherent intradyne transmission. This is another

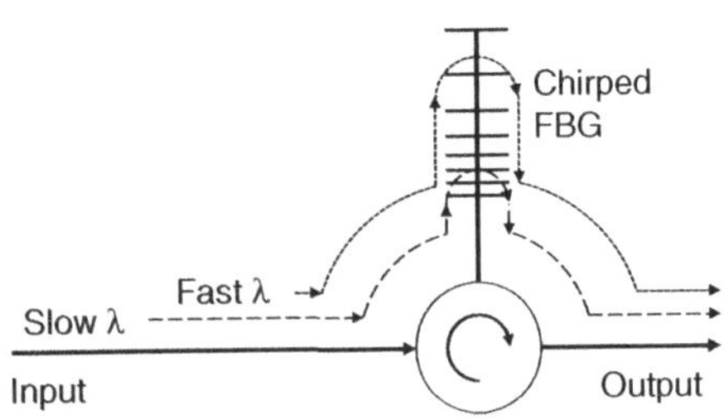

FIGURE 3.59 Setup with chirped FBG as CD compensator.

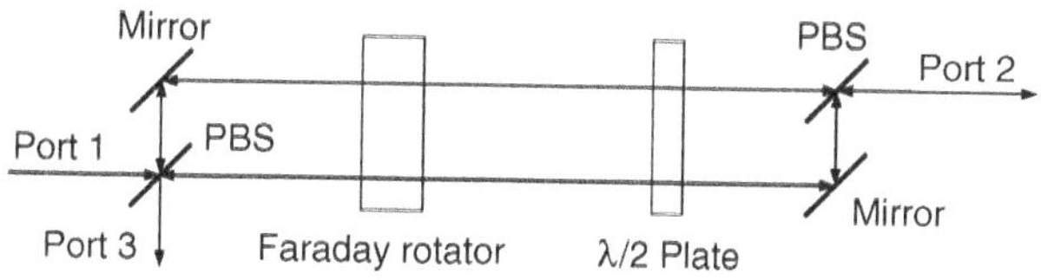

FIGURE 3.60 Block diagram of an optical circulator.

advantage of FBGs over DCF, since DCF can severely decrease the performance of coherent intradyne systems. The use of channelized FBGs, however, can become a problem if gridless systems (i.e., with flexible channel grid) are to be used (e.g., for combined 100G/400G WDM transmission).

3.2.5.3 Self-Phase Modulation (Soliton Effect) The effect of self-phase modulation has been described in Section 2.2.1.1 already. In long-haul transmission, SPM interworks with CD. The result can be constructive or destructive, depending on the sign of the CD (normal or anomalous CD). For $D > 0$, which is the case for almost all transmission fibers (refer to Fig. 2.6), partial cancellation of the SPM and CD effects can be achieved, for $D < 0$, the detrimental effects of both accumulate. This effect is shown in Fig. 3.61 [125]. Note that DCF have $D < 0$ (see Table 3.9). This is the reason why DCF must be launched at comparatively low power levels.

The compensating effect of SPM depends on the power level and the pulse shape, where the allowable power level in turn depends on the pulse shape. CD compensation through SPM can work perfectly for Solitons, that is, hyperbolic-secant-shaped RZ pulses with suitably high pulse energy (refer to Chapter 2, Section "Solitons"). For general RZ pulses, the compensation effect is better, the smaller the duty cycle is. In addition, the allowable power level also increases with decreasing duty cycle. This effect is shown in Fig. 3.62. Higher power levels lead to an increase of the maximum number of spans, or maximum link length. One drawback of RZ (apart from the necessity of an additional pulse carver) is the required bandwidth that also increases with decreasing duty cycle (also see Fig. 5.7).

With the exception of soliton pulses, SPM is not sufficient as the only means for CD compensation. It must be combined with in-line or digital (pre/post) compensation. For in-line compensation, RZ pulse shaping gives somewhat better total (reach) performance. In particular, less dependence on the specific dispersion map (required

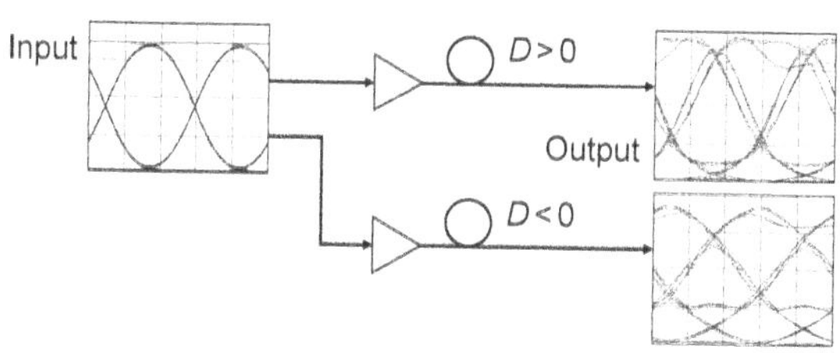

FIGURE 3.61 Combined effect of SPM and CD for $D > 0$ (top) and $D < 0$ (bottom).

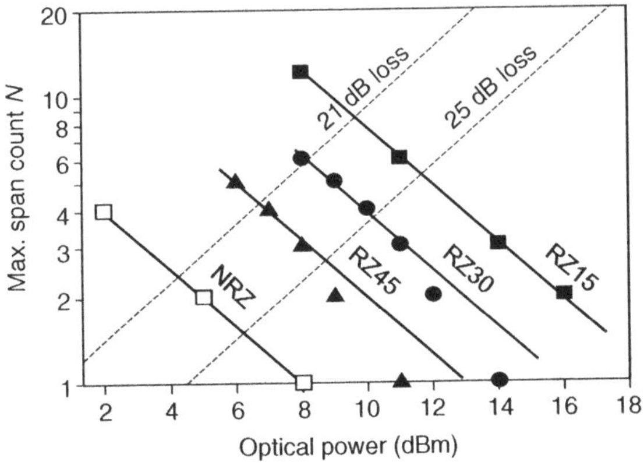

FIGURE 3.62 Effect of NRZ versus RZ.

degree of optimum undercompensation) is achieved with RZ shaping. In general, RZ pulse shaping requires undercompensation [144].

3.2.5.4 TODC Tunable optical dispersion compensation (TODC) was necessary for the first generations of directly detected and self-coherent 40 Gb/s systems. These systems had to run on links optimized for 10 Gb/s transport (i.e., DUCS). Forty gigabits per second systems can work on such links with appropriate predistortion at the transmitter. In addition, the receiver CD allowance is significantly smaller as compared to 10 Gb/s systems (depending on the modulation scheme, the CD allowance may decrease according to the square of the Baud-rate increase). Since CD compensation for 10 Gb/s systems is designed for the typical CD allowance of a 10 Gb/s IM/DD receiver (say, ±600 ps/nm), 40 Gb/s receivers require additional CD compensation (a DPSK 40 Gb/s receiver may have CD allowance in the range of ±50 ps/nm). In the absence of suitable digital compensation (lack of sufficiently powerful DSPs, no access to the electrical field), this additional compensation had to be provided by tunable optical devices.

Many TODCs are based on thermally tunable chirped FBGs [145,146]. The respective devices can have tuning rang up to 1600 ps/nm and low insertion loss of <3 dB. However, a broad range of alternative devices has been developed in the first decade of the twenty-first century.

One such TODC is described in Ref. [147]. It is based on a waveguide grating router (WGR) that is implemented in a planar lightwave circuit (PLC), together with a variable-curvature (deformable) mirror. WGRs are also known as AWGs and described in Section 3.2.6.3. The principle setup is shown in Fig. 3.63.

The mirror is tuned (deformed) according to the required amount of CD compensation, or group delay. The mirror curvature controls the illumination of the WGR light paths and hence the group delay of the device. The PLC-WGR TODC demonstrated a tuning range of more than ±500 ps/nm. The group delay ripple was less than ±6 ps over 70 GHz of bandwidth, and the insertion loss, including the circulator, was ~10 dB (dispersion-setting dependent).

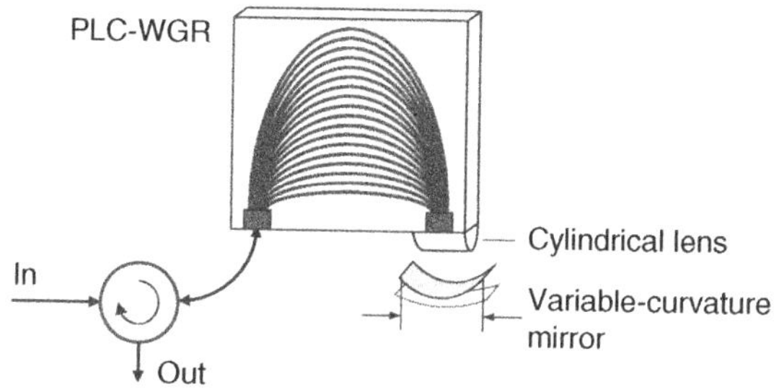

FIGURE 3.63 TODC based on PLC-WGR and deformable mirror.

A TODC based on a virtually imaged phased-array (VIPA) and spatial light modulator (SLM) or flexible 3D mirror has been proposed in [148,149]. The setup is shown in Fig. 3.64.

Similar setups have been used with either flexible curved mirrors (shown in Fig. 3.64), or with a SLM to provide tunability without moving parts. The optical input is fed into a VIPA using a cylindrical focusing lens. The VIPA spatially disperses the wavelength components, and another lens focuses the spatially dispersed components at the Fourier plane. To achieve tunable CD compensation, a flexible mirror or a liquid-crystal SLM with quadratic phase distribution that varies with the wavelength (spatial position) can be used.

The insertion loss for the TODC including circulator and polarization controller is ~15 dB. The tuning range is 1700 ps/nm. The TODC can work simultaneously on different DWDM channels spaced by the free spectral range of the VIPA (here 50 GHz).

Further TODC were based on higher-order mode fibers and switchable fiber gratings [150], and multicavity etalon all-pass filter [151].

3.2.6 Passive WDM Filters and Couplers

Passive optical filters are an essential part of cost-effective WDM transmission systems. They are used as multiplex, demultiplex or add–drop filters and as noise-bandwidth limiters. Fixed add–drop filters were originally used for WDM rings or linear add–drop structures ("buses") where only a subset of the WDM channels had to be terminated at a given site. As indicated in Fig. 3.65, they often consist of cascaded WDM filters realized in different technologies.

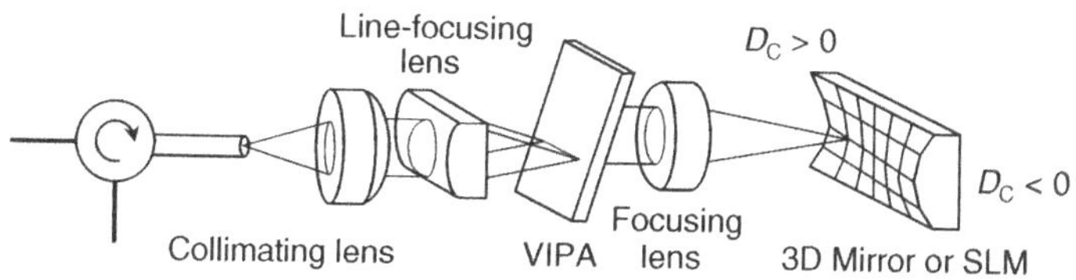

FIGURE 3.64 VIPA-based TODC.

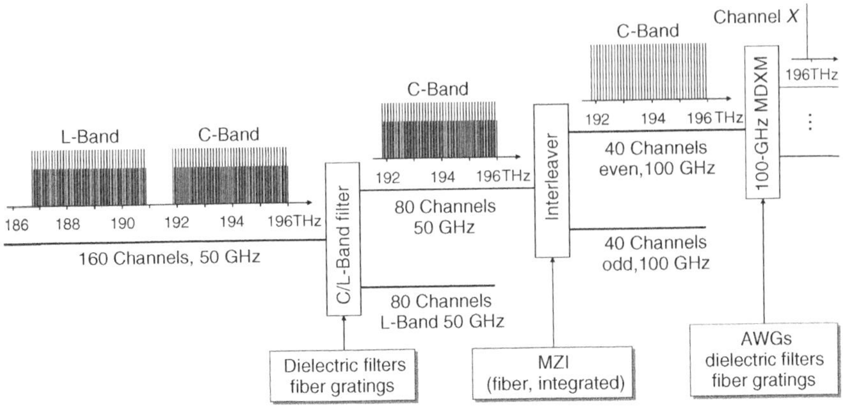

FIGURE 3.65 Cascaded filters for a 50 GHz, C + L-band WDM systems.

Figure 3.65 shows the filters and wavelength-band assignment for a 160 channels, 50 GHz WDM system. Today there is a strong trend to replacing passive static filters by flexible reconfigurable filters (ROADMs).

3.2.6.1 Fused Couplers Fused couplers are used for power splitters, directional couplers, monitoring taps and similar applications. In most cases, they are wavelength-flattened in order not to show any wavelength dependence. Basic element is the 2×2 coupler, shown in Fig. 3.66. Derivatives are 1×2, $1 \times N$, and $2 \times N$ couplers, respectively.

The 2×2 coupler consists of two fibers that are fused together over a certain length z in a defined way (i.e., with a certain curvature). By imprinting a certain periodicity in the coupling region, wavelength selectivity can be achieved. Since always two fibers are fused, the 1×2 coupler is just a standard 2×2 coupler with one port unused. $1 \times N$ and $2 \times N$ couplers result from cascading 2×2 couplers.

The transfer function of a 2×2 coupler is given by

$$\begin{pmatrix} E_{01}(f) \\ E_{02}(f) \end{pmatrix} = \mathrm{e}^{-j\beta z l} \begin{pmatrix} \cos kz & j \sin kz \\ j \sin kz & \cos kz \end{pmatrix} \begin{pmatrix} E_{i1}(f) \\ E_{i2}(f) \end{pmatrix}. \tag{3.34}$$

Here, z is the coupling length, and k the coupling coefficient that depends on waveguide shape, coupling length and refractive index profile. With i and j denoting input and output, respectively, the power transfer function can be derived as

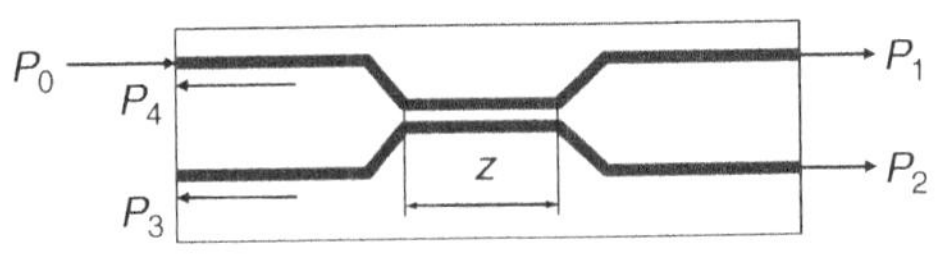

FIGURE 3.66 Fused fiber coupler.

$T_{ij}(f) = |E_{0j}|^2/|E_{ii}|^2$. For a 3 dB coupler (a 1×2 coupler with same power at the output ports), one gets $T_{11}(f) = T_{12}(f) = 0.5$. An important characteristic of the 3 dB coupler is the phase difference of $\pi/2$ between the two output signals. This phase difference is used, for example, in DQPSK delay demodulation [152].

Commercial $1 \times 2/2 \times 2$ couplers are often described by some performance parameters. The coupling ratio CR or splitting ratio is defined as the power from any single output divided by total output power:

$$\mathrm{CR} = 10 \log_{10}\left(\frac{P_2}{P_1 + P_2}\right). \tag{3.35}$$

The excess loss α_E is the logarithmic ratio of total input power and total output power:

$$\alpha_E = 10 \log_{10}\left(\frac{P_0}{P_1 + P_2}\right). \tag{3.36}$$

The insertion loss α_I is given as any single output power divided by total input power:

$$\alpha_I = 10 \log_{10}\frac{P_{1,2}}{P_0}. \tag{3.37}$$

The isolation loss α_{ISO}, which is also known as (near-end) cross talk or directivity, is defined as the input power at one port divided by the power that is reflected to the other input port:

$$\alpha_{ISO} = 10 \log_{10}\left(\frac{P_0}{P_3}\right). \tag{3.38}$$

Finally, the return loss α_R is given by the ratio of input power at one port and power reflected to same port:

$$\alpha_R = 10 \log_{10}\frac{P_0}{P_4}. \tag{3.39}$$

Wavelength-selective directional couplers can be made from fibers with different core diameters and refractive indices. These fibers are matched at only one wavelength and can hence be used as band-pass filters [153]. The principle of wavelength-selective fused directional couplers is shown in Fig. 3.67.

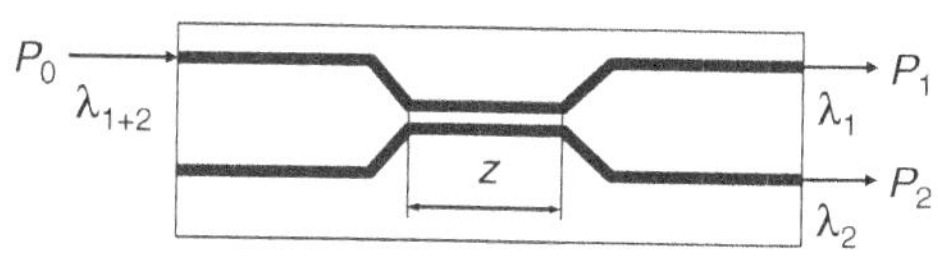

FIGURE 3.67 Wavelength-selective fused fiber coupler.

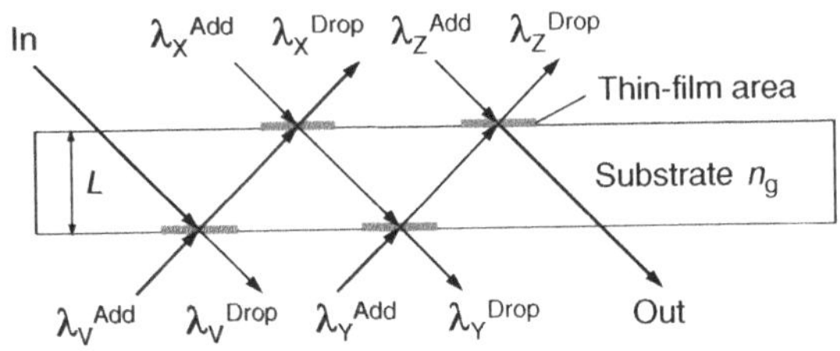

FIGURE 3.68 Thin-film WDM filter.

For the wavelength-selective directional coupler, a different isolation parameter is defined. The isolation loss is now given as the power of the desired wavelength at a given output port divided by the power of other wavelength at the same port:

$$\alpha_{\text{ISO}} = 10 \log_{10}\left(\frac{P_{\lambda 1,\text{Port1}}}{P_{\lambda 2,\text{Port1}}}\right). \tag{3.40}$$

In multiport filters, this is also referred to as adjacent-channel isolation and given as the ratio of the desired wavelength at one port, divided by the power of the two adjacent wavelengths at the same port. The *total isolation* in a multiport device (e.g., an AWG) is then given by the power of the desired wavelength divided by the accumulated power of all other wavelength (and in both polarizations).

3.2.6.2 Thin-Film Filters Optical filters based on dielectric thin films are relevant components for WDM multiplexing, demultiplexing and add–drop. It is one of the oldest technologies used for WDM systems. Thin-film filters (TFF) are based on Fabry–Pérot filters (FPFs), the principle is shown in Fig. 3.68.

A FPF is essentially a resonator. The resonator cavity is bounded on either end by a partial mirror. The structure is also called an Etalon if the mirrors are fixed (otherwise, an interferomcter). Light of frequencies different from the resonant frequency is mostly reflected. Nonresonant light that enters the resonator undergoes destructive interference. Resonant frequencies that try to reflect undergo destructive interference from light already in the cavity going out through one mirror. Therefore, nearly all light at resonant frequency enters the cavity, and exits through the opposite mirror.

The TFF shown in Fig. 3.68 follows the FPF principle. Here, layers of dielectric material are used to form the mirrors, causing multiple reflections inside the substrate (the Etalon). Reflection or transmission occurs at the thin-film dielectric areas, depending on the wavelength. Hence, the add–drop locations depend on wavelength, and add–drop channels can therefore be assigned different (individual) fibers.

The transfer function of an FPF is given by [154]

$$H_{\text{FPF}}(f) = \frac{I_{\text{out}}}{I_{\text{in}}} = \frac{(1-R)^2}{(1-R)^2 + 4R\sin^2(2\pi f n_g L/c)}, \tag{3.41}$$

R is the reflectivity of the mirrors, L is the distance between the mirrors (i.e., the thickness of the Etalon substrate), and n_g is the group index of the material. Further, I_{in} and I_{out} are the input and output intensity, respectively.

Relevant parameters of an FPF (as with any other filter) are the finesse and the FSR. The finesse is a measure for the quality of the filter, analogous to Q in electrical theory. Higher reflectivity of the thin-film mirrors leads to higher finesse.

Resonant wavelengths of an FPF are given by $\lambda = 2Ln/m$, where n is the refractive index of the substrate, and $m = 1, 2, 3, \ldots$ The distance between peaks in the filter response is called free spectral range. The FRS is given by

$$\mathrm{FSR} = \Delta f = \frac{c}{n_g L}. \tag{3.42}$$

The finesse can be expressed as the ratio of FSR and FWHM, $\mathrm{FSR}/\mathrm{FWHM} = \pi\sqrt{R}/(1 - R)$, where FWHM is the full width at half maximum, and R is the reflectance of the mirrors.

3.2.6.3 Arrayed Waveguide Gratings

Arrayed waveguide gratings, sometimes also referred to as optical phased-array (Phasor), phased-array waveguide grating (PAWG), or WGR are relevant DWDM multiplexers/demultiplexers [155,156]. This is mainly due to the fact that AWGs can de-/multiplex a high number of WDM channels with relatively low loss.

Figure 3.69 shows a schematic representation of an N:M AWG. The device consists of two slab waveguide star couplers (or free propagation regions, FPR) that are connected by a waveguide array with equal length difference ΔL between adjacent array waveguides.

Light coming from an input waveguide is diffracted and coupled into the arrayed waveguides by the first FPR. The optical path-length difference ΔL between adjacent array waveguides equals an integer (m) multiple of the central wavelength λ_0 of the demultiplexer. As a consequence, the field distribution at the input aperture will be reproduced at the output aperture at the second FPR. Therefore, at the center wavelength the light focuses in the center of the image plane, provided that the input waveguide is centered in the input plane. If the input wavelength deviates from

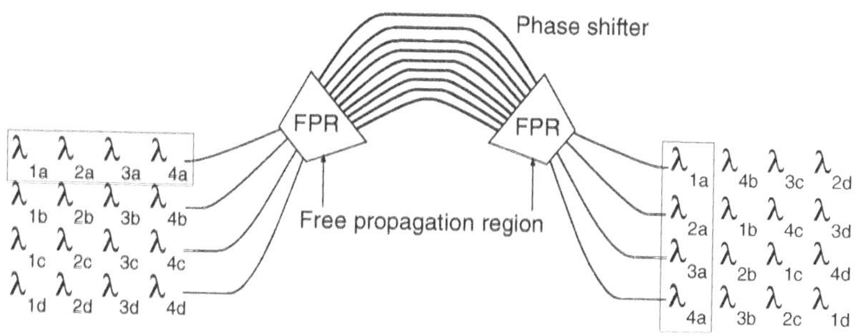

FIGURE 3.69 N:M AWG as column–row converter.

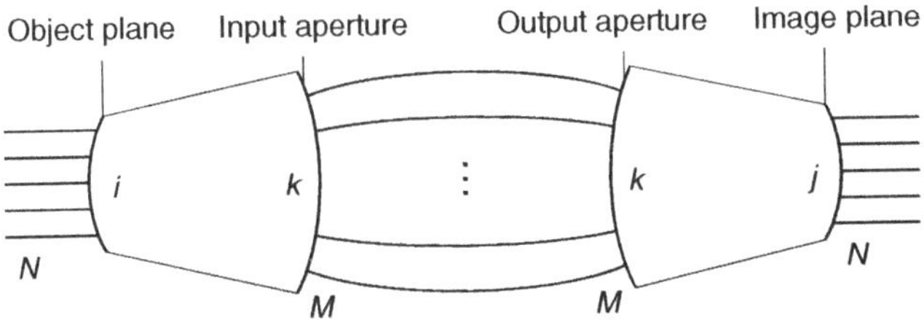

FIGURE 3.70 Numbering of input/output fibers and arrayed waveguides of an AWG.

the central wavelength, phase changes occur in the array branches. Due to the constant path-length difference between adjacent waveguides, this phase change increases linearly from inner to outer array waveguides, which causes the wavefront to be tilted at the output aperture. Consequently, the focal point in the image plane in the second FPR is shifted away from the center. Therefore, spatial separation of different wavelengths can be provided by respective positioning of the output waveguides.

The relative phases for input i, output j, and array waveguide k (with $k = 1, 2, \ldots, m$) according to Fig. 3.70 is given by [157]

$$\phi_{ijk} = \frac{2\pi}{\lambda}\left(n_1 d_{ik}^{\text{in}} + n_2 k\,\Delta L + n_1 d_{kj}^{\text{out}}\right). \tag{3.43}$$

Here, $d_{ik}^{\text{in}} = d_i^{\text{in}} + k\delta_i^{\text{in}}$ and $d_{kj}^{\text{out}} = d_j^{\text{out}} + k\delta_j^{\text{out}}$ are the differences in distance between input i and array waveguide k and array waveguide k and output j, respectively. Then, the relative phase can be expressed by

$$\begin{aligned}\phi_{ijk} &= \frac{2\pi n_1}{\lambda}\left(d_i^{\text{in}} + d_j^{\text{out}}\right) + \frac{2\pi k}{\lambda}\left(n_1\delta_i^{\text{in}} + n_2 k\,\Delta L + n_1\delta_j^{\text{out}}\right). \\ &= \phi + \Delta\phi\end{aligned} \tag{3.44}$$

Obviously, important design parameters of an AWG chip are the optical distances and the refractive indices of the waveguides.

With $n_1\delta_i^{\text{in}} + n_2 k\,\Delta L + n_1\delta_j^{\text{out}} = p\lambda_j$ and integer p, the phase difference $\Delta\varphi$ can also be written as $\Delta\phi = 2\pi kp$. λ_j will add in phase at output j, and hence be present at output j. According to $n_1\delta_i^{\text{in}} + n_2 k\,\Delta L + n_1\delta_j^{\text{out}} = (p+1)\lambda_j'$, λ_j' will also be present at output j. The difference in frequency between two wavelengths that can be routed to the same output port (for $\Delta p = 1$) is referred to as the FSR of the filter. With $p\lambda = pc/f = (p+1)c/f'$, the FSR is given by

$$\text{FSR} = f' - f = \frac{c}{n_2\,\Delta L}. \tag{3.45}$$

When properly designed, AWGs can route one wavelength in the C-band and one wavelength in the L-band to each of the output ports. In addition, further wavelengths in the S-, U-, E- and O-band can also be routed to the respective ports. The respective devices are referred to as cyclic AWGs. The cyclic characteristic can be used, for example, for wavelength-band enhancements (i.e., capacity upgrades) of WDM systems, or in systems that make use of single fiber working (e.g., WDM-PONs, refer to Section 6.1.4). In ITU Recommendation G.698.3, cyclic AWGs with up to 48 ports (100 GHz in the C-band) have been standardized.

The transfer function for an input waveguide near the optical axis can be approximated as

$$T_x = \sum_{k=1}^{M} A_k \exp\left(j\phi_k + j\frac{2\pi}{\lambda_x} k n_2 \,\Delta L\right), \tag{3.46}$$

M is the number of arrayed waveguides, and φ_k and A_k are the phase and the attenuation of the kth arrayed waveguide, respectively. The absolute square yields the transmission spectrum, and the argument describes delay and chromatic dispersion of the device.

Mostly depending on the shape of the slab FPR, AWGs can be designed with different per-channel transmission spectrum characteristics. Two specific characteristics are commonly produced, AWGs with Gaussian or FT (flat-top) transmission. FT AWGs result from Gaussian-shaped AWGs by truncating the transmission peak. When using AWGs for WDM systems, this characteristic has to be considered. Generally, Gaussian-shaped AWGs have narrower passband, lower insertion loss, and less phase ripple. For many applications, this makes Gaussian AWGs the better choice for demultiplexers, whereas FT AWGs are the preferred choice for multiplexers (or combiners).

As integrated waveguide devices, AWGs show strong temperature dependence. The temperature-induced wavelength shift can be expressed as [158,159]

$$\frac{\partial\lambda}{\partial T} = \frac{\lambda}{nL}\left(\frac{\partial(nL)}{\partial T}\right) = \frac{\lambda}{nL}\left(n\frac{\partial L}{\partial T} + L\frac{\partial n}{\partial T}\right) = \lambda\left(\frac{1}{L}\frac{\partial L}{\partial T} + \frac{1}{n}\frac{\partial n}{\partial T}\right) = \lambda\left(\alpha + \frac{1}{n}\frac{\partial n}{\partial T}\right). \tag{3.47}$$

For silica, $\partial n/\partial T$ is in the range of $7.5 \times 10^{-6}/°$, and for silicon, $\alpha = 2.63\,\text{ppm}/°$, respectively. A silica-on-silicon device hence has a wavelength drift (redshift) of $d\lambda/dT = 12\ \ 12\,\text{pm}/°$. A 50 GHz DWDM device ($\sim$400 pm channel spacing) is completely transparent every 400 pm, and opaque in between. The device becomes a wavelength stop if the temperature changes by $\sim$17°. Hence, AWGs are often temperature stabilized.

Temperature stabilization, or athermalization, is usually done via mechanical or material compensation. In mechanical compensation, the point at which the signal (input fiber) enters the device is mechanically adjusted via a fiber-coupling element whose thermal expansion is designed such as to compensate the temperature drift of the AWG chip. In material compensation, the waveguides are designed to be inherently athermal.

TABLE 3.10 Experimental Parameters of AWGs

Channel spacing	100 GHz	50 GHz	25 GHz	25 GHz
No. of channels	32	64	128	256
On-chip loss	2.1 dB	2.8 dB	3.5 dB	2.7 dB
3 dB bandwidth	49 GHz	19 GHz	11 GHz	14 GHz
Total isolation	28 dB	27 dB	16 dB	33 dB

Commercial AWGs are available for 32 and 40 channels (100 GHz), and 64, 80, and 96 channels (50 GHz). Experimental devices with very high port count and WDM grid down to 12.5 GHz have been produced. Table 3.10 summarizes some experimental results.

For total insertion loss, the fiber coupling loss has to be considered. In addition, a loss penalty for temperature drift must be taken into account, even if the device is stabilized or compensated. Typical total insertion loss of 100 and 50 GHz AWGs is in the range of 5 and 6 dB, respectively.

3.2.6.4 Fiber Bragg Gratings Fiber Bragg gratings are a third class of filters that are relevant for WDM systems. Generally, a Bragg grating is a periodic perturbation in the propagating medium (periodic variation of the refractive index). In FBGs, this perturbation is written into the fiber by means of a UV laser [160,161].

An FBG is shown in Fig. 3.71. Λ is the period of the grating (or perturbation). The Bragg phase-matching condition is satisfied for two waves with β_0 and β_1 if $|\beta_0 - \beta_1| = 2\pi/\Lambda$. If a wave with β_0 propagates through the grating, its energy is coupled onto a scattered wave traveling in the counterdirection at the same wavelength if $|\beta_0 - (-\beta_0)| = 2\beta_0 = 2\pi/\Lambda$.

With $\beta_0 = 2\pi n_{\text{eff}}/\lambda_0$, the Bragg condition results

$$n_{\text{eff}}\Lambda = \lambda_0/2, \tag{3.48}$$

λ_0 is referred to as Bragg wavelength, and n_{eff} is the average refractive index of the material (in an FBG, of the core region with the imprinted grating). Other wavelengths than λ_0 can pass the grating.

Fiber Bragg gratings can be produced from photosensitive (Ge-doped) single-mode fibers. The grating is written into the fiber with UV lasers, where regions with higher UV intensity impact produce higher refractive index. The required change of the refractive index is in the range of $\Delta n \approx 10^{-4}$.

Advantages of FBGs include low loss (down to 0.1 dB), ease of coupling to transmission fibers, polarization insensitivity, and high cross talk suppression of

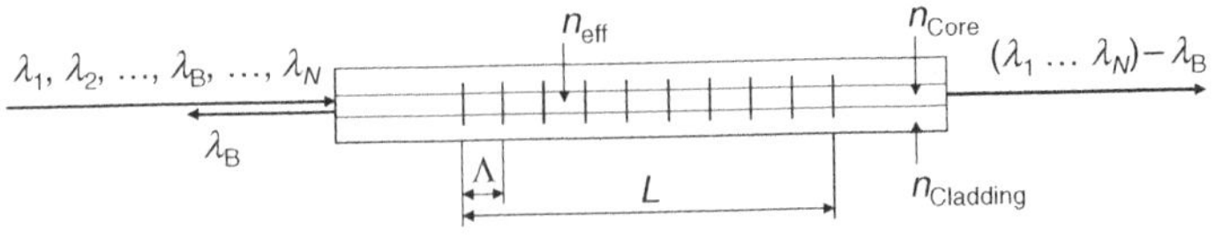

FIGURE 3.71 Fiber Bragg grating.

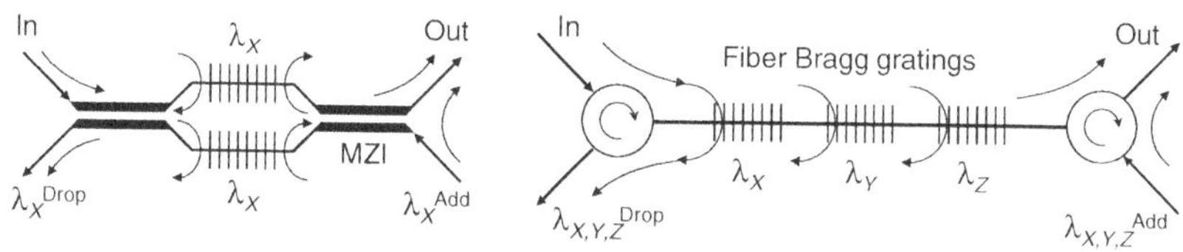

FIGURE 3.72 Fiber Bragg gratings used for add–drop multiplexers.

~40 dB. The typical temperature coefficient is in the range of $\sim 1.2 \times 10^{-2}$ nm/°C. Therefore, FBGs must be operated in temperature-controlled environment, or be temperature compensated. An additional advantage of FBGs is their low cost.

Two kinds of FBGs are produced, with short period (Bragg grating $\Lambda \approx 0.5\ \mu$m) or with long period ($\Lambda \approx 100$–$1000+\ \mu$m) [162,163]. The advantages, specially low loss, apply to short-period FBGs. Due to their characteristics, these devices are well suited as filters or CD compensation devices (refer to Section 3.2.5.2) in WDM systems. Since the Bragg (add–drop) wavelengths are reflected into the same fiber, FBGs must be combined with suitable means for separation of the Bragg wavelengths and the remaining WDM spectrum. Circulators or interferometers can be used, two different add–drop constellations are shown in Fig. 3.72.

The configuration on the right-hand side of Fig. 3.72 shows cascaded FBGs. In this case, all drop wavelengths are guided into a common fiber, and all added wavelengths have to be added via another common fiber.

Long-period FBGs can be used, for example, for EDFA gain equalization. Modes propagating in the core couple to lossy cladding modes, effectively producing a wavelength-selective loss element. Several FGBs or chirped FBGs can be used to obtain the desired loss profile.

3.2.6.5 Interleavers (Mach–Zehnder Interferometer) Interleavers are periodic filters used for combining or separating (deinterleaving) two WDM signals (multiplex sections) with the same channel spacing and a grid offset of half the spacing. They can be used, for example, to build a 50 GHz WDM multiplex section out of two 100 GHz multiplex sections. The principle function is shown in Fig. 3.73.

Due to the required periodicity, MZIs are suitable devices for interleavers. Similar to MZIs used as modulators (refer to Fig. 3.24), MZIs used as interleavers introduce a delay ΔL to one of the branches in the interferometric section of the device. Also, similar to the MZI application as delay demodulators for DQPSK receivers, the 2×2 input and output couplers have to be considered due to $\pi/2$ shift that is introduced between the output ports (refer to Section 3.2.6.1). The complete structure of an MZI interleaver is shown in Fig. 3.74. The signal coming from Input 1 and going to Output 1

FIGURE 3.73 Interleaver functionality.

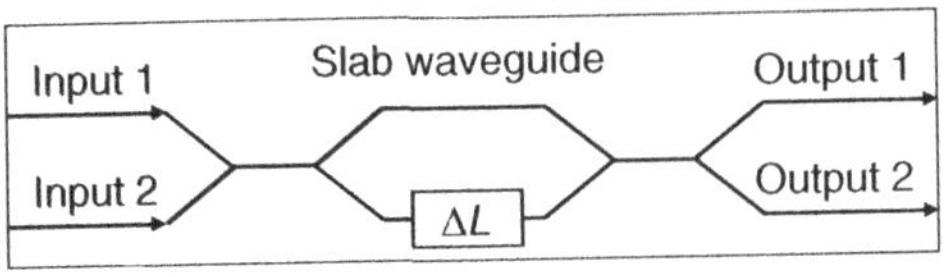

FIGURE 3.74 MZI as (de)interleaver.

through the upper arm acts as reference. Then, the signal coming from Input 1 and going through the lower arm to the upper output experiences a phase lag of

$$\frac{\pi}{2} + \beta\,\Delta L + \frac{\pi}{2} = \pi + \beta\,\Delta L. \tag{3.49}$$

At the lower output (Input 1, going through the lower arm), the phase difference is

$$\frac{\pi}{2} + \beta\,\Delta L - \frac{\pi}{2} = \beta\,\Delta L. \tag{3.50}$$

For $\beta\,\Delta L = k\pi, k = (2n + 1)$, the signals in the upper output add in phase. At Output 2, the phase difference is $(2n + 1)\pi$ that is out of phase, hence there is no signal at the output. Opposite behavior is achieved for $\beta\,\Delta L = 2n\pi$.

Taking the 2×2 input and output couplers into account, the transfer function of the MZI can be derived:

$$\begin{pmatrix} T_{11}(f) \\ T_{12}(f) \end{pmatrix} = \begin{pmatrix} \sin^2(\beta\,\Delta L/2) \\ \cos^2(\beta\,\Delta L/2) \end{pmatrix}. \tag{3.51}$$

For use as an interleaver, λ_1 and λ_2 (according to Fig. 3.73) are chosen to coincide with the minima and maxima of the transfer function. With $\beta = 2\pi n_{\text{eff}}/\lambda$, the delay ΔL and the corresponding phase lag $\beta\,\Delta L$ can be expressed:

$$\Delta L = \frac{m_i \lambda_i}{2n_{\text{eff}}}, \quad \beta\,\Delta L = \frac{2\pi n_{\text{eff}}}{\lambda_i}\frac{m_i \lambda_i}{2n_{\text{eff}}} = m_i \pi. \tag{3.52}$$

For $\lambda_1 = 2\,\Delta L n_{\text{eff}}/m_i, m_i$ odd, Output 1 has a signal whereas Output 2 has no signal. For $\lambda_2 = 2\,\Delta L n_{\text{eff}}/m_i, m_i$ even, the output assignment is vice versa. Hence, the device acts as a deinterleaver. Operation as interleaver is similar.

3.2.6.6 Tunable Filters Tunable filters have been investigated since the early WDM days [164]. Advantages of tunable filters include remote reconfigurability (possibly controlled by a control plane), and reduction of system items and spare parts. Tunability of WDM filters can be based on one out of the following mechanisms (where WDM in most cases refers to DWDM):

- Fabry–Pérot interferometer
- Liquid-crystal (FP) filter

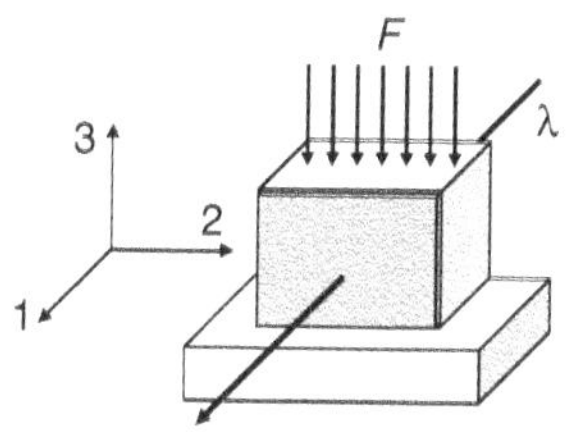

FIGURE 3.75 Schematic of tunable TFF with active substrate.

- MEM-tunable devices
- FBG, temperature- or strain-tuned
- AOTF, acousto-optic tunable filter
- EOTF, electro-optical tunable filter
- AWG (thermally tuned)
- Mach–Zehnder interferometer (cascaded, with heaters)
- Tunable ring-resonator filters

So far, only few of these technologies have been commercialized, and only few have the potential for low cost and small form factor, for example, to support future low-cost WDM-PONs. The majority of the literature [165–176] indicates that relevant contenders for low-cost tunable filters come out of the first four technologies listed above.

FP interferometers are the basis of most tunable filters. They are based on TFF or liquid-crystal etalons similar to the ones used for many fixed DWDM filters (see Section 3.2.6.2). An overview of tunable TFF is given in Ref. [170]. The etalons are either angle-tuned, linearly tuned (i.e., a variable TFF addressed by translation), or tuned with active substrates or active layers. A simplified schematic diagram of an active-substrate TFF-FP interferometer is shown in Fig. 3.75.

The basic idea of the tuning scheme shown in Fig. 3.75 is to transfer a mechanical strain of the substrate to the spacer thickness of a DWDM band-pass filter deposited at its surface, and to achieve a change in the resonant wavelength of the all-dielectric FP structure.

The mechanical strain can be induced either thermally, or via the piezoelectric effect. For thermal tuning, two effects are relevant: the increase of the refractive index created by rising temperature, and the differential thermal expansion between the layers and the substrate. These effects can be used to partly compensate temperature sensitivity or, at the opposite, to maximize thermal sensitivity for thermal tuning. If temperature is kept constant, a mechanical deformation of the substrate, that is, a strain, can also be used to induce a shift in the wavelength: this strain can be produced for instance by a compressive force applied on a standard substrate, or by an electric field applied on a piezoelectric substrate.

Thermally tuned TFF are commercially available, for example, for WDM overlay in G-PON systems [174,175]. They were designed for stacked (X)G-PONs where

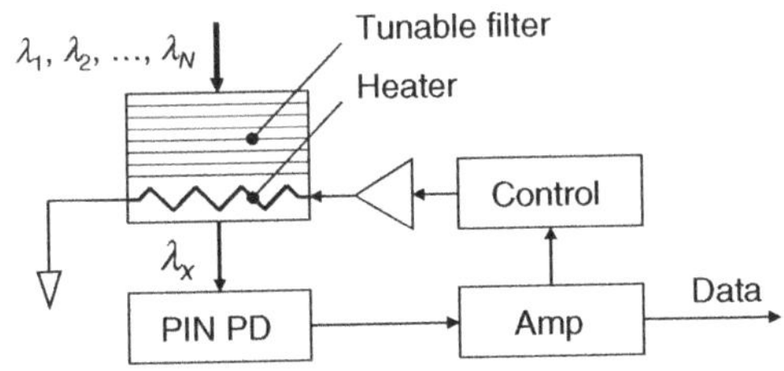

FIGURE 3.76 Schematic of tunable TFF for G-PON WDM overlay.

only few wavelengths can be added. The filters have limited tuning range (~5 nm, or 600 GHz). Tuning is performed with a heater, tuning power is <200 mW. A schematic diagram of this filter is shown in Fig. 3.76.

An advantage of the WDM overlay filter is the compactness that allows integration with the photodiode subassembly. According to the manufacturer, broader tuning ranges contradict the small size and power consumption.

FP etalons based on liquid crystals (LCFP) are an alternative to TFF FP [166]. As compared to other FP filters, main advantages of LCFP filters include low driving voltage (i.e., low power consumption), broad tuning range, high resolution, and also low insertion loss. LCFP filters can also be used in spectroscopy and laser radar. So far, LC filters have been used in certain ROADMs.

Nematic liquid crystals have optical anisotropy, or birefringence. They exhibit double refraction: light polarized parallel to the director sees a different refractive index than light polarized perpendicular to the director. Nematic liquid crystals also show dielectric anisotropy, the dielectric constants parallel and perpendicular to the nematic director are not the same. The dielectric anisotropy introduces body torque on the molecules in the presence of an external field, which in turn gives rise to the director reorientation. Under the external field, the director of the liquid crystal with a positive dielectric anisotropy tends to be aligned parallel to the external field, while the director of the liquid crystal with a negative dielectric anisotropy tends to be aligned perpendicular to the external field.

By filling an etalon cavity with nematic liquid crystals, the refractive index will change with applied voltage. The wavelength of the transmission peak is tunable with different voltages applied. The structure of an LCFP etalon is shown in Fig. 3.77.

A polarizer needs to be placed parallel to the alignment direction of the liquid crystal. For WDM transport systems, polarization dependence generally is

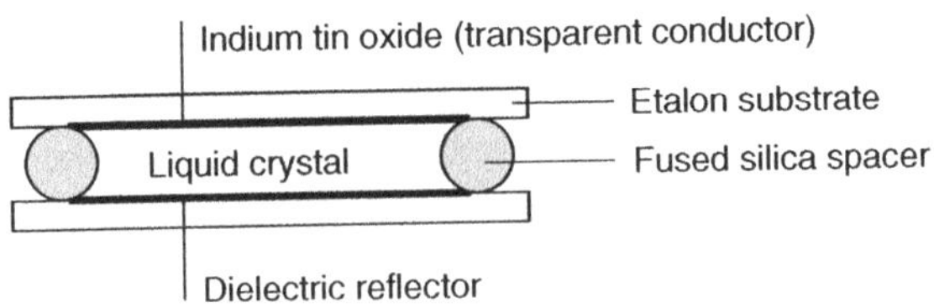

FIGURE 3.77 Schematic of liquid-crystal filter.

considered critical. Several methods have been discussed to make polarization-insensitive LCFP devices. One is to split the input light (with unknown polarization) into two components with orthogonal polarization. Each component is passing through a tunable LCFP filter with a direction of the liquid crystal aligned to the polarization direction. After passing through the LCFP filters, the two components are recombined through another calcite crystal. An alternative is to use two neighboring LC pixels to match the two polarized components of one channel. The two pixels need to have alignment layers perpendicular to each other, and LCD production technology is a cost-effective solution to achieve this.

Fiber Bragg gratings are an alternative to FP etalons when it comes to tunable filters. They can be designed to cover WDM grids from 200 GHz down to less than 25 GHz [169]. They have low-cost potential, but lack integration potential. FBGs consist of periodic refractive-index variations along a piece of SMF (refer to Section 3.2.6.4), and they can be thermally or strain-tuned. Heating can be achieved through a metal layer that covers the fiber and which is traversed by an electric current. The reflected central wavelength then varies linearly with temperature, and 3 dB bandwidth and central peak transmission do not change significantly. For DWDM with grids of 1.6–0.2 nm, the temperatures ranges to be covered can easily be achieved. If strain is uniform along the FBG, it can also be used for tuning. Then, strain does not affect BW and peak transmission of the reflected wavelengths.

Volume holographic gratings (VHG) are similar to FBGs except that the recording medium for the grating is not a (single-mode) fiber but a volume medium. Consequently, the incident and diffracted light are not confined to the modes of the fiber, but can be assigned to any mode that can propagate in the bulk material. This opens new possibilities for device design and for example allows removal of the circulator that is essential for FBGs. VHGs as means for tunable filters have been described in Ref. [171]. Wide tuning range of 1510–1590 nm and low insertion loss of ~1 dB can be achieved.

Figure 3.78 shows a schematic representation of the angle-tunable volume holographic filter. The Bragg wavelength of a holographic grating is determined by $\lambda_B = 2n_{eff}\Lambda\cos\theta$, where θ is the angle of incidence inside the holographic material. By changing θ, the Bragg wavelength can be tuned continuously. The tuning range can cover the entire C-band and is limited mainly by the higher insertion loss at larger angle θ. A critical task is to collect the drop signal into a fiber during angle tuning without an expensive tracking mechanism or feedback control system.

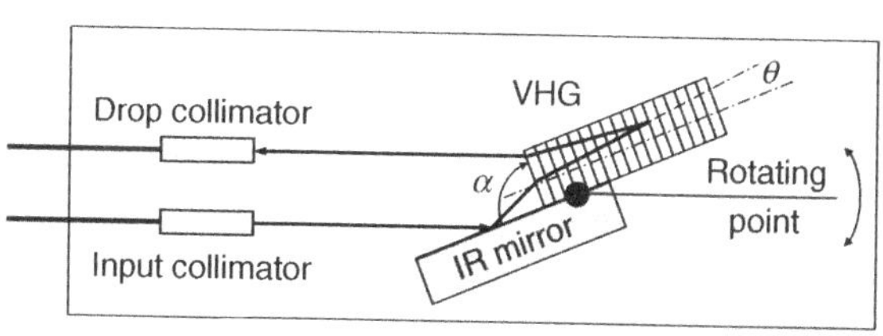

FIGURE 3.78 Schematic of tunable volume holographic grating.

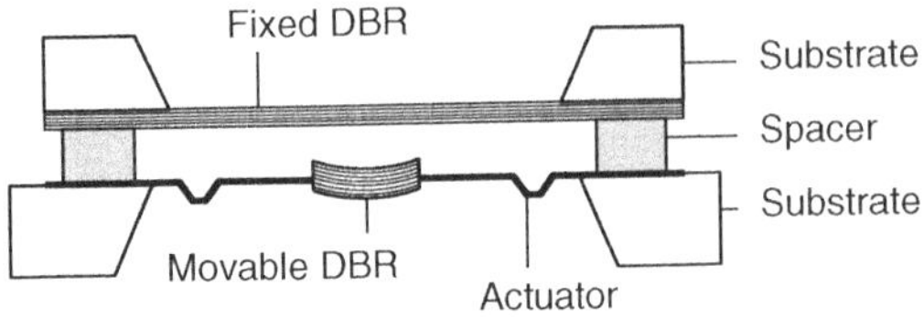

FIGURE 3.79 Schematic of MEM-tunable filter.

This can be achieved with a self-reflector architecture that combines the reflective holographic grating with a wideband IR mirror, as shown in Fig. 3.78.

A retroreflector consists of two mirrors with a fixed angle between the mirror surfaces. This also applies to the structure shown in Fig. 3.78, where one mirror is replaced by the VHG. At the Bragg wavelength, this VHG works as a reflective mirror with an effective depth into the material. When rotating the mirror/grating structure around the crossing point of the mirror and grating surfaces at the effective reflection depth, the drop signal beam is fixed spatially while the wavelength is tuned by the angle θ. With the addition of a temperature sensor and suitable control system, the device can be made athermal by compensating θ to offset temperature-induced variations of the grating period.

MEM-tunable devices are promising for low-cost tunable filters. The design principle is similar to the one described for tunable MEM-VCSELs (see Section 3.1.1.5). MEM-tunable filters are described, for example, in Refs [165,176]. Several MEM-tunable optical filters based on electrostatic actuation or thermal tuning have been reported. Electrostatic actuation is adequate for low power operation, but its tuning range is limited and its actuation voltage is relatively high. Also, electrostatic actuation intrinsically has a nonlinear behavior. Thermal actuation can achieve a wide tuning range but it consumes more power and its response time is slow. A more recent alternative is magnetic actuation [165]. A schematic diagram of one such filter is shown in Fig. 3.79.

The filter consists of a FP cavity with two dielectric DBRs, similar to other cavity-type MEM-tunable devices. One DBR is formed on the fixed part of the filter, the other is located on the actuator. The two DBRs are separated by a spacer. Since the actuator consists of suspended conductor bridges, it receives Lorentz force by electrical current flowing through it in a magnetic field. The magnetic field is applied in plane with the actuator. Displacement of the actuated DBR changes the cavity length and hence allows tuning. This design allows very wide tuning range in excess of 200 nm and also very low power consumption in the range of 25 μW.

Various derivatives of MEM-tunable filters have been described. One of these approaches is based on tunable microcavities [176]. It is applicable to both VCSELs and tunable filters and has low-cost potential. The basic idea is shown in Fig. 3.80.

The device shown in Fig. 3.80 is based on a micromachined two-chip concept with a movable GaAs DBR mirror membrane on one chip and a fixed DBR mirror on a second chip, which together form the microcavity. The mirror membrane is movable due to four flexible suspension beams. Deflection of the membrane as a

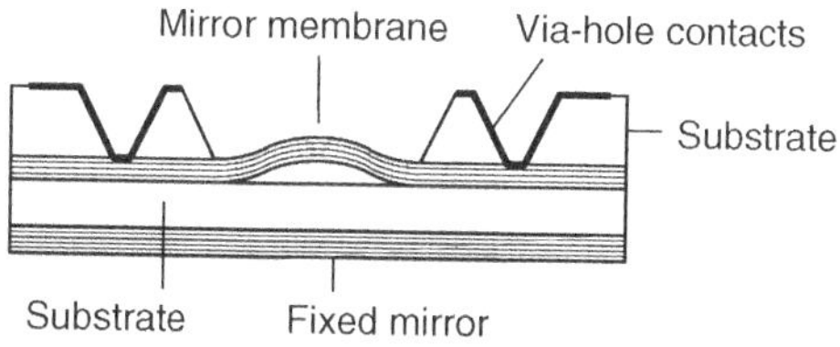

FIGURE 3.80 Schematic of tunable microcavities.

consequence of electrothermal heating is achieved by injecting a small electrical current (<4 mA, equaling 10 mW of electrical power) through the suspension beams. The second chip can be a simple dielectric DBR on a plane glass substrate for filter applications (in case of a tunable VCSEL, it is one-half of the VCSEL consisting of DBR plus active region).

Monolithic approaches are preferable in terms of fabrication costs. However, the two-chip concept allows implementing long cavities. For filter applications, a longer cavity automatically reduces the bandwidth, which is important for DWDM with small channel spacing. Usually, a two-chip concept has the disadvantage that assembly costs are high. Assembly can be done by placing the membrane chip directly on top of the second chip. A permanent connection between both chips can then be obtained by bonding, gluing, or using flip-chip technique.

A *comparison* on the most relevant characteristics of the tunable-filter technologies described herein is provided in Table 3.11.

In general, low-cost, low energy consumption, and channel spacing and bandwidth suitable for full-band DWDM tuning (C- or L-band) must be achieved. From this, TFF-FP, LCFP, MEM-FP, and VHG filters can be identified as suitable technologies.

3.2.6.7 Wavelength Switching Devices Optical switching matrices for WDM networking need to fulfill certain requirements. They must have suitable optical characteristics such as low insertion loss, high isolation, and low back reflection. In addition, they need to be highly reliable, fast (at least, for certain application), compact, low power consuming, and scalable. Techniques that are generally suitable for optical switching matrices include but are not restricted to MEMS, some MEMS-like technologies (thermocapillary, piezoelectric), liquid-crystal technology (LCT), bubbles, thermo- and acousto-optics, and holograms [177–183].

MEMS refer to arrays of tiny tilting mirrors sculpted from semiconductor materials such as silicon. In *2D-MEMS* arrays, the mirrors can only flap up and down incorporating a single axis [184]. 2D arrays were used in first-generation all-optical switches because of maturity. A 2D-MEMS array is schematically shown in Fig. 3.81.

In *3D-MEMS* arrays, the mirrors can be tilted in any direction. 3D-MEMS arrays can be used for large-scale optical cross-connects (OXCs), particularly if groups of wavelengths (wavelength bands) are to be switched from one fiber to another. A 3D-MEMS assembly for fiber (or wavelength band) switching is shown in Fig. 3.82.

TABLE 3.11 Comparison of Relevant Parameters of Tunable Filters

	Insertion Loss	Channel Spacing/ Isolation	Bandwidth (3 dB)	Tuning Range	Tuning Speed	Power Consumption
TFF-FP	3 dB	50 GHz/30 dB	<0.5 nm	~40 nm	ms	Medium
LCFP	3 dB	50 GHz/30 dB	<0.5 nm	~50 nm	μs	Low
MEM-FP	1 dB	50 GHz/30 dB	<0.5 nm	~60 nm	100 μs	Low
FBG	0.1 dB (w/o circulator)	25 GHz/22 dB	<0.2 nm	<10 nm	ms	Medium
VHG	3 dB	100 GHz/25 dB	<0.7 nm	~40 nm	ms?	Low
AOTF	4 dB	200 GHz/30 dB	<1.5 nm	~60 nm	μs	Medium
EOTF	4 dB	200 GHz/25 dB	<1.5 nm	~50 nm	ms	Medium
Cascaded MZI	19 dB (LiNbO3)	25 GHz/22 dB	<0.2 nm	<10 nm	50 ns	Medium
AWG (tunable)	5 dB	50 GHz/22 dB	<0.2 nm	~40 nm	10 ms	Medium
Ring resonator	3 dB	50 GHz/30 dB	~0.2 nm	~25 nm	ms	Medium

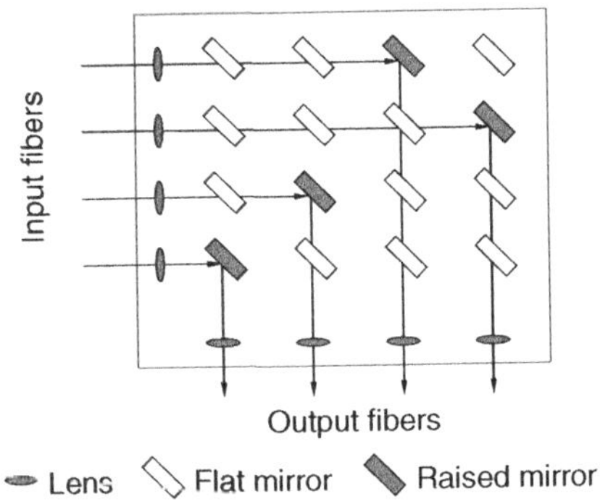

FIGURE 3.81 2D-MEMS array used for fiber cross-connect.

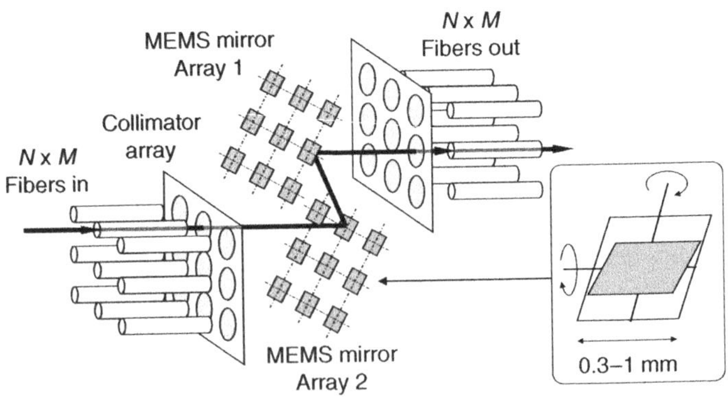

FIGURE 3.82 3D-MEMS used for multidimensional fiber cross-connect.

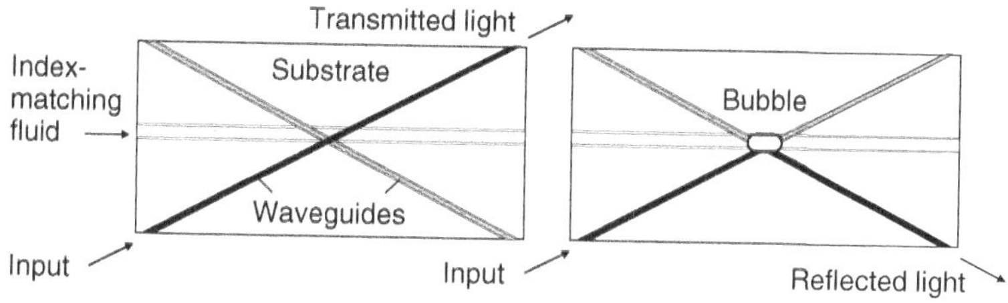

FIGURE 3.83 Principle of a bubbles switch.

As MEMS is based on mechanical movement, it offers the best isolation performance available. For the same reason, MEMS cannot be used for drop-and-continue applications. It was therefore not used in early ROADMs that used the broadcast-and-select architecture. In newer ROADMs, MEMS can be used in the add–drop sections.

Bubbles are formed by printer ink pens and have surfaces that act like mirrors. They have been used for optical switches produced by Agilent, reusing parts of their Desk Jet printer technology [179]. The technology is suitable for ROADMs and medium-sized switches. It lacks 3D capability. The principle of a (2D) bubbles switch is shown in Fig. 3.83.

In an *LCT* switch, light is passed through liquid crystals that alter their transmission characteristics when an electric current is applied [178]. Most LCT devices depend on the polarization properties of the light for switching. LCT allows low power consumption. An LCT switch with PBS and polarization beam combiners (PBC) and two LC cells for polarization-independent switching is shown in Fig. 3.84.

LCT devices can throttle the amount of light that passes through them. They can hence be used as variable attenuators or power splitters with configurable splitting ratio. This enables drop-and-continue application. For these reasons, LCT is used in Degree-2 ROADMs that are based in integrated planar lightwave circuits (iPLC) (see Fig. 3.85) [185].

LCT can be combined with electroholographic technology for increasing switching speeds. With holograms, electrically energized Bragg gratings are created inside

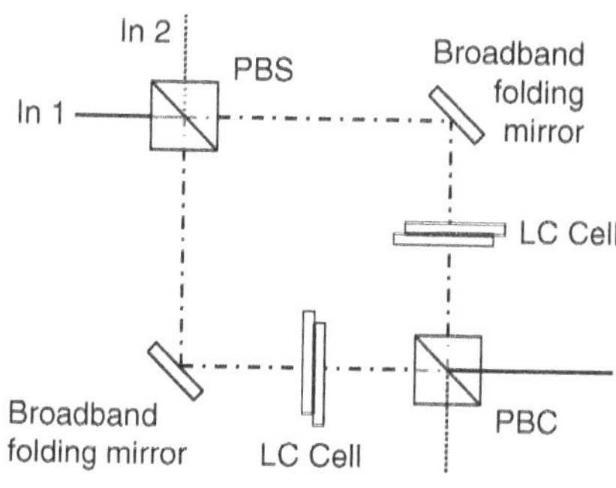

FIGURE 3.84 Liquid-crystal switch.

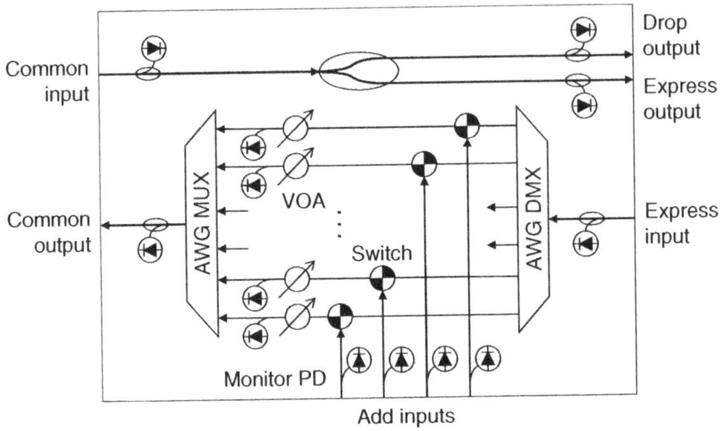

FIGURE 3.85 Block diagram of an iPLC ROADM module.

a crystal. When voltage is applied, the grating deflects the light to an output port. With no voltage, the light passes straight through. Holograms compete with 3D-MEMS on scalability but are better suited to switching individual wavelengths rather than groups of wavelengths. They achieve nanosecond switching speed. Like LCT, holograms support drop-and-continue applications.

Liquid crystal on silicon (LCoS) is the latest addition to LC switching technologies [186,187]. It allows matrix switches, where one switching element for one WDM channel comprises of many matrix subelements. Variable numbers of these subelements can be combined to achieve varying channel bandwidth without attenuation dips between the subchannel elements. This principle is shown in Fig. 3.86. It can be used for so-called gridless ROADMs (i.e., ROADMs that are not bound to a specific ITU WDM grid, for example, 50 GHz).

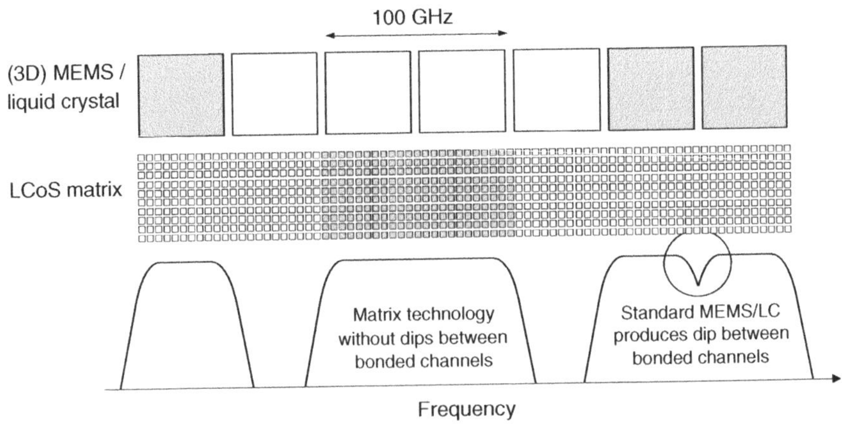

FIGURE 3.86 LCoS gridless matrix WSS.

TABLE 3.12 Comparison of Relevant Parameters of Switching Technologies

	Bubbles	LCT	(3D) MEMS	LCoS
Scalability	Medium	Medium	Very high	High
Insertion loss	~5 dB	3–6 dB	3–7 dB	5–10 dB
PMD/PDL		Low	Low	Low
Power consumption	Medium	Low	Medium	Medium
Reliability	Good	Good	Moving parts	Good

Further optical switch technologies include thermo-optic and acousto-optic switches. Thermo-optic switches consist of passive splitters. When these are heated, their refractive index is changed to alter the way in which they divide wavelengths between either of two output ports. They are hence wavelength sensitive and need temperature control. Thermo-optics are well suited to integration. Polymers are a suitable material because they keep the heat localized at the switch. As a waveguide approach, thermo-optics eliminate beam alignment issues created by free-space approaches but are susceptible to cross talk, polarization dependence, and loss issues [180].

Acousto-optic switches use sound waves to deflect light that passes through crystals or fused couplers. Like thermo-optic switches, this is a waveguide approach that can be used for wavelength-selective switches (WSS) [182].

A *comparison* of the most relevant switching technologies is given in Table 3.12.

Today, most WDM switching devices use derivatives of LCT or MEMS technology.

3.2.7 ROADMs and OXCs

3.2.7.1 ROADM Structures Reconfigurable optical add–drop multiplexers are a key component for flexible WDM transport networks. They are also the main reason why large-matrix OXCs have never been installed in large quantities though being available. Generally, ROADMs can be based on the filter-switch-fabric-filter type shown in Fig. 3.87a [188,189]. This is also the basic structure of larger OXCs. First commercial ROADMs, however, were based on reconfigurable wavelength blockers (WBs), together with power splitters for the add and drop paths. This scheme is shown in Fig. 3.87b. Today, it has no relevance anymore. The next ROADM technologies were WSS based on MEMS or iPLC. These are shown in different configurations in Fig. 3.87c and d, respectively.

ROADMs based on iPLC technology typically have a degree of 2, that is, they can mainly be used in WDM rings or linear add–drop configurations. An iPLC ROADM module consists of a pair of AWGs, per-wavelength switches, monitoring diodes, and VOAs. The AWGs are required to separate the WDM wavelengths and to allow their individual switching and power-level adaptation by means of the VOA array. The switch array performs the reconfigurable add functionality. The drop path is

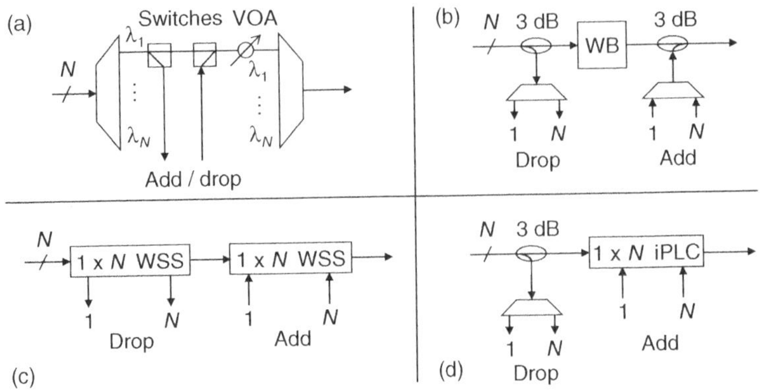

FIGURE 3.87 Basic ROADM implementations. (a) Discrete mux-switch-demux. (b) Wavelength blocker. (c) Wavelength-selective switch. (d) Integrated planar lightwave circuit.

implemented via static passive power split. The block diagram of an iPLC ring ROADM with East + West termination is shown in Fig. 3.88 [185].

Advantages of iPLC-based ROADMs include the integrated waveguide technology that avoids free-space optics with the related problems regarding cost, light coupling, and so on. A disadvantage is the potential polarization dependence of planar waveguide circuits. From the block diagram in Fig. 3.88, an insertion loss of one ROADM module of at least 10 dB (driven by the AWGs) can be derived. An East-plus-West ring ROADM consists of two modules plus two dedicated AWGs for the drop paths.

*3.2.7.2 **Multidegree WSS*** Early ROADM implementations, for example, based on iPLC, led to Degree-2 devices suitable for applications in rings or linear add–drop

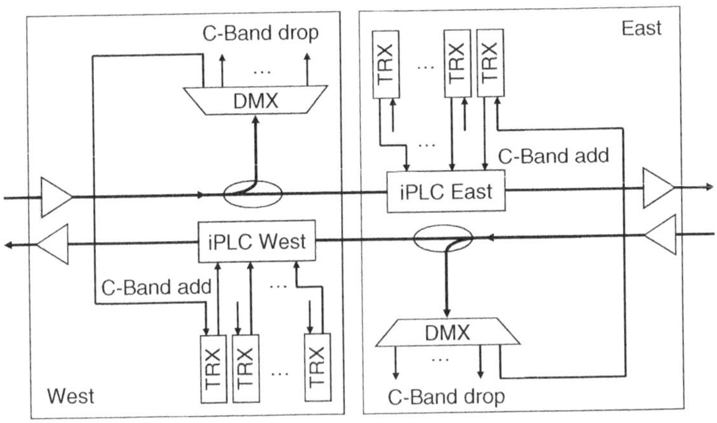

FIGURE 3.88 Block diagram of an iPLC-based Degree-2 ring ROADM. DMX: demultiplexer.

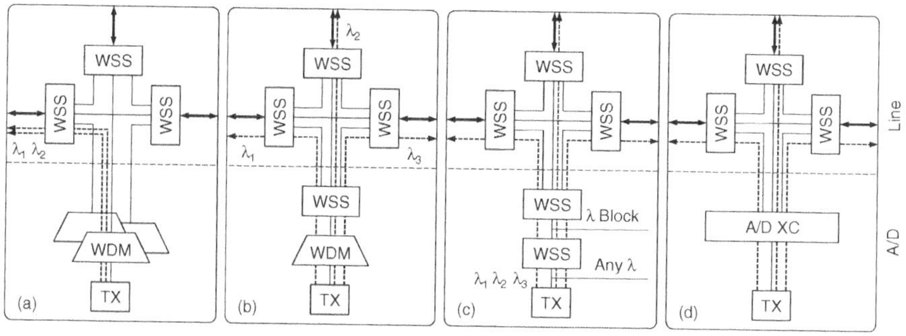

FIGURE 3.89 Basic ROADM functionality. (a) Colored per-degree add–drop. (b) Directionless add–drop. (c) Directionless and colorless add–drop. (d) Directionless, colorless, and contentionless add–drop.

configurations. In meshed networks, multidegree (MD)-ROADMs are required. Unless based on filter-switch-fabric-filter architectures (Fig. 3.87a), these devices are based on multidegree wavelength-selective switches (MD-WSS). This started with MEMS components that enabled Degree-4 (four network ports plus local add–drop) and evolved via Degree-8 to Degree-20+ [190].

In parallel, the functionality of the MD-ROADMs evolved from colored add–drop, where each add–drop port of a WSS is terminated on a static filter (e.g., an AWG) via directionless and colorless ROADMs to contentionless devices (see Fig. 3.89a–d).

In a *basic WSS-based MD-ROADM*, each network fiber (line, trunk, degree), which is to be terminated, requires dedicated add–drop filters. This increases cost, and still no flexibility for an add–drop port with regard to the outgoing direction can be provided. A client device hence needs to be connected to the correct add–drop filters. This problem is solved in *directionless* MD-ROADMs. An additional WSS (see Fig. 3.89b) connects the add–drop ports of the line-terminating WSSs. An add–drop AWG is connected to this additional WSS. Hence, drop signals can now be selected from any of the terminated lines, and add signals be directed respectively. One problem of this configuration is that the AWG add–drop ports are still colored, that is, specific add–drop wavelengths have to be connected to specific ports. In addition, due to the use of a single AWG (in the add–drop path), each wavelength can only be terminated once, leading to a certain degree of wavelength blocking.

In the *colorless, directionless* MD-ROADM, a further WSS is used to provide uncolored add–drop ports (see Fig. 3.89c). Now, add–drop ports are colorless, which allows connecting any add–drop wavelength at any port. The colorless, directionless combination allows restoration switching. Obvious disadvantages are the limited number of add–drop ports and the wavelength blocking that already occurred in the directionless, single-AWG configuration. Hence, without $1 \times 20+$ WSS, this configuration is suitable for smaller nodes with limited add–drop traffic and little risk of wavelength blocking only.

The wavelength-blocking limitation is overcome in *contentionless* MD-ROADMs. Through the use of additional WSSs and $N \times N$ (e.g., $N = 8$) switches,

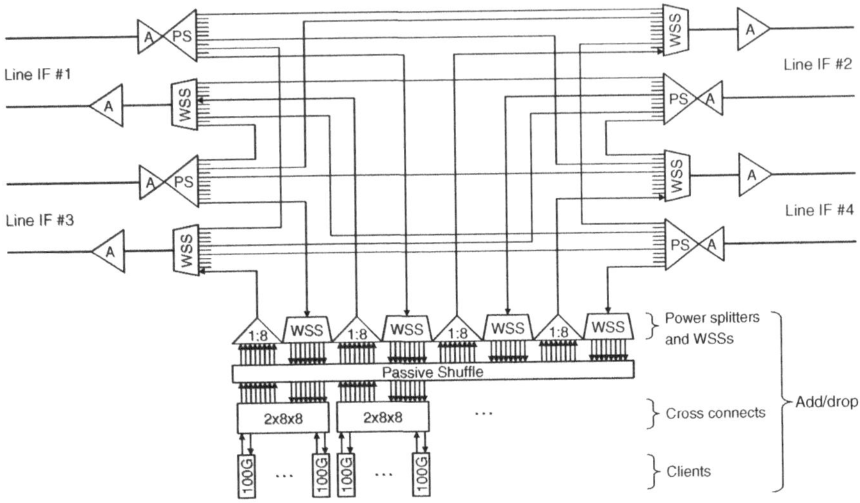

FIGURE 3.90 Block diagram of Degree-*N* directionless, colorless, contentionless ROADM. PS: power splitter, A: amplifier (EDFA).

wavelength blocking is avoided, thus allowing termination of the same wavelengths coming from/going to different lines. The lowest-layer switches (Fig. 3.89d) have $N \times N$ capability, where N is the number of lines to be terminated. A more detailed view of a Degree-N colorless and directionless ROADM is given in Fig. 3.90. This configuration is based on WSSs in the outgoing (line) directions and power splitters for the incoming directions. The add–drop part consists of WSS/power-splitter combinations, $N \times M$ switches, and a passive shuffle (i.e., patches) connecting all WSS/power-splitter combinations to all switches. Figure 3.90 shows one example of a contentionless add–drop path. Similar configurations are shown, for example, in Refs [190–192].

WSS-based ROADMs, like other filter structures, exhibit severe bandwidth narrowing when being cascaded. This effect is shown in Fig. 3.91 as an effect of the number of cascaded devices and the ROADM filter order.

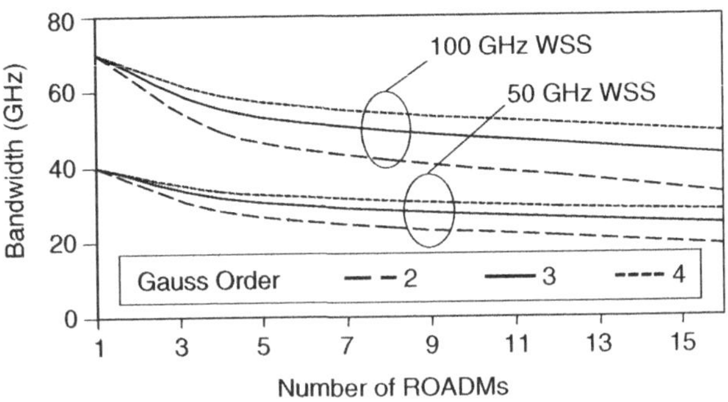

FIGURE 3.91 Bandwidth narrowing due to cascaded WSS.

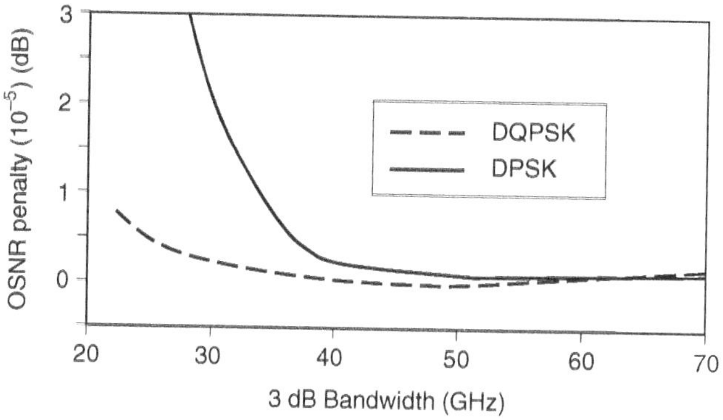

FIGURE 3.92 Penalty on OSNR for 43 Gb/s transmission as function of bandwidth narrowing.

The bandwidth narrowing of cascaded filters must be considered in (long-haul) transmission at high bit rates. Depending on the filter order and number of cascaded filters, penalties on the receive-end OSNR must be applied [193]. These penalties depend on the spectral allocation of the respective signals and their modulation, for example, DPSK versus DQPSK (see Fig. 3.92). Consequently, the effect can be counteracted by proper spectral shaping [194].

The negative effects of bandwidth narrowing in cascaded ROADMs can also be tackled with gridless ROADM technologies (see Section 3.2.6.7). Then, optical channel bandwidths can be adapted to the needs of the specific WDM channel, without heavily sacrificing spectral efficiency in the network.

3.2.7.3 Optical Cross-Connects Optical cross-connects are the generalization of ROADMs. Depending on implementation, they can scale to very high port counts without wavelength blocking. Conceptually, they are older than ROADMs but were never deployed in large quantities due to cost restrictions.

For cost-effective implementations, OXCs can be divided in several layers: fiber layer, wavelength bands, and individual wavelengths [195,196]. Each layer can be switched independently. Switching on the lowest (fiber) layer releases the switching fabric for the higher layers. Likewise, switching wavelength bands bypasses the wavelength switch and releases it from unnecessary traffic. A layered OXC (or MD-ROADM) is shown in Fig. 3.93.

Depending on actual size, the fiber cross-connect can be replaced by a manual patch frame. In large installations (e.g., in international internet exchanges), replacing such patch frames by automated fiber-handling robots [197,198] makes sense in order to suppress misconnections. The wavelength-band cross-connect can be based on (2D, 3D) MEMS, the wavelength XC can be based on any of the technologies described in Section 3.2.6.7.

Several ways of implementing large, blocking-free switch matrices for (wavelength) XCs exist [195,196,199,200]. Two solutions are shown in Fig. 3.94.

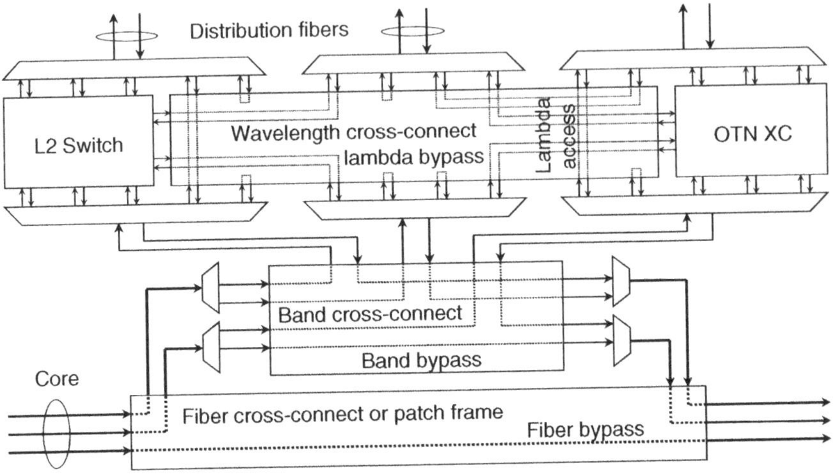

FIGURE 3.93 Structure of layered OXC/MD-ROADM.

The OXC structure shown in Fig. 3.94 (left) is basically identical to the one shown in Fig. 3.87a. The respective ROADM configuration has local add–drop ports and also VOAs for aligning add–drop signals with pass-through signals with regard to per-channel power levels. The $NM \times NM$ switch matrix, like in digital cross-connects, is most often implemented in a three-stage Clos architecture (see Fig. 3.95). Clos matrices can scale to large cross-connect sizes. They consist of switches that have significantly smaller port count than the resulting cross-connect. They are inherently blocking free [201].

The OXC structure shown in Fig. 3.94 (right) is an alternative implementation for large OXCs. It consists of cascaded power splitters, (fiber) switches, power combiners, and tunable filters. In addition, it also uses (transparent) wavelength

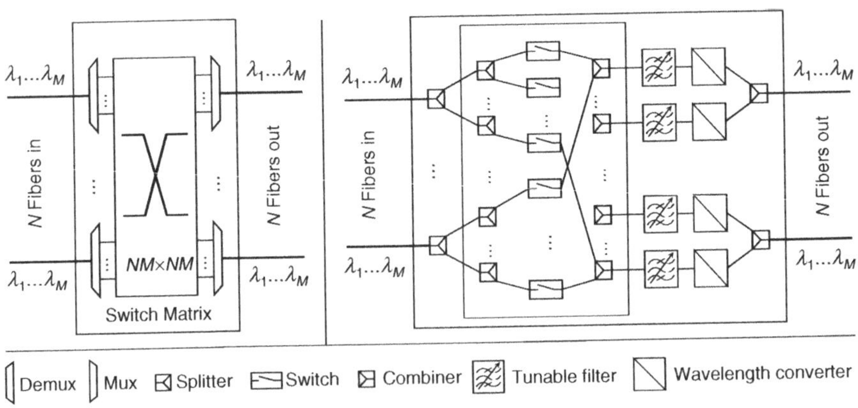

FIGURE 3.94 OXC structures. Filtered with $NM \times NM$ switch matrix (left), with power splitter/combiner and switch array, tunable filters, and additional wavelength converters (right).

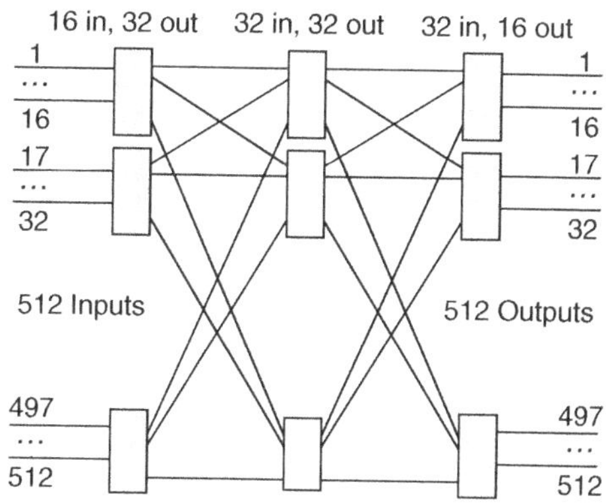

FIGURE 3.95 Three-stage, 512 × 512 Clos switch matrix.

converters. Ideally, the converters are transparent to the channel protocols and bit rates. This leads to all-optical converters, for example, based on SOAs or FWM [202–205]. Though under investigation since the early WDM days, no such devices are available commercially today. Hence, converters must be based on 3R regenerators, involving O/E/O conversions. They help avoiding wavelength blocking for those cases where the same incoming wavelength is to be connected to the same outgoing fiber. The OXC shown in Fig. 3.94 (left) can also be complemented by these wavelength converters.

3.3 RECEIVERS

In this chapter, we focus on the photodetectors (photodiodes) and the associated active electronics only. Other parts of a WDM receiver (such as optical preamplifiers or WDM or noise filters) have been covered in other chapters of this book.

3.3.1 Photodiodes

In high-speed WDM transmission, almost exclusively photodiodes are used as photodetectors [206–210]. Two types are commonly used that are described hereinafter.

3.3.1.1 PIN Photodiodes A p–n junction of suitable semiconductor material can be used as a high-speed photodetector. To improve responsivity, a lightly doped intrinsic semiconductor is put between the p- and n-type semiconductors, hence the term p-i-n or PIN photodiode. To achieve absorption in the intrinsic region, the p- and n-regions are kept transparent to the desired wavelength detection band. This leads to a double heterojunction structure of the diode. The structure of a front-illuminated PIN photodiode is shown in Fig. 3.96, together with the energy model.

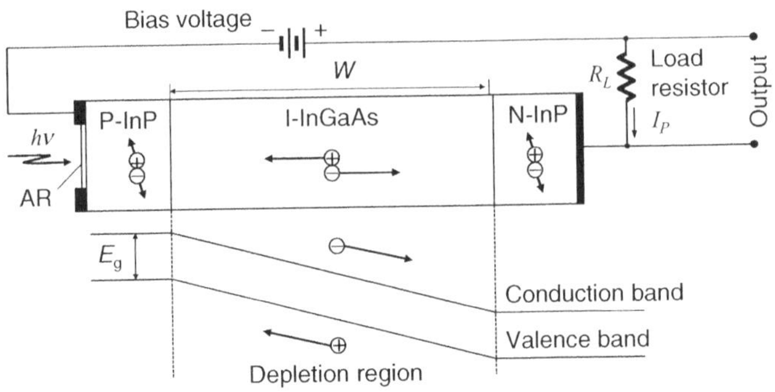

FIGURE 3.96 Structure and energy model of a PIN photodiode.

The *primary* photocurrent resulting from absorption over the entire width W is given by

$$I_{\mathrm{p}} = \frac{q}{h\nu} P_0 \left(1 - \mathrm{e}^{-\alpha_{\mathrm{s}}(\lambda) W}\right)(1 - R_{\mathrm{f}}) = RP_0. \tag{3.53}$$

P_0 is the incident optical power, $\alpha_{\mathrm{s}}(\lambda)$ is the signal (wavelength)-dependent absorption coefficient of the depletion region, q is the electron charge, $h\nu$ is the photon energy, and R_{f} is the entrance-facet reflectivity, respectively [211]. The (external) quantum efficiency η is defined as the ratio of photogenerated electron–hole pairs and incident photons:

$$\eta = \frac{I_{\mathrm{P}}/q}{P_0/h\nu}. \tag{3.54}$$

Now, the responsivity R of the (PIN) photodiode can be defined as

$$R = \frac{I_{\mathrm{P}}}{P_0} = \frac{\eta q}{h\nu} \quad (\mathrm{A/W}). \tag{3.55}$$

The responsivity of the semiconductor materials is a function of wavelength. Figure 3.97 shows the responsivity of InGaAs and Ge that are the relevant materials covering the relevant WDM transmission bandwidths [212]. For wavelengths $\lambda < 900\,\mathrm{nm}$, Si can be used.

For PIN photodiodes, there is a trade-off between efficiency (responsivity, and hence S/N) and bandwidth. High bandwidth of a PIN diode is determined by small transit time t_{T}, small RC of the receiver (consisting of photodiode and amplifier), and small diffusion time of the photocarriers outside the depletion region. The transit time through the depletion region is given by $t_{\mathrm{T}} = W/v_{\mathrm{d}}$, with v_{d} the drift velocity (which is in the range of 10^5 m/s). Obviously, for fast transit W needs to be *short*. Diffusion outside the depletion region is relatively slow, but does not depend on the

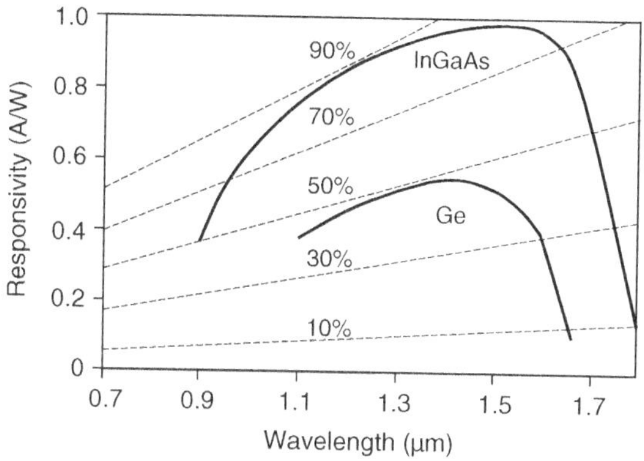

FIGURE 3.97 Spectral responsivity of Ge and InGaAs photodiodes.

width W. The bandwidth that results from the RC time constant of the receiver according to Fig. 3.98 is given by $B = 1/2\pi R_T C_T$.

The left part of Fig. 3.98 shows the photodiode connected to a load resistor R_L and an amplifier. The right part of the figure shows the equivalent circuit with an ideal diode (current source) and an amplifier. R_D and C_D are the photodiode resistance and capacitance, R_S is an additional series resistor, and R_A and C_A are the amplifier parameters, respectively [213]. With $R_D \rightarrow \infty$, the receiver time constant is then given via $R_T = R_S || R_L$ and $C_T = C_D + C_A$. Since $C_D = \varepsilon A/W$, with A the p–n junction area and ε the permittivity, the width W now needs to become large. The same holds for high responsivity, but it contradicts fast transit through the depletion region. Hence, W must be chosen carefully for high-speed diodes.

3.3.1.2 Avalanche Photodiodes Avalanche photodiodes (APDs) are a derivative of PIN diodes that provide an inherent gain through a controlled avalanche effect. APDs are doped such that an avalanche region with very high field strength results when the diode is reverse biased with high voltage. This region then shows an internal current gain due to impact ionization (the avalanche effect). In case of Si ADPs, the reverse bias is in the range of 100–200 V, providing current gain around 100. Bias voltage for Ge and InGaAs are significantly lower. Generally, higher reverse bias leads to higher gain (which includes noise generation). The structure of a reach-through APD is shown in Fig. 3.99, together with the electrical field strength along the device [212,214].

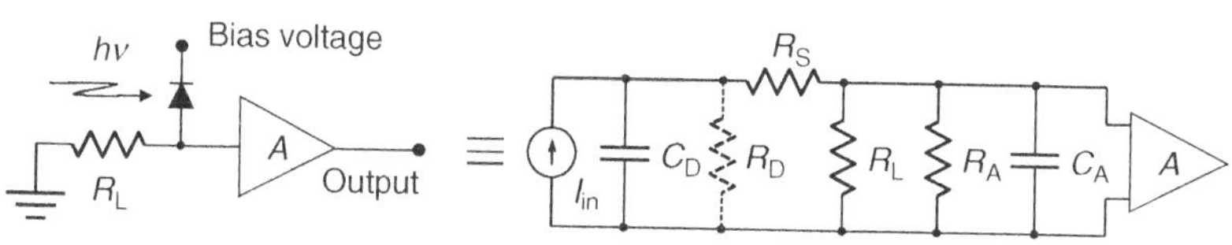

FIGURE 3.98 Photodiode receiver (left) and equivalent circuit (right).

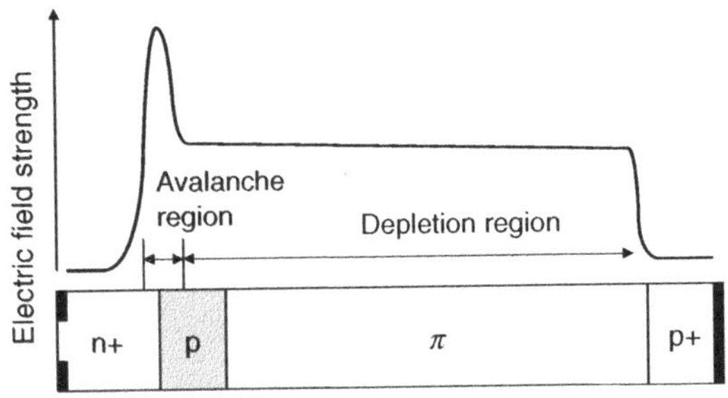

FIGURE 3.99 Structure of an APD, and electrical field strength along its structure.

APDs for WDM applications can be based on the same semiconductor material that is used for PIN diodes. Si-APDs detect in the visible and near-infrared (which commonly is not used for WDM). Germanium (Ge) detects infrared up to 1.7 μm, but has high multiplication noise. InGaAs can be used up to >1.6 μm, and has less multiplication noise than Ge. It is normally used as the absorption region of heterostructure diodes, typically involving InP as substrate and also as the multiplication layer. The resulting structure is shown in Fig. 3.100.

The responsivity of an APD can be expressed by the gain M (set to 1 for PIN diodes):

$$R_{\mathrm{APD}} = \frac{\eta q}{h\nu} M = R_{\mathrm{PIN}} M. \tag{3.56}$$

M is given by $M = I_{\mathrm{M}}/I_{\mathrm{P}}$, with I_{M} the average value of the total multiplied output current, and I_{P} the primary photocurrent. Since the noise figure $F(M)$ increases with M, there exists an optimum value of M that maximizes S/N. For sinusoidally modulated signals $s(t)$ with modulation index $m = 1$, optimum gain M is given by

$$M_{\mathrm{opt}}^{x+2} = \frac{2qI_{\mathrm{L}} + 4k_{\mathrm{B}}T/R_{\mathrm{L}}}{xq(I_{\mathrm{P}} + I_{\mathrm{D}})}. \tag{3.57}$$

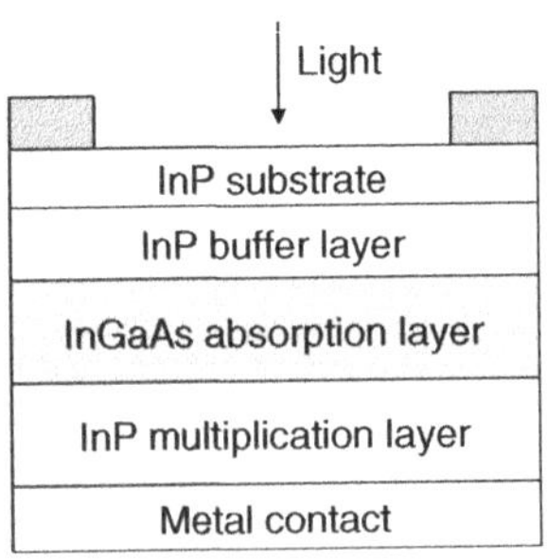

FIGURE 3.100 Detailed structure of an InP/InGaAs APD.

TABLE 3.13 Comparison of Photodiodes

		Ge	InGaAs
Wavelength range, λ (nm)	PIN, APD	900–1650	1000–1700
Responsivity, R (A/W)	PIN	0.4–0.5	0.7–0.95
Quantum efficiency, η (%)	PIN	50–55	60–70
Avalanche gain, M	APD	50–200	10–40
Dark current, I_d (nA)	PIN	50–500	0.5–5
	APD	50–500	5–50
Rise time, τ_r (ns)	PIN	0.1–0.5	0.05–0.5
	APD	0.5–0.8	0.1–0.5
Bandwidth, B (GHz)	PIN	0.5–3	1–40
	APD	1.5 max.	1–5
Bias voltage, V_B (V)	PIN	5–10	5–6
	APD	20–40	20–30
k-factor, k_A	APD	0.7–1.0	0.5–0.7

The gain-dependent noise figure is given by $F(M) \approx M^x$. The excess noise factor (which is produced by the multiplied shot and bulk dark current noise) of APDs is low enough to permit a gain-bandwidth product in excess of 100 GHz for the InP/InGaAs system, and up to 400 GHz for InGaAs on silicon. Therefore high-speed operation is possible, and commercial devices are available to bit rates of 10 Gb/s.

A comparison of relevant photodiode parameters is given in Table 3.13 [212,215].

3.3.2 Electronic Amplifiers

Photodiode detectors are current sources, with the current produced being proportional to the light intensity illuminating them. As a result, photodiodes are ideally suited to transimpedance amplifiers (TIAs). TIAs are current/voltage converters, where the output voltage is proportional to the input current (similarly as at an impedance). Since the output signal of the TIA usually is inadequate for following circuits (like CDR), the TIA is followed by a limiting amplifier (LA). The LA must be able to boost the signal swing produced by the TIA to a level required by the following circuitry, for example, 400 mV. A block diagram of an optical receiver with photodiode, load resistor R_L, TIA, and LA is shown in Fig. 3.101.

LAs often consist of multiple gain cells. For high-speed (WDM) applications, the resulting multistage amplifiers are usually integrated, using HEMT (high electron-mobility transistors) or HBT (heterojunction bipolar transistors) technologies [216–219].

3.3.2.1 TIA Transimpedance amplifiers are used to convert the photodiode current into voltage [220–222]. The amplifier must have low input impedance and generate either a 50 Ω signal to drive a coaxial cable or a voltage signal for further amplification. Since the most linear amplification is current amplification by a bipolar transistor, amplification may be provided before impedance conversion.

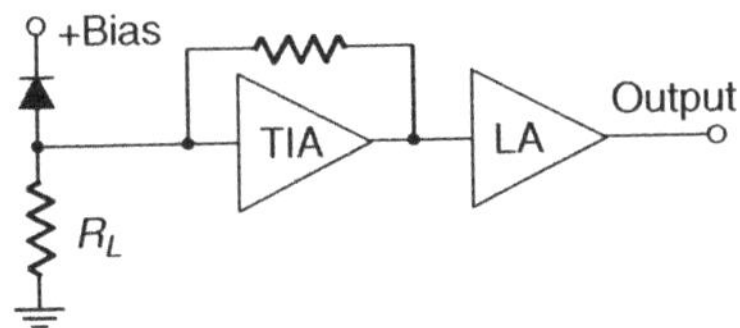

FIGURE 3.101 Optical receiver with electrical amplifiers.

The transimpedance can consist of a resistor R_f. However, in order to avoid gain peaking, and to improve stability, a parallel capacitor C_f may also be required. The resulting TIA block diagram is shown in Fig. 3.102, together with the photodiode equivalent circuit (and neglecting the other resistors and capacitors shown in Fig. 3.98).

The frequency-dependent transimpedance $Z_T(\omega)$ of the receiver circuit shown in Fig. 3.102 is given by [221]

$$Z_T(\omega) = \frac{V_{out}}{I_{in}} = \frac{Z_f}{1 + (1/A(\omega))(1 + (Z_f/Z_D))} \approx \frac{R_f}{1 + j\omega R_f C_f}. \tag{3.58}$$

Here, $Z_D = C_D| |R_D$. The approximation on the right-hand side of Eq. (3.58) holds for $A_0 \to \infty$ and $\omega_T \to \infty$, where ω_T is the transit frequency of the op-amp. At f_T, the open-loop gain has fallen to unity, and f_T = closed-loop voltage gain × bandwidth [223]. The op-amp gain $A(\omega)$ is given by

$$A(\omega) = \frac{A_0}{1 + j\omega A_0/\omega_T}. \tag{3.59}$$

Neglecting C_f, the input impedance of the TIA is given by

$$Z_{in} = \frac{V_-}{I_{in}} = \frac{R_f}{1 + A_0}. \tag{3.60}$$

The TIA according to Fig. 3.102 has two relevant noise sources, the thermal noise of the transimpedance resistor R_f, and the voltage noise connected at the V_{ref} input (see Fig. 3.103) [224].

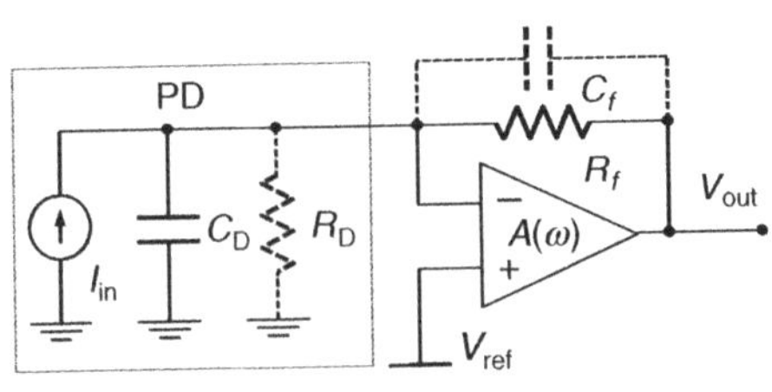

FIGURE 3.102 Photodiode connected to transimpedance amplifiers.

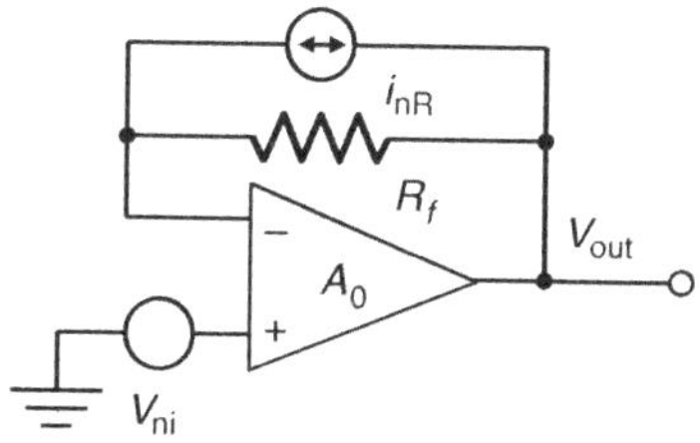

FIGURE 3.103 Transimpedance amplifier noise sources.

Since both noise sources are uncorrelated, the mean output voltage noise $\overline{v_{out}^2}$ is given by

$$\overline{v_{out}^2} = \overline{v_{ni}^2} + \overline{i_{nR}^2} \cdot R_f^2 = \overline{i_{eq}^2} \cdot R_f^2. \tag{3.61}$$

$\overline{v_{ni}^2}$ and $\overline{i_{nR}^2} \cdot R_f^2$ are the mean voltage noise contributions according to Fig. 3.103. They can be expressed as

$$\overline{v_{ni}^2} = \frac{4kT\gamma}{g_m} \cdot \Delta f + \frac{K}{C_{FET}} \cdot \frac{\Delta f}{f}, \quad \overline{i_{nR}^2} = \frac{4kT}{R_f} \cdot \Delta f. \tag{3.62}$$

$\overline{i_{nR}^2}$ is given by the well-known thermal or Johnson noise within a given bandwidth Δf. This noise is spectrally white and Gaussian distributed. The expression for $\overline{v_{ni}^2}$ in Eq. (3.62) consists of two terms. The first term (proportional to $4kT\,\Delta f$) can be viewed as the thermal noise of a resistor equal to γ/g_m, placed in series with the input gate of the op-amp. Here, g_m is the FET transconductance, and γ is a constant noise factor. The second term (proportional to $1/f$) represents the pink $1/f$ or flicker noise of the op-amp [225]. K is a constant in the range of $10^{-25}\,V^2$ F, and C_{FET} contains the FET capacitance and the channel width and length, respectively.

Flicker noise occurs in almost all electronic devices, however, it is higher in MOSFETs as compared to JFETs or bipolar transistors. Flicker noise is often characterized by the corner frequency f_C between the region dominated by the $1/f$ flicker noise and the higher frequency white noise. This is shown in Fig. 3.104. MOSFETs can have f_C in the gigahertz range.

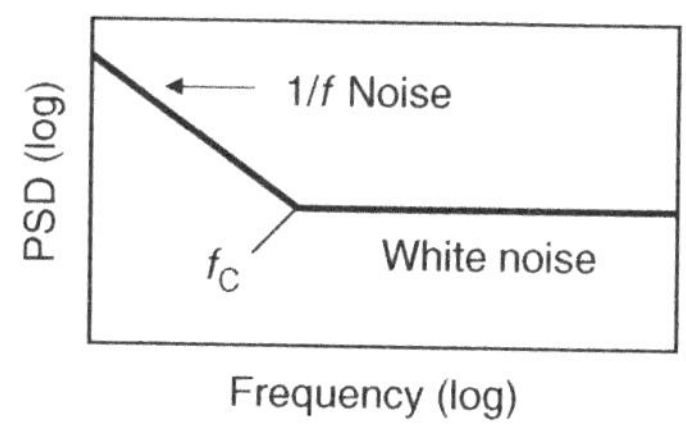

FIGURE 3.104 Simple flicker noise representation.

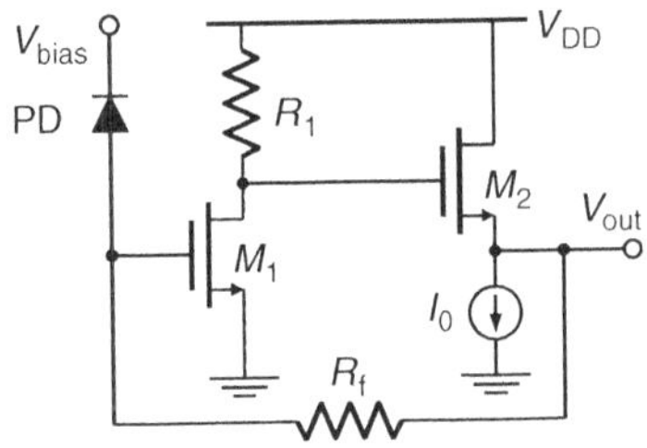

FIGURE 3.105 Simple transistor-based TIA.

With the expressions given in Eq. (3.62), $\overline{i_{eq}^2}$ can be written as

$$\overline{i_{eq}^2} = \frac{4kT}{R_f} \cdot \left[1 + \frac{\gamma}{g_m R_f} + \frac{K}{C_{FET} R_f f}\right] \cdot \Delta f. \tag{3.63}$$

An example of a simple transimpedance amplifier based on two (field-effect) transistors is shown in Fig. 3.105 [221,222,226]. Similar circuits can be found for HBTs.

R_1 controls the open-loop gain, it must be reduced to decrease peaking. The transimpedance is approximately given by R_f. Low R_f increases both peaking and bandwidth. Therefore, the circuit may have to be complemented by a capacitor as shown in Fig. 3.102.

For an analytic treatment of the design of maximally flat high-sensitivity transimpedance amplifiers, refer to Ref. [227].

3.3.2.2 Limiting Amplifier In coherent intradyne fiber-optic receivers (see Section 5.3.5.1), the TIA can directly be followed by analog/digital converters (ADC), given it provides sufficient gain to drive the ADC. In directly or self-coherently detected receivers, the TIA output is inadequate for the CDR circuit. Therefore, a postamplifier has to be interposed between the TIA and the CDR. In binary transmission (per in-phase or quadrature component, hence including DQPSK), the postamplifier is implemented as a *limiting amplifier* (see Fig. 3.101).

Main task of the LA is the amplification of an input with variable amplitude (~10–30 mV) to a fixed-amplitude (~400–500 mV) output. The LA needs to have sufficiently high bandwidth, low noise, and low offset voltage.

The core of an LA consists of multiple cascaded gain stages [219,228,229]. This core has to be complemented by an offset cancellation circuit as shown in Fig. 3.106.

Without further feedback, multistage LAs tend to gain peaking. Figure 3.107 shows the block diagram of an LA comprising six gain stages with interleaving active feedback and DC-offset cancellation. The interleaving active feedback effectively suppresses gain peaking while maintaining broadband characteristics.

The complex-frequency transfer function $H_{LA}(s)$ of the LA is given by [219,229]

$$H_{LA}(s) = \frac{A^6(s)}{1 + A^2(s)A_f(s) + A^4(s)A_f^2(s)}. \tag{3.64}$$

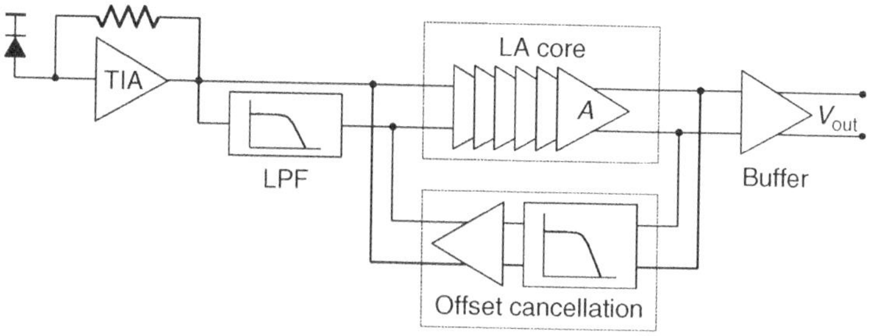

FIGURE 3.106 *N*-Stage limiting amplifier with offset cancellation.

The frequency-dependent gain functions $A(s)$, $A_f(s)$ of the forward gain stages and active feedback stages are given by, respectively,

$$A(s) = \frac{G_m R}{1 + sRC}, \quad A_f(s) = \frac{G_{mf} R}{1 + sRC}, \tag{3.65}$$

where G_m and G_{mf} represent the transconductances of the gain and feedback stages, respectively. R and C are the load resistance and capacitance of the differential pairs, respectively. As a simplification, the gain and feedback stages are assumed to be single-pole stages. Then, the overall gain is flattened since the interleaving active feedback stages cause pole splitting.

3.3.3 Pluggable Transceivers for WDM

Pluggable transceivers for WDM systems became available around 2002. The idea was to reduce cost and form factor of (WDM) transceivers and to unify WDM transponder design. With pluggable transceivers, only the different pluggables need to be spared, whereas a single unified transponder or muxponder card can be used for the different wavelengths.

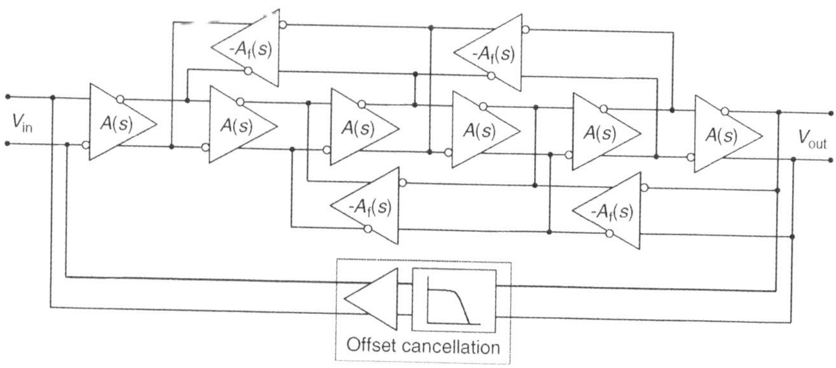

FIGURE 3.107 Limiting amplifier with interleaved active feedback.

FIGURE 3.108 SFP cage (left), and SFP top/bottom view (right).

Pluggable WDM transceiver development started with small form factor pluggables (SFPs). SFPs were originally developed as uncolored (850, 1310, or 1550 nm) GbE interfaces. Today, CWDM and DWDM SFPs with different optical specifications and covering bit rates of up to 5 Gb/s are available. In the form of the so-called SFP+, bit rates up to 11 Gb/s are covered as well. An SFP and its PCB-mounted cage are shown in Fig. 3.108.

Colored SFPs do not play an important role in WDM core networking anymore. The main reason is the bit rate limitation, further reasons include lack of tunability. They are still in use in certain WDM backhaul and access applications. SFPs still play an important role as muxponder local interfaces as well as switch/router interfaces. Here, they are being complemented by compact SFPs (CSFPs) that have two single fiber-working transceivers within the same form factor. These transceivers use two wavelengths for upstream and downstream, for example, 1310 and 1490 nm. CWDM options are available as well.

For bit rates around 10 Gb/s (OC-192/STM-64, 10GbE, OTU2), 10-Gb/s small form factor pluggables (XFPs) have been developed. Cage and pluggable are shown in Fig. 3.109.

As compared to SFPs, XFPs have larger form factor, which is mainly to allow accommodation of an external modulator (MZM or EAM). Because of the relevance of 10 Gb/s as a channel bit rate for WDM networking, XFPs are meanwhile available as tunable devices. These XFPs mainly use multisection DBR lasers (see Section 3.1.1.5).

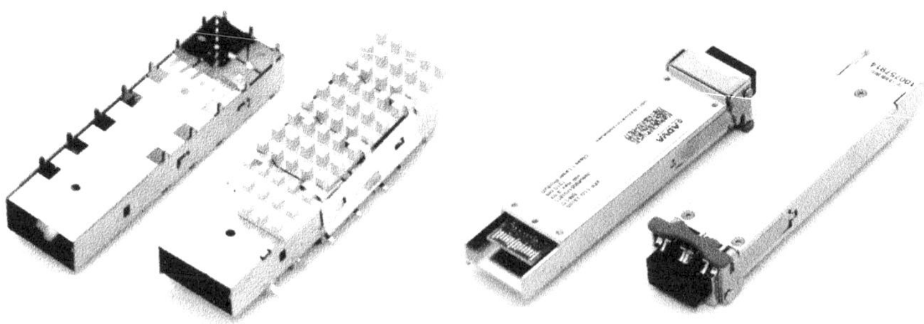

FIGURE 3.109 Cage with heat sink for XFPs (left), and XFP bottom/top view (right).

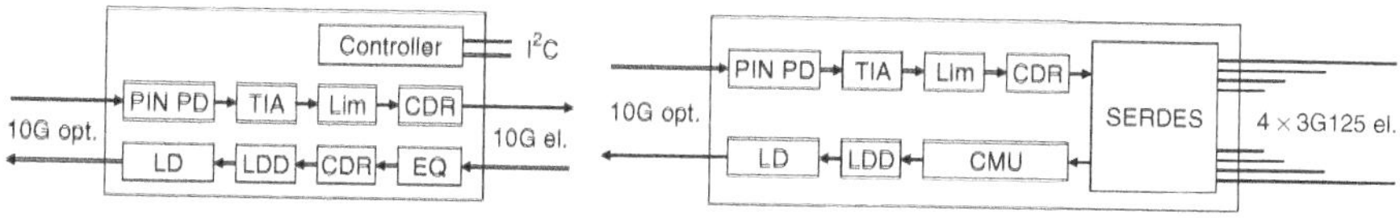

FIGURE 3.110 Simplified block diagrams of an XFP (left), and a XENPAK (right). CDR: clock and data recovery, CMU: clock multiplier unit, LDD: LD driver.

XFPs are not the only (WDM) pluggables that were developed for bit rates around 10 Gb/s. Specially for 10GbE, so-called XENPAKs were also built. XENPAKs have even larger form factor, but they include a serializer/deserializer (SERDES) for the 4×3.125 Gb/s 4-lane 10GbE interface. Block diagrams for XFPs and XENPAKs are shown in Fig. 3.110.

For WDM signals with 8–11 Gb/s, the SFP+ can also be used. It has the same form factor as standard SFPs, but does not include the clock data recovery (CDR). The respective functionality has thus to be implemented in the host device. Tunable SFP+ are available.

Relevant parameters of some WDM pluggables are summarized in Table 3.14.

For 100GbE (and 40GbE), so-called CFPs have been developed, where the letter C stands for "100." So far, standard CFPs are not available with general WDM remote (line) interfaces, however, the CFP standards LR-4 (long reach, four lanes) and ER-4 (extended reach, four lanes) make use of so-called *LAN WDM* (four wavelengths spaced 800 GHz at 1295.6, 1300.1, 1304.6, and 1309.1 nm) to multiplex the four lanes of ~25 Gb/s each in the wavelength division onto a common fiber [230]. A CFP module and its transceiver block diagram are shown in Fig. 3.111.

CFPs are available in a number of standardized and vendor-specific variants. The IEEE 802.3ba standard as of 2010 includes the LR-4 and ER-4 variants that use LAN WDM on single SMFs, the SR-10 variant (10×10 Gb/s on MMF ribbon cables), and electrical variants (CR-10, based on twinax cables). An additional variant, LR-10, is available as an MSA (multisource agreement) industry standard. LR-10 also uses single SMF and WDM. Table 3.15 summarizes relevant parameters of optical 100GbE CFP interfaces.

TABLE 3.14 SFP, XFP Parameters

	Tunable XFP	DWDM SFP (LR)	CWDM SFP
Linewidth (kHz)	~500	<1000	<1000
Extinction ratio	~9	~8	~8
Tx power (dBm)	+1 to +4	+4 to +7	0 to +5
Rx sensitivity (dBm)	−22 to −23	−28 (2G7)	−28/−30 (2G7/1G2)
CD allowance (ps/nm)	1600	2400 (2G7)	1600/2400 (2G7/1G2)
Bit rates (Gb/s)	9.9–11.3	0.15–2.7	0.1–2.7
Power dissipation	3.5–4.0	~1.5	1.0–1.5

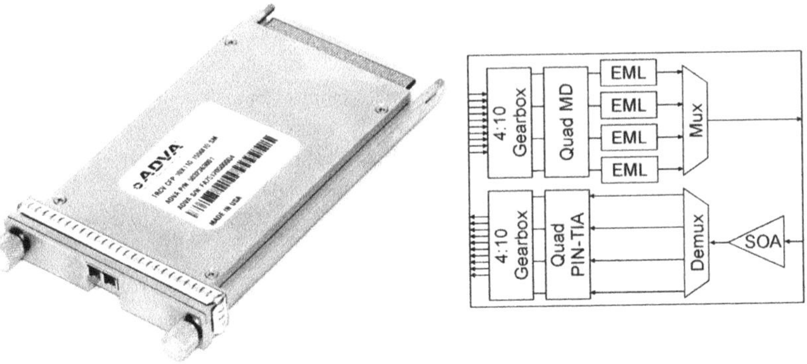

FIGURE 3.111 LR-4/ER-4 CFP module view (left), and transceiver block diagram (right).

TABLE 3.15 CFP Parameters (for 100 GbE)

	SR-10	LR-10	LR-4	ER-4
No. of fibers (per direction)	10 × MMF	1 × SMF	1 × SMF	1 × SMF
No. of wavelengths (per direction)	1	10	4	4
Bit rate per lane (Gb/s)	10	10	25	25
Max. reach (km)	0.1 /0.15	2 /10	10	40

Apart from the IEEE and MSA standard variants, proprietary CFP variants that support 50 GHz DWDM are available as well [231]. These modules make use of duobinary modulation (also refer to Fig. 5.30). They must be accommodated on transponder/muxponder cards that perform signal processing such as monitoring, FEC, and the gearbox required for 100 Gb/s client signals. With FEC, maximum reach is in the range of 500 km. Primary applications include metropolitan area and interdata-center WDM networking.

3.4 DIGITAL ELECTRONICS

3.4.1 SERDES

While on the optical link usually a data stream is transmitted in a serial manner, the electronic processing is performed in a parallel way. That means multiple bits are processed simultaneously. At the interface between the parallel electronics and the serial data format a serializer is used, while in the other direction, the serial data stream is converted to multiple parallel data lines at a lower data rate using a deserializer. As an additional function, the deserializer also often recovers the clock of the serial signal and converts the signal from the analog domain to the digital data domain by making the decision between "zero" and "one" levels. This function is also called CDR.

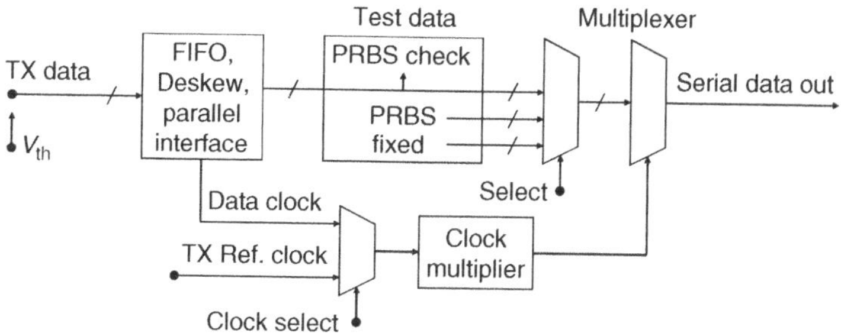

FIGURE 3.112 Schematic of a serializer IC. Crossed lines signify multilane (parallel) interfaces.

Recently, multilevel modulation formats (see Section 5.3.3) have been introduced in commercial transmission systems. Here, multiple information bits are transported in one symbol at a lower symbol rate, such that multiple bits are transported in parallel. At the interface between the parallel electronic data format and the parallel transmission format with typically less parallel data streams, a so-called "gearbox" IC is used, which converts the parallel data formats.

3.4.1.1 Serializer The serializer IC converts a parallel data stream to a serial data stream, which can then be transmitted over the optical link. The schematic diagram of a serializer is shown in Fig. 3.112. The parallel data in electronic format are received. This parallel interface can be any of the standardized interfaces such as SFI 4.2 and SFI 5.1 [232]. To align the parallel data lanes and account for different lengths of the cables of waveguides, a deskew function is included in this block. This deskew can be based on an additional lane in the parallel data interface, which carries samples of the data lanes. This additional lane is called the deskew channel. The data lanes are then aligned by delaying individual lanes such that the data bits fit to the samples in the deskew channel.

To allow testing of the individual channels and individual building blocks of the transmitter, test data can be monitored or generated in the serializer IC. For example, in the schematic in Fig. 3.112, the error rate on each lane can be tested in PRBS checkers. The data source (for instance a framing IC) generates a repetitive PRBS pattern on each lane, the serializer locks onto the received patterns and counts the differences between expected and received bits. Likewise, a PRBS signal can be generated to test the line side, or a fixed pattern can be generated for test purposes. Through a selector, either the data stream or one of the generated test patterns is sent to the multiplexer unit, where the parallel lanes are multiplexed into a serial data stream. The clock of the output signal can come from a variety of sources. It can be derived from the clock of the parallel interface, or it is externally fed into the serializer IC. The external clocks usually have a lower rate than the required line clock. Therefore, a clock multiplier is used to obtain the line rate clock for the multiplexer.

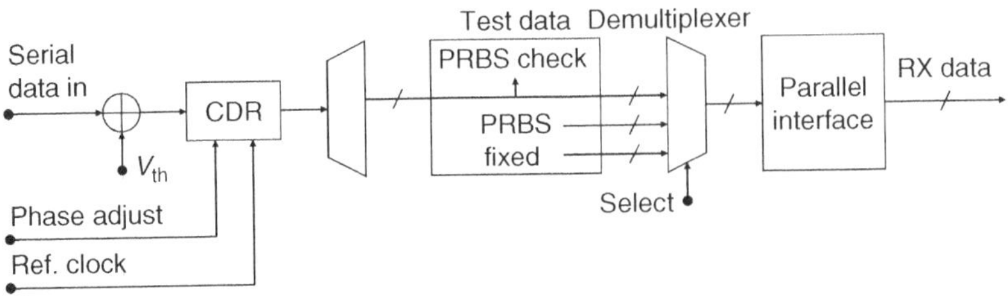

FIGURE 3.113 Schematic of a deserializer IC. Crossed lines signify multilane (parallel) interfaces.

3.4.1.2 Deserializer On the receive side, the serial data stream is detected and converted to a number of parallel streams at lower bit rate in order to enable further processing. For typical direct-detection systems, the deserializer also comprises one of the relevant functions in the transmitter–receiver chain, the clock and data recovery, which is depicted by the CDR block in Fig. 3.113. The CDR recovers the clock of the received signal and samples the signal in the middle of the bit period. By applying a decision threshold, the analog sample is converted into a digital "0" or "1." In coherent intradyne receivers with digital implementation of the main building blocks, CDR and deserializer are separated (refer to Section 5.4).

The serial digital data are then demultiplexed into parallel lanes. If PRBS data were sent over the link, the error rate can be measured in a PRBS checker unit. Similar to the serializer IC, PRBS or fixed pattern test data can be generated in the deserializer IC to test the interface between deserializer and the following electronics. Through a selector, the line data or the test data are sent to the parallel interface block, which is clocked with a fraction of the recovered received clock. In this interface block, depending on the interface specification, additional lanes might be added to enable synchronization of the parallel lanes.

Clock Recovery. To always sample the received signal at the optimum time position, the clock frequency of the signal must be recovered and the correct phase relation must be set. In general, a phase-locked loop (PLL) is used (see Fig. 3.114), where the phase of the incoming signal is compared to the phase of a locally generated clock from a voltage-controlled oscillator (VCO). A voltage corresponding to the phase difference is low-pass filtered and then fed back to control the frequency of the VCO.

As phase detector, often an Alexander or bang-bang phase detector, is used [233]. Here, the signal is sampled twice during a bit period, the 0/1 decision is made for each sample, and consecutive samples are compared. If one of the samples (sample 2 in Fig. 3.115, left) lies in the center of the signal eye, the alternate sample times are at

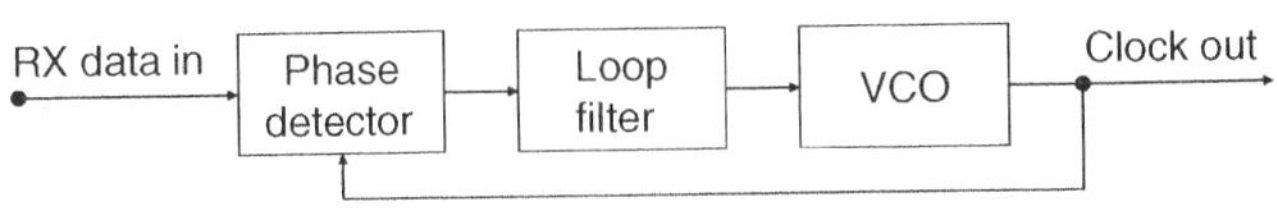

FIGURE 3.114 Schematic of a phase-locked loop.

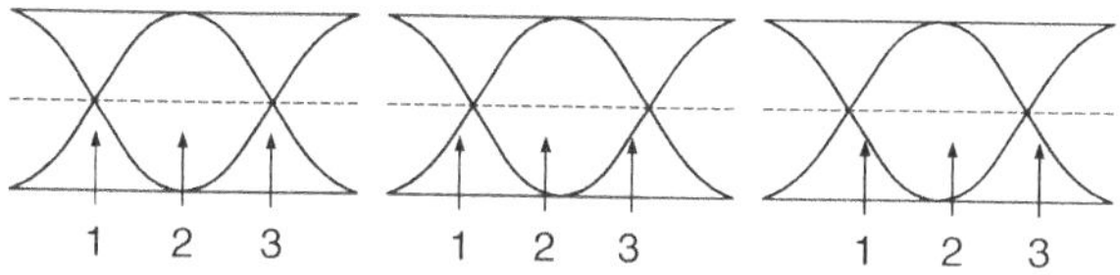

FIGURE 3.115 Sample phase relations for Alexander phase detection. The dashed line depicts the decision threshold.

the bit transitions and their data values are random compared to the center sample. If the data sample time is too early (Fig. 3.115, middle), the following sample 3 has consistently the same data value, and if the data sample time is too late, the previous sample 1 sees the same data value. Comparing the logical sample values therefore allows the VCO to be accelerated or slowed down.

If the eye-shape of the received signal is distorted, it is sometimes required to sample the data at a time offset from the mid-point between the transitions. In this case, an additional phase adjust input to the CDR can be used to shift the sampling point. This shift can be set fixed during manufacturing of the optical receiver, or it is variable and controlled by some feedback of the signal quality. If FEC is used, the count of corrected errors is a good quality parameter. Alternatively, the frame structure of some transmission formats (e.g., SONET/SDH, OTN, see Section 8.1) contain checksum bytes, which can be used to estimate the error rate of the received signal and, in turn, to adjust the sampling phase.

Slicing. To obtain the optimum bit error rate for a received signal, it is important to make the 0/1 decision based on an optimum setting of the decision threshold. This decision-making is sometimes referred to as "slicing." Depending on the impairments the optical signal has experienced during propagation over the fiber, the optimum decision threshold can vary widely. If accumulated optical noise is the main signal impairment, the variance of the 1 rail is usually larger than the variance of the 0 rail, as explained in Section 5.3.1.2. The optimum decision threshold in this case is lower than the middle between the 0 and 1 rails. If, on the other hand, the 0 rail has larger variations, as for instance seen for optical duobinary modulation (see Section 5.3.5.1), the decision threshold needs to move to higher levels. Figure 3.116 shows, as an example, how the receiver decision threshold impacts the bit error rate for different noise distribution functions. As explained in Section 5.3.1.2, the noise distribution of a signal dominated by optical noise is close to a noncentral chi-squared function. A good estimate of the BER can be obtained assuming a Gaussian function, but the optimum decision threshold differs for both functions.

In a practical CDR, the decision threshold is set at the middle of the eye. However, to enable a variable decision threshold, the signal eye can be shifted up or down by adding a constant voltage in front of the CDR. This voltage can be modified based on some feedback of the signal error rate. If FEC is used, the count of corrected 1-errors (i.e., a 1 that has been detected erroneously as a 0) and 0-errors (a 0 detected as 1) can be used to shift the signal in the correct direction and thus balance the number of counted 1- and 0-errors. Although for different receive-end probability distributions

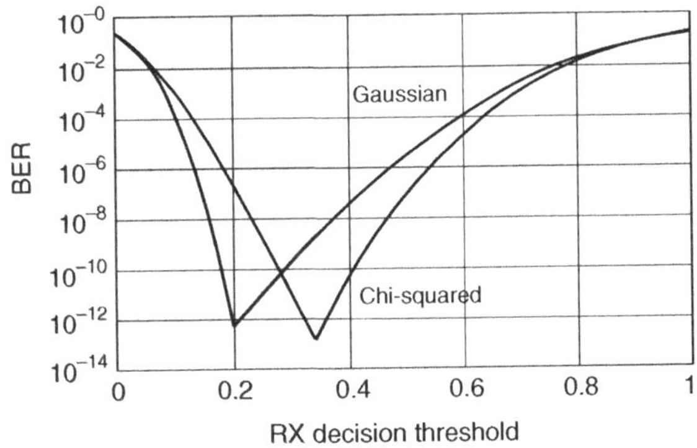

FIGURE 3.116 Bit error rate as a function of the receiver decision threshold for different noise distributions.

of 1s and 0s an equal number of 1- and 0-errors is not necessarily optimal for the overall error rate, it is at least close and therefore a common optimization scheme.

If no FEC is present in the system, a threshold optimization still can be achieved by a slicer with a main and two auxiliary thresholds. The auxiliary thresholds are placed above and below the main threshold and produce (in comparison with the center threshold) 1 and 0 errors assuming the center threshold is error free. After balancing the number of 1 and 0 errors at low levels the center between those levels is a good position for the main threshold.

3.4.2 Forward Error Correction*

The bit error rate is one of the most useful measures of transmission quality in digital transmission systems (refer to Section 8.2.2). For typical optical transmission, a BER of 10^{-12} needs to be obtained. Some network operators require a BER of 10^{-15} or even 10^{-16} for certain services to ensure that the errors introduced in the optical transmission line are negligible compared to other error sources. For a data rate of 10 Gb/s, less than one error will occur per day at a BER of 10^{-15}. This makes testing difficult.

An optical signal with a data rate of 10 Gb/s, whose BER is determined by optical noise, requires an OSNR of at least 20 dB for a BER of 10^{-12}. This would limit the reach of typical long-haul transmission systems to less than 1000 km. In order to increase the transmission distance or to allow other transmission impairments, FEC can be applied to the signal. FEC allows the operation of the optical transmission at a higher BER, while maintaining payload data integrity.

The possibility to reduce OSNR requirements and thus lower system costs or increase the transmission distance has made various FEC codes part of almost all medium and long-reach WDM transmission systems. The gain in OSNR, however, comes at the price of higher complexity of the electronics for encoding and decoding, and an increased latency for the data. An added benefit (as discussed in Section

*Section 3.4.2 is contributed by Helmut Grießer.

"Slicing") is that FEC can also provide the information required for a simple threshold control.

In general, FEC denotes the method of adding redundant symbols to the data that allow detecting and/or correcting errors at the receiving end without having to retransmit (parts of) the data. The error-correcting capabilities of specific codes depends on the type of distortions introduced during transmission.

When characterizing a specific FEC, usually simple optical-channel models are used due to the complexity of precise channel models. Suitability of a channel model depends on two aspects:

- Use of reliability information for decoding (soft decision) versus hard decision (after slicer).
- Channel is considered as memoryless (i.e., subsequent modulation symbols are independent) or not (e.g., burst-error channel).

For the common assumption of a memoryless channel without reliability information and a common threshold optimization that aims to balance the number of 1- and 0-errors (see Section "Slicing"), this leads to the so-called binary symmetric channel (BSC) that is fully characterized by its average BER.

Alternatively, an additive white Gaussian noise (AWGN) channel can be considered (again memoryless) that allows considering reliability information for the decoding process (soft-decision decoding) and that is a good fit for coherent receivers.

Nonlinear fiber effects like FWM and sequence-estimation receivers for electrical distortion equalization (using the Viterbi algorithm, see next section) can cause short error bursts. The error-correcting code should be able to cope with such bursts.

The following sections give a brief introduction into the basic properties of FEC codes in the context of WDM systems. For further reading, a textbook on FEC [234] is recommended.

3.4.2.1 FEC Basics The functional concept of FEC is simple: in addition to the payload (called *information* hereinafter), redundant data (that depends on the information) is transmitted over the noisy channel. If at the receiver some of the transmitted data is incorrect, this redundancy can help detecting and/or correcting the errors.

Often, coding is organized in isolated blocks (block code). Then, k information symbols are extended by $n - k$ additional symbols for error correction, leading to a code word of length n. This leaves the information symbols untouched (systematic encoding), which saves complexity and avoids error propagation at the decoder when retrieving the transmitted information. Depending on the code, symbols are most often binary, that is, bits. A very simple code is the single parity-check code (SPC). Redundancy is just a single bit, extending the information bits to achieve even parity. Obviously, the SPC can detect a single error (single bit flip) in the block, but is unable to find the erroneous bit and completely fails if two bits are wrong in the same block.

The error-correcting capability of a code is determined by the number of symbols in which two valid code words differ (also known as Hamming distance d [235]). The parity of an SPC code ensures a minimum distance of 2 between all possible code words and allows the detection of a single symbol error. With $d = 3$ and a single error, the received data can be attributed unambiguously to a valid code word, that is, every single error can be corrected. In general, a code allows to detect $d - 1$ or correct $t = \lfloor (d-1)/2 \rfloor$ errors (with $\lfloor \cdot \rfloor$ denoting the integer part). A class of single-error correcting codes (i.e., $d = 3$) is the Hamming code with the property that every possible received word can be attributed without ambiguity to a valid code word (perfect code).

Almost all codes used in practice are linear codes, that is, all linear combinations of code words are again valid code words. This property allows defining a code by a small set of code words that define a basis and simplifies encoding and decoding of the code. A Hamming (7, 4) code with code words of length 7 can encode 4 information bits and correct a single error. The code can be defined with a matrix $\mathbf{G}$ of four linearly independent code words (basis), with $\mathbf{G}$ also providing a specific encoding method to obtain the coded code word $\mathbf{c}$ from the uncoded code word $\mathbf{u}$, $\mathbf{c} = \mathbf{u} \cdot \mathbf{G}$ (note that all arithmetic with binary codes is done modulo-2, that is, $1 + 1 = 0$):

$$\mathbf{G} = \begin{pmatrix} 1 & 0 & 0 & 0 & 0 & 1 & 1 \\ 0 & 1 & 0 & 0 & 1 & 0 & 1 \\ 0 & 0 & 1 & 0 & 1 & 1 & 0 \\ 0 & 0 & 0 & 1 & 1 & 1 & 1 \end{pmatrix}. \tag{3.66}$$

The correction capabilities of a code do not change for linear combinations of the rows of $\mathbf{G}$ (different basis of the same code and different encoding rule) and permutations of the columns of $\mathbf{G}$ (equivalent code). Therefore, any linear code can be transformed into its equivalent systematic representation.

For the length-7 Hamming code, decoding is achieved by means of a parity-check matrix $\mathbf{H}$ with the syndrome $\mathbf{s} = \mathbf{c} \cdot \mathbf{H}^{\mathrm{T}} = \mathbf{0}$ for all valid code words $\mathbf{c}$ and $\mathbf{s} \neq \mathbf{0}$ for erroneous code words ($\mathbf{0}$ is the all zero vector). A parity-check matrix $\mathbf{H}$ matching $\mathbf{G}$ above is given by

$$\mathbf{H} = \begin{pmatrix} 0 & 1 & 1 & 1 & 1 & 0 & 0 \\ 1 & 0 & 1 & 1 & 0 & 1 & 0 \\ 1 & 1 & 0 & 1 & 0 & 0 & 1 \end{pmatrix}. \tag{3.67}$$

Since the Hamming code is perfect, each single error can be paired with a unique syndrome, and a simple decoding algorithm can determine the error by means of a syndrome lookup table (see Table 3.16). More effective decoding algorithms exist for cyclic codes.

TABLE 3.16 Syndrome Lookup Table for Example Hamming Code

Syndrome	(0 1 1)	(1 0 1)	(1 1 0)	(1 1 1)	(1 0 0)	(0 1 0)	(0 0 1)
Error position	1	2	3	4	5	6	7

3.4.2.2 Cyclic Codes A code is called cyclic if an arbitrary cyclic shift of any code word is again a valid code word. This allows efficient implementations with shift register circuitry and an elegant algebraic description via polynomials over finite fields. The *Reed–Solomon* (RS) code is a conceptually simple, yet very powerful cyclic code. If the information is given as a polynomial of degree $<k$, then the code word can be represented as the values of the polynomial at n different positions. For any k correct values of the polynomial, the original polynomial and hence the information can be recovered. Reed–Solomon codes allow the correction of up to $t = \lfloor (n-k)/2 \rfloor$ errors by means of an efficient algebraic bounded minimum distance (BMD) decoder, which can correct errors up to half of the minimum distance of the code. A disadvantage of Reed–Solomon codes is that the code word length is limited to the cardinality (number of elements) of the symbol alphabet. The Reed–Solomon code of length 255 from ITU-T Recommendation G.709 is based on byte symbols (not bits) and errors are corrected on a byte level also. This is very efficient for bursty channels, but in general binary codes are better for memoryless binary channels.

Assuming that the probability of undetected false decoding is negligible (i.e., the decoder can either correct or detect a failure, leaving the received data untouched), the code symbol error rate after decoding P_{out} can be estimated quite accurately from the input code symbol error rate P_{in} according to ITU-T Recommendation G.975:

$$P_{\text{out}} = \sum_{i=t+1}^{n} \frac{i}{n} \binom{n}{i} P_{\text{in}}^{i} (1 - P_{\text{in}})^{n-i}. \tag{3.68}$$

Here, n denotes the symbol length of the code and $t = \lfloor (n-k)/2 \rfloor$ is the number of symbol errors that can be corrected with algebraic bounded minimum decoding.

For independent bit errors (e.g., DP-QPSK modulation from Section 5.4 with Gray mapping [236]), the relation between the bit error probability p and the code symbol error probability P before and after decoding is well defined (q denotes the cardinality of the symbol alphabet):

$$P = 1 - (1-p)^{q}. \tag{3.69}$$

Codes can be *interleaved*, that is, instead of transmitting one code word after the other, the first symbol of several code words is transmitted followed by the second symbol of all code words, and so on (block interleaving). Interleaving does not change the correcting capabilities for memoryless channels, but is a simple yet effective way to enhance burst-error correction capabilities. The FEC according to ITU-T Recommendation G.709 defines the bytewise block interleaving of 16 code words of the Reed–Solomon code RS(255, 239). Since a single code can correct up to 8 arbitrary bytes, the interleaving of 16 code words allows the correction of up to 16×8 symbols in a row. This corresponds to a binary burst-error correcting capability (i.e., maximum length of a single error burst that can be corrected irrespective of its position in the frame) of $(16 \times 8 - 2) \times 8 + 2 = 1010$ bits.

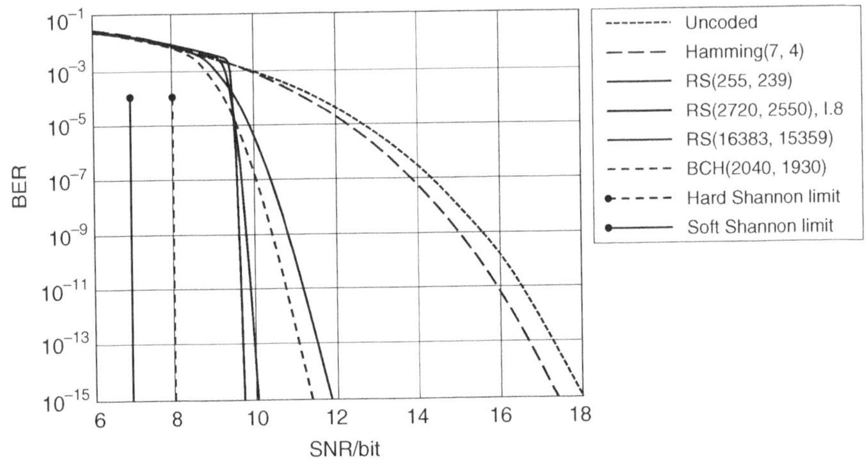

FIGURE 3.117 Performance of various algebraic codes in comparison with the hard and soft Shannon limit for 6.7% overhead and intensity modulation (OOK).

BCH (Bose–Chaudhuri–Hocquenghem) codes are a generalization of Reed–Solomon codes that allow the construction and efficient algebraic decoding of longer binary codes.

Figure 3.117 shows a comparison of the performance of all code classes introduced so far for OOK on an AWGN channel that can be transformed into a BSC with the following correspondence if optimum threshold decision is performed at the receiver:

$$p = \frac{1}{2}\operatorname{erfc}\left(\sqrt{\frac{1}{2}\frac{\text{SNR}}{\text{bit}}}\right) = Q\left(\sqrt{\frac{\text{SNR}}{\text{bit}}}\right). \tag{3.70}$$

Q and erfc are the Q-function and the complementary error function, respectively. They are defined in Section 5.3.

For a line rate, which is increased by a factor n/k, the SNR is reduced by the same factor, as a proportionally wider noise bandwidth needs to be taken into account (see Section 4.2.1). However, with a higher line rate, a larger BER can be tolerated after decoding. This, in turn, results in a lower SNR requirement. The SNR gain, or net effective coding gain (NECG) that can be achieved with FEC is usually given in decibel and is defined as

$$\text{NECG}_{\text{dB}} = 10\log_{10}\left(\frac{\text{SNR}_{\text{Uncoded}}(\text{BER}_{\text{ref}})}{\text{SNR}_{\text{FEC}}(\text{BER}_{\text{ref}})}\right) + 10\log_{10}\left(\frac{k}{n}\right). \tag{3.71}$$

The NECG depends on the reference bit error rate and therefore should always be specified with the corresponding reference BER.

Depending on the limiting noise source in the receiver, the FEC-enabled reduction in (electrical) SNR transforms into a reduction of the required OSNR (for ASE-limited systems) or P_{in} (for thermal- or shot-noise-limited systems). If optical amplification leads to an ASE-limited system, the noise variance before the FEC decoder is proportional to the OSNR at the receiver (cf. Eq. (4.24)). Hence the gain in OSNR is equal to the electrical coding gain. For thermal-noise-limited systems without optical amplification and a PIN receiver (see Section 3.3.1.1), the noise variance is proportional to the square of the optical input power (cf. Eq. (4.15)), and therefore the gain with respect to the optical input power is only half of the electrical coding gain. Finally, for purely shot-noise-limited receivers, the noise variance is proportional to P_{in} (cf. Eq. (4.19)) and therefore the gain with respect to the optical input power for APD receivers is typically reduced by 10–30% compared to the electrical coding gain (depending on the dominance of the shot noise vs. thermal noise) [237].

The short Hamming code that was introduced to explain the FEC concept is below 1 dB of NECG even for a large overhead of more than 40%. Significant gains can only be achieved with longer codes. The RS code from ITU-T Recommendation G.709 achieves a NECG of 6.2 dB for a reference BER of 10^{-15} with only 6.7% overhead. Longer RS codes (e.g., the RS code of length 2720 over 10 bit symbols from ITU-T Recommendation G.975.1) show an increasing NECG (see Fig. 3.117). It is also apparent that the additional gain gets smaller and does not approach the hard Shannon limit whereas decoding complexity increases significantly. The main reason is the symbol-based code construction that does not match the binary nature of the channel very well. The binary BCH code in Fig. 3.117 performs better than the RS(255, 239) (which has the same binary length), although the overhead is lower. A problem with BCH codes is that for constant code rate, their relative usable minimum distance decreases with increasing code length (a property that is known as *asymptotically bad* in coding literature) [238]. Therefore, BCH codes are also not able to approach the Shannon limit, especially since decoding complexity grows quadratically with the length.

For many years, the approach of the classic coding theory was to find "*good*" codes that would approach the Shannon limit with feasible decoding complexity. Later, a different strategy gained momentum that aimed at constructing very long codes by combining shorter ones (accepting a lower minimum distance), but with the benefit of being able to decode them with limited complexity via its component decoders.

3.4.2.3 Code Concatenation and Iterative Decoding Code concatenation is a form of code combination that performs additional encoding of the same symbols of several code words with a second code, and leads to so-called product codes if both codes are defined over the same (often binary) alphabet.

A very simple example of a product code with binary Hamming component codes is shown in Fig. 3.118. 16 bits of data are arranged in four rows of four bits each (shaded area). In each row, three redundant ("parity") bits are added such that all rows are valid code words of the Hamming code. Then, three rows are added such

0	0	0	0	0	0	0
0	0	0	0	0	0	0
0	0	0	0	0	0	0
0	0	0	0	0	0	0
0	0	0	0	0	0	0
0	0	0	0	0	0	0
0	0	0	0	0	0	0

0	0	0	0	0	0	0
0	0	0	0	0	0	0
0	0	0	1	1	0	0
0	0	1	1	0	0	0
0	0	0	0	0	0	0
0	0	0	0	0	0	0
0	0	0	0	0	0	0

FIGURE 3.118 Example for encoding product code with Hamming (7,4,3) component codes for each row and column (left, the all zero code word), and a potential received code word with four bit errors (right).

that each column is a valid code word of the same Hamming code (since it is a linear code these three rows will automatically be valid code words).

The resulting bit matrix is a code word (Fig. 3.118 assumes the all-zero code word for simplicity) of the product code and transmitted sequentially, for instance column-by-column. At the receiver, the serial bit stream of 49 bits is rearranged again into a 7×7 matrix. A simple example shall demonstrate the benefit of iterative (i.e., repeated) decoding via the component codes. If bit errors have occurred during transmission, for example, the four errors marked with a one in the right-hand side of Fig. 3.118, a first row-by-row decoding will rather introduce two additional errors, see left-hand side of Fig. 3.119. A subsequent columnwise decoding in the next step will remove all single errors in the columns and add a third one in the center column (Fig. 3.119, right). A second row-by-row decoding will finally remove remaining errors.

Iterative decoding cannot correct all arrangements of four errors (if the four errors mark the corners of a rectangle, the algorithm fails and rather adds five additional errors) and thus does not achieve decoding up to half of the minimum distance of the code (which is only the square of the minimum distance of the component code). On the other hand, iterative decoding can correct many error patterns with more than four errors. This property of the iterative decoder leads to good performance for high bit error rates and simultaneously to a bad asymptotic performance (a property known as *error-flaring* in literature). Through careful design of code and decoder it is

0	0	0	0	0	0	0
0	0	0	0	0	0	0
1	0	0	1	1	0	0
0	0	1	1	0	0	1
0	0	0	0	0	0	0
0	0	0	0	0	0	0
0	0	0	0	0	0	0

0	0	0	0	0	0	0
0	0	0	0	0	0	0
0	0	0	1	0	0	0
0	0	0	1	0	0	0
0	0	0	0	0	0	0
0	0	0	0	0	0	0
0	0	0	1	0	0	0

FIGURE 3.119 Iterative decoding of the Hamming product code: after row decoding (left, no correction but added two additional errors) and after column decoding (right, only errors in one column remain). A second row decoding step in an iterative decoding algorithm will remove the residual errors completely.

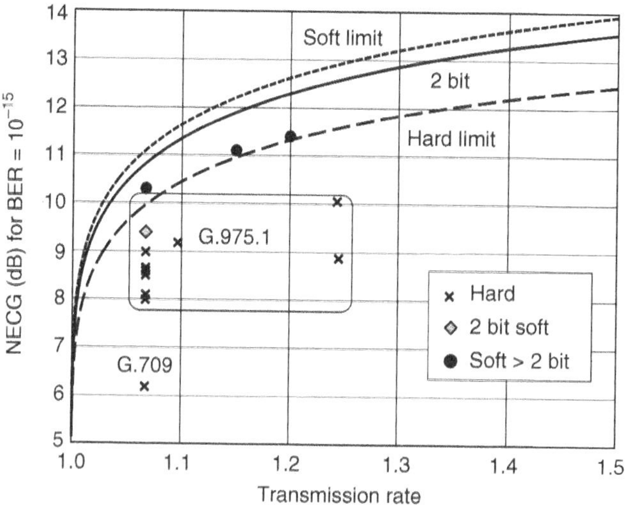

FIGURE 3.120 Net effective coding gain of various hard- and soft-decision codes for a target BER of 10^{-15} in comparison to the fundamental limits (Shannon bound).

possible to push the kink in the BER curve below the target residual BER (e.g., 10^{-15}).

Figure 3.120 gives an overview of the NECG for several second-generation FEC codes that were proposed in G.975.1 (almost all of them being concatenated codes) in comparison to the standard G.709 Reed–Solomon code. The bullets represent examples for the third generation of FEC codes with soft-decision decoding.

Table 3.17 summarizes the achievable NECGs of the G.975.1 FEC codes.

All the algorithms introduced so far have assumed the decision between 0 and 1 to be made in the CDR at the input of the receiver. This is referred to as hard-decision decoding. The algorithms then operate on binary signals. An improvement of the decoding capability can be made when the reliability of the received signal (given through the analog data signal) is used instead. A signal value close to the decision threshold would, for instance, be more unreliable than a signal value close to the ideal levels. This so-called soft-decision decoding can improve the maximum NECG by approximately 1 dB, but requires more complex receiver structures [239].

Iterative decoding with soft decision not only considers soft values at the input of the decoder but also provides soft values to the following decoding stage (soft-in/soft-out), and is known as Turbo decoding. With very long concatenated codes (e.g., product codes) and Turbo-like iterative decoding, a performance close to the soft Shannon limit can be achieved. This comes at the cost of increased complexity for soft-decision decoding, and growing latency.

A different code construction that also allows approaching the Shannon limit is defined via its sparse parity check matrix **H** and is known as low-density parity check (LDPC) code. All nonzero symbols in each row of **H** define a check for the corresponding code positions. Equally, all nonzero symbols in a column of **H** define

TABLE 3.17 Enhanced FEC Codes According to G.975.1 with NECG for a BER of 10^{-15}

I.2	Concatenated RS(255, 239)/CSOC (convolutional self-orthogonal code with $n_0/k_0 = 7/6$, $J = 8$) (25% OH)	NECG = 8.88 dB
I.3	Concatenated BCH(3860, 3824) and BCH(2040, 1930) code	NECG = 8.99 dB
I.4	Concatenated RS(1023,1007) and BCH (2047, 1952) code	NECG = 8.67 dB
I.5	Concatenated RS(1901,1855) and extended Hamming (512, 502) × (510, 500) product code	NECG = 8.5 dB (hard) NECG = 9.4 dB (2 bit soft)
I.6	LDPC(32640, 30592) (low-density parity check) code	NECG = 8.02 dB
I.7	Two orthogonally concatenated BCH (7, 11, 25% OH)	NECG = 8.09dB, 9.19dB, 10.06 dB
I.8	Long Reed–Solomon (2720, 2550) code	NECG = 8.0 dB
I.9	Two interleaved extended BCH(1020, 988) codes	NECG = 8.6 dB (estimation), code has floor below 2×10^{-14}

the checks that provide information to decode the corresponding code position. Decoding is again done iteratively with soft-in/soft-out stages that alternate between calculating the soft-information for each check and combining the soft-information of all checks for every symbol. The complexity of iterative decoding of LDPC codes is growing only linearly with the length, an important property for getting close to the Shannon limit.

3.4.3 Electronic Distortion Compensation

Electronic distortion compensation (e.g., CD compensation) is an alternative to analog optical compensation. Generally, electronic compensators can be implemented as digital or analog filters. Here, we give an overview of relevant digital filter concepts. An example of an analog electronic PMD compensator is described in Ref. [240].

Filters that compensate phase distortion need to be applied to the electrical field, rather than intensity. Hence, they can only be implemented in the transmitter, acting as precompensator and feeding an I/Q modulator, or in coherent receivers. Coherent receivers as used for 100 Gb/s WDM transmission are described in more detail in Section 5.4. In addition, MLSE can also be applied to intensity signals.

Besides PMD compensation as applied to coherent dual-polarization receivers, CD compensation is of primary concern. It can be performed efficiently in transmitters or coherent receivers. Digital implementation in a DSP and using DACs (digital/analog converters) that feed an I/Q modulator on the transmit side is shown in Fig. 3.121.

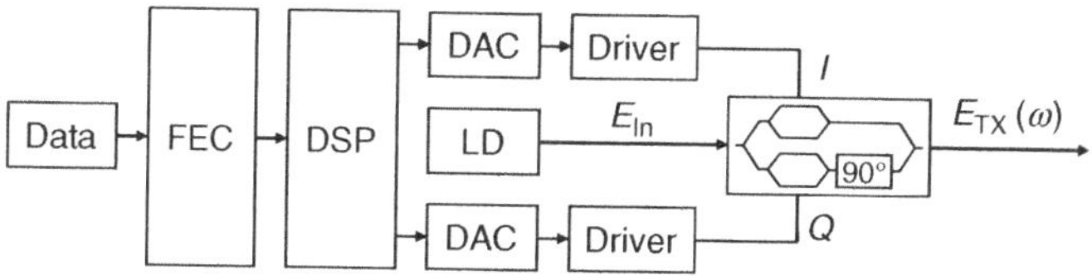

FIGURE 3.121 Electronic transmitter-side predistortion.

If the receive-end field signal $E_{Rx}(t)$ is affected by accumulated CD $\beta_2 L$, with β_2 the fiber dispersion and L the length of the link, then the predistorted optical signal $E_{Tx}(t)$ is given by calculating the transmitted signal spectrum from the *desired* received signal spectrum

$$E_{Tx}(\omega) = E_{Rx}(\omega)\exp\left(-j\frac{\beta_2\omega^2 L}{2}\right). \tag{3.72}$$

and taking the inverse Fourier transform of $E_{Tx}(t)$. The predistorted signal can be generated using a dual-drive MZM [241] or, as shown in Fig. 3.121, a double-nested MZM [242].

The amount of precompensation (i.e., the filter settings) can be optimized based on feedback on the quality of the received far-end signal, derived from Q-factor measurements or FEC error analysis.

According to Ref. [243], the nonlinear tolerance is decreased due to intrachannel effects when electronic precompensation as shown in Fig. 3.121 is applied.

For compensation of residual CD (i.e., on compensated links), FIR filters or FFE are suitable. This is also known as time-domain equalization (TDE). An N-tap FFE is shown in Fig. 3.122.

For residual CD compensation, the number N of equalizer coefficients is low, typically $N < 10$. For total-link CD compensation, the number of taps or coefficients can be approximated:

$$N \approx DL\frac{c}{f_c^2}f_{\text{Symb}}^2. \tag{3.73}$$

DL is the total link dispersion, c is the speed of light, f_c is the carrier frequency, and f_{Symb} is the Baud rate, respectively. For $DL = 17{,}000$ ps/nm (~1000 km of SSMF)

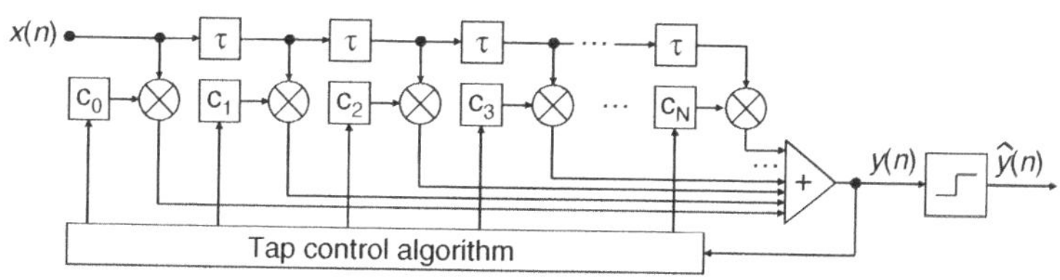

FIGURE 3.122 N-tap feed-forward equalizer.

and a Baud rate of 28 GBd, a tap number of $N > 100$ results. This tap number is to be applied per physical lane, that is, fourfold in 100 Gb/s DP-QPSK transmission.

The delay τ of the FFE taps is usually chosen as half the symbol period, $\tau = T_s/2$. Unlike filters with feedback, FFEs are always stable. Specially as receive-end filters, they amplify noise as well as the signal that is to be processed.

In scenarios that require flexibility, for example, with respect to link lengths, a mechanism must be used to set the tap weights. This can be done with the least mean squares (LMS) algorithm. LMS is an approximate steepest-descent algorithm that does not require knowledge of the channel [244,245]. It uses the error $e(n)$ between $y(n)$ and $\hat{y}(n)$:

$$e(n) = \hat{y}(n) - \vec{x}(n)^{\mathrm{T}}\vec{c}(n), \tag{3.74}$$

$\vec{x}(n)^T, \vec{c}(n)$ are the transposed input vector at time instances $n - (N + 1)$ to n, and the actual filter-coefficients vector, respectively. The LMS minimizes the error by modifying the filter coefficients according to

$$\vec{c}(n+1) = \vec{c}(n) + \mu \cdot e(n) \cdot \vec{x}(n). \tag{3.75}$$

Here, μ is a factor that determines speed and stability of the coefficients adaptation, and $\vec{c}(n+1)$ is the updates coefficients vector, respectively. At each time instance, the error is calculated, and the filter coefficients are updated.

Filters with high tap count are used today. However, for such tap numbers, *frequency-domain equalization* (FDE) can be more efficient [246]. Then, the filter is implemented in the frequency domain that requires conversion of the signal that is to be equalized via fast-Fourier transform (FFT). FDE then requires complex multiplications and additions for performing FFT/IFFT, frequency-domain multiplications, and the required verlap-add/save mechanism (additions only). Both the length N_{FFT} of the FFT and the FFT Radix are parameters to be optimized for minimization of effort.

The length N_{FFT} can be optimized because for short lengths, too few signal samples per filtering process are processed. In turn, for very large N_{FFT}, the effort increases with $N_{\mathrm{FFT}} \cdot \log_2(N_{\mathrm{FFT}})$.

Typically, Radix-2 FFT is used, meaning the length N_{FFT} is a power of 2 [247]. However, more efficient FFTs can be implemented at the cost of some constraints regarding N_{FFT}. This can be seen from comparing the numbers of real multiplications required for Radix-2 and Radix-4 (i.e., N_{FFT} is a power of 4) FFTs:

$$N_{\mathrm{Mult,Rad\text{-}2}} = \frac{3}{2} N_{\mathrm{FFT}} \log_2(N_{\mathrm{FFT}}), \quad N_{\mathrm{Mult,Rad\text{-}4}} = \frac{9}{8} N_{\mathrm{FFT}} \log_2(N_{\mathrm{FFT}}). \tag{3.76}$$

The higher Radix FFT requires 25% less multiplications. Even more efficient FFT algorithms, using (modified) Split-Radix lengths have been developed [248,249].

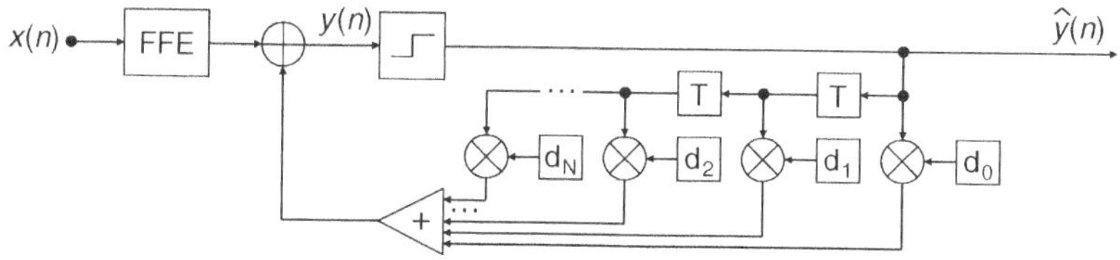

FIGURE 3.123 *N*-tap decision-feedback equalizer.

The FFE shown in Fig. 3.122 can be implemented in the analog or digital domain. Analog filters can be fast and low in power consumption. However, most filters are implemented in the digital domain due to ease of filter design.

An alternative for reducing filter (compute) effort is the addition of feedback to time-domain (FFE) filters. This results in *decision-feedback equalizers* (DFE), or infinite impulse response (IIR) filters. An example where feedback is derived from behind the decider is shown in Fig. 3.123.

When feedback is derived from behind the decider, the DFE only feeds back noise-free samples. Unlike in FFEs, there is no noise amplification. However, error bursts can be generated when the DFE feeds back a decision error. Unlike FFEs, DFEs can also get unstable. Hence, careful feedback-coefficients setting is required.

When there is intersymbol interference (ISI) in the channel, multiple received samples can be used to make decisions on a sequence of bits. This leads to MLSE [244,250–252]. MLSE can be applied to both, field and intensity signals [253,254].

MLSE selects *most likely sequences*. Along a set of sampled symbols $\bar{x}$, the most likely sequence $\bar{S}$ is selected according to $\max_S p(\bar{x}|\bar{S})$. Although the samples are correlated due to ISI, the noise is not. The total probability is then given by the product of the individual probabilities:

$$p(\bar{x}|\bar{S}) = \prod_{k=0}^{\infty} p(x_k|\bar{S}). \tag{3.77}$$

Path selection is usually done according to log-likelihood, and for limited sequence length:

$$\ln p(\bar{x}|\bar{S}) = \sum_{k=0}^{\infty} \ln p(x_k|\bar{S}), \quad \Gamma_{k_1}^{k_2}(\bar{S}) = -\sum_{k=k_1}^{k_2} \ln p(x_k|\bar{S}). \tag{3.78}$$

Each Γ is called a path length. It gives the probability of getting from time instance k_1 to k_2 using the path $\bar{S}$. The most probable path length must be selected with a computationally efficient algorithm. Efficient calculation can be performed with the *Viterbi algorithm*. It can be combined with histogram-based adaptive blind channel estimation (Fig. 3.124).

MLSE looks at all paths along a trellis graph. The Viterbi algorithm works left to right through this graph. At each state, only the best of the incoming paths, the

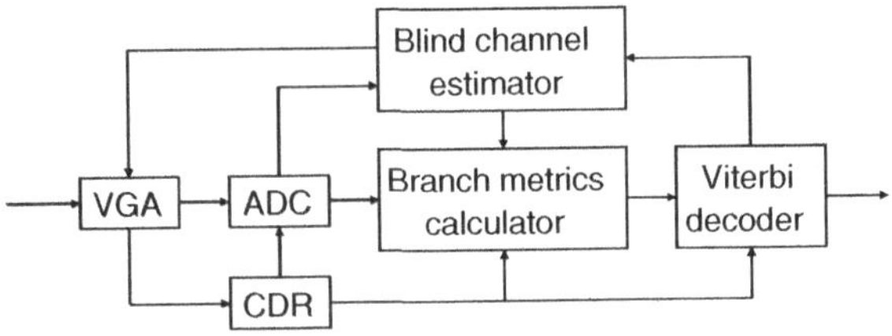

FIGURE 3.124 MLSE with blind channel estimation. VGA: variable gain amplifier.

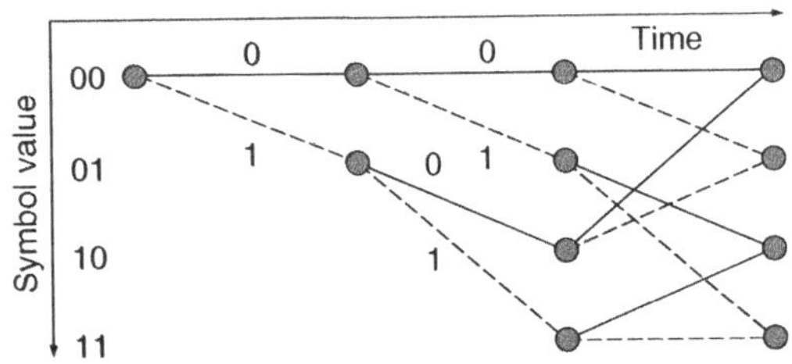

FIGURE 3.125 Simple trellis example for quaternary modulation as used Viterbi-based MLSE.

survivor, needs to be stored. This either yields an unambiguous result over the length of the sequence (all survivors form a closed path), or the most probable path is chosen. A trellis example for quaternary modulation (four possible symbol states) is shown in Fig. 3.125.

MLSE leads to optimal reception with respect to block-error probability. For AWGN, it can be simplified because the path length reduces to the Euclidian distance between the noisy sample and the mean for the sequence. This simplification cannot be used in direct detection, because the actual symbol probabilities are not Gaussian anymore.

REFERENCES

1. S.O. Kasap, Optoelectronics, Prentice Hall, 1999.
2. L.A. Coldren and S.W. Corzine, Diode Lasers and Photonic Integrated Circuits, Wiley, 1995.
3. J. Seufert et al., DFB laser diodes in the wavelength range from 760 nm to 2.5 μm, Spectrochim. Acta A: Mol. Biomol. Spectrosc., Vol. 60, No. 14, December 2004, pp. 3243–3247.
4. G.P. Agrawal and N.K. Dutta, Long-Wavelength Semiconductor Lasers, Van Nostrand Reinhold, New York, 1986.
5. L. Bjerkan et al., Measurement of laser parameters for simulation of high-speed fiberoptic systems, IEEE J. Lightwave Technol., Vol. 14, May 1996, pp. 839–850.

6. I. Tomkos et al., Measurement of laser rate equations parameters for simulating the performance of directly modulated 2.5 Gb/s metro area transmission systems and networks, Lasers and Electro-Optics Society Annual Meeting (LEOS 2000), Vol. 2, November. 2000, pp. 692–693.
7. U. Malzahn, Grundlagen der Laserdiodenansteuerung: Laserdioden verstehen und sicher ansteuern, Photonik, Vol. 5, 2003, pp. 2–4.
8. J. Shen and X. Cai, Compact laser-diode driver provides protection for precision-instrument use, EDN Europe, Vol. 53, May 2008, p. 60.
9. J. Williams, Current sources for fiber-optic lasers: a compendium of pleasant current events, EDN, August 22, 2002, pp. 69–78.
10. H. Soda et al., GaInAsP/InP surface emitting injection lasers, Jap. J. Appl. Phys., Vol. 18, No. 12, 1979, pp. 2329–2330.
11. F. Koyama et al., Room temperature cw operation of GaAs vertical cavity surface emitting laser, Trans. IEICE, Vol. E71, No. 11, 1988, pp. 1089–1090.
12. J. Buus and E.J. Murphy, Tunable lasers in optical networks, IEEE J. Lightwave Technol., Vol. 24, No. 1, January 2006, pp. 5–11.
13. L.A. Coldren, Tunable Semiconductor Lasers: A Tutorial (Invited), OFC2003, Atlanta, March 2003, Paper ThF1. Available at http://www.ece.ucsb.edu/Faculty/Coldren/papers/TSL.pdf.
14. L.A. Coldren et al., Tunable semiconductor lasers: a tutorial, IEEE J. Lightwave Technol., Vol. 22, No. 1, January 2004, pp. 193–202.
15. G. Busico, Introduction to DS-DBR, Oclaro technical slides, 2009, Contact: giacinto.busico@oclaro.com.
16. S. Dahlfort et al., Tunable transmitters and receivers for WDM PON, FSAN NG-PON2 Workshop, San Francisco, February 2010.
17. J.P. Wilde et al., Frequency noise characterization of a widely tunable narrow-linewidth DFB laser array source, NFOEC2009, San Diego, March 2009, Poster Session I, JWA33.
18. A. Nevsky et al., A narrow-line-width external cavity quantum dot laser for high-resolution spectroscopy in the near-infrared and yellow spectral ranges, Appl. Phys. B, Vol. 92, No. 4, 2008, pp. 501–507.
19. M.G. Thompson et al., InGaAs quantum-dot mode-locked laser diodes, Invited Paper, IEEE J. Sel. Top. Quantum Electron., Vol. 15, No. 3, 2009, pp. 661–672.
20. A.J. Ward et al., Widely tunable DS-DBR laser with monolithically integrated SOA: design and performance, IEEE J. Sel. Top. Quantum Electron., Vol. 11, No. 1, 2005, pp. 149–156.
21. Y. Liu, et al., Directly-modulated DS-DBR tunable laser for uncooled C-band WDM system, OFC2006, Anaheim, March 2006.
22. C.C. Renaud et al., 100 GHz spaced 10 Gb/s WDM over 10 °C to 70 °C using an uncooled DBR laser, OFC2006, Anaheim, March 2006, OWI76.
23. M.J. Wale et al., Technology routes for cost-effective, high performance, next-generation access components, EPIC Workshop, FTTH Council Europe, Copenhagen, February 2009.
24. M. Jiang, et al., Error Free 2.5 Gb/s transmission over 125 km conventional fiber of a directly modulated widely tunable vertical cavity surface emitting laser, OFC2001, Anaheim, March 2001, TuJ3-1.

25. P. Debernardi et al., Modal properties of long-wavelength tunable MEMS-VCSELs with curved mirrors: comparison of experiment and modeling, IEEE J. Quantum Electron., Vol. 44, No. 4, April 2008, pp. 391–399.
26. ChemOptics: Preliminary Technical Specifications Tunable Laser Module, January 2009, Contact: 104-11 Moonji-Dong, Yusung-Gu, Deajeon 305-380, South Korea, URL: www.chemoptics.co.kr, E-mail: sales@chemoptics.co.kr.
27. G.W. Yoffe et al., Widely-tunable 30 mW laser source with sub-500 kHz linewidth using DFB array, 21st LEOS Annual Meeting, Newport Beach, November 2008, ThBB2.
28. K. Achyut et al., WDM Technologies: Active Optical Components, Elsevier Science, San Diego, 2002.
29. C. Rolland et al., High speed and low loss, bulk electroabsorption waveguide modulators at 1.3 mm, IEEE Photonics Technol. Lett., Vol. 3, October 1991, pp. 894–896.
30. D.A.B. Miller et al., Band-edge electroabsorption in quantum well structures: the quantum-confined Stark effect, Phys. Rev. Lett., Vol. 53, 1984, pp. 2173–2175.
31. O. Sahlen, Optimization of DFB lasers integrated with Franz-Keldysh absorption modulators, IEEE J. Lightwave Technol., Vol. 12, 1994, pp. 969–976.
32. D. Delprat et al., High performance DFB laser-electroabsorption modulator based on the identical active layer approach and application to 10 Gb/s transmission over 125 km of standard fiber, ECOC'97, 1997, pp. 167–170.
33. A. Ramdane et al., Monolithic integration of multiple-quantum-well lasers and modulators for high-speed transmission, IEEE J. Sel. Top. Quantum Electron., Vol. 2, 1996, pp. 326–335.
34. K. Wakita et al., Bluechirp electroabsorption modulators with very thick quantum wells, IEEE Photonics Technol. Lett., Vol. 8, 1996, pp. 1169–1171.
35. D. Delprat et al., Integrated multiquantum well distributed feedback laser electroabsorption modulator with a negative chirp for zero bias voltage, Electron. Lett., Vol. 33, 1997, pp. 53–55.
36. Y.K. Park et al., Dispersion-penalty-free transmission over 130-km standard fiber using a 1.55-mm, 10-Gb/s integrated EA/DFB laser with low-extinction ratio and negative chirp, IEEE Photonics Technol. Lett., Vol. 8, 1996, pp. 1255–1257.
37. A. Naughton et al., Error-free 10 Gb/s duobinary transmission over 215 km of SSMF using a hybrid photonic integrated reflective modulator, OFC2012, Los Angeles, March 2012.
38. W.-J. Choi et al., Low insertion loss and low dispersion penalty InGaAsP quantum-well high-speed electroabsorption modulator for 40-Gb/s very-short-reach, long-reach, and long-haul applications, IEEE J. Lightwave Technol., Vol. 20, No. 12, 2002, pp. 2052–2056.
39. W.R. Cook and H. Jaffe, Piezooptic and electrooptic constants, in K.-H. Heilwege (ed.), Landolt–Börnstein New Series, Group III Vol. 11, Springer, 1979.
40. A. Yariv and P. Yeh, Optical Waves in Crystals, John Wiley & Sons, Inc., 1984.
41. F.C.A. Pockels, Lehrbuch der Kristalloptik, Teubner, Leipzig, 1906.
42. J. Kerr, On rotation of the plane of the polarization by reflection from the pole of a magnet Phil. Mag., Vol. 3, 1877, p. 321, Phil. Mag., Vol. 5, 1878, p. 161.
43. A. Banerjee et al., Wavelength-division multiplexed passive optical network (WDM-PON) technologies for broadband access: a review [Invited], J. Opt. Netw., Vol. 4, No. 11, November 2005, pp. 737–758.

44. CIP White Paper: WDM-PON Technologies. Available at http://www.ciphotonics.com/New_PDFs/WPON_White_Paper_v1%200.pdf.
45. M. Fujiwara et al., Impact of back reflection on upstream transmission in WDM single fiber loopback access networks, IEEE J. Lightwave Technol., Vol. 24, No. 2, 2006, pp. 740–746.
46. S.H. Cho et al., Reducing the back reflection-induced penalty by using a receiver with an optimized decision level in a reuse single fiber bidirectional WDM-PON, Proceedings of ECOC 2007, 2007, Paper WeP124.
47. J.H. Lee et al., Improved reflection tolerance by near-threshold operation of DFB seed light for an 80 km reach loop-back WDM-PON, Proceedings of OFC2009, 2009, Paper JThA71.
48. J.C. Cartledge and G.S. Burley, The effect of laser chirping on lightwave system performance, IEEE J. Lightwave Technol., Vol. 7, No. 3, 1989, pp. 568–573.
49. G. Berry, 100 °C, 10 Gb/s directly modulated InGaAsP DFB lasers for uncooled Ethernet applications, Proceedings of OFC2002, 2002, Paper ThF1.
50. P.A. Morton et al., 38.5 km error free transmission at 10 Gb/s in standard fibre using a low chirp, spectrally filtered, directly modulated 1.55 mm DFB laser, Electron. Lett., Vol. 33, 1997, pp. 310–311.
51. S. Mohrdiek et al, 10 Gb/s standard fiber transmission using a directly modulated 1.55 mm quantum well DFB lasers, IEEE Photonics Technol. Lett., Vol. 7, No. 11, 1995, pp. 1357–1359.
52. M.D. Feuer et al., Electronic dispersion compensation for a 10 Gb/s link using a directly modulated laser, IEEE Photonics Technol. Lett., Vol. 15, No. 12, 2003, pp. 1788–1790.
53. B. Wedding, New method for optical transmission beyond dispersion limit, Electron. Lett., Vol. 28, No. 14, 1992, pp. 1298ff.
54. R. Heidemann et al., 10-Gb/s transmission and beyond, Proc. IEEE, Vol. 81, No. 11, 1993, pp. 1558ff.
55. B. Wedding et al., 10 Gb/s optical transmission up to 253 km via standard singlemode fiber using the method of dispersion supported transmission, IEEE J. Lightwave Technol., Vol. 12, No. 10, 1994, pp. 1720–1727.
56. D. Mahgerefteh et al., Chirp managed laser and applications, IEEE J. Sel. Top. Quantum Electron., Vol. 16, No. 5, 2010, pp. 1126–1139.
57. P.A. Morton et al., High speed, low chirp, directly modulated 1.55 μm DFB laser sources for 10 Gb/s local distribution, presented at OFC1996, 1996, Paper TuH6.
58. S. Chandrasekhar et al., Chirp-managed laser and MLSE-RX enables transmission over 1200 km at 1550 nm in a DWDM Environment in NZDSF at 10 Gb/s without any optical dispersion compensation, IEEE Photonics Technol. Lett., Vol. 18, No. 14, 2006, pp. 1560–1562.
59. X. Zheng et al., Transmission from 0–360 km (6120 ps/nm) at 10 Gb/s without optical or electrical dispersion compensation using digital pulse shaping of a chirp managed laser, presented at OFC2009, San Diego, 2009, Paper OThE5.
60. R.-J. Essiambre et al., Capacity limits of optical fiber networks, IEEE J. Lightwave Technol., Vol. 28, No. 4, 2010, pp. 662–701.
61. T. Hayashi et al., Low-crosstalk and low-loss multi-core fiber utilizing fiber bend, OFC/NFOEC2011, Los Angeles, 2011, Paper OWJ3.

62. B. Zhu et al., Seven-core multicore fiber transmissions for passive optical network, Opt. Express, Vol. 18, No. 11, 2010, pp. 11117–11122.
63. S. Inao et al., Multicore optical fiber, Optical Fiber Communication, 1979 OSA Technical Digest Series (Optical Society of America, 1979), Paper WB1.
64. B. Rosinski et al., Multichannel transmission of a multicore fiber coupled with VCSEL, IEEE J. Lightwave Technol., Vol. 17, 1999, pp. 807–810.
65. B. Yao et al., Reduction of crosstalk by hole-walled multi-core fibers, OFC2012, Los Angeles, March 2012, Paper OM2D.5.pdf.
66. O. Ziemann et al., POF Handbook: Optical Short Range Transmission Systems, 2nd ed., Springer, 2008.
67. L. Möllers et al., Plastic optical fiber technology for reliable home networking: overview and results of the EU project POF-ALL, IEEE Commun. Mag., Vol. 47, No. 8, 2009, pp. 58–68.
68. G. Giaretta et al., High-speed (11 Gb/s) data transmission using perfluorinated graded-index polymer optical fibers for short interconnects (<100 m), IEEE Photonics Technol. Lett., Vol. 12, No. 3, 2000, pp. 347–349.
69. P.S.J. Russell, Photonic crystal fibers, IEEE J. Lightwave Technol., Vol. 24, No. 12, 2006, pp. 4729–4749.
70. R.F. Cregan et al., Single-mode photonic band gap guidance of light in air, Science, Vol. 285, No. 5433, 1999, pp. 1537–1539.
71. A. Bjarklev et al., Photonic crystal fibres, Kluwer Academic Publishers, Boston, MA, 2003.
72. P.J. Roberts et al., Ultimate low loss of hollow-core photonic crystal fibers, Opt. Express, Vol. 13, No. 1, 2005, pp. 236–244.
73. B.J. Mangan et al., Low loss (1.7 dB/km) hollow core photonic bandgap fiber, Proceedings of OFC2004, Los Angeles, USA, March 2004, Paper PDP24.
74. The Fiber Optic Association Guide to Fiber Optics and Premises Cabling. Available at http://www.thefoa.org/tech/ref/contents.html#contents.
75. R.J. Rapp, Submarine cables: critical infrastructure supplier perspective, 34th COLP Conference, Washington, May 2010.
76. E. Gjesteland, Technical solution and implementation of the Svalbard fibre cable, Telektronikk, Vol. 3, 2004, pp. 140–152.
77. R.H. Stolen and J.E. Bjorkholm, Parametric amplification and frequency conversion in optical fibers, IEEE J. Quantum Electron., Vol. QE-18, 1982, pp. 1062–1071.
78. R. Tang et al., Noise-figure limit of fiber-optical parametric amplifiers and wavelength converters: experimental investigation, Opt. Lett., Vol. 29, No. 20, 2004, pp. 2372–2374.
79. J. Hansryd et al., Fiber-based optical parametric amplifiers and their applications, Invited Paper, IEEE J. Sel. Top. Quantum Electron., Vol. 8, No. 3, 2002, pp. 506–520.
80. J. Kakande et al., Detailed characterisation of a fiber-optic parametric amplifier in phase-sensitive and phase-insensitive operation, Opt. Express, Vol. 18, No. 5, 2010, pp. 4130–4137.
81. N.A. Olsson and J.P.Van der Ziel, Characteristics of a semiconductor laser pumped Brillouin amplifier with electronically controlled, IEEE J. Lightwave Technol., Vol. 5, No. 1, 1987, pp. 147–153.

82. E. Desurvire, Erbium-Doped Fiber Amplifiers, John Wiley & Sons, Inc., New York, 1994.

83. B.P. Pal, Guided Wave Optical Components and Devices, Academic Press, Burlington, US, 2006.

84. G. Keiser, Optical Fiber Communications, 4th ed., Tata McGraw-Hill Education Pvt. Ltd., New Delhi, 2009.

85. D.K. Mynbaev and L.L. Schiner, Fiber Optics Communications Technology, Pearson Education, New Delhi, 2003.

86. R.J. Mears et al., Low-threshold tunable CW and Q-switched fiber laser operating at 1.55 μm, Electron. Lett., Vol. 22, 1986, pp. 59–160.

87. R.J. Mears et al., Low-noise Erbium-doped fiber amplifier at 1.54 μm, Electron. Lett., Vol. 23, 1987, pp. 1026–1028.

88. S.W. Harun et al., Optimization of the 1050 nm pump power and fiber length in single-pass and double-pass thulium doped fiber amplifiers, Prog. Electromagn. Res. B, Vol. 14, 2009, pp. 431–448.

89. S.S.-H. Yam and J. Kim, Ground state absorption in thulium-doped fiber amplifier: experiment and modeling, IEEE J. Sel. Top. Quantum Electron., Vol. 12, No. 4, 2006, pp. 797–803.

90. R.C. Schimmel et al., Characterisation and modelling of praseodymium doped fibre amplifiers, Proceedings of the 2001 IEEE/LEOS Symposium Benelux Chapter, Brussels, December 2001, pp. 133–136.

91. M. Yamada et al., Advanced PDFA for 1.3 μm amplification using newly developed Pr3+-doped PbF2/InF3-based fluoride fiber, CLEO/Pacific Rim'97, 1997, pp. 48–49.

92. M. Karasek, Numerical analysis of Pr^{3+}-doped fluoride fibre amplifier, IEEE Photonics Technol. Lett., Vol. 4, 1992, pp. 1266–1269.

93. A. Hirano, Optical amplifiers and their standardization in ITU-T & IEC, ITU-T Workshop on IP/Optical, 2002, Available at http://www.itu.int/itudoc/itu-t/workshop/optical/s8-p02r.html.

94. A.R. Bahrampour et al., Analysis of gain dynamics in EDFA and transient response of optical fiber inverter using a simplified model, Scientia Iranica, Vol. 7, No. 1, January 2000, pp. 32–40.

95. A. Mori et al., Ultra-wideband tellurite-based Raman fibre amplifier, Electron. Lett., Vol. 37, No. 24, November 2001, pp. 1442–1443.

96. E.S. Son and Y.C. Chung, Pump-to-signal crosstalk in Raman amplifier systems using polarization-multiplexed pumps lasers, Proc. SPIE, Vol. 4906, 2002, pp. 35–42.

97. C. Han Kim et al., Experimental study on the effect of codirectional Raman gain on system's performance, Opt. Express, Vol. 15, No. 10, 2007, pp. 6146–6151.

98. A.H. Beshr et al., Different pump configurations for discrete Raman amplifier, Int. J. Sci. Eng. Res., Vol. 2, No. 11, 2011, ISSN 2229-5518.

99. Red-C data sheet, Hybrid Raman-EDFA Module 2 and 3-Pump, 2012. Available at http://www.red-c.com/_Uploads/dbsAttachedFiles/HybridRamanEDFA.pdf.

100. P.B. Hansen et al., Rayleigh scattering limitations in distributed Raman pre-amplifiers, IEEE Photonics Technol. Lett., Vol. 10, No. 1, 1998, pp. 159–161.

101. W. Zhang et al., An analytical expression of equivalent noise figure for distributed fiber Raman amplifiers with Rayleigh scattering, Opt. Commun., Vol. 199, November 2001, pp. 231–236.

102. X. Zhou et al., Theoretical investigation of fiber Raman amplifier with dynamic gain control, OSA/OFC2001, 2001, Paper WDD17-1.
103. H. Wei et al., All optical method to achieve gain-clamping in broadband distributed fiber Raman amplifiers, Optica Applicata, Vol. 34, No. 3, 2004, pp. 365–372.
104. Xiang Zhou et al., Submicrosecond transient control for a forward-pumped Raman Fiber amplifier, IEEE Photonics Technol. Lett., Vol. 17 No. 10, 2005, pp. 2059–2061.
105. G.P. Agrawal, Fiber-Optic Communication Systems, John Wiley & Sons, Inc., New York, 1992.
106. L.H. Spiekman et al., Transmission of 8 DWDM Channels at 20 Gb/s over 160 km of standard fiber using a cascade of semiconductor optical amplifiers, IEEE Photonics Technol. Lett., Vol. 12, No. 6, 2000, pp. 717–719.
107. Y. Sun et al., Error free transmission of 32 × 2.5 Gb/s DWDM channels over 125 km using cascaded in-line semiconductor optical amplifiers, Electron. Lett., Vol. 35, No. 21, 1999, pp. 1863–1865.
108. T. Akiyama and M. Sugawara, Recent progress in quantum-dot semiconductor optical amplifiers, Invited Paper, Photonics for Space Environments XI, in E.W. Taylor (ed.), Proc. of SPIE, Vol. 6308, 2006.
109. Special mini-issue on dispersion compensation, IEEE J. Lightwave Technol., Vol. 12, No. 10, 1994.
110. K. Thyagarajan et al., A Novel Design of a Dispersion Compensating Fiber, IEEE Photonics Technol. Lett., Vol. 8, No. 11, 1996, pp. 1510–1512.
111. I.C. Goyal et al., Dispersion compensating fibers, ICTON 2002, Warsaw, April 2002, Paper Mo.A.5.
112. L. Grüner-Nielsen et al., Dispersion-compensating fibers, Invited, IEEE J. Lightwave Technol., Vol. 23, No. 11, 2005, pp. 3566–3579.
113. N.M. Litchinitser et al., Fiber Bragg gratings for dispersion compensation in transmission: theoretical model and design criteria for nearly ideal pulse recompression, IEEE J. Lightwave Technol., Vol. 15, No. 8, 1997, pp. 1303–1313.
114. I. Riant et al., Chirped fiber Bragg gratings for WDM chromatic dispersion compensation in multispan 10-Gb/s transmission, IEEE J. Sel. Top. Quantum Electron., Vol. 5, No. 5, 1999, pp. 1312–1324.
115. D. van den Borne et al., 42.8-Gb/s RZ-DQPSK transmission with FBG-based in-line dispersion compensation, IEEE Photonics Technol. Lett., Vol. 19, No. 14, 2007, pp. 1069–1071.
116. T. Saito et al., Prechirp technique for dispersion compensation for a high-speed long-span transmission, IEEE Photonics Technol. Lett., Vol. 3, No. 1, 1991, pp. 74–76.
117. N. Henmi et al., Prechirp technique as a linear dispersion compensation for ultrahigh-speed long-span intensity modulation directed detection optical communication systems, IEEE J. Lightwave Technol., Vol. 12, No. 10, 1994, pp. 1706–1719.
118. R.-M. Mu et al., Dynamics of the chirped return-to-zero modulation format, IEEE J. Lightwave Technol., Vol. 20, No. 1, 2002, pp. 47–57.
119. A. Royset et al., Linear and nonlinear dispersion compensation of short pulses using midspan spectral inversion, IEEE Photonics Technol. Lett., Vol. 8, No. 3, 1996, pp. 449.
120. A. Yariv et al., Compensation for channel dispersion by nonlinear optical phase conjugation, Opt. Lett., Vol. 4, No. 2, 1979, pp. 52ff.

121. R.M. Jopson et al., 10-Gb/s 360-km transmission over normal-dispersion fiber using mid-system spectral inversion, OFC'93, San Jose, 1993, Paper PD3.

122. R.-D. Li et al., Dispersion compensation with phase-sensitive optical amplifiers, IEEE J. Lightwave Technol., Vol. 12, No. 3, 1994, pp. 541ff.

123. H. Kubota and M. Nakazawa, Partial soliton communication system, Opt. Commun., Vol. 87, No. 1, 2, 1992, pp. 15ff.

124. G.P. Agrawal, Nonlinear Fiber Optics, 2nd ed., Academic Press, San Diego, 1995.

125. F. Forghieri, R. Tkach, and A. Chraplyvy, Fiber nonlinearities and their impact on transmission systems, in I. Kaminov and T. Koch, (eds), Optical Fiber Telecommunications IIIa, Academic Press, San Diego, 1997.

126. G.P. Agrawal, Fiber-Optic Communication Systems, John Wiley & Sons, Inc., New York, 1992.

127. H. Taga et al., Performance evaluation of the different types of fiber-chromatic-dispersion equalization for IM-DD ultralong-distance optical transmission, IEEE J. Lightwave Technol., Vol. LT-12, No. 9, 1994, p. 1616.

128. L.F. Mollenauer, Ultra-long-haul, dense WDM using dispersion-managed solitons in an all-Raman system, in M.N. Islam, et al. (eds), Raman Amplifiers for Telecommunications 2, Springer Series in Optical Sciences, Vol. 90/2, Springer, New York, 2004, pp. 627–672.

129. K. Aikawa et al., High performance dispersion and dispersion slope compensating fiber modules for non zero dispersion shifted fibers, Fujikura Tech. Rev., 2003, pp. 5–10.

130. Corning® Vascade® Optical Fibers, Product Information, PI1445, Issue August 2010, www.corning.com/opticalfiber.

131. J.-P. Elbers and C. Glingener, Efficient design of high-capacity dense wavelength-division multiplex systems, Int. J. Electron. Commun., Vol. 55, No. 5, 2001, pp. 295–304.

132. L. Grüner-Nielsen et al., Dispersion-compensating fibers, Invited Paper, IEEE J. Lightwave Technol., Vol. 23, No. 11, 2005, pp. 3566–3579.

133. C. Xie, Suppression of inter-channel nonlinearities in WDM coherent PDMQPSK systems using periodic-group-delay dispersion compensators, ECOC 2009, Vienna, September 2009, Paper P4.08.

134. I. Riant et al., Chirped fiber Bragg gratings for WDM chromatic dispersion compensation in multispan 10-Gb/s transmission, IEEE J. Sel. Top. Quantum Electron., Vol. 5, No. 5, 1999, pp. 1312–1324.

135. D.v.d. Borne et al., Cost effective 10.7-Gb/s long-haul transmission using fiber Bragg gratings for in-line dispersion compensation, OFC2007, Anaheim, March 2007, Paper OThS5.

136. D.v.d. Borne et al., 42.8-Gb/s RZ-DQPSK transmission with FBG-based in-line dispersion compensation, IEEE Photonics Technol. Lett., Vol. 19, No. 14, 2007, pp. 1069–1071.

137. X. Shu et al., Single-reflection-band fiber Bragg gratings with channelized linear and nonlinear dispersion and their applications, OFC2007, Anaheim, March 2007, Paper JThA17.

138. C. Scheerer et al., Influence of filter group delay ripples on system performance, ECOC 1999, Nice, 1999, pp. I-410–I411.

139. T.N. Nielsen et al., Penalties associated with group delay imperfections for NRZ, RZ and duo-binary encoded optical signals, ECOC 1999, Nice, 1999, pp. I-388–I-389.

140. X. Liu et al., Impact of group-delay ripple in transmission systems including phase-modulated formats, IEEE Photonics Technol. Lett., Vol. 16, No. 1, 2004, pp. 305–307.

141. Y. Painchaud et al., Low-penalty cascade of low-ripple FBG-based, dispersion compensators, ECOC 2006, Cannes, September 2006, Paper Th4.2.7.

142. C. Xie, Suppression of inter-channel nonlinearities in WDM coherent PDM-QPSK systems using periodic-group-delay dispersion compensators, ECOC 2009, Vienna, September 2009, Paper P4.08.

143. E.H. Turner and R.H. Stolen, Fiber Faraday circulator or isolator, Opt. Lett., Vol. 6, No. 7, 1982, pp. 322ff.

144. G. Bellotti et al., Dependence of self-phase modulation impairments on residual dispersion in 10 Gb/s based terrestrial transmission using standard fiber, IEEE Photonics Technol. Lett., Vol. 11, No. 7, 1999, pp. 824–826.

145. J. Sun et al., Thermally tunable dispersion compensator in 40-Gb/s system using FBG fabricated with linearly chirped phase mask, Opt. Express, Vol. 14, No. 1, 2006, pp. 44–49.

146. ClearSpectrum™: TDC Tunable Dispersion Compensator, Datasheet, TeraXion, 2004.

147. C.R. Doerr et al., 40-Gb/s colorless tunable dispersion compensator with 1000-ps/nm tuning range employing a planar lightwave circuit and a deformable mirror, OFC 2005, Anaheim, March 2005, PDP5.

148. I.P Kaminow, T. Li, and A.E. Willner, Optical Fiber Telecommunications, A. Components and Subsystems, Academic Press, San Diego, 2008.

149. G.-H. Lee et al., 1700 ps/nm tunable dispersion compensation for 10 Gbit/s RZ lightwave transmission, Electron. Lett., Vol. 42, No. 13, June 2006, pp. 768–769.

150. S. Ramachandran et al., Tunable dispersion compensators utilizing higher order mode fibers, IEEE Photonics Technol. Lett., Vol. 15, No. 5, 2003, pp. 727–729.

151. L.M. Lunardi et al., Tunable dispersion compensation at 40-Gb/s using a multicavity etalon all-pass filter with NRZ, RZ, and CS-RZ modulation, IEEE J. Lightwave Technol., Vol. 20, No. 12, 2002, pp. 2136–2144.

152. A. Ohta et al., 43-Gbps RZ-DQPSK transponder for long-haul optical transmission system, Yokogawa Techn. Rep., English Ed., No. 46, 2008, pp. 3–6.

153. J.M. Senio and M.Y. Jamro, Optical Fiber Communications: Principles and Practice, 3rd ed., Pearson Education Ltd., Harlow, 2009.

154. G. Hernandez, Fabry–Perot Interferometers, University Press, Cambridge, 1986.

155. C. Dragone, An NxN optical multiplexer using a planar arrangement of two star couplers, IEEE Photonics Technol. Lett., Vol. 3, No. 9, 1991, pp. 812–815.

156. C. Dragone, Low-loss wavelength routers for WDM optical networks and high-capacity IP routers, IEEE J. Lightwave Technol., Vol. 23, No. 1, 2005, pp. 66–79.

157. R. Ramaswami et al., Optical Networks. A Practical Perspective, 3rd ed., Morgan Kaufmann, Burlington, MA, 2010.

158. L. Grave de Peralta et al., Temperature-insensitive reflective arrayed-waveguide grating multiplexers, IEEE Photonics Technol. Lett., Vol. 16, No. 3, 2004, pp. 831–833.

159. T. Saito et al., Temperature-insensitive (Athermal) AWG modules, Furukawa Rev., No. 24, 2003, pp. 29–33.

160. R. Delmdahl, Compact excimer lasers in FBG production, The Laser User, Issue 63 2011, p. 8.

161. G.N. Smith et al., Demonstration of inscription and ablation of phase masks for the production of 1st, 2nd and 3rd order FBG gratings using a femtosecond laser, Proc. SPIE, Vol. 7205, 2009, pp. 1–11.

162. A.M. Vengsarkar et al., Long-period fiber Bragg gratings as band-rejection filters, IEEE J. Lightwave Technol., Vol. 14, No. 1, 1996, pp. 58–65.

163. M. Harurnoto et al., Gain-flattening filter using long-period fiber gratings, IEEE J. Lightwave Technol., Vol. 20, No. 6, 2002, pp. 1027–1033.

164. D. Sadot and E. Boimovich, Tunable optical filters for dense WDM networks, IEEE Commun. Mag., Vol. 36, No. 12, December 1998.

165. H.-K. Lee et al., A wide-range linearly tunable optical filter using Lorentz force, IEEE Photonics Technol. Lett., Vol. 16, No. 9, September 2004.

166. H. Zhang et al., Novel tunable liquid crystal Fabry–Perot filters for fiber-optical system, Proc. SPIE, Vol. 4583, 2001, doi: 10.1117/12.445110 (W. Gu, J. Zhou, J.-Y. Pan (eds), Metro and Access Networks, Vol. 64).

167. D. Hohlfeld et al., Tunable thermo-optic filter for WDM applications, Proc. IEEE, 2002, pp. 564–567.

168. T. Nakazawa et al., Ti:LiNbO$_3$ acousto-optic tunable filter (AOTF), Fujitsu Sci. Tech. J., Vol. 35, No. 1, July 1999, pp. 107–112.

169. M. Ferchichi et al., Design of temperature-strain tunable UDWDM, DWDM, WDM FBG filter for passive optical network access, 2nd ICTON—Mediterranean Winter 2008, Marrakech, Morocco, December 2008, Fr1A.5.

170. M. Lequime., Tunable thin-film filters: review and perspectives, Proc. SPIE, Vol. 5250 February 2004, doi: 10.1117/12.516026 (Advances in Optical Thin Films, ISBN 9780819451347).

171. F. Havermeyer et al., Volume holographic grating-based continuously tunable optical filter, SPIE J. Opt. Eng., Vol. 43, No. 9, September 2004.

172. Fiber Fabry–Perot Tunable Filter | FFP-TF2, Data Sheet, Micron Optics, Inc. 2008. Available at http://www.micronoptics.com/uploads/library/documents/Datasheets/Micron%20Optics%20FILTERS%20-%20FFP-TF2%20%20-%200812.1.pdf.

173. Thin-film Variable Filters, Data Sheet, Lightwaves2020, 2010. Available at http://www.lightwaves2020.com/publications/lightvision/newsletter-Feb-2010-4p.pdf.

174. R. Murano et al., Tunable 2.5 Gb/s Receiver for Wavelength-Agile DWDM-PON, OFC2008, San Diego, March 2008.

175. R. Murano et al., Tunable GPON receivers enable phased migration to 1 Gb/s per subscriber, OFC2009, San Diego, March 2009.

176. F. Riemenschneider et al., Low-cost electrothermally tunable optical microcavities based on GaAs, IEEE Photonics Technol. Lett., Vol. 14, No. 11, November 2002. pp. 1566–1568.

177. C. Marxer et al., Vertical mirrors fabricated by deep reactive ion etching for fiber optic switching applications, IEEE J. Microelectromech. Syst., Vol. 6, No. 3, 1997, pp. 277–285.

178. J.-C. Chiao, Liquid-Crystal Optical Switches, Photonics in Switching OSA Topical Meetings, Invited Paper, June 2001.

179. S. Hengstler et al., Laser-activated optical bubble switch element, IEEE/LEOS International Conference on Optical MEMS, 2003, Paper WB3.

180. M. Hoffman et al., Thermooptical digital switch arrays in silica on silicon with defined zero voltage state, IEEE J. Lightwave Technol., Vol. 16, No. 3, 1998, pp. 395–400.
181. A.J. Agranat et al., The electroholographic optical switch, Proceedings of ECOC'99, Vol. I, September 1999, pp. 334–335.
182. D.A. Smith et al., Multiwavelength performance of an apodized acousto-optic switch, IEEE J. Lightwave Technol., Vol. 14, No. 9, 1996, pp. 2044–2051.
183. F. Dorgeuille et al., Novel approach for simple fabrication of high-performance InP-switch matrix based on laser-amplifier gates, IEEE J. Lightwave Technol., Vol. 14, No. 9, September 1996.
184. M.C. Wu, Current Trends in Optical MEMS, Tutorial, OFC2005, 2005.
185. JDSU Data Sheet: 40 Channel Integrated PLC Reconfigurable Optical Add/Drop Multiplexer. Available at www.jdsu.com/ProductLiterature/40plcroadm_ds_cms_ae_020806 .pdf.
186. P. Evans et al., LCOS-based WSS with true integrated channel monitor for signal quality monitoring applications in ROADMs, OFC2008, 2008, Paper OWC3.pdf.
187. Y. Sakurai et al., LCOS-based gridless wavelength blocker array for broadband signals at 100 Gbps and beyond, OFC2012, Los Angeles, March 2012, Paper OTh3D.
188. P. Roorda and B. Collings, Evolution to colorless and directionless ROADM architectures, OFC/NFOEC2008, 2008, NWE2.pdf.
189. Evolution of ROADM, ZTE Technologies, Issue 63, No. 3, 2005.
190. S. Perrin, Building a Fully Flexible Optical Layer with Next-Generation ROADMs, Heavy Reading White Paper, October 2011. Available at www.jdsu.com/ProductLiterature/HR-ROADM-2011-Final.pdf.
191. R. Jensen et al., Colourless, directionless, contentionless ROADM Architecture using low-loss optical matrix switches, ECOC2010, Torino, September 2010, Mo.2.D.2.
192. A. Devarajan et al., Colorless, directionless and contentionless multi-degree ROADM architecture for mesh optical networks, Second International Conference on Communication Systems and Networks (COMSNETS), January 2010.
193. S. Tibuleac and M. Filer, Transmission impairments in DWDM Networks with ROADMs, IEEE J. Lightwave Technol., Vol. 28, No. 4, 2010, pp. 557–568.
194. H. Griesser et al., Quaternary duobinary modulation for superchannels with sub-nyquist subcarrier spacing, OFC2012, 2012, OTh1B.4.
195. A.A.M. Saleh and J.M. Simmons, Architectural principles of optical regional and metropolitan access networks, IEEE J. Lightwave Technol., Vol. 17, No. 12, 1999, pp. 2431–2448.
196. H. Liu et al., The design of a novel multistage OXC for large-scale optical networks, Proc. SPIE, Vol. 4910, 2002, pp. 197–202.
197. M. Mizukami et al., Development of automated optical fiber cross-connect equipment, NTT Tech. Rev., Vol. 2, No. 11, November 2004.
198. M. Mizukami et al., 200 × 200 automated optical fiber cross-connect equipment using a fiber-handling robot for optical cabling systems, OFC2005, 2005, Paper OFP5.
199. L. Zong et al., Low crosstalk structure for integrated OXC/OADM in WDM optical transport networks, Opt. Commun., Vol. 195, 2001, pp. 179–186.

200. L. Wosinska et al., Large-capacity strictly nonblocking optical cross-connects based on microelectrooptomechanical systems (MEOMS) switch matrices: reliability performance analysis, IEEE J. Lightwave Technol., Vol. 19, No. 8, 2001, pp. 1065–1075.

201. C. Clos, A study of non-blocking switching networks, Bell Syst. Tech. J., Vol. 32, No. 2, March 1953, pp. 406–424.

202. J. Leuthold, 100 Gb/s all-optical wavelength conversion with integrated SOA delayed-interference configuration, Electron. Lett., Vol. 36, No. 13, June 2000, pp. 1129–1130.

203. C. Joergensen et al., All-optical wavelength conversion at bit rates above 10 Gb/s using semiconductor optical amplifiers, Invited Paper, IEEE J. Sel. Top. Quantum Electron., Vol. 3, No. 5, 1997, pp. 1168–1180.

204. V.G. Ta'eed et al., All optical wavelength conversion via cross phase modulation in chalcogenide glass rib waveguides, Opt. Express, Vol. 14, No. 23, November 2006, pp. 11242–11247.

205. S. H. Wang et al., All-optical wavelength conversion using multi-pump Raman-assisted four-wave mixing, OFC2007, Anaheim, March 2007, Paper OWQ1.pdf.

206. J.E. Bowers and Y.G. Wey, High-speed photodetectors, Chapter 17, in Handbook of Optics, 2nd ed., M. Bass (Editor in Chief), McGraw-Hill, New York, 1995.

207. S. Katsutoshi et al., Development of high-speed photodiode modules for optical communications, Yokogawa Tech. Rep. English Ed., No. 24, 1997, pp. 32–35.

208. H.-G. Bach, Ultra-broadband photodiodes and balanced detectors towards 100 Gb/s and beyond, Invited Paper, Proc. SPIE, Vol. 6014, 2005, pp. 1–13.

209. H.-G. Bach, Ultra high-speed photodetectors and photoreceivers for telecom and datacom also aiming at THz applications, Proceedings of ECIO 2007, Copenhagen, April 2007, Paper FB0.

210. M.S. Islam and M.C. Wu, Recent advances and future prospects in high-speed and high-saturation-current photodetectors, Invited Paper, Proc. SPIE, Vol. 5246, 2003, pp. 448–457.

211. T. Viswanathan, Telecommunication Switching Systems and Networks, Prentice Hall of India, New Delhi, 2006 (26th Printing).

212. M. Azadeh, Fiber Optics Engineering, Springer, Dordrecht, Heidelberg, 2009.

213. J.M. Senior and M.Y. Jamro, Optical Fiber Communications: Principles and Practice, 3rd ed., Pearson Education Ltd., Harlow, 2009.

214. K. Ogasawara et al., The effect of depletion layer thickness in avalanche photodiodes for measurement of low-energy electrons, Nucl. Instrum. Methods Phys. Res. A, Vol. 566, No. 2, October 2006, pp. 575–583.

215. A. Rogalski, Fundamentals of Infrared Detector Technologies, CRC Press, Boca Raton, 2011.

216. K.W. Kobayashi, State-of-the-art 60 GHz, 3.6 K-Ohm transimpedance amplifier for 40 Gb/s and beyond (Invited), IEEE Radio Frequency Integrated Circuits (RFIC) Symposium, June 2003, pp. 55–58.

217. J.-M. Lee et al., Design and fabrication of wideband transimpedance amplifier by using InGaAs/InP HBT technology, J. Kor. Phys. Soc., Vol. 45, December 2004, pp. S906–S908.

218. H.-Y. Huang et al., A 10-Gb/s inductorless CMOS limiting amplifier with third-order interleaving active feedback, IEEE J. Solid-State Circuits, Vol. 42, No. 5, 2007, pp. 1111–1120.

219. H.-L. Chen et al., Inductorless CMOS receiver front-end circuits for 10-Gb/s optical communications, Tamkang J. Sci. Eng., Vol. 12, No. 4, 2009, pp. 449–458.
220. G.H. Rieke, Detection of Light: From the Ultraviolet to the Submillimeter, 2nd ed., Cambridge University Press, Cambridge, 2003.
221. H. Zimmermann, Silicon Optoelectronic Integrated Circuits, Springer, Berlin, 2004.
222. P. Müller and Y. Leblebici, CMOS Multichannel Single-Chip Receivers for Multi-Gigabit Optical Data Communications, Springer, Dordrecht, 2007.
223. J. Bird, Electrical and Electronic Principles and Technology, 4th ed., Newnes, Oxford, UK, 2010.
224. T. Wang and B. Erhman, Compensate Transimpedance Amplifiers Intuitively, Texas Instruments Application Report SBOA055A, Revised March 2005.
225. B. Razavi, Design of Analog CMOS Integrated Circuits, McGraw-Hill, 2001.
226. B. Razavi, Design of Integrated Circuits for Optical Communications, McGraw-Hill, New York, 2003.
227. M.B. Das et al., Designing optoelectronic integrated circuit (OEIC) receivers for high sensitivity and maximally flat frequency response, IEEE J. Lightwave Technol., Vol. 13, No. 9, 1995, pp. 1876–1884.
228. Q. Xi et al., A 0.8–3 GHz RF-VGA with 35 dB dynamic range in 0.13 μm CMOS, J. Semicond., Vol. 33, No. 1, January 2012, pp. 015009-1-015009-6.
229. J. Han et al., A 20-Gb/s transformer-based current-mode optical receiver in 0.13-μm CMOS, IEEE Trans. Circuits Syst. II: Express Briefs, Vol. 57, No. 5, 2010, pp. 348–352.
230. R. Arima et al., Demonstration of world-first 112 Gb/s 1310 nm LAN-WDM optical transceiver for 100GbE and 100GbE over OTN applications, OFC/NFOEC 2010, San Diego, March 2010, Paper PDPD3.
231. ADVA AG Optical Networking: Specification of 100-Gb/s metro card, 2012.
232. OIF Electrical Interface Implementation Agreements. Available at http://www.oiforum.com/public/impagreements.html#EI.
233. J.D.H. Alexander, Clock recovery from random binary signals, Electron. Lett., Vol. 11, No. 22, 1975, pp. 541ff.
234. S. Lin and D.J. Costello, Error Control Coding, 2nd ed., Prentice Hall, Englewood Cliffs, 2004.
235. R.W. Hamming, Error detecting and error correcting codes, Bell Syst. Tech. J., Vol. 29, No. 2, 1950, pp. 147–160.
236. F. Gray, Pulse code communication, March 17, 1953 (filed November 1947), U.S. Patent 2,632,058.
237. L.G. Kazovsky et al., Broadband Optical Access Networks, John Wiley & Sons, Inc., Hoboken, 2011.
238. F.J. MacWilliams and N.J.A. Sloane, The Theory of Error Correcting Codes, North-Holland, Amsterdam, 1977.
239. T. Sugihara, Reality check: implementation of soft-decision FEC in a DSP LSI, WS1: DSP & FEC: towards the Shannon limit, ECOC2009, Vienna, September 2009.
240. H. Jiang et al., SiGe equalizer IC for PMD mitigation and signal optimization of 40 Gb/s transmission, OFC2005, 2005, Paper OWO2.

241. R.I. Killey et al., Electronic dispersion compensation by signal predistortion using digital processing and a dual-drive Mach–Zehnder modulator, IEEE Photonics Technol. Lett., Vol. 17, No. 3, 2005, pp. 714–716.

242. S.J. Savory, Electronic signal processing in optical communications, Invited Paper, Proc. SPIE, Vol. 7136, November 2008 pp. 2C-1–2C-9.

243. A. Klekamp et al., Nonlinear limitations of electronic dispersion pre-compensation by intrachannel effects, OFC2006, 2006, Paper OWR1.pdf.

244. J. Proakis, Digital Communications, 4th ed., McGraw Hill, 2000.

245. B. Widrow and M.E. Hoff, Adaptive switching circuits. IRE WESCON Convention Record, Vol. 4, Los Angeles, 1960, pp. 96–104.

246. J. Leibrich and W. Rosenkranz, Frequency domain equalization with minimum complexity in coherent optical transmission systems, OFC2010, San Diego, March 2010, Paper OWV1.

247. J.W. Cooley and J.W. Tukey, An algorithm for the machine calculation of complex Fourier series, Math. Comput., Vol. 19, April 1965, pp. 297–301.

248. M. Vetterli and P. Duhamel, Split-radix algorithms for length-p^m DFTs, IEEE Trans. Acoust., Speech Signal Process., Vol. 37, No. 1, 1989, pp. 57–64.

249. S.G. Johnson and M. Frigo, A modified split-radix FFT with fewer arithmetic operations, IEEE Trans. Signal Process., Vol. 55, No. 1, 2007, pp. 111–119.

250. J.F. Hayes, The Viterbi algorithm applied to digital data transmission, IEEE Commun. Mag., Vol. 13, No. 2, March 1975, re-published in IEEE Commun. Mag., 50th Anniversary Commemorative Issue, May 2002, pp. 26–32.

251. G.D. Forney, Maximum likelihood sequence estimation of digital sequences in the presence of intersymbol interference, IEEE Trans. Inform. Theory, Vol. 18, No. 3, May 1972. pp. 363–378.

252. G.D. Forney, The Viterbi algorithm, Proc. IEEE, Vol. 61, No. 3, March 1973. pp. 268–278.

253. A. Färbert et al., Performance of a 10.7 Gb/s receiver with digital equaliser using maximum likelihood sequence estimation, ECOC2004, Stockholm, 2004, PDTH415.

254. T. Kupfer, Measurement of the performance of 16-states MLSE digital equalizer with different optical modulation formats, OFC2008, San Diego, February 2008, PDP13.

4

NONFIBER-RELATED EFFECTS

In Chapter 2, transmission effects related to the transmission fiber itself were discussed. In addition to these effects, the components along a link cause further distortions of the wavelength division multiplexing (WDM) signals. Two of these effects not covered so far are linear cross talk and (accumulated) noise. They are discussed in more detail in this chapter.

4.1 LINEAR CROSS TALK

Linear cross talk originates in WDM terminals (receivers) from imperfect demultiplexing of the multichannel signal into individual channels or faulty routing of certain channels prior to detection in single-channel receivers. We refer to these effects as linear cross talk in order to separate them from cross talk caused by fiber or amplifier nonlinearity. An overview of this effect can also be found in ITU-T G-series Supplement 39.

Two cross talk mechanisms can be separated:

- *Interchannel Cross Talk*: Ratio of total power in the disturbing channels to that in the wanted channel. Wanted and disturbing channels have different wavelengths.
- *Interferometric or Intrachannel Cross Talk*: Ratio of the disturbing power [not including amplified spontaneous emission (ASE) noise] to the wanted power *within* a single channel (wavelength).

Wavelength Division Multiplexing: A Practical Engineering Guide, First Edition.
Klaus Grobe and Michael Eiselt.

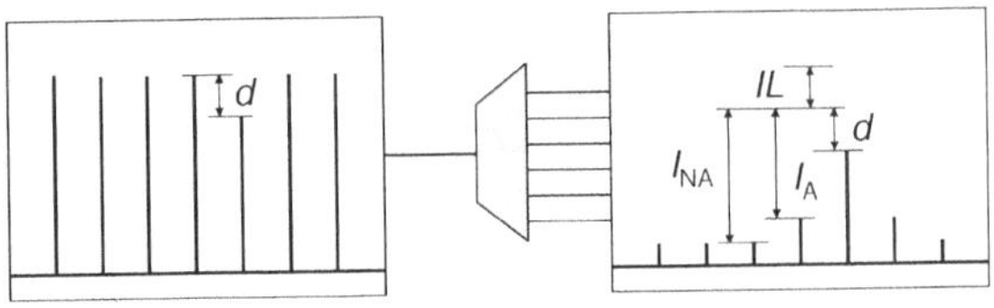

FIGURE 4.1 Demultiplexer example.

Interchannel cross talk occurs when several dense wavelength division multiplexing (DWDM) channels are demultiplexed in a receive-end terminal or in an optical add–drop multiplexer (OADM). The worst case for any channel under consideration is for its power to be at the minimum and the power of all other channels to be at the maximum. The maximum allowable difference between channels (which is to be specified by the respective system vendor) is denoted d (dB). At the demultiplexer output port under consideration, the disturbing channels have been attenuated with respect to the wanted channel by the (unidirectional and adjacent) filter isolation I (dB).

In practical demultiplexers (e.g., AWGs), the isolation of the channels adjacent to the wanted channel I_A is smaller than the isolation of the nonadjacent disturbing channels I_{NA}, $I_A < I_{NA}$. This situation is illustrated in Fig. 4.1. Here, IL denotes the filter insertion loss.

The interchannel cross talk C_C in a k-channel WDM system is given by

$$C_C = d + 10\log_{10}\left(2 \times 10^{-I_A/10} + (k-3) \times 10^{-I_{NA}/10}\right)\ (\text{dB}). \qquad (4.1)$$

The cross talk penalty P_C for the channel under consideration and a single disturbing channel, with r the *linear* extinction ratio (ER), yields

$$P_C = 10\log_{10}\left(1 - 10^{C_C/10}\frac{r+1}{r-1}\right)\ (\text{dB}). \qquad (4.2)$$

This equation does not yet include eye-opening reductions due to transmitter eye closure or fiber transmission effects. These effects can be included by considering an *effective* extinction ratio r' that takes extinction ratio and eye closure into account:

$$r' = \frac{(r+1) + 10^{(-E/10)}(r-1)}{(r+1) - 10^{(-E/10)}(r-1)}. \qquad (4.3)$$

Here, E is the eye closure penalty in decibels. If the extinction ratio is 6 dB, then $r = 3.98$ results. With an additional eye closure penalty of 3 dB, $r' = 1.86$ results.

The optical penalty is plotted as a function of single-interferer interchannel cross talk for a variety of assumptions regarding eye closure and ER in Fig. 4.2. Penalties for actual systems lie somewhere below the highest curve.

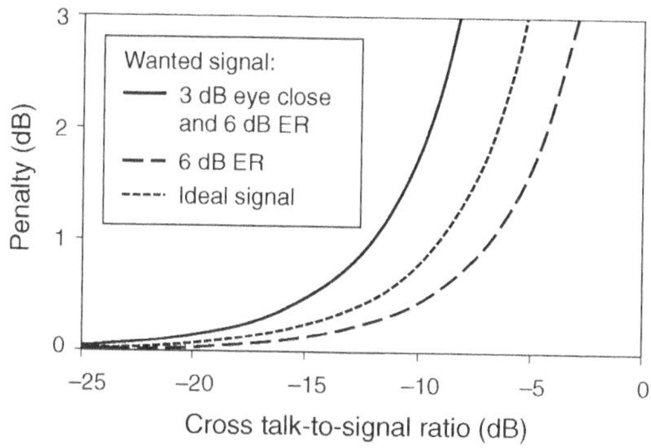

FIGURE 4.2 Optical penalty versus interchannel cross talk from single interferer.

For a large number of disturbing channels with equal amplitude and uncorrelated data, the interchannel cross talk becomes noise-like, and a Gaussian approximation can be used. The resulting cross talk must be convolved with the noise distribution of the receiver or the ASE generated along the transmission link to produce an effective penalty. This yields [1,2]

$$P_C = -5 \log_{10}\left(1 - \frac{10^{2C_C/10}}{k-1} Q^2 \left(\frac{r+1}{r-1}\right)^2\right) \text{ (dB)}, \tag{4.4}$$

where $Q = \sqrt{2}\,\text{erfc}^{-1}(2 \times \text{BER})$ (see Section 5.3.1).

For multiple interferers, the penalty becomes smaller than the single-interferer ideal signal penalty shown in Fig. 4.2 for $C_C \approx -4$ dB and higher for larger cross talk. The penalty can also depend on RZ versus NRZ line coding and the relative bit rates of the interfering signals.

The required filter isolation can be derived according to ITU-T G-series Supplement 39:

- Specify a system-specific value for P_C. As an example, we choose $P_C = 0.5$ dB.
- Derive C_C from P_C. The model required is somewhere between that of two interfering signals with large difference between I_A and I_{NA} through to a Gaussian model when $I_A \approx I_{NA}$ and k is large. In our example, a worst case of $C_C = -15$ dB follows.
- Specify the system-specific value for d. This parameter can cover broad ranges, for example, 2–5 dB. Here, we use $d = 3$ dB. Also, specify channel count k. Here, we set $k = 40$.
- With $C_C = d - I + 10 \log 10 (k-1)$, the (worst-case) result $I = 31.8$ dB can be derived.

Interferometric cross talk occurs when the disturbing channel and the wanted channel are at the same nominal wavelengths. Obviously, the disturbing channel cannot be suppressed with WDM filters. Interferometric cross talk can occur:

- In an OADM, where a wavelength is incompletely dropped before the new signal is added.
- In an optical terminal multiplexer, where a (mistuned) transmitter emits at the wavelength of another channel. In particular, this can occur if groups of channels are combined with power splitters rather than filters. This effect is referred to as *transmit-side cross talk* in ITU-T Recommendation G.692.
- In an OXC or multidegree ROADM with poor isolation, where light from more than one trunk fiber can reach the respective receive port.
- In any component, subsystem, or network element with more than one light path that connects to the receiver. [This can again include optical cross-connects (OXCs) or reconfigurable optical add–drop multiplexers (ROADMs).] This effect is also called multipath interference (MPI).
- Intentionally, which meanwhile is also referred to as cross talk attack [3].

Interferometric cross talk behaves differently compared to interchannel cross talk between two WDM signals spaced such that their beat frequency falls within the electrical receiver bandwidth. Now, optical fields rather than the related intensities generate the cross talk. Consequently, lower cross talk levels lead to particular penalties.

For a single interferer, the cross talk can be modeled using a bounded probability density function (PDF). The cross talk penalty, including imperfect extinction ratio and considering an average-power decision threshold, becomes [4]

$$P_\mathrm{I} = 10 \log_{10} \left(\frac{(r-1)/(r+1)}{[(r-1)/(r+1)] + 10^{C_\mathrm{I}/10} - 4\sqrt{(r/r+1) \times 10^{C_\mathrm{I}/10}}} \right) \text{ (dB)}. \tag{4.5}$$

For optimized decision threshold, one gets

$$P_\mathrm{I} = -10 \log_{10} \left(1 - 2 \left(\frac{(1+\sqrt{r})\sqrt{10^{C_I/10}(r+1)}}{r-1} \right) \right) \text{ (dB)}. \tag{4.6}$$

The interferometric cross talk penalty is shown in Fig. 4.3 for a single interferer.

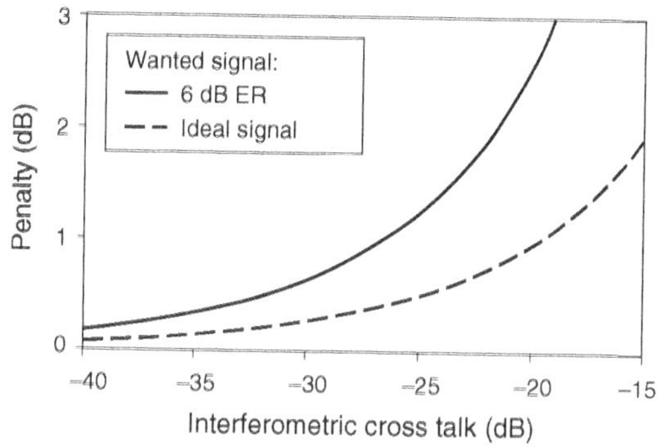

FIGURE 4.3 Optical penalty versus interferometric cross talk (single interferer, optimum decision threshold, and bounded model).

For multiple interferers, the related PDF again becomes approximately Gaussian. For a PIN receiver, the optical cross talk penalty for average-power decision threshold is given by [2]

$$P_\mathrm{I} = -5\log_{10}\left(1 - 4 \times 10^{C_\mathrm{I}/10} Q'^2 \frac{1 + (1/r)}{(1 - (1/r))^2}\right)\,(\mathrm{dB}). \tag{4.7}$$

For optimized decision threshold, one gets

$$P_\mathrm{I} = -5\log_{10}\left(1 - 2 \times 10^{C_I/10} Q^2 \left(\frac{r+1}{r-1}\right)^2 + \left(10^{C_I/10}\right)^2 Q^4 \left(\frac{r+1}{r-1}\right)^2\right)\,(\mathrm{dB}). \tag{4.8}$$

Here, $Q' = \sqrt{2}\,\mathrm{erfc}^{-1}\,(4 \times \mathrm{BER})$ and $Q = \sqrt{2}\,\mathrm{erfc}^{-1}\,(2 \times \mathrm{BER})$. For $\mathrm{BER} = 10^{-12}$, $Q' \approx 6.94$ and $Q \approx 7.03$ follows. Interferometric multi-interferer cross talk may have to be reduced by 2–4 dB compared to single-interferer cross talk for penalties ranging 0.5–1 dB.

Note that multi-interferer interferometric cross talk can be regarded a highly unlikely event that only results from severe faults or malicious attacks.

Alternative approaches to calculating intrachannel or interferometric cross talk have been reported in Refs [5–8].

4.2 NOISE IN OPTICAL TRANSMISSION SYSTEMS

Fiber and other components' loss leads to decreasing signal power levels. On long transmission links, this loss has to be compensated by optical amplifiers that can almost transparently amplify a multitude of WDM signals. Like all amplifiers, optical amplifiers add noise, called amplified spontaneous emissions [9]. Together with the unavoidable noise sources in receivers, this decreases the signal-to-noise ratio (SNR), which is a relevant measure for the achievable bit error rate (BER).

The SNR could be increased by increasing the signal power level. However, this approach is severely limited by nonlinear effects (in fibers and amplifiers) and by laser safety requirements. Therefore, other ways to improve the BER have to be utilized. These can make use of advanced modulation and detection techniques (as discussed in Section 5.3) and improved forward error correction (FEC) (see Section 3.4.2). In addition, lowest-noise amplifiers (Raman amplifiers, possibly parametric amplifiers) (see Section 3.2.4) can be used. Nonetheless, noise, and hence loss, is one of the ultimately limiting factors in long-haul transmission.

As long as the transmission is (at least, mainly) limited by noise, the resulting BER can be derived analytically (see Section 5.3). The BER is then often described as a function of the Q-factor, which can be derived from the statistical distribution of the received symbols. High-speed long-haul fiber-optic transmission in addition is disturbed by cross talk, dispersion, and nonlinear effects, which altogether cannot be treated analytically anymore. Together, these effects can only be approximated, as a penalty on the Q-factor.

In the following, the Q-factor analysis of different optical receivers [PIN, avalanche photodiodes (APDs), and added optical preamplifier (OPA)] is given, followed by a brief analysis of the noise figures of (chains of) optical amplifiers [Erbium-doped fiber amplifier (EDFA) and EDFA with added Raman amplifier].

4.2.1 Noise in Optical Receivers

In order to derive the SNR that a photodiode generates, the primary noise sources must be considered. These are quantum (shot) noise and dark current noise. Quantum noise arises from the statistical nature of the production and collection of photo-generated electrons upon optical illumination. Its statistics follow a Poisson process.

Dark current noise is generated by the current that continues to flow through the bias circuit in the absence of light. Dark current is the combination of *bulk dark current*, which consists of thermally generated electrons and holes in the p–n junction, and the *surface dark current*, which is generated by ohmic conduction (i.e., the bias voltage) and surface defects. Dark current usually is no significant noise source at high bandwidths for PIN photodiodes, but it can become significant in APDs.

Further relevant noise sources in a photodiode receiver include the thermal noise of the electronic receiver [10] and the ASE from optional optical preamplifiers. Consequently, PIN-based, APD-based, and optically preamplified receivers must be considered separately.

The photocurrent I_p in a photodetector is proportional to the incident optical power P_{in} and is given by $I_P(t) = RP_{in}(t)$ (see Eq. 3.53). Due to noise, I_p is time dependent even if $P_{in} = \text{const}$.

We now consider modulated optical power $P_{in}(t)$ falling on a PIN photodetector:

$$P_{in}(t) = P_0[1 + ms(t)]. \tag{4.9}$$

Here, $s(t)$ is the electrical signal that is to be transmitted and m is the modulation index. The primary photocurrent then gets (where for PIN photodiodes, $M = 1$)

$$i_{\text{Ph}} = \frac{\eta q}{h\nu} M P_{\text{in}}(t) = I_{\text{P}}[\text{DC}] + i_{\text{P}}(t)[\text{AC}]. \tag{4.10}$$

The root mean square (RMS) *signal* current is given by

$$\langle i_{\text{s}}^2 \rangle = \sigma_{\text{s}}^2 = \langle i_{\text{P}}^2 \rangle M^2. \tag{4.11}$$

For sinusoidal signals, σ_{P}^2 is given by $\langle i_{\text{P}}^2 \rangle = \sigma_{\text{P}}^2 = m^2 I_{\text{P}}^2/2$.

The total photodetector *noise* (first, without optical preamplifier) is given by the sum of the individual contributions from quantum noise, bulk dark current noise, and surface dark current noise. The total RMS photodetector noise current is

$$\langle i_{\text{N}}^2 \rangle = \sigma_{\text{N}}^2 = \left\langle i_{\text{Q}}^2 \right\rangle + \langle i_{\text{DB}}^2 \rangle + \langle i_{\text{DS}}^2 \rangle = 2q(I_{\text{P}} + I_{\text{D}})BM^2F(M) + 2qI_{\text{L}}B. \tag{4.12}$$

$\langle i_{\text{Q}}^2 \rangle$, $\langle i_{\text{DB}}^2 \rangle$, and $\langle i_{\text{DS}}^2 \rangle$ are the contributions from quantum noise (Q), bulk dark (DB) current noise, and surface dark (DS) current noise, respectively. Again, for PIN diodes, $M^2F(M) = 1$ [M and $F(M)$ get relevant for APDs], and I_{D} and I_{L} are the bulk dark current and the surface current, respectively. B is the effective noise bandwidth of the receiver. It depends on both the intrinsic photodetector bandwidth and other receiver components such as the electrical low-pass filter (LPF). In Eq. (4.12), the shot noise contributions are given by (with $M^2F(M) = 1$) [10–12]

$$\sigma_{\text{S}}^2 = 2qRP_{\text{in}}B. \tag{4.13}$$

In addition to noise sources of the photodetector, thermal noise of the amplifier that is connected to the photodiode might have to be considered [10,11]:

$$\langle i_{\text{T}}^2 \rangle = \sigma_{\text{T}}^2 = \frac{4k_{\text{B}}TB}{R_{\text{L}}}. \tag{4.14}$$

R_{L} is the matched load resistance of the amplifier, $k_{\text{B}} = 1.38 \times 10^{-23}$ J/K is the Boltzmann constant, T is the temperature, and B is the electronic (effective noise) receiver bandwidth.

Now, the SNR can be derived from Eqs. (4.11 and 4.12) as

$$\frac{S}{N} = \frac{\langle i_{\text{P}}^2 \rangle M^2}{2q(I_{\text{P}} + I_{\text{D}})BM^2F(M) + 2qI_{\text{L}}B + 4k_{\text{B}}TB/R_{\text{L}}}. \tag{4.15}$$

For weak photodetector input signals, thermal noise dominates shot noise. Then the thermal noise limit of a PIN-based receiver, considering the thermal noise of the load resistor only, is given by

$$\mathrm{SNR}_{\mathrm{PIN,therm}} = \frac{R_{\mathrm{L}}\sigma_{\mathrm{P}}^2}{4k_{\mathrm{B}}TB}. \tag{4.16}$$

This SNR has to be complemented (i.e., decreased) by the noise figure F_{N} of the electronic amplifier that follows the load resistor.

For large input signals, total noise is dominated by shot noise. The respective noise limit for a PIN receiver is then given by

$$\mathrm{SNR}_{\mathrm{PIN,Shot}} = \frac{\sigma_{\mathrm{P}}^2}{2qI_PB}. \tag{4.17}$$

For APDs, the factor $M^2F(M)$ in Eq. (4.18) has to be considered. Since the noise figure $F(M)$ increases with avalanche gain M, there exists an optimum value of M that maximizes the SNR. For sinusoidally modulated signals with modulation index $m = 1$, $F(M) \approx M^x$ and $0 \leq x \leq 1$, one gets

$$M_{\mathrm{opt}}^{x+2} = \frac{2qI_L + 4k_BT/R_L}{xq(I_{\mathrm{P}} + I_{\mathrm{D}})}. \tag{4.18}$$

SNR in APD receivers is dominated by shot noise. The total shot noise (to be inserted in Eq. (4.17) for SNR calculation) is given by

$$\langle i_{\mathrm{N}}^2 \rangle = 2q(I_{\mathrm{P}} + I_{\mathrm{D}})BM^2F(M) + 2qI_{\mathrm{L}}B. \tag{4.19}$$

The excess noise factor $F(M)$ is given by [12]

$$F(M) = kM + (1-k)(2 - 1/M) \tag{4.20}$$

The range for the factor k is given by $0 < k < 1$. InGaAs detectors typically have $k = 0.7$.

As mentioned, typical PIN receivers are dominated by thermal noise. They can be combined with OPAs in order to increase the sensitivity. The OPA will generate beat noise between signal–spontaneous and spontaneous–spontaneous emissions (signal–ASE and ASE–ASE, respectively). When the OPA gain G is large, signal–ASE beat noise dominates over thermal and shot noise of the receiver. An optical receiver with OPA is shown in Fig. 4.4.

The most relevant noise contributions in a preamplified receiver are the signal–ASE beat noise $\sigma_{\mathrm{signal-ASE}}^2$ and the shot noise σ_{S}^2 as given by

$$\sigma_{\mathrm{signal-ASE}}^2 = 4R^2\,GP_{\mathrm{in}}P_{\mathrm{n}}(G-1)B, \quad \sigma_{\mathrm{S}}^2 = 2qR[GP_{\mathrm{in}} + P_{\mathrm{n}}(G-1)B_{\mathrm{o}}]B. \tag{4.21}$$

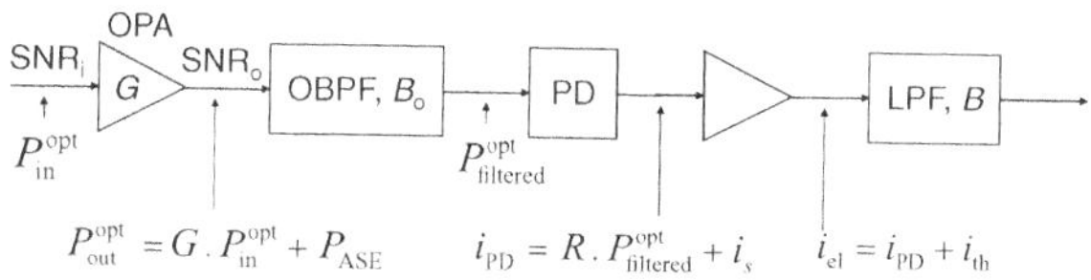

FIGURE 4.4 Optical receiver with optical preamplifier (OPA).

Here, B_o is the optical bandwidth according to Fig. 4.4, and P_n is given via the spontaneous noise power P_N, which is given by

$$P_N = n_{sp}h\nu(G-1)B_o = P_n(G-1)B_o. \tag{4.22}$$

h is the Planck's quantum, ν is the optical frequency, and n_{sp} is the population inversion factor, defined via the atomic populations N_1 and N_2 of the ground and excited states, respectively:

$$n_{sp} = N_2/(N_2 - N_1) \tag{4.23}$$

The population inversion factor has a value of $n_{sp} = 1$ for amplifiers with full population inversion, and is >1 for incomplete population inversion.

Since signal–ASE beat noise dominates, input SNR$_i$ and output SNR$_o$ according to Fig. 4.4 can be approximated. Then, the noise figure F_N of the preamplifier can be derived. This assumes a shot noise-limited source. Therefore, (far-end) laser noise is ignored, and receiver thermal noise is negligible. The signal/noise ratios are as follows:

$$\begin{aligned} \mathrm{SNR_i} &= \mathrm{SNR_{shot}} = \frac{\eta P_{in}}{2h\nu B} = \frac{(RP_{in})^2}{2qRP_{in}B}, \\ \mathrm{SNR_o} &= \frac{(GRP_{in})^2}{N_{signal-ASE} + N_{ASE-ASE} + N_{shot}} \approx \frac{(GRP_{in})^2}{4R^2GP_{in}P_n(G-1)B}. \end{aligned} \tag{4.24}$$

Dividing input SNR$_i$ and output SNR$_o$ yields the OPA noise figure:

$$F_N = \frac{\mathrm{SNR_i}}{\mathrm{SNR_o}} = \frac{(RP_{in})^2 \cdot 4R^2GP_{in}P_n(G-1)B}{2RqP_{in}B \cdot (GRP_{in})^2} \approx 2n_{sp}. \tag{4.25}$$

For large gain G, an approximate noise figure $F_N \approx 2n_{sp}$ results. Large gain requires high population inversion; hence, ideally $n_{sp} \approx 1$ and, consequently, the optimum OPA noise figure yields $F_{N,opt} \approx 2$. An ideal low-noise OPA has a noise figure of (only) 3 dB, and it helps avoiding the thermal noise limit of the photodetector.

4.2.2 Receiver Sensitivities

Sensitivity of an optical receiver is defined as the minimum average power that is required to achieve a BER of 10^{-12}. For OOK (see Section 5.3.1.1), we can write

$$\bar{P}_{\text{rec}} = (P_0 + P_1)/2 \approx P_1/2. \tag{4.26}$$

For BER $= 10^{-12}$, a Q-factor of $Q = 7$ is required (see Section 5.3.1.1). With $P_1 = Q \cdot (\sigma_0 + \sigma_1)/MR$ and $Q = (I_1 - I_0)/(\sigma_0 + \sigma_1) \approx I_1/(\sigma_0 + \sigma_1)$, one yields

$$\bar{P}_{\text{rec}} = Q \cdot (\sigma_0 + \sigma_1)/2MR. \tag{4.27}$$

With the expressions for thermal and shot noise:

$$\sigma_0^2 = \sigma_{\text{th}}^2, \quad \sigma_1^2 = \sigma_{\text{th}}^2 + \sigma_{\text{shot}}^2, \quad \sigma_{\text{shot}}^2 = 2qR2\bar{P}_{\text{rec}}BM^2F(M), \tag{4.28}$$

we can derive the sought-after required power [12]:

$$\bar{P}_{\text{rec}} = \frac{Q}{R} \cdot [qBF(M)Q + \sigma_{\text{th}}/M]. \tag{4.29}$$

APD gain reduces the effective thermal noise, but increases the effective shot noise. If we assume $\lambda = 1.55\ \mu\text{m}$, $R = 1.25$ A/W, $\eta = 1$, $T = 300$ K, $B = R_B/2$, $R_B = 10^9$ b/s, $F_N = 3$ dB (low-noise electrical front-end amplifier), and $R_L = 100\ \Omega$, then one gets

$$\sigma_{\text{th}}^2 = \frac{4k_BT}{R_L} F_N B = 1.656 \times 10^{-10} B\,(\mu\text{A})^2. \tag{4.30}$$

For PIN diodes with electrical front-end amplifier only, $M = 1$. Therefore,

$$\bar{P}_{\text{rec}} = Q \cdot \sigma_{\text{th}}/R = -26\ \text{dBm}. \tag{4.31}$$

For APD with $M = 10$ and $F(M) = 1.3$, one accordingly gets

$$\bar{P}_{\text{rec}} = Q \cdot \sigma_{\text{th}}/(RM) = -36\ \text{dBm}. \tag{4.32}$$

An APD gives an advantage of 10 dB in terms of sensitivity. In most practical devices, the sensitivity gain is close to these 10 dB.

For optical preamplifiers, one can neglect thermal noise:

$$Q = (I_1 - I_0)/(\sigma_0 + \sigma_1) \approx I_1/(\sigma_0 + \sigma_1) \approx I_1/\sigma_1. \tag{4.33}$$

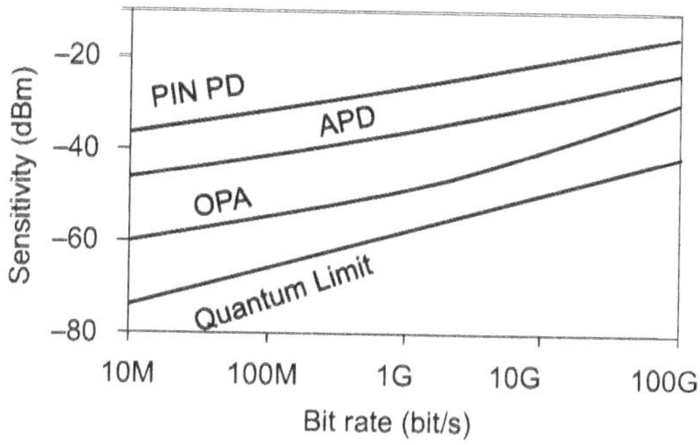

FIGURE 4.5 Optical receiver sensitivities.

With $I_1 = RGP_1$ and $\sigma_1^2 = 4R^2GP_1P_n(G-1)B$, one gets the required Q-factor:

$$Q = \frac{\sqrt{GP_1}}{2\sqrt{P_n(G-1)B}}. \tag{4.34}$$

For large gain G, this leads to $\bar{P}_{rec} = 2Q^2P_nB$, with $P_n = n_{sp}h\nu$ [see Eq. (4.22)]. For $n_{sp} = 1$, $\lambda = 1.55\,\mu m$, $R = 1.25$ A/W, $\eta = 1$, $T = 300$ K, $B = R_B/2$, and $R_B = 10^9$ b/s, we get

$$\bar{P}_{rec} = 98 \cdot h\nu \cdot B \approx -50\,\text{dBm}. \tag{4.35}$$

An optically preamplified receiver has a minimum sensitivity of 98 photons per bit. This result is valid for on–off keying (OOK). Practical receivers do not achieve these sensitivities. They are subject to penalties that depend on implementation. Results (for OOK) are compared in Fig. 4.5.

The achievable sensitivities heavily depend on the modulation and detection scheme. Sensitivity increases from direct detection via heterodyne detection to homodyne detection (see Section 5.6). In Ref. [13], the equivalence between optically preamplified and asynchronous heterodyne detections has been derived. Quantum-limited sensitivity of nine photons per bit is reached for coherent homodyne PSK reception (see Section 5.3.2).

4.2.3 Noise Figures and OSNR in Optical Amplifier Chains

Optical amplifiers can be described in terms of quantum-mechanical field operators, rate equations, photon multiplication, and field beating. The first two are quantum-mechanical treatments, while the third and fourth are semiclassical treatments [14,15]. The first and the last treatments consider the wave-like aspect of light and deal with the electric field at optical frequencies; the other approaches consider the particle-like nature of light and deal with photon number.

The quantum-mechanical treatments are exact. They determine the minimum amount of noise that an optical amplifier must add to the amplified signal not to violate quantum-mechanical commutation rules. The well-known result for the minimum noise figure of 3 dB is thus obtained [16]. However, these approaches become too complex for practical cases where realistic fiber links with cascaded optical amplifiers have to be considered.

The reduction to a semiclassical description is commonly used in electronics and photonics. The standard field beating theory [15,17] is based on the analysis, at the amplifier output, of beating terms generated by the square law process of photodetection of the amplified signal field and the ASE field.

The beatings between the signal and itself, signal and ASE, and ASE and itself correctly give the DC components and the dominant noise term of the photocurrent. However, the exact output noise expression is not obtained from the beatings. Shot noise terms are added heuristically *a posteriori*. Hence, the field beating theory can be regarded as inconsistent from a noise analysis viewpoint because at optical frequencies, quantum noise predominates over thermal noise. Nonetheless, it is the standard model for the description of practical WDM systems. In Ref. [14], an exact semiclassical, wave-like description is given.

For any long-haul transmission or WDM network with a multitude of optical amplifiers, it is necessary to know the receive-end optical signal-to-noise ratio (OSNR) that results from the amplifiers and the lossy components along the link under consideration. The OSNR is the signal/noise ratio that is delivered to the photodetector. It has obvious impact on the achievable SNR after photodetection. Typically, the OSNR is derived based on a simplified semiclassical view.

EDFAs produce amplified spontaneous emissions. The effect of ASE on amplifier chains is described by noise figures F_N, as already introduced in Eq. (4.25):

$$F_N = \frac{\mathrm{OSNR_i}}{\mathrm{OSNR_o}}. \tag{4.36}$$

The input SNR can be simplified for a shot noise-limited *source* [11]:

$$\mathrm{OSNR_i} = \frac{P_{\mathrm{in}}}{2h\nu B}. \tag{4.37}$$

This expression is similar to the one already obtained for an OPA [see Eq. (4.24)].

The output SNR can then be approximately described as follows [11]:

$$\mathrm{OSNR_o} = \frac{P_{\mathrm{Signal,out}}}{P_{\mathrm{Noise,out}}} = \frac{GP_{\mathrm{Signal,in}}}{4n_{\mathrm{sp}}h\nu(G-1)B} = \frac{P_{\mathrm{Signal,in}}}{F_N 2h\nu B}. \tag{4.38}$$

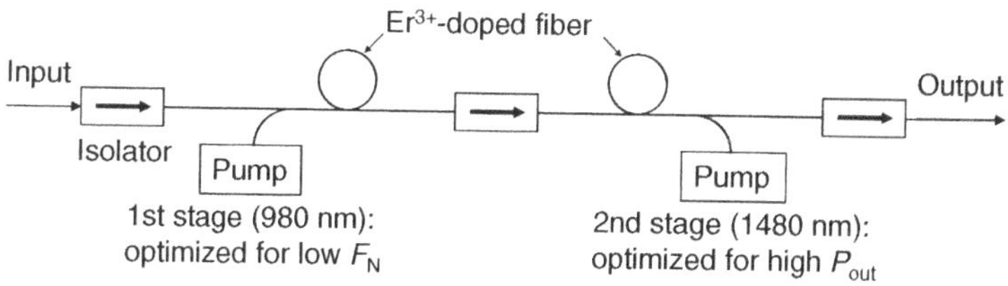

FIGURE 4.6 Dual-stage EDFA.

Using Eqs. (4.37) and (4.38), the following expression for the noise figure can be found:

$$F_{\mathrm{N}} = \frac{P_{\mathrm{ASE}}}{h\nu B_{\mathrm{o}} G} = 2n_{\mathrm{sp}}\frac{(G-1)}{G} \approx 2n_{\mathrm{sp}}. \tag{4.39}$$

In the first part of Eq. (4.39), the ASE power P_{ASE} has been used:

$$P_{\mathrm{ASE}} = 2n_{\mathrm{sp}}h\nu(G-1)B_{\mathrm{o}}. \tag{4.40}$$

Equation (4.39) shows that the OSNR is always decreased by at least 3 dB (or a factor of 2), even for ideal amplifiers with inversion factor $n_{\mathrm{sp}} = 1$. The noise figure can be better than 3 dB, but then the amplifier is driven well below its nominal saturation gain G.

Optical amplifiers can be cascaded. This is possible with or without lossy elements between the gain stages. A co- and counterpumped dual-stage EDFA is shown in Fig. 4.6.

For this configuration, the well-known Friis formula for the total noise figure F_{tot} of cascaded amplifiers applies [18,19]:

$$F_{\mathrm{tot}} = F_1 + (F_2 - 1)/G_1. \tag{4.41}$$

For sufficiently large gain G_1 of the first amplifier stage, the total noise figure is dominated by the first stage. The first EDFA stage should be pumped with 980 nm, as shown in Fig. 4.6. Pumping at 980 nm achieves better inversion compared to 1480 nm pumping, leading to noise figures of 980-nm-pumped EDFAs that are typically better by 1 dB [20].

Often, dual-stage EDFAs are used to accommodate lossy components (e.g., CD compensators and switches) between the two stages. This is shown in Fig. 4.7.

The midstage (MS) components can have loss L that cannot be neglected. Then, Eq. (4.41) needs to be modified accordingly. The result can be derived from the cascading formula, assuming the midstage block has noise figure $F_{\mathrm{MS}} = L$ and gain $G_{\mathrm{MS}} = 1/L$:

$$F_{\mathrm{tot}} = F_1 + (F_2 L - 1)/G_1. \tag{4.42}$$

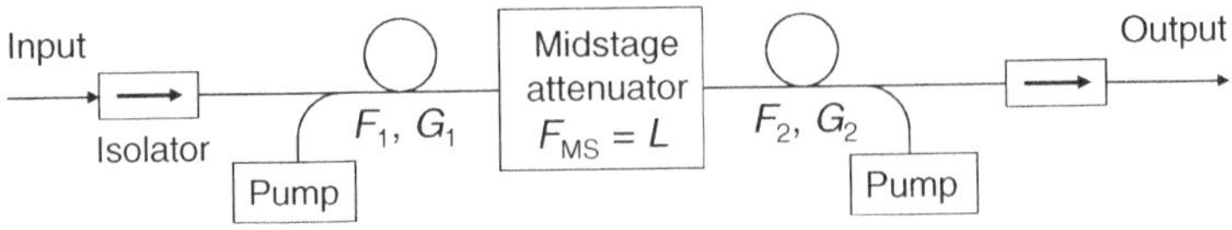

FIGURE 4.7 Dual-stage EDFA with midstage attenuator.

It becomes obvious that for small inserted loss L, the total noise figure is still dominated by the first amplifier stage. For large L, however, F_{tot} tends toward the sum of the noise figures of the two amplifier stages.

The generalization toward cascaded amplifiers in long-haul links is straightforward, combining Eqs. (4.41) and (4.42). The general configuration is shown in Fig. 4.8. Amplifiers have total noise figures (in case they accommodate lossy components in their midstage access) F_{Ni}, gains G_i, and the intermediate spans have attenuations A_i (mostly determined by αL_i, with α being the loss coefficient and L_i the respective span length).

The noise figure of a chain of amplifiers along the link is then given by

$$F_{\text{total}} \approx \frac{F_1}{A_1} + \frac{F_2}{A_1 G_1 A_2} + \cdots + \frac{F_n}{A_n G_1 A_1 \cdots G_{n-1} A_{n-1}}. \quad (4.43)$$

If $G_i \gg 1/A_i$, only the first amplifier dominates the receive-end noise behavior (which is the well-known result of Friis' formula [19]). If $G_i \leq 1/A_i$—which is the case especially for many metro and regional WDM systems—then F_{total} can be decreased significantly by the respective spans [16]. If all amplifiers have equal gain G_i, and all links have equal loss A_i with $G_i = 1/A_i$, one gets

$$F_{\text{total}} = nF_N/A_i. \quad (4.44)$$

The total loss figure is given by the sum of all contributions divided by the (mean) span loss.

The OSNR is often used when expressed in decibels. Useful expressions can be derived for both the receivers with an OPA and the cascades of amplifiers.

For the case of a single OPA, refer to Fig. 4.4. The linear OSNR can be expressed as

$$\frac{1}{\text{OSNR}_{\text{lin},B_{\text{o,ref}}}} = \frac{P_{\text{ASE}}}{G \cdot P_{\text{in}}^{\text{opt}}} = \frac{F_N h\nu B_{\text{o,ref}}}{P_{\text{in}}^{\text{opt}}}. \quad (4.45)$$

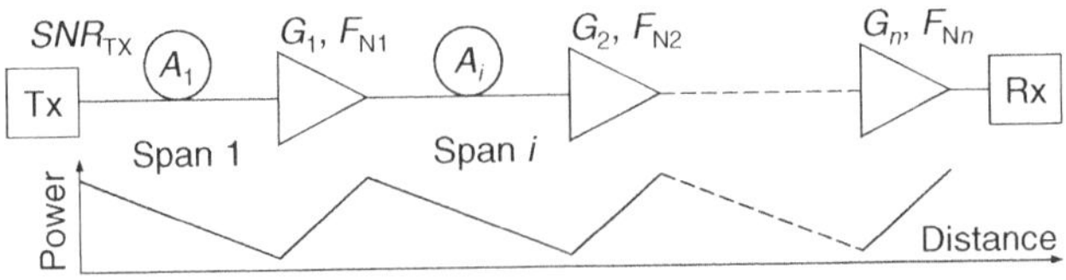

FIGURE 4.8 Long-haul link with cascaded amplifiers.

This definition is related to a reference bandwidth of $B_{\text{o,ref}} = 12.5$ GHz (or 0.1 nm). Expressed in decibels, this yields

$$\text{OSNR}_{\text{dB},B_{\text{o,ref}}} = 58\ \text{dBm} - 10\log\left(\frac{F_{\text{N}}}{P_{\text{in}}^{\text{opt}}}\right). \tag{4.46}$$

With the usual restriction to signal–ASE beat noise, the linear OSNR can also be related to the Q-factor:

$$Q = \frac{i_1 - i_0}{\sigma_1 + \sigma_0} \approx \sqrt{\frac{B_{\text{o,ref}}}{B}\frac{\left(\sqrt{P_1} - \sqrt{P_0}\right)^2}{G \cdot P_{\text{in}}^{\text{opt}}}\text{OSNR}_{\text{lin}}}. \tag{4.47}$$

Likewise, output OSNR of a single optical line amplifier (not a preamplifier) is given by

$$\begin{aligned}\text{OSNR}_{\text{o,lin}} &= \frac{P_{\text{Signal,out}}}{P_{\text{Noise,out}}} = \frac{GP_{\text{Signal,in}}}{4n_{\text{sp}}h\nu(G-1)B} = \frac{P_{\text{Signal,in}}}{F_{\text{N}}2h\nu B},\\ \text{OSNR}_{\text{o,dB}} &= P_{\text{Signal,in,dBm}} - F_{\text{N}} - 10\log_{10}(2h\nu B) = P_{\text{Signal,in,dBm}} - F_{\text{N}} + 58\ \text{dBm}.\end{aligned} \tag{4.48}$$

Finally, for the accumulated OSNR in the chains of optical amplifiers, one gets with Eq. (4.44)

$$\text{OSNR}_{0.1\ \text{nm}} = P_{\text{out,Ch}} - A - F_{\text{N}} - 10\log(n) + 58. \tag{4.49}$$

A is the mean span loss (now in decibels) and n is the number of spans (or amplifiers). As before, the OSNR is derived for the reference bandwidth or 0.1 nm. Obviously, the OSNR in long-haul systems (and assuming $G = 1/A$) drops logarithmically with increasing number of spans (Fig. 4.9).

Limits of the maximum amplifier number, depending on the detection scheme, were already derived in Ref. [16]. In ultralong-haul systems with low-noise design,

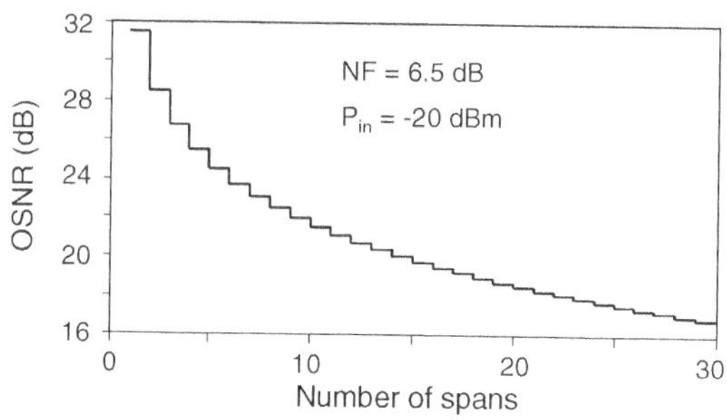

FIGURE 4.9 Development of OSNR in long chains of (equidistantly spaced) amplifiers.

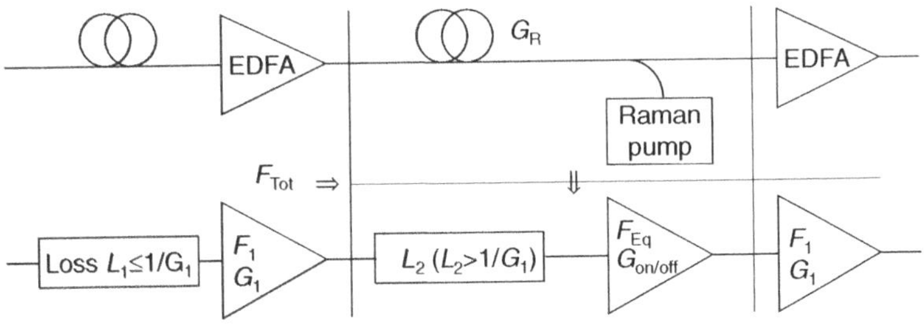

FIGURE 4.10 Multispan link with EDFAs and Raman preamplifiers, and equivalent model.

several hundred amplifiers can be cascaded. In regional systems with irregular span lengths (and spans with $A_i > 1/G_i$), the number of amplifiers is much smaller, often not even approaching 20.

The noise limit of EDFAs with respect to either very long single spans or very long multispan links can be improved by using the nonlinear effect of stimulated Raman scattering (SRS) along the transmission fiber (see Section 3.2.4.2). In these scenarios, Raman amplifiers can be used as low-noise preamplifiers that improve the resulting total noise figure. A multispan configuration with backward-pumped Raman amplifiers is shown in Fig. 4.10.

In this configuration, SRS acts as a *distributed* preamplifier to the following EDFA. The per-channel power can be prevented to reach a certain critical OSNR limit, as shown in Fig. 3.52. This results in improved OSNR (or noise figure) of the combined system of Raman preamplifier and EDFA. Often, the distributed fiber Raman amplifier (FRA) is described via an *equivalent* noise figure F_{Eq} and the Raman on/off gain $G_{\mathrm{on/off}}$ of a *lumped* Raman amplifier [21–25]. These are given by

$$
\begin{aligned}
F_{\mathrm{Eq}} &= F_{\mathrm{N,FRA}} \exp(-\alpha_S L), \\
G_{\mathrm{on/off}} &= G_R(L)\exp(\alpha_S L) = \exp(g_R P_{\mathrm{P_0}} L_{\mathrm{eff}}).
\end{aligned}
\tag{4.50}
$$

$F_{\mathrm{N,FRA}}$ is the noise figure of the distributed FRA, considering the distributed fiber loss, and $G_{\mathrm{R}}(z)$ and g_{R} are the length-dependent Raman gain and the Raman gain coefficient, respectively. P_{P0} is the launched pump power (i.e., at $z = L$) and L_{eff} is the effective fiber length at the pump wavelength. α_{S} is the fiber loss coefficient at signal wavelengths.

The on/off gain is explained in Fig. 4.11.

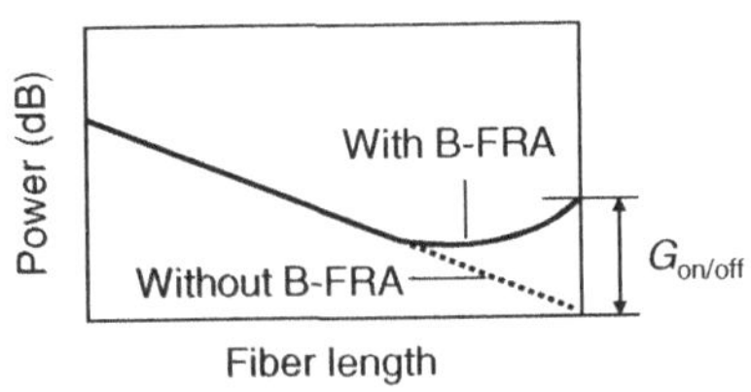

FIGURE 4.11 On/off Raman gain. B-FRA: backward-pumped FRA.

Depending not only on the fiber characteristics such as loss and mode field diameter but also on the patch and splice losses between fiber segments, the typical value for F_{Eq} is

$$F_{\mathrm{Eq}} \approx -2.5\,\mathrm{dB}. \tag{4.51}$$

The associated Raman on/off gain is in the range of $G_{\mathrm{on/off}} = 15\text{–}20\,\mathrm{dB}$.

The noise figure $F_{\mathrm{N,FRA}}$ of the distributed FRA has been derived in Ref. [24]:

$$F_{\mathrm{N,FRA}} = 1 + \frac{G_R - 1}{G_R} + 2\alpha_S D_{\mathrm{inv}}, \quad D_{\mathrm{inv}} = \int_0^L \frac{1}{G_R(z)}\,\mathrm{d}z. \tag{4.52}$$

The quantum limit of a 3 dB noise figure for very high gain is satisfied. Forward pumping schemes achieve lowest noise figures due to the minimization of D_{inv}.

The total noise figure F_{tot} of combined EDFA and Raman preamplification is given by

$$F_{\mathrm{tot}} = F_{\mathrm{Eq}} + \frac{(F_{\mathrm{EDFA}}/\alpha_c) - 1}{G_{\mathrm{on/off}}}. \tag{4.53}$$

Here, α_c is the (coupling) loss between the fiber output (at the end of L_2) and the EDFA.

In similar ways, the receive-end OSNR in very long multispan transmission can be improved by Raman-only amplification, using forward- and backward-pumped distributed FRAs. In Ref. [9], the noise figures for forward-pumped Raman amplification have been derived. Values of $F_{\mathrm{N,FRA}} = 4\text{–}5\,\mathrm{dB}$ can be achieved.

Further work on Raman noise figures can be found in Refs [25,26].

One aspect to be considered when using FRAs is laser safety, due to high pump power levels. This leads to the requirement of suitable automatic power shutdown (APS) mechanisms.

REFERENCES

1. H. Takahashi et al., Impact of crosstalk in an arrayed-waveguide multiplexer on $N \times N$ optical interconnection, IEEE J. Lightwave Technol., Vol. 14, No. 6, 1996, pp. 1097–1105.
2. F. Liu et al., Experimental verification of a new model describing the influence of incomplete signal extinction ratio on the sensitivity degradation due to multiple interferometric crosstalk, IEEE Photonics Technol. Lett., Vol. 11, No. 1, 1999, pp. 137–139.
3. Y. Peng et al., High-powered optical attack propagation in transparent optical networks, Optoelectron. Commun., 2011, doi: 10.1117/2.1201111.003933.

4. P.J. Legg et al., Solution paths to limit interferometric noise induced performance degradation in ASK/direct detection lightwave networks, IEEE J. Lightwave Technol., Vol. 14, No. 9, 1996, pp. 1943–1954.
5. L.G.C. Cancela and J.J.0. Pires, Impact of intrachannel crosstalk on the performance of direct-detection DPSK optical systems, QELS'05, 2005, Paper JThE48.
6. S. Yamamoto, Influence of intrachannel crosstalk with frequency dependence on signal degradation in optical switch network, IEEE J. Lightwave Technol., Vol. 27, No. 24, 2009, pp. 5716–5722.
7. M.R.G. Leiria and A.V.T. Cartaxo, Performance assessment of transparent optical networks with incoherent homodyne crosstalk and amplified spontaneous emission noise, 4th IASTED International Multi-Conference Wireless and Optical Communications, Banff, Canada, July 2004, pp. 851–856.
8. S.D. Dods and R.S. Tucker, Analysis of incoherent homodyne crosstalk in optical networks, J. Opt. Commun., Vol. 23, No. 5, 2002, pp. 178–181.
9. M. Premaratne, Analytical characterization of optical power and noise figure of forward pumped Raman amplifiers, Opt. Express, Vol. 12, No. 18, 2004, pp. 4235–4245.
10. S.B. Alexander, Optical Communication Receiver Design, SPIE Tutorial Texts in Optical Engineering, Vol. TT22, SPIE Publications, Bellingham, 1997.
11. G. Kweon, Noise figure of optical amplifiers, J. Korean Phys. Soc., Vol. 41, No. 5, 2002, pp. 617–628.
12. G.P. Agrawal, Fiber-Optic Communication Systems, John Wiley & Sons, Inc., New York, 1992.
13. O.K. Tonguz and R.E. Wagner, Equivalence between preamplified direct detection and heterodyne receivers, IEEE Photonics Technol. Lett., Vol. 3, No. 9, 1991, pp. 835–837.
14. S. Donati, Noise in an optical amplifier: formulation of a new semiclassical model, IEEE J. Quantum Electron., Vol. QE-33, No. 9, 1997, pp. 1481–1488.
15. E. Desurvire, Erbium-Doped Fiber Amplifiers, John Wiley & Sons, Inc., New York, 1994.
16. R. Loudon, Theory of noise accumulation in linear optical-amplifier chains, IEEE J. Quantum Electron., Vol. QE-21, No. 7, 1985, pp. 766–773.
17. N.A. Olsson, Lightwave systems with optical amplifiers, IEEE J. Lightwave Technol., Vol. 7, No. 7, 1989, pp. 1071–1082.
18. H.A. Haus, The noise figure of optical amplifiers, IEEE Photonics Technol. Lett., Vol. 10, No. 11, 1998, pp. 1602–1604.
19. H.T. Friis, Noise figures of radio receivers, Proceedings of the Institute of Radio Engineers (IRE), Vol. 32, No. 7, July, 1944, pp. 419–422 (Correction, Vol. 32, No. 12, p. 729, December 1944).
20. I.P. Kaminow and T.L. Koch, Optical Fiber Telecommunications III, Academic Press, San Diego, 1997.
21. Red-C data sheet, Hybrid Raman-EDFA Module 2 and 3-Pump, 2012. Available at http://www.red-c.com/_Uploads/dbsAttachedFiles/HybridRamanEDFA.pdf
22. P.B. Hansen et al., Rayleigh scattering limitations in distributed Raman pre-amplifiers, IEEE Photonics Technol. Lett., Vol. 10, No. 1, 1998, pp. 159–161.
23. P. Galliona et al., Noise in distributed Raman amplification, Proc. SPIE, Vol. 6781, 2007. pp. 67810U-1–67810U-15.

24. B. Bristiel et al., New model of noise figure and RIN transfer in fiber Raman amplifiers, IEEE Photonics Technol. Lett., Vol. 18, No. 8, 2006, pp. 980–982.
25. W. Zhang et al., An analytical expression of equivalent noise figure for distributed fiber Raman amplifiers with Rayleigh scattering, Opt. Commun. 199, November 2001, pp. 231–236.
26. P.M. Krummrich et al., System performance improvements by codirectional Raman pumping of the transmission fiber, Proceedings of the 27th European Conference Optical Communications, ECOC'01, Amsterdam, 2001, Paper Tu.A.1.4.

5

MODULATION FORMATS FOR WDM

5.1 BASIC MODULATION

The per-channel vector field $\overrightarrow{E}(t)$ that is transmitted in the fiber can be expressed as

$$\overrightarrow{E}(t) = \begin{bmatrix} A_x(t) \cdot e^{j(\omega_x(t)t+\phi_x(t))} \\ A_y(t) \cdot e^{j\left(\omega_y(t)t+\phi_y(t)\right)} \cdot e^{j\phi_{xy}} \end{bmatrix}. \tag{5.1}$$

Here, x and y refer to two orthogonal polarization states. A_i, ω_i, and φ_i are the per-polarization amplitudes, instantaneous frequencies, and phases, respectively. They all can depend on time, that is, carry information. ψ_{xy} is a delay between both polarizations; this too may be time dependent. This gives options for the parameters of the (electric) field, which can be modulated in order to carry payload information.

If $|A(t)|^2 = A_x^2(t) + A_y^2(t)$ is modulated, it is called intensity modulation. It can be detected noncoherently in direct detection (DD) receivers. The respective transmission scheme is called IM-DD (intensity modulation with direct detection). It is also referred to as on–off keying (OOK). However, this term also holds for coherently detected unipolar 2ASK.

If $A(t)$ is used for modulation, we have amplitude-shift keying (ASK). ASK can use just two amplitudes (2ASK) or multiple amplitude levels [multilevel ASK (M-ary

Wavelength Division Multiplexing: A Practical Engineering Guide, First Edition.
Klaus Grobe and Michael Eiselt.

ASK)], and it can be detected coherently or noncoherently. IM-DD is the special case of directly detected, binary, unipolar ASK.

If $\omega(t)$ is modulated, it is referred to as frequency-shift keying (FSK). FSK can use two frequencies [binary FSK (BFSK), or 2FSK], or multiple frequencies (although this is not common in fiber optics), and it can be detected coherently or noncoherently.

If $\varphi(t)$ is chosen to carry the information, it is called phase-shift keying (PSK). PSK works as binary (BPSK) or multilevel (M-ary PSK) modulation. It can use coherent or noncoherent (NC) receivers. Noncoherent PSK requires differential precoding that leads to differential PSK (DPSK).

The combination of (multilevel) ASK and PSK leads to quadrature amplitude modulation (QAM). QAM plays an important role in ultrahigh-speed wavelength division multiplexing (WDM) transmission. It is preferentially detected coherently.

For all modulation schemes named so far, it is possible to use full response or partial response pulses. Full response pulses have duration or width, measured as full width at half maximum (FWHM), not exceeding the symbol duration. For partial response (PR), the pulse spreads over several symbols, causing intentional intersymbol interference (ISI). This is achieved by precoding and narrowband filtering. PR affects bandwidth (BW) efficiency and chromatic dispersion tolerance. In general, the ISI effects can be eliminated with receive-end maximum-likelihood sequence estimation (MLSE). The simplest form of PR in WDM transmission is optical duobinary, ODB. Like QAM, PR can be extended into the complex symbol space, which then leads to M-ary quadrature partial response, or QPR [1,2].

If the orthogonally polarized x- and y-planes are used for modulation, it leads to polarization-shift keying (PolSK). If both planes are used independently for multiplexing, it is called polarization-domain multiplexing (PDM), or orthogonally polarized (OP) or dual-polarization (DP) modulation, respectively. A special variant of DP modulation is polarization interleaving, where pairs of adjacent WDM channels get orthogonal polarization.

The next parameter that determines the modulation scheme is the pulse shape. It refers to the function that describes the pulse (e.g., raised cosine, cosine, and Gaussian), and to the length or duty cycle. The latter refers to a single-symbol time slot and is not to be confused with the full- versus partial response characteristic. It mainly splits into nonreturn-to-zero (NRZ), and return-to-zero (RZ) pulses, where RZ can have various duty cycles.

Further parameters that have influence on the resulting modulation scheme include the application of chirp or strong asymmetric filtering, leading to vestigial or single sidebands.

An overview on resulting modulation schemes that are relevant for WDM transmission (or have been investigated extensively) is given in Fig. 5.1.

Chirped modulation [e.g., chirped RZ-OOK (CRZ-OOK)] and vestigial sideband (VSB) and single sideband (SSB) modulations are not listed here for clarity, and due to lack of importance.

The basic modulation schemes ASK, FSK, PSK, and consequently the derived complex and partial response schemes (ODB, quaternary PSK (QPSK), QAM, and

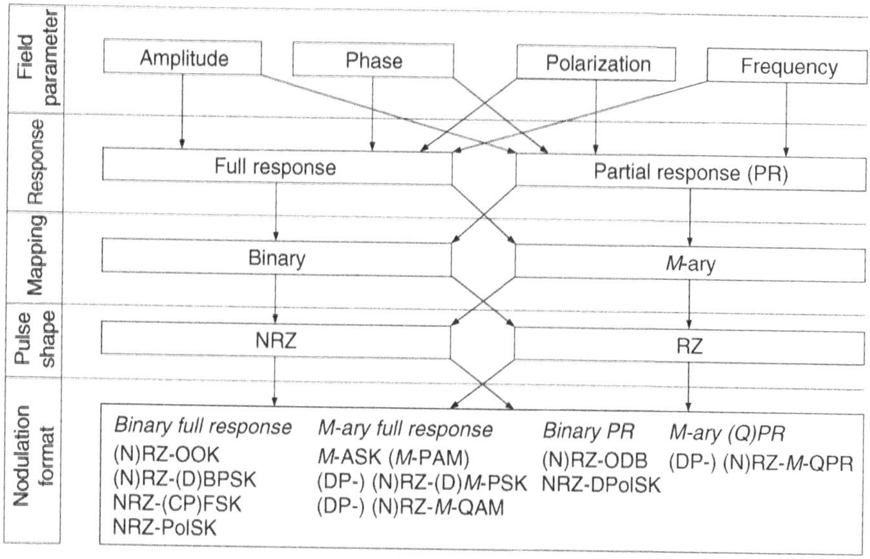

FIGURE 5.1 Overview on fiber-optic modulation and pulse-shaping schemes.

QPR) can be detected noncoherently or coherently. Noncoherent detection refers to direct detection or asynchronous heterodyne detection using envelope detectors or (self-coherent) delay demodulators. Coherent detection refers to heterodyne or intradyne detection with synchronous demodulators, and homodyne detection, respectively. The possible combinations are listed in Table 5.1.

In Table 5.1, the detection schemes relevant for WDM are highlighted (*bold-face italic*). The schemes listed for phase hold for BPSK and QPSK, including the relevance. Noncoherent higher-order PSK detection gets increasingly difficult, and it has no relevance for WDM.

TABLE 5.1 Detection Schemes for Basic Modulation Schemes

Detection/Modulation	Amplitude	Frequency	Phase
Direct detection (DD)	***IM-DD***	BFSK/DD w/delay demodulator	***DPSK/DD w/delay demodulator***
Heterodyne incoherent (asynchronous)	w/Envelope detector	w/Envelope detector, single or dual filter	DPSK w/delay demodulator
Heterodyne coherent (synchronous)	w/Synchronous Demodulator	w/Synchronous demodulator, single or dual filter	w/Synchronous demodulator
Intradyne coherent (synchronous)	***w/Synchronous Demodulator***	–	***w/Synchronous demodulator***
Homodyne	w/OPLL	–	w/OPLL

OPLL: optical phase-locked loop.

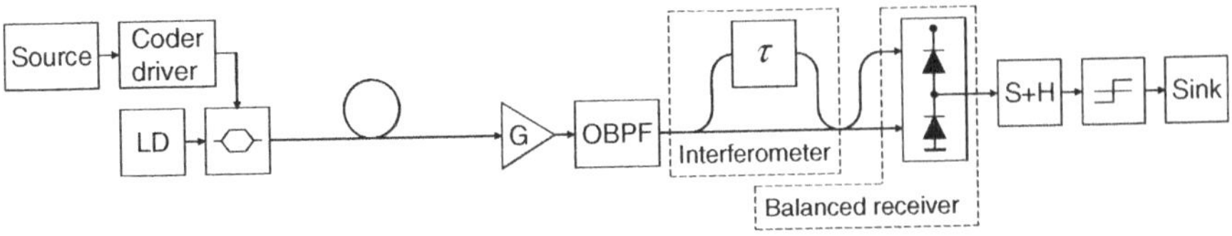

FIGURE 5.2 Direct detection for IM-DD, binary FSK, and DPSK.

The detection schemes listed for amplitude also hold for ODB.

Figure 5.2 shows the generic fiber-optic direct detection scheme. It can be applied to IM-DD, BFSK, and differentially precoded binary PSK (DPSK). For IM-DD, $\tau = 0$. For DPSK, $\tau = T$, with T the bit (or symbol) duration. For BFSK, τ is a fraction of T. For BFSK with $\Delta f = 1/(2T)$, optimum $\tau = T/2$ results [3,4]. In the BFSK case, the interferometer in Fig. 5.2 acts as a frequency discriminator, with delay τ tuned to optimum separation. As indicated in Table 5.1, this structure is relevant for both IM-DD and DPSK.

The electrical receiver shown here consists of a sample and hold circuit (S + H) (i.e., an integrator with sampling unit) and a decider (slicer). The S + H circuit is an approximation of the matched filter (MF, that is, ideally, S + H is replaced by MF plus sampler).

Figure 5.3 shows the generic structure of fiber-optic heterodyne noncoherent or asynchronous detection [4]. Again, $\tau = 0$ for OOK, $\tau = T$ for DPSK, and τ is a fraction of T for BFSK, respectively. Figure 5.3 also shows a polarization-diversity receiver that is one way of avoiding orthogonality between received-signal and local oscillator (LO) polarizations. The incoming signal is split in a polarization beam splitter (PBS). The LO is split in a 3 dB power splitter. One branch is $\lambda/4$ polarization shifted, and both signal and LO branches are combined in further 3 dB couplers and fed into two photodiodes (PDs). This structure compensates any polarization rotations of the incoming signal. Alternatively, polarization scrambling (of the LO) can be used with single-branch detectors that leads to 3 dB penalty [5]. A further alternative is active polarization alignment of the LO to the incoming signal. This polarization tracking and alignment must be endless, that is, it must allow an arbitrary number of polarization rotations in one direction. Tracking and alignment must also be able to follow fast rotations in the order of several 10 krad/s [6,7].

The output of the noncoherent detector is also used for automatic frequency control (AFC).

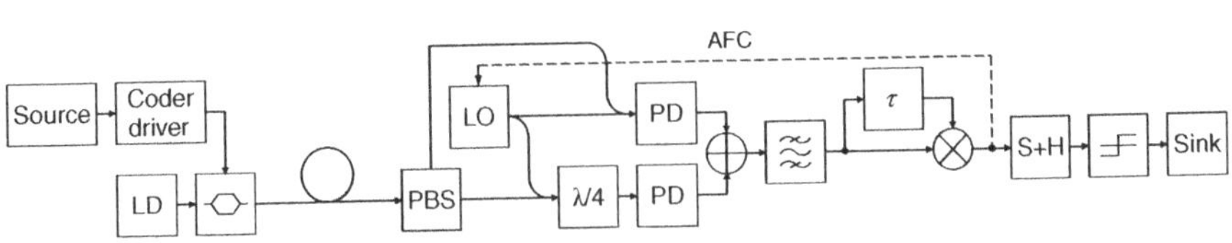

FIGURE 5.3 Heterodyne asynchronous (noncoherent) detection for OOK, 2FSK, and DPSK.

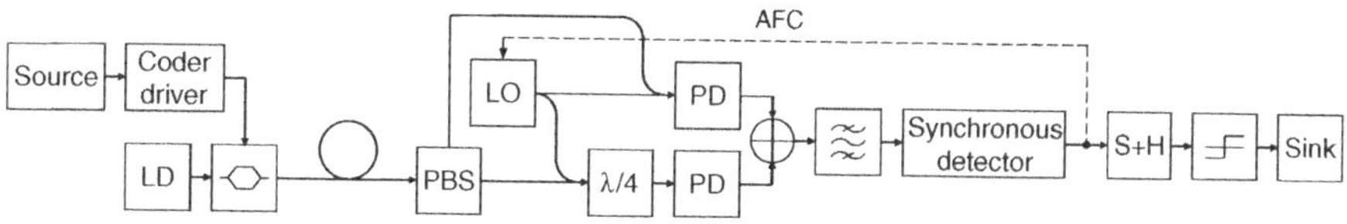

FIGURE 5.4 Heterodyne synchronous (coherent) detection for 2ASK, 2FSK, and BPSK.

Figure 5.4 shows the generic structure for heterodyne synchronous (coherent) detection. Different synchronous detectors have to be implemented for ASK, PSK, and FSK. For optimum FSK detection, a two-filter synchronous detector must be implemented [8]. Similar structures for synchronous intradyne detection result. The structure shown here is also discussed for heterodyne QPSK in WDM-PON (see Section 6.1.4.3).

With coherent detection, the combination of amplitude and phase leads to the QAM, including the relevance stated in Table 5.1. This also covers the partial response variant QPR (which may have to be complemented by MLSE).

An important parameter of modulation schemes, in particular multilevel schemes, is the location of the different symbol states in the complex symbol space. It is graphically represented in constellation diagrams, where the respective inphase (I) and quadrature (Q) components of each symbol state are shown. For PolSK and DP modulation, the two-dimensional symbol space has to be extended to four dimensions to account for the polarization domain. In Fig. 5.5, some constellation diagrams are shown. These schemes have been used or investigated extensively in fiber-optic transmission, in particular for 10, 40, and 100 Gb/s WDM transmission.

The modulation schemes in the lower row are dually polarized.

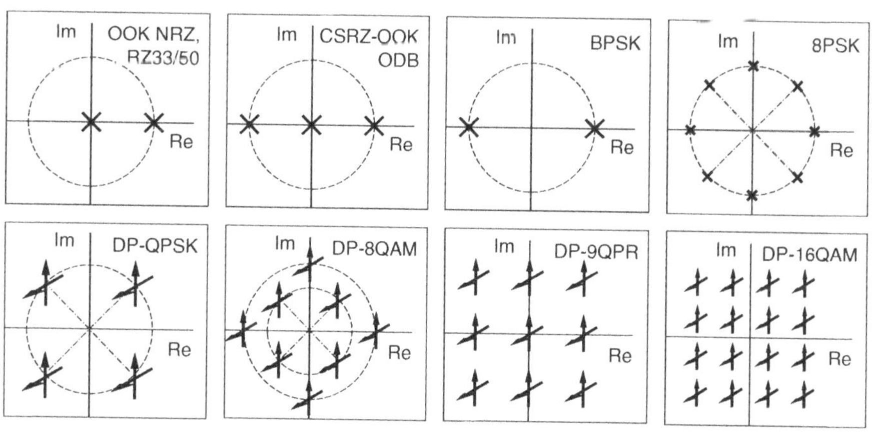

FIGURE 5.5 Some fiber-optic modulation schemes' constellation diagrams.

Up to and including 10 Gb/s WDM transmission, NRZ-OOK (IM-DD) has been the most important modulation scheme. For 40 Gb/s, ODB, DPSK, and DQPSK (differential quaternary PSK) are relevant, and for 100 Gb/s and beyond, most effort is spent on DP-QPSK and DP-QAM with intradyne detection.

5.2 PULSE SHAPING

In digital transmission, pulse shape must be considered. In fiber-optic transmission, cosine NRZ or RZ can be used, where RZ typically splits into only three dedicated duty cycles, namely 33%, 50%, and 67% (which are referred to as RZ33, RZ50, and RZ67 or CSRZ (carrier-suppressed RZ), respectively). These pulse shapes are applicable to most modulation schemes, for example, DP-QPSK. In addition, in the 1980s and 1990s, much effort was spent on *Solitons* that are a special (very low-duty cycle and high pulse energy) RZ variant for OOK, with hyperbolic-secant shape [9–14].

As compared to NRZ, RZ pulse shaping has a slight impact on the receive-end optical signal-to-noise ratio (OSNR) (because a better matched filter can be implemented at the receiver), and it has severe impact on both, the resulting required optical bandwidth and the resistance against polarization-mode dispersion (PMD) and nonlinear fiber transmission effects.

In fiber-optic transmission, RZ pulse shaping usually is applied through a dedicated external Mach–Zehnder modulator (MZM), which for RZ50 is driven with a sinusoidal frequency equaling the bit rate, and a bias at the 3 dB point. For RZ33 and RZ67, the MZM is biased at peak and null points, respectively, and driven with $2V_{\pi}$ at a frequency equaling *half* the bit rate. RZ67 is often referred to as CSRZ. For OOK, it leads to an additional phase modulation between succeeding pulse (Fig. 5.6). This effect must be considered in CSRZ-DPSK demodulation.

Effects of RZ pulse shaping on nonlinear fiber transmission will be discussed later. However, RZ coding also has an effect on the optical bandwidth. It is obvious through Fourier transform that bandwidth is increasing if pulse duration is decreasing, which is the case for RZ as compared to NRZ. For a given symbol rate R_S, bandwidth differs for different duty cycles. It is also severely dependent on the bandwidth definition. In Ref. [15], bandwidths containing 90% and 99% of the optical power have been considered. The resulting bandwidths for RZ33, RZ50, and RZ67 are summarized in Table 5.2 for binary OOK, ODB, and *M*-ary PSK or QAM, respectively.

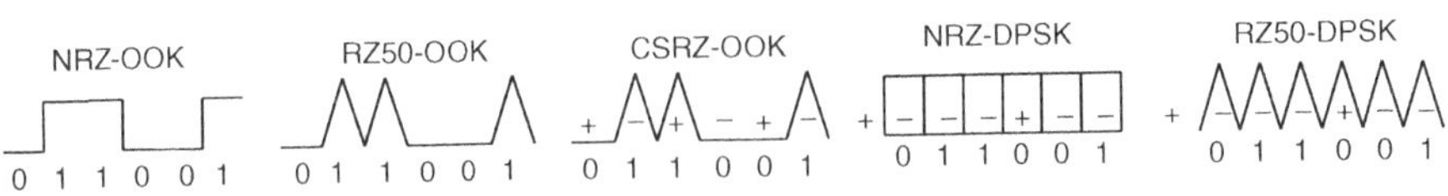

FIGURE 5.6 NRZ, RZ50, and RZ67 (CSRZ) for OOK and DPSK.

TABLE 5.2 Comparison of RZ Bandwidths

	RZ33		RZ50		RZ67	
	90% BW	99% BW	90% BW	99% BW	90% BW	99% BW
Binary OOK	$2.00/T_S$	$2.75/T_S$	$2.00/T_S$	$2.84/T_S$	$1.66/T_S$	$3.69/T_S$
ODB	$2.06/T_S$	$2.59/T_S$	$1.94/T_S$	$2.72/T_S$	$1.34/T_S$	$3.28/T_S$
M-ary PSK, QAM	$2.00/T_S$	$2.97/T_S$	$1.75/T_S$	$2.84/T_S$	$1.47/T_S$	$3.69/T_S$

From Table 5.2 it is clear that bandwidth definition is relevant, and has to be considered in detail for any given WDM transmission problem. It also gets clear that RZ67 has the lowest bandwidth for 90%-power bandwidth definition, whereas RZ33 has lowest bandwidth for 99%. It is also apparent that ODB leads to the lowest bandwidth per symbol rate for RZ67 and 90%, and RZ33 and 99%, respectively. Examples of optical power spectra, normalized to the bit rate, are shown in Fig. 5.7.

There are several methods to generate optical RZ signals, for example, by directly modulating a semiconductor laser with an electrical RZ data signal, by *pulse carving* of an optical NRZ signal by means of a dedicated MZM, or by generating an optical pulse train (i.e., a periodic RZ signal) first and modulating it with a NRZ data signal. The pulse train can again be generated with a pulse carver (in which case, the order of the pulse carver and the data modulator is reversed as compared to the second option) [16].

The pulse carver options are frequently used because of simplicity and because various duty cycles can be realized by appropriate combination of the bias voltage and the modulation amplitude of the MZM. The optical NRZ data signal can be

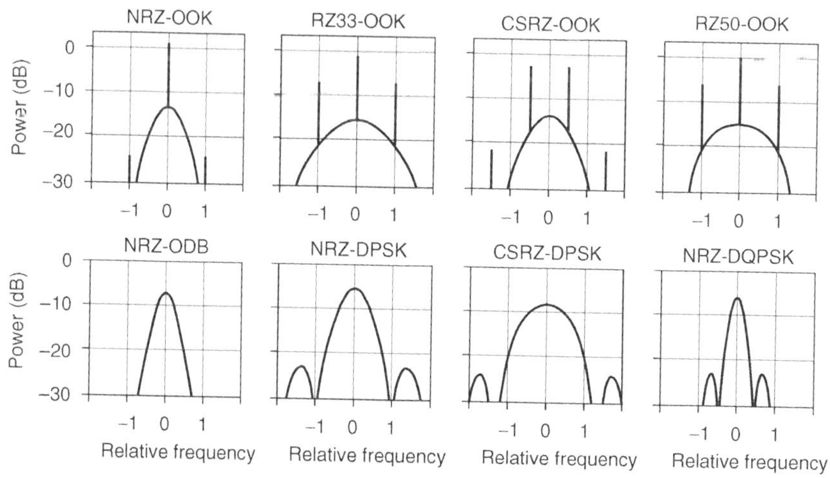

FIGURE 5.7 Effect of modulation, pulse shaping, and duty cycle on optical bandwidth.

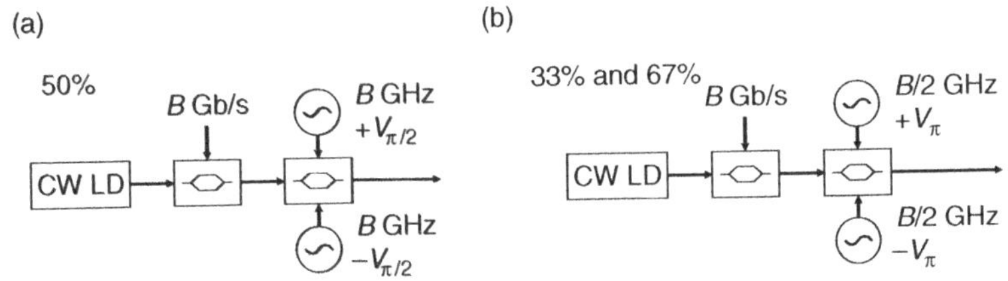

FIGURE 5.8 RZ modulators: 50% (a), 33% and 67% (b).

generated from a continuous wave (CW) laser with either another MZM or an electroabsorption modulator.

Three easily implementable duty cycles for RZ modulation are 33%, 50%, and 67%, respectively. Possible MZM implementations for these duty cycles are shown in Fig. 5.8. RZ pulse carving is achieved with push–pull MZM.

The driving voltage of the pulse-carving MZM is given by

$$V_{\mathrm{m}}(t) = V_{\mathrm{bias}} + V_{\mathrm{RF}}(t) = V_{\mathrm{bias}} + V_{\mathrm{RF}} \cos(2\pi f_{\mathrm{mod}} t + \varphi_{\mathrm{m}}). \tag{5.2}$$

Here, V_{bias} is the DC bias voltage, V_{RF} is the RF modulation amplitude, f_{mod} is the RF modulation frequency, and ϕ_{m} is the phase shift. The optical power transfer function of the MZM can be written as [16]

$$T(t) \propto \cos^2\left[\frac{\pi V_m(t)}{2V_\pi} + \frac{\theta}{2}\right] = \cos^2\left[\frac{\pi V_{\mathrm{bias}}}{2V_\pi} + \frac{\pi V_{\mathrm{RF}}(t)}{2V_\pi} + \frac{\theta}{2}\right]. \tag{5.3}$$

Here, θ is the intrinsic phase shift of the MZM without the driving voltage, and V_π is the π-phase-shift voltage of the MZM. If $V_{\mathrm{bias}} = V_{\mathrm{max}}$, by definition the MZM is DC-biased at its maximum optical transmission. If $V_{\mathrm{bias}} = V_{\mathrm{min}}$, the MZM is DC-biased at its minimum optical transmission. The MZM can also be driven in a balanced push–pull way.

In the case of chirp-free push–pull modulation of a two-arm $LiNbO_3$ MZM, an electrical peak-to-peak modulation of V_π is split into $+V_{\pi/2}$ and $-V_{\pi/2}$ to obtain an RZ50 format (Fig. 5.8). Alternatively, pulse carving can be realized using a single-arm MZM by applying peak-to-peak modulation of V_π at the single arm to obtain the RZ50 format.

The generation of the three different RZ duty cycles depends on the RZ modulator frequency, the electrical peak-to-peak modulation voltage, and the modulator bias.

For RZ33, the input signal to the MZM is an optical NRZ signal with a symbol rate of $1/T_{\mathrm{s}}$ (where T_{s} is the symbol duration). The MZM is DC-biased at its maximum optical transmission by $V_{\mathrm{bias}} = V_{\mathrm{max}}$, and RF-modulated by a sinusoidal signal, with a frequency of $f = 1/(2T_{\mathrm{s}})$ and an amplitude of V_π ($2V_\pi$ peak-to-peak). Then, the amplitude of the optical field $E_{\mathrm{RZ33}}(t)$ of the MZM output is proportional to

$$E_{\mathrm{RZ33}}(t) \propto \cos\left[\frac{\pi}{2}\cos\left(\pi\frac{t}{T_{\mathrm{b}}}\right)\right] e_{\mathrm{NRZ}}(t), \tag{5.4}$$

where $e_{NRZ}(t)$ is the optical field of the input NRZ signal. The optical output power of the MZM then becomes [16]

$$P_{RZ33} \propto E_{RZ33}(t)E_{RZ33}(t)^* \propto \left[\cos\left[\frac{\pi}{2}\cos\left(\pi\frac{t}{T_b}\right)\right]e_{NRZ}(t)\right]^2. \tag{5.5}$$

Due to suppression of the optical carrier, RZ67 is also called CSRZ. It has better robustness against chromatic dispersion than RZ modulation with 33% or 50% duty cycle because of lower bandwidth (Fig. 5.7). To generate CSRZ, the MZM is DC biased at its minimum optical transmission by $V_{bias} = V_{min}$, and modulated by a sinusoidal RF signal with a frequency of $f = 1/(2T_s)$, and a phase shift, $\phi_m = \pi/2$. The RF modulation amplitude is V_π ($2V_\pi$ peak-to-peak), corresponding to the half-wave voltage of the MZM. The amplitude of the optical field at the output of the MZM, $E_{RZ67}(t)$, is proportional to

$$E_{RZ67}(t) \propto \sin\left[\frac{\pi}{2}\sin\left(\pi\frac{t}{T_b}\right)\right]e_{NRZ}(t). \tag{5.6}$$

The output power of the MZM is proportional to $E_{RZ67}(t)E_{RZ67}(t)^*$, and given by

$$P_{RZ67} \propto E_{RZ67}(t)E_{RZ67}(t)^* \propto \left[\sin\left[\frac{\pi}{2}\sin\left(\pi\frac{t}{T_b}\right)\right]e_{NRZ}(t)\right]^2. \tag{5.7}$$

To generate RZ50, the MZM is DC biased at its 3 dB optical transmission by $V_{bias} = V_{3\,dB}$, and modulated by a sinusoidal RF signal, with a frequency of $f = 1/(T_s)$ (Fig. 5.8). The RF modulation amplitude is $V_\pi/2$ (V_π peak-to-peak). The amplitude of the optical field at the output of the MZM, $E_{RZ50}(t)$, is proportional to

$$E_{RZ50}(t) \propto \cos\left[\frac{\pi}{4}+\frac{\pi}{4}\cos\left(\frac{2\pi t}{T_b}\right)\right]e_{NRZ}(t). \tag{5.8}$$

The output power of the MZM is proportional to $E_{RZ50}(t)E_{RZ50}(t)^*$, which gives

$$P_{RZ50} \propto E_{RZ50}(t)E_{RZ50}(t)^* \propto \left[\cos\left[\frac{\pi}{4}+\frac{\pi}{4}\cos\left(\frac{2\pi t}{T_b}\right)\right]e_{NRZ}(t)\right]^2. \tag{5.9}$$

Table 5.3 summarizes the key figures of the three RZ duty cycles. f_{mod} is the modulation frequency, V_{mod} is the peak-to-peak modulation voltage ($2V_{RF}$), and V_{bias} describes the bias condition, where V_{min} and V_{max} are the bias points at transmission

TABLE 5.3 Modulation Figures of RZ Formats at 43 Gb/s [16]

RZ-	33%	50%	67% (CSRZ)
f_{mod} (GHz)	21.5	43	21.5
V_{mod}	$2V_\pi$	V_π	$2V_\pi$
V_{bias}	V_{max}	$V_{3\,dB}$	V_{min}
Phase shift	0,0,0	0,0,0	0,π,0

minimum (leading to carrier suppression) and maximum, respectively, and $V_{3\,\mathrm{dB}}$ is the conventional MZM bias point that is also used for NRZ data modulation. "Phase shift" describes the phase shift between consecutive RZ pulses (i.e., modulated data symbols) (also refer to Fig. 5.6).

Comparing RZ33, RZ50, and RZ67 from the viewpoint of components effort, slight advantages for RZ33 and RZ67 can be identified because of lower bandwidth demand for the pulse carvers.

5.3 MODULATION FORMATS

Closed analytical calculation of the receive-end bit error rate (BER) P_B of a WDM transmission system including fiber nonlinearity, time variance [via PMD and polarization-dependent loss (PDL)/PDG (polarization-dependent gain)], dispersion, and noise is not possible. Hence, a simplified AWGN (added white Gaussian noise) system model for the derivation of P_B must be used, for example, Refs [5,17–22]. Effects of nonlinearity, dispersion, and cross talk can then be considered approximately via penalties on the receive-end OSNR.

Further, we consider optimal (matched filter) reception and single polarization only. Generally, noise can be considered unpolarized, whereas the payload (WDM) signals have unknown state of polarization (SOP), which depend on transmitter design, link length, PDL/PDG, and (higher order) PMD. For direct detection, a penalty caused by the second polarization (i.e., payload signal is not linearly polarized and the receiver does not use a corresponding polarizer) in the range of 0.3–0.4 dB results [23]. For coherent detection, a similar penalty results from polarization-diverse detection [5].

In the AWGN model, the received signal vector $\mathbf{r}$ is given by $\mathbf{r} = \mathbf{s}_i + \mathbf{n}$, where the signal vector $\mathbf{s}_i = (a_{i1}, a_{i2}, \ldots, a_{iN})$ is deterministic. The elements of the noise vector $\mathbf{n} = (n_1, n_2, \ldots, n_N)$ are Gaussian random variables with zero-mean and variance $\sigma^2 = N_0/2$. N_0 is the single-sided noise density.

The noise vector probability density function (PDF) is given by

$$p_\mathbf{n}(\mathbf{n}) = \frac{1}{(\pi N_0)^{N/2}} \exp\left(\frac{-\|\mathbf{n}\|^2}{N_0}\right). \tag{5.10}$$

The elements of the observed receive-end vector $\mathbf{r} = (r_1, r_2, \ldots, r_N)$ are independent Gaussian random variables. Their PDF is given by

$$p_\mathbf{r}(\mathbf{r}|\mathbf{s}_i) = \frac{1}{(\pi N_0)^{N/2}} \exp\left(-\frac{\|\mathbf{r} - \mathbf{s}_i\|^2}{N_0}\right). \tag{5.11}$$

The BER P_B for binary, synchronous heterodyne (coherent) detection is given by

$$P_\mathrm{B} = Q\left(\frac{\|\mathbf{s}_1 - \mathbf{s}_2\|}{2\sigma}\right) = Q\left(\frac{d}{2\sigma}\right) = Q\left(\frac{\|\mathbf{s}_1 - \mathbf{s}_2\|}{\sqrt{2N_0}}\right) = Q\left(\sqrt{\frac{\eta E_\mathrm{b}}{2N_0}}\right) = Q\left(\sqrt{\frac{d^2}{E_\mathrm{b}}} \cdot \sqrt{\frac{E_\mathrm{b}}{2N_0}}\right). \tag{5.12}$$

For homodyne detection, the noise contribution is halved. The BER depends on the distance $d = ||\mathbf{s}_1 - \mathbf{s}_2||$ between the signal states, normalized by the noise standard deviation per dimension (I and Q, x and y in PolSK). $\eta = d^2/E_\mathrm{b}$ is a scale-invariant measure of the power efficiency, with E_b the per-bit energy. BER performance depends on the signal-to-noise *ratio* E_b/N_0 rather than on absolute strengths. For given E_b/N_0, performance is better for higher power efficiency η.

Q is the quality function or Q-function, which is commonly used in digital-transmission description. It can be derived from the complementary error function (erfc) [19,20]:

$$\mathrm{erfc}(x) = 2Q\left(\sqrt{2}x\right) = \frac{2}{\sqrt{\pi}}\int_x^\infty e^{-y^2}dy. \tag{5.13}$$

Following, BERs for the AWGN channel are derived for several modulation formats. This is split into coherent (homodyne and heterodyne, where applicable) and noncoherent (direct band-pass-filtered or self-coherent) detection. In Ref. [4], the equivalence between heterodyne asynchronous and optically preamplified direct detection has been derived.

5.3.1 Amplitude Modulation

Amplitude modulation is linear modulation, which can directly be seen from Eq. (5.1). For practical considerations, it can be split into *binary intensity modulation* [i.e., IM-DD, often also referred to as (NRZ) OOK], and bipolar multilevel ASK. IM-DD is the standard modulation format for WDM channel bit rates up to 10 Gb/s. Multilevel ASK is not relevant for WDM, but its quadrature extension QAM is.

5.3.1.1 On–Off Keying (OOK) In practically all cases, OOK is detected directly using a band-pass filter, as indicated in Fig. 5.2. The scheme is then referred to as IM-DD. Its constellation diagram is shown in Fig. 5.9.

Due to added noise, the receive-end probability density is spread according to Fig. 5.10. The PDFs for the two symbol states are not equal. For direct detection, they are also not Gaussian anymore due to band-pass filtering. Hence, the Gaussian description is an approximation. The PDFs are described by mean values and variances of i_0, σ_0 and i_1, σ_1, respectively.

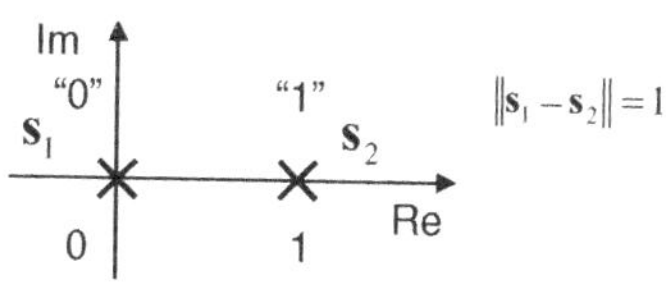

FIGURE 5.9 Constellation of IM-DD and unipolar 2ASK (OOK).

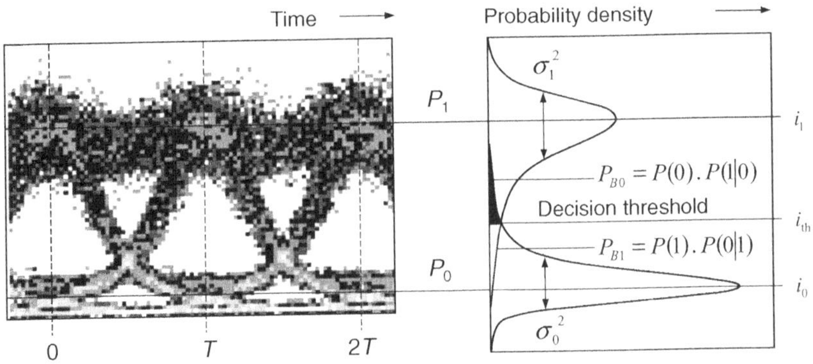

FIGURE 5.10 Symbol detection in binary OOK.

Bit errors occur if noisy symbols fall into the wrong region with regard to the decider threshold i_{th}. Hence, the BER is given by

$$P_{\mathrm{B}} = \Pr(1) \cdot \Pr(0|1) + \Pr(0) \cdot \Pr(1|0) = \frac{1}{2}[\Pr(0|1) + \Pr(1|0)]. \quad (5.14)$$

Pr(·) stands for probability (·). The derivation of the BER P_{B} can be found in the literature, for example, in Refs [5,17,18]. For noncoherent IM-DD, the BER results as follows:

$$P_{\mathrm{B}} = \frac{1}{2}\exp\left(-\frac{E_{\mathrm{b}}}{4N_0}\right). \quad (5.15)$$

The block diagram of an NRZ-OOK (IM-DD) system can be derived from Fig. 5.2 and is shown in Fig. 5.11. The transmitter consists of a CW laser diode (LD), and an external modulator. The receiver consists of preamplifier, optical band-pass filter (OBPF), and direct detection receiver. The OBPF is necessary for both, WDM channel separation and as a noise-bandwidth limiter. The direct detection receiver consists of a S + H unit that approximates a matched filter, followed by a decision unit. Not shown here are means for receive-end clock data recovery (CDR).

With signal field strength E_{Sig}, the receive-end photodiode current is given by

$$i = |E_{\mathrm{Sig}} + n|^2 = |E_{\mathrm{Sig}}|^2 + 2\,\mathrm{Re}(E_{\mathrm{Sig}}n^*) + |n|^2. \quad (5.16)$$

The total receiver noise is dominated by the beat of signal and noise n, $\mathrm{Re}(E_{\mathrm{S}} \cdot n^*)$.

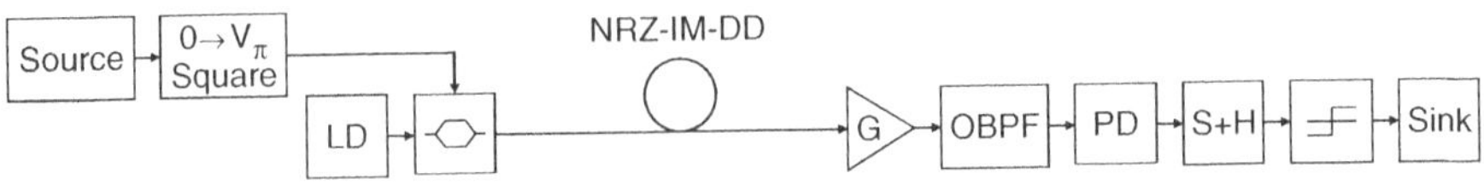

FIGURE 5.11 NRZ-OOK with external modulation and direct detection (DD).

TABLE 5.4 Q-Factor and Resulting BER

log(BER)	Q_F	Q_F (dB)
−3	3.1	9.8
−4	3.7	11.4
−5	4.3	12.7
−6	4.8	13.6
−7	5.2	14.3
−8	5.6	15.0
−9	6.0	15.6
−10	6.4	16.1
−11	6.7	16.5
−12	7.0	16.9
−13	7.3	17.3
−14	7.7	17.7
−15	7.9	18.0

As unipolar 2ASK, OOK can also be detected coherently. With the definitions given in Fig. 5.9, and with $E_b = 0.5\left(\|\mathbf{s}_1{}^2 + \mathbf{s}_2\|^2\right)$ and power efficiency $\eta = 2$, the BER for coherent heterodyne OOK can be stated. In addition, one also gets with Ref. [17]:

$$P_B = Q\left(\sqrt{\frac{E_b}{2N_0}}\right) = \frac{1}{2}\operatorname{erfc}\left(\sqrt{\frac{E_b}{4N_0}}\right) = \frac{1}{2}\operatorname{erfc}\left(\frac{Q_F}{\sqrt{2}}\right), \quad Q_F = \frac{i_1 - i_0}{\sigma_1 + \sigma_0}. \qquad (5.17)$$

The first part of Eq. (5.17) uses the Q-function expression, which can be found frequently in the literature. The second part uses the complementary error function and, in its argument, the so-called Q-factor or Q parameter Q_F [17]. Note that Eq. (5.17) is sometimes also used as an approximation for asynchronous detection, instead of Eq. (5.15). The deviation between the two is below 1 dB.

Table 5.4 lists the achievable BER in dependence of the Q-factor.

Q-function and Q-factor must not be confused.

As mentioned earlier, homodyne detection has 3 dB better signal to-noise ratio (SNR), compared to synchronous heterodyne detection. Hence, the BER for homodyne ASK yields

$$P_B = Q\left(\sqrt{\frac{E_b}{N_0}}\right). \qquad (5.18)$$

OOK is used for bit rates up to 10–11 Gb/s. Due to various transmission impairments, it is not used for ultrahigh-speed transmission. However, transmission of 100GbE signals up to 40 km, using a four-fold parallel-lane approach using OOK and so-called LAN WDM, has been defined in IEEE 802.3ba (see Section 3.3.3).

5.3.1.2 *Detailed Derivation of OSNR Tolerance of Directly Detected OOK*

The error rate of the received signal is impacted by semideterministic effects like chromatic dispersion (CD), PMD, etc. as well as by random effects like the accumulated optical noise from optical amplification or cross talk from other channels [filter cross talk, four-wave mixing (FWM), and XPM (cross-phase modulation)]. The semideterministic effects can vary randomly, for example, by changes in the environment (temperature and mechanical stress of the fiber). However, these changes are relatively slow (compared to the bit period), such that compensators can be used to mitigate at least the main part of these impacts. The random effects cannot be mitigated and therefore are the main factors that determine the error rate of the received signal.

The major random effect that determines the bit error rate of the received signal is optical noise, which is measured by the OSNR. The other random effects, like cross talk, can be minimized by proper transmission system design, that is, choosing proper channel spacing or a sufficient filter extinction ratio. Remaining effects can be translated into an effective reduction of the OSNR. It is therefore educative to determine the impact of optical noise on the BER of an optical signal.

We now consider an amplitude-modulated signal that is directly detected on a photodiode. The Q-factor is a good measure for signal quality. It was already introduced in Eq. (5.17), where it was called Q_F to distinguish it from the Q-function. It is defined as

$$Q = \frac{\bar{i}_1 - \bar{i}_0}{\sigma_0 + \sigma_1}. \tag{5.19}$$

In this equation, $\bar{i}_1$ and $\bar{i}_0$ are the average levels of the one and the zero currents of the signal after photo detection, respectively, while σ_1 and σ_0 are the standard deviations of the respective currents. The direct conversion of the Q-factor to the BER using the relation

$$\text{BER} = 0.5 \exp(-(Q^2)/2) \tag{5.20}$$

is exact only if the noise on both the zero and one rails is Gaussian distributed and the optimum decision threshold is used. This is not the case for optical noise, as the Gaussian distributed noise amplitude in the optical domain is converted to a noncentral chi-square distribution during the nonlinear process of detection in the photodiode [24]. Still, Eq. (5.20) can be used as an approximation with a good precision for an optimum decision threshold. It should be noted, however, that the optimum decision threshold itself, obtained for a Gaussian distribution, is different than that for the chi-square distribution.

In addition to optical noise, electrical noise, generated in the first amplification stage after the photodiode, contributes to the overall noise. On the other hand, the signal quality is reduced due to a nonfinite extinction ratio X_r, which is defined as

$$X_r = \frac{P_{avg}(1)}{P_{avg}(0)} = \frac{\bar{i}_1}{\bar{i}_0}. \tag{5.21}$$

A reduced extinction ratio not only reduces the eye-opening $m_1 - m_0$ for a constant signal power, but also a higher power in the zero-level increases the noise on the zero-rail, as will be explained below.

For a signal with finite extinction ratio that is impacted by optical noise as well as electrical receiver noise, the following terms contribute to the Q-factor according to Eq. (5.19). Shot noise is neglected, as it is small compared to the other noise contributions.

Eye-Opening Reduced by Finite Extinction Ratio. The received optical power is converted to a current in the photodiode. The current is proportional to the instantaneous optical power with a proportionality constant R (responsivity) (see Section 3.3.1.1). For an average optical power P_{opt}, the average currents for the zero and one bits are

$$\bar{i}_0 = 2 \cdot R \cdot P_{opt} \cdot \frac{1}{1+X_r}, \quad \bar{i}_1 = 2 \cdot R \cdot P_{opt} \cdot \frac{X_r}{1+X_r}. \tag{5.22}$$

The eye opening is therefore reduced by a factor $(X_r - 1)/(X_r + 1)$ compared to infinite extinction ratio.

Beat Noise Between Signal and Optical Noise. The noise power incident on the receiver photodiode is described by the OSNR. The OSNR is measured by taking into account the noise in two orthogonal polarizations within an optical bandwidth of 0.1 nm. This measurement bandwidth is equivalent to $B_m = 12.5\,\mathrm{GHz}$ for a wavelength of 1550 nm. The optical noise power density N'_{opt} in each polarization for a given OSNR is

$$N'_{opt} = \frac{P_{opt}}{2 \cdot \mathrm{OSNR} \cdot B_m}. \tag{5.23}$$

Before the receiver, an optical filter is located, which reduces the bandwidth of the optical noise to B_o. The beat between signal and noise power detected on a photodiode results in noise in the (electrical) baseband. The spectrum of this baseband noise is rectangular, with a width of half the optical bandwidth, but only the spectral part within the electrical receiver bandwidth contributes to the electrical noise. This is shown in Fig. 5.12. The resulting variance of the electrical current depends on the instantaneous signal power and is expressed as

$$\sigma^2_{\text{S-ASE},x} = 4R \cdot N'_{opt} \cdot \bar{i}_x \cdot B_e, \quad (x = 0, 1). \tag{5.24}$$

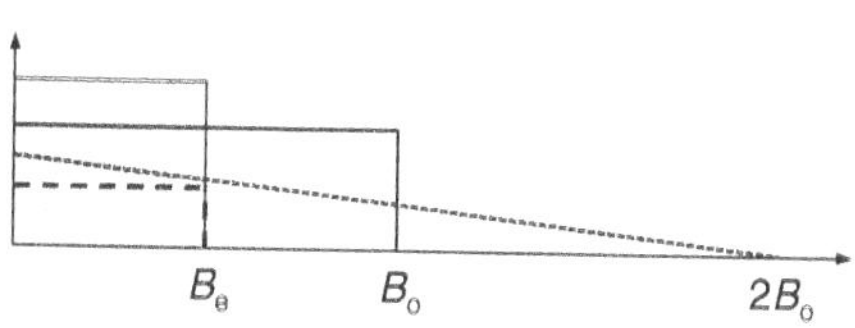

FIGURE 5.12 Baseband noise spectrum with different contributions: solid, signal-ASE beat noise; dotted, ASE–ASE beat noise; dashed, thermal noise. The electrical receiver bandwidth is depicted as a shaded rectangle. $B_o > B_e$ is assumed.

Beat Noise Between Optical Noise Components at Different Frequencies. Also the beat between different spectral components of the optical noise received on the photodiode results in noise in the baseband. The spectrum of this electrical noise is triangular, as it results from a convolution of the (rectangular) noise spectrum with itself. The width of this triangular spectrum depends on the bandwidth of the optical filter before the receiver, but only the spectral part passing the electrical filter contributes to the electrical noise. This noise contribution is also shown in Fig. 5.12. The resulting variance of the electrical current is expressed in Eq. (5.25). Please note that only the noise with the same polarization can interact. The overall noise depends on the number of polarizations incident to the receiver. Normally, both polarizations are present ($p = 2$), but if a polarizer is used before the receiver, the noise in only one polarization needs to be considered ($p = 1$), although the OSNR has been measured for two polarizations:

$$\sigma^2_{\text{ASE-ASE}} = p \cdot R^2 \cdot N'^2_{\text{opt}} \cdot (2B_\text{o} - B_\text{e}) \cdot B_\text{e}. \tag{5.25}$$

Thermal Receiver Noise. Noise generated in the electrical amplifier following the photodiode also contributes to the received signal quality. The equivalent thermal noise (i_{th}) referred to the amplifier input is often given as the equivalent root mean square (RMS) current per unit bandwidth in pA/$\sqrt{\text{Hz}}$. The noise variance is

$$\sigma^2_{\text{th}} = i^2_{\text{th}} \cdot B_\text{e}. \tag{5.26}$$

Usually, for an optical receiver the thermal noise is not given as a separate parameter, but the back-to-back sensitivity is specified. The thermal noise can be estimated from this value, as for this case the signal quality is only impacted by thermal noise and the Q-factor is obtained as

$$Q_{\text{bb}} = \frac{2 \cdot R \cdot P_{\text{opt}}}{2 \cdot \sqrt{i^2_{\text{th}} \cdot B_\text{e}}}. \tag{5.27}$$

The receiver sensitivity is normally given for a BER of 10^{-12}. According to Eq. (5.20), this corresponds to a Q-factor of 7 (17 dB). The electrical bandwidth can be estimated from the maximum bit rate of the receiver as $B_\text{e} = 0.7 \cdot R_\text{b}$. With a photodiode responsivity of 0.7, and a receiver sensitivity $P_{\text{opt,sens}}$, the thermal noise is then obtained as

$$i^2_{\text{th}} \approx \frac{P^2_{\text{opt,sens}} \cdot \left(\frac{A}{W}\right)^2}{70 \cdot R_\text{b}}. \tag{5.28}$$

As an example, the sensitivity of a 10 Gb/s receiver might be −24 dBm for a BER of 10^{-12}. This would result in an equivalent thermal noise current of 4.8 pA/$\sqrt{\text{Hz}}$.

Combining Eqs. (5.19) and (5.24) through Eq. (5.26), the Q-factor of an OOK modulated signal with direct detection is obtained from the equation

$$Q = \frac{\overline{i_1} - \overline{i_0}}{\sqrt{\sigma_{\text{th}}^2 + \sigma_{\text{S-ASE},0}^2 + \sigma_{\text{ASE-ASE}}^2} + \sqrt{\sigma_{\text{th}}^2 + \sigma_{\text{S-ASE},1}^2 + \sigma_{\text{ASE-ASE}}^2}}. \tag{5.29}$$

Using the Eq. (5.21) through Eq. (5.27) and Eq. (5.29) can be expanded to the somewhat lengthy Eq. (5.30)

$$Q = \frac{2[(X_r - 1)/(X_r + 1)]}{\begin{aligned}&\sqrt{\left[\left(P_{\text{opt,sens}}^2\right)/\left(49 \cdot P_{\text{opt}}^2\right)\right] + [(4B_e)/(\text{OSNR} \cdot B_m(X_r + 1))] + \left[(p(2B_o - B_e)B_e)/\left(4\text{OSNR}^2 \cdot B_m^2\right)\right]}\\ &+\sqrt{\left[\left(P_{\text{optp,sens}}^2\right)/\left(49 \cdot P_{\text{opt}}^2\right)\right] + [(4B_e)/(\text{OSNR} \cdot B_m \cdot (1 + {}^1/_{X_r}))] + \left[(p(2B_o - B_e)B_e)/\left(4\text{OSNR}^2 \cdot B_m^2\right)\right]}\end{aligned}}. \tag{5.30}$$

Commonly, the optical bandwidth is much larger than the electrical bandwidth ($B_o \gg B_e$) and the extinction ratio is considered as large ($X_r \to \infty$). The signal quality is only determined by optical noise, if the optical input power to the receiver, P_{opt}, is much larger than the receiver sensitivity. For two polarizations incident to the receiver, the following equation for Q is obtained:

$$Q \approx \frac{2\text{OSNR}\left[(B_m)/\left(\sqrt{B_o B_e}\right)\right]}{1 + \sqrt{1 + 4\text{OSNR}[(B_m)/(B_o)]}}. \tag{5.31}$$

For very large OSNR values (e.g., >20 dB), Eq. (5.31) can be further approximated as

$$Q^2 \approx \text{OSNR}\frac{B_m}{B_e}. \tag{5.32}$$

In this approximation, the square of the Q-factor is proportional to the OSNR. This approximation is often used, as the Q-factor in decibels is calculated as $Q_{\text{dB}} = 20 \log_{10}(Q)$ and therefore is the same as the OSNR in decibels, with a bandwidth related offset. In this case, the Q-factor scales with the OSNR with 1 dB/dB.

5.3.1.3 *Multilevel Amplitude Modulation*

Multilevel amplitude modulation has been investigated for 40 Gb/s WDM transmission. In most cases, this refers to *bipolar* ASK, rather than multilevel intensity modulation.

The simplest form of bipolar ASK is 2ASK (Fig. 5.13). This constellation diagram is identical to DPSK or BPSK, respectively.

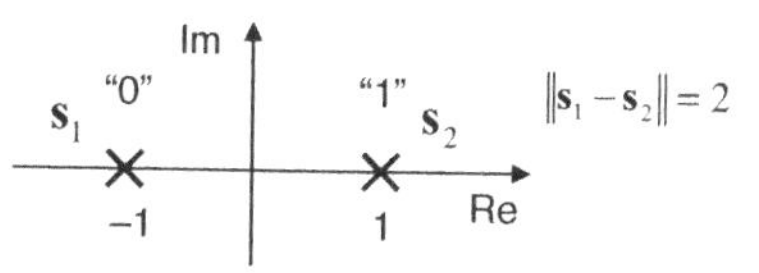

FIGURE 5.13 Constellation diagram of bipolar 2ASK or BPSK.

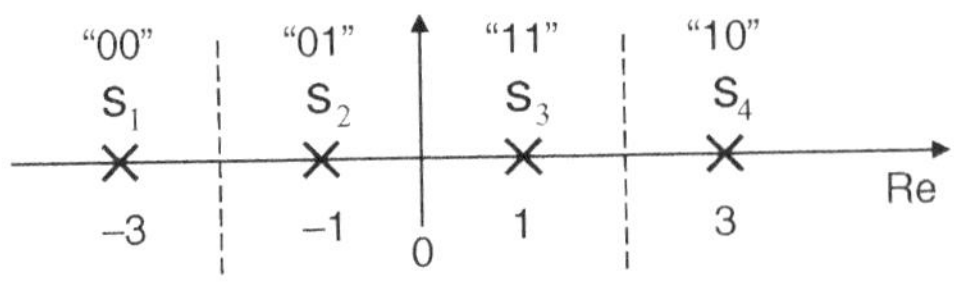

FIGURE 5.14 Constellation diagram example for bipolar 4ASK (DPSK-2ASK).

The BER of 2ASK is derived similarly to OOK. For NC, that is, preamplified directly detected or heterodyne asynchronous) and coherent homodyne (C) 2ASK, this yields, respectively

$$P_{B,NC} = \frac{1}{2}\exp\left(-\frac{E_b}{2N_0}\right), \quad P_{B,C} = Q\left(\sqrt{\frac{2E_b}{N_0}}\right). \tag{5.33}$$

Coherent homodyne bipolar 2ASK (like homodyne BPSK) achieves the best BER performance of all modulation schemes, or sensitivity of 9 photons/bit for $P_B = 10^{-9}$. For unipolar ASK, the numbers get worse by 3 dB, due to nonantipodal symbol spacing.

Multilevel (bipolar) ASK is shown in Fig. 5.14. The constellation diagram shown here can also be referred to as DPSK-2ASK, depending on the transmitter implementation.

Since M-ary ASK creates a multilevel signal, M-1 decision thresholds are required at the receiver. For optimum performance, a maximum-likelihood (ML) detector is required. The respective coherent receiver is shown in Fig. 5.15. $\psi(t)$ denotes the LO signal, r_1 is the decision variable, and $\hat{m}$ is the detected and decided output signal, respectively.

Symbol errors happen if the noise $n_1 = r_1 - \mathbf{s}_m$ exceeds in amplitude one-half of the distance between symbols adjacent in the symbol space. For the outer symbols $\mathbf{s}_1$ and $\mathbf{s}_M$ (the ones with $||\mathbf{s}|| = \pm 3$), errors can happen only in one direction. For the different symbol error probabilities one gets

$$\begin{aligned} P_e(\mathbf{s}_1) &= \Pr\left(n_1 = r_1 - \mathbf{s}_1 > \sqrt{E_g}\right), \\ P_e(\mathbf{s}_m) &= \Pr\left(|n_1| = |r_1 - \mathbf{s}_m| > \sqrt{E_g}\right) \quad \text{for} \quad 1 < m < M, \\ P_e(\mathbf{s}_M) &= \Pr\left(n_1 = r_1 - \mathbf{s}_M < -\sqrt{E_g}\right). \end{aligned} \tag{5.34}$$

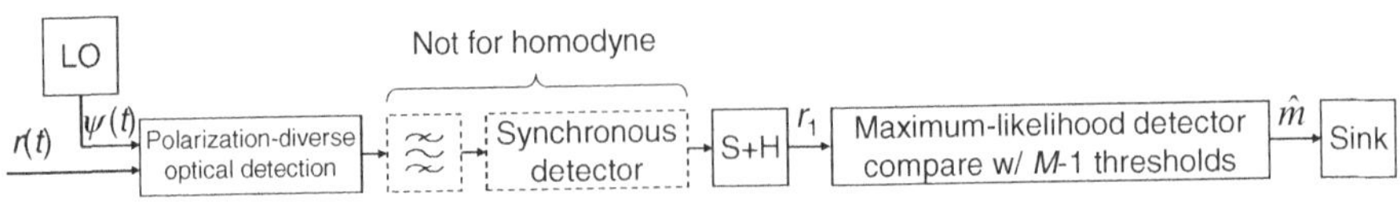

FIGURE 5.15 Coherent ML receiver for multilevel ASK.

The mean symbol error probability P_S can be derived as

$$P_S(M) = \frac{1}{M}\sum_{m=1}^{M} P_e(\mathbf{s}_m) = \frac{M-2}{M}\Pr\left(|n_1| > \sqrt{E_g}\right) + \frac{1}{M}\Pr\left(n_1 > \sqrt{E_g}\right) + \frac{1}{M}\Pr\left(n_1 < -\sqrt{E_g}\right).$$

$$= \frac{2(M-1)}{M}\Pr\left(n_1 > \sqrt{E_g}\right) = \frac{2(M-1)}{M}\int_{\sqrt{E_g}}^{\infty} p_{n_1}(n)dn = \frac{2(M-1)}{M} Q\left(\sqrt{\frac{2E_g}{N_0}}\right). \quad (5.35)$$

With the symbol energy $E_s = (\log_2 M)E_b = \left[(M^2-1)/(3)\right]E_g$, the symbol error probability of M-ary bipolar ASK [which is also referred to as M-PAM (pulse-amplitude modulation)] with homodyne detection gets

$$P_S(M) = \frac{2(M-1)}{M} Q\left(\sqrt{\frac{6\log_2 M}{M^2-1}\frac{E_b}{N_0}}\right). \quad (5.36)$$

For M-ary ASK (as well as for M-ary PSK and M-ary QAM), the BER follows from P_S:

$$P_B \approx P_S/\log_2 M \quad \text{for} \quad P_S \ll 1. \quad (5.37)$$

Today, M-ary ASK does not play a major role in WDM transmission. The concept, however, has been extended into the complex symbol space, leading to QAM. M-ary QAM is an important modulation scheme for ultrahigh-speed transmission.

5.3.2 Phase Modulation

5.3.2.1 Phase-Shift Keying PSK has first been considered for ultralong-haul WDM transmission in the 1990s. Advantages were seen in the constant-envelope characteristic (for cosine roll-off NRZ, making is less susceptible to certain fiber nonlinear effects), and the superior OSNR.

The BER of homodyne BPSK has already been stated (for bipolar 2ASK). It is given by

$$P_B = Q\left(\sqrt{\frac{2E_b}{N_0}}\right). \quad (5.38)$$

This BER follows from Eq. (5.12) with the antipodal BPSK symbol distance

$$||\mathbf{s}_1 - \mathbf{s}_2|| = 2. \quad (5.39)$$

The block diagram of a homodyne BPSK system (with differential precoding for receive-end phase unambiguity) is shown in Fig. 5.16. The receiver uses polarization diversity that is required in coherent detection when the principal states of polarization (PSPs) of the received signal are unknown.

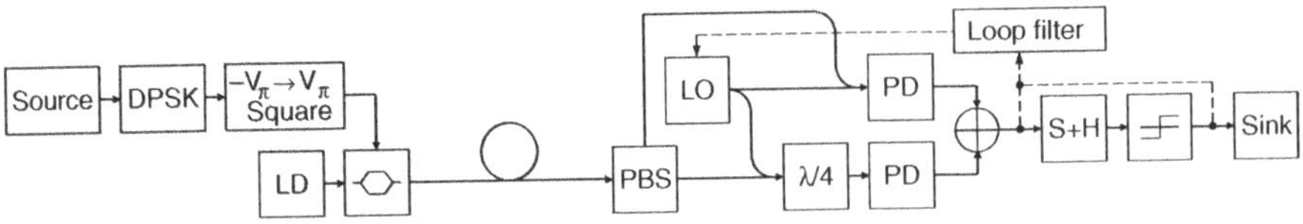

FIGURE 5.16 Optical homodyne BPSK system. PBS: polarization beam splitter, LO: local oscillator.

Differential precoding as shown in Fig. 5.16 can be used in order to avoid receive-end phase ambiguity. This precoding requires some limited, dedicated hardware effort. It can also lead to a small system penalty (typically <1 dB), but avoids error bursts due to phase jumps.

Although achieving optimum BER, the structure shown in Fig. 5.16 is not used anymore. The analog loop-filter implementation caused high effort. Today, one uses intradyne receivers, where the phase-estimation problem has been shifted into the digital domain.

Since coherent fiber-optic transmission almost vanished after the invention of the Erbium-doped fiber amplifier (EDFA), most of the early work concentrated on noncoherent binary PSK with differential precoding and delay demodulation, DPSK. Only after year 2000, focus turned back onto coherent PSK, now with intradyne detection.

5.3.2.2 DPSK with Direct Detection Similarly to noncoherent OOK, one gets for the BER of noncoherently detected DPSK

$$P_{\mathrm{B}} = \frac{1}{2} \exp\left(-\frac{E_{\mathrm{b}}}{2N_0}\right). \tag{5.40}$$

This BER is worse as compared to homodyne BPSK by ~3.5 dB. It is thus in the same range as the BER for heterodyne synchronous BPSK and homodyne unipolar 2ASK detection.

The system block diagram of a DPSK system again can be derived from Fig. 5.2 and is shown in Fig. 5.17. It comprises an external (phase) modulator in the transmitter, which is suitable for ultrahigh-speed modulation up to >40 Gb/s. Between NRZ source and modulator driver, a differential precoder should be implemented in order to avoid error propagation after detection. The receiver consists of preamplifier, OBPF (WDM demultiplexing and noise bandwidth

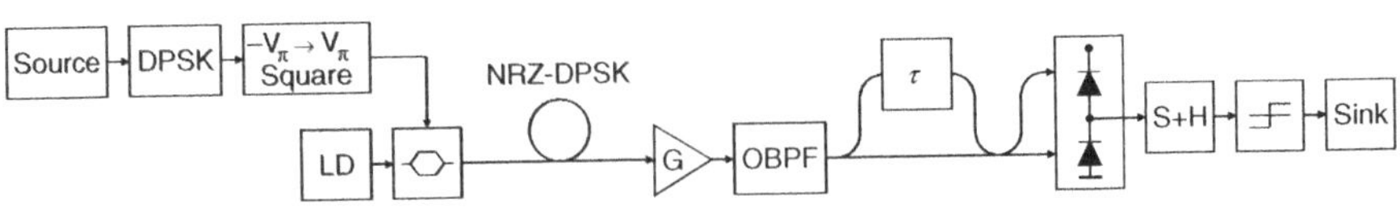

FIGURE 5.17 NRZ-DPSK with direct detection and self-coherent delay demodulation.

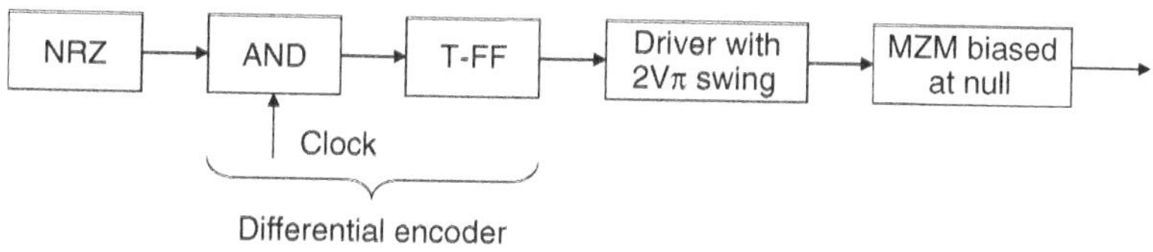

FIGURE 5.18 Differential precoder for DPSK.

limitation), and a delay demodulator with balanced receiver, which is followed by direct detection. As stated earlier, $\tau = T$.

A possible implementation of the DPSK precoder is given in Fig. 5.18. The input to the precoder is a unipolar NRZ signal. This signal is processed, together with the clock, in an AND gate and a toggle flip-flop (T-FF).

The resulting DPSK precoding and phase is listed in Table 5.5 for an 8 bit sequence.

The delay demodulator in Fig. 5.17 consists of a Mach–Zehnder interferometer (MZI), where the delay of one branch is tuned to the bit duration T. This delay demodulator compares successive bits and thus demodulates the differential precoding. In practical implementations, the MZI must be temperature controlled in order to maintain the correct delay. Mismatch of the interferometer by 0.1 mm path-length can lead to ~15° of phase error. The penalty induced by phase error is shown in Fig. 5.19 for NRZ-DPSK.

TABLE 5.5 DPSK Precoder

	0	1	2	3	4	5	6	7	8
NRZ	–	0	1	1	0	0	1	0	1
DPSK	1	0	0	0	1	0	0	1	1
Phase	$+E$	$-E$	$-E$	$-E$	$+E$	$-E$	$-E$	$+E$	$+E$

1: no change of DPSK signal; 0: change of DPSK signal.

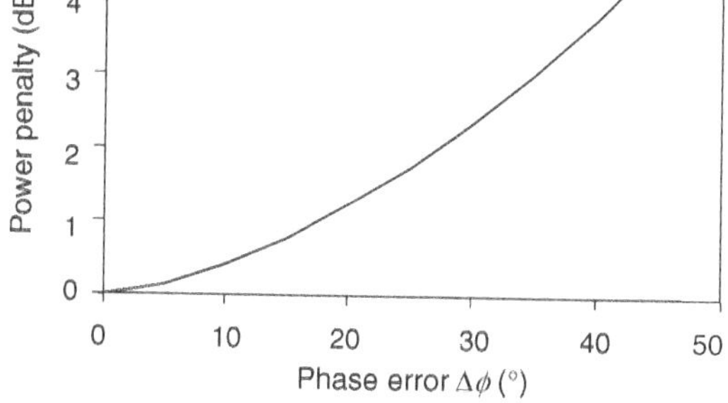

FIGURE 5.19 Power penalty due to phase error in NRZ-DPSK.

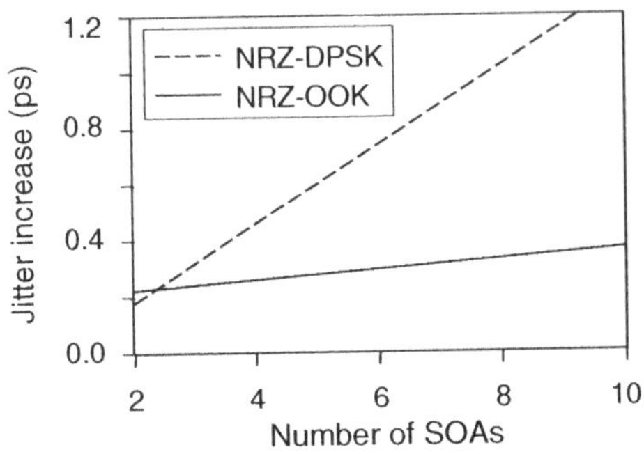

FIGURE 5.20 Jitter increase due to added phase noise for NRZ-OOK and NRZ-DPSK.

In certain DPSK implementations [adaptive DPSK (ADPSK)], the MZI delay can be varied in order to account for certain transmission effects like accumulated filter bandwidth.

One problem of DPSK is its sensitivity to added phase noise. The rise time jitter of a demodulated DPSK signal can be correlated to the phase noise, which is generated in long transmission links as nonlinear phase noise though AM-PM conversion of EDFA ASE (amplified spontaneous emissions), and which is also generated by semiconductor optical amplifiers (SOAs). As an example, Fig. 5.20 shows the jitter increase of a demodulated DPSK signal, as compared to NRZ-OOK, as a function of the number of traversed SOAs.

From Fig. 5.20, it becomes apparent that PSK is more susceptible to added phase noise than OOK. Today, the most relevant source for phase noise is nonlinear conversion of ASE into phase noise in chains of EDFAs in long-haul transmission [25,26]. For this reason, power levels in long-haul XPSK transmission must usually be kept lower as compared to typical values for intensity modulation.

5.3.2.3 (D)XPSK The most relevant implementation of multilevel PSK is DQPSK. It has been used for both noncoherent and coherent systems, and it is the basis of the quasi-standardized 100 Gb/s transmission scheme. The BERs for NC and coherent homodyne (C) detection are the same as for BPSK, for the same per-bit energies E_b:

$$P_{B,NC} = \frac{1}{2}\exp\left(-\frac{E_b}{2N_0}\right), \quad P_{B,C} = Q\left(\sqrt{\frac{2E_b}{N_0}}\right). \tag{5.41}$$

The reason for the good BER performance is that QPSK can be regarded the combination of two BPSK systems: the first in the inphase component of a complex signal, and the second in the orthogonal quadrature component. In an ideal receiver, both components stay orthogonal and hence do not disturb each other.

The noncoherent version is used for 40 Gb/s WDM transmission, where it is often combined with RZ pulse coding for better tolerance against fiber nonlinearity (and

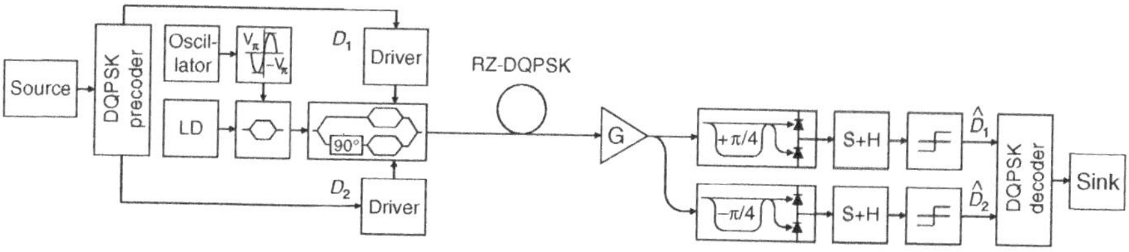

FIGURE 5.21 RZ-DQPSK with self-coherent delay demodulation. D: decision circuit.

somewhat better tolerance against PMD) (refer to Table 2.2). Figure 5.21 shows the block diagram of an RZ-DQPSK transmission system.

The transmitter consists of a pulse carver (driven at Baud rate for RZ50) and a double-nested MZM for the quadrature modulation. The two branches of the double-nested modulator are delayed against each other by 90° in order to provide the orthogonal I and Q components. Other modulator structures can be used, for example, a push–pull modulator followed by a phase modulator [27,28], and DQPSK can also be generated with directly modulated chirp-managed lasers (CML) [29].

The transmitter requires a DQPSK precoder, which performs differential precoding and parallelization into the I and Q components. In real transmission systems, this is complemented by means for forward error correction (FEC), performance monitoring (PM), and OTN (Optical Transport Network) mapping and framing.

The receiver consists of two DPSK demodulators, which are separated by a 3 dB coupler. This makes use of the basic property of 3 dB couplers of inducing a 90° phase shift between the two output ports (refer to Section 3.2.6.1). Combined with delay demodulation, a 3 dB coupler can thus be used to demultiplex the I and Q components without penalty. The delay demodulators in the two branches have to be tuned to $T_S \pm \pi/4$, with T_S being the symbol duration.

Typically, there is a positive impact on OSNR from RZ coding, which results from better PMD and SPM (self-phase modulation) performance, and also better receive-end matched filter approximation aided by the RZ coding. On the downside, RZ leads to worse performance with regard to XPM when it comes to WDM pulse collisions in long-haul transmissions.

The general structure of a noncoherent M-ary DPSK receiver has been given in Ref. [27]. It is based on an extension of the receiver shown in Fig. 5.21 by a direct detection branch with multilevel decision capability. However, M-ary DPSK for $M > 4$ suffers higher penalties than M-ary QAM and plays no major role in WDM transmission.

The general receiver structure for coherent (heterodyne) M-PSK is shown in Fig. 5.22. $\psi(t)$ is the LO signal, which is combined with the receive signal with polarization diversity or polarization multiplex. The combining element gives access to the resulting I and Q components. This can be achieved in so-called 90° hybrids (see Section 5.4).

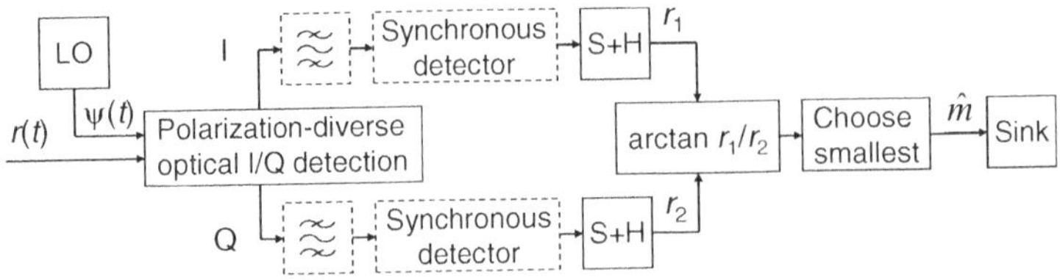

FIGURE 5.22 Coherent detection of M-PSK.

The detector compares the phase of the observation vector to $M-1$ thresholds. Due to the circular symmetry of the signal space, one gets for the symbol error probability

$$P_S(M) = 1 - P_C(M) = 1 - \frac{1}{M}\sum_{m=1}^{M} P_c(\mathbf{s}_m) = 1 - P_c(\mathbf{s}_1) = 1 - \int_{-\pi/M}^{\pi/M} p_{\hat{\varphi}}(\varphi)d\varphi. \tag{5.42}$$

Here, $P_C(M) = \Pr(\text{no error detected on I and Q branches})$, and

$$p_{\hat{\varphi}}(\varphi) \approx \sqrt{\frac{2\,E_s}{\pi N_0}}\cos(\varphi)\exp\left(-\frac{E_s}{N_0}\sin^2\varphi\right), \quad |\varphi| \leq \frac{\pi}{2}. \tag{5.43}$$

The symbol error probability for homodyne detection and for $M \geq 4$ yields

$$P_S(M) \approx 2Q\left(\sqrt{\frac{2(\log_2 M)E_b}{N_0}}\sin\left(\frac{\pi}{M}\right)\right), \quad P_B \approx P_S/\log_2 M \quad \text{for} \quad P_S \ll 1. \tag{5.44}$$

P_B can be derived from Eq. (5.37). For coherent QPSK, the BER equals the one for coherent BPSK for the same E_b. For higher order PSK, increasing penalties result.

5.3.3 QAM-Combined Amplitude and Phase Modulation

QAM is the combination of phase and amplitude modulation. The term is often applied to rectangular symbol constellations in the complex symbol space, however, other constellations based on multiple PSK stars or even nonsymmetric constellations exist.

Although QAM with differential precoding can be detected self-coherently (noncoherently) with I and Q delay demodulators and a multilevel direct detection branch, the clear focus lies on coherent (intradyne) detection.

The generic structure of a coherent M-QAM receiver is shown in Fig. 5.23. Similar to the coherent M-PSK receiver shown in Fig. 5.22, it consists of two branches (I and Q). Both branches use ML multilevel detectors.

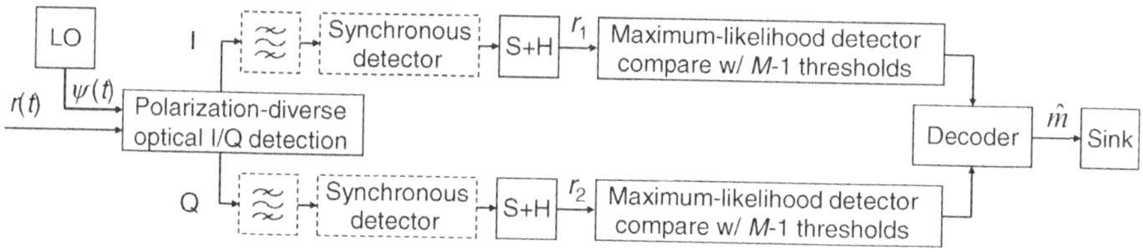

FIGURE 5.23 Coherent detection of *M*-QAM.

M-QAM can be viewed as the combination of two modulations in the I and Q branches. No bit errors occur if no errors are detected in neither the I nor the Q branches. Due to the orthogonality of I and Q, the symbol error rate can be expressed as $P_S(M) = 1 - P_C(M)$, where $P_C(M)$ is the joint probability of no errors being detected in both branches.

Considering the symmetry in the signal space, P_C can be expressed as

$$P_C = \Pr(\text{no error on I}) \cdot \Pr(\text{no error on Q}) = \Pr(\text{no error on I})^2 = \left(1 - P_S\left(\sqrt{M}\right)\right)^2. \tag{5.45}$$

$P_S\left(\sqrt{M}\right)$ is the symbol error rate of bipolar $\sqrt{M}$-ASK as given earlier. The symbol error rate for QAM with homodyne detection therefore follows as

$$P_S(M) = 4\left(1 - \frac{1}{\sqrt{M}}\right) Q\left(\sqrt{\frac{3\log_2 M}{M-1}\frac{E_b}{N_0}}\right). \tag{5.46}$$

P_B again follows from Eq. (5.37). Comparisons between the different modulation schemes with regard to achievable BER are given later. In general, *M*-ary QAM achieves better results than *M*-ary PSK beyond $M = 4$ (where both are identical).

Today, the focus in long-haul, ultrahigh-speed transmission lies on dual-polarization QAM with digital intradyne receivers. For 100 Gb/s transport in the 50 GHz WDM grid, this leads to NRZ-DP-4QAM (or NRZ-DP-QPSK). For higher transmission speeds on the same grid (200 and 400 Gb/s), NRZ-DP-16QAM can be used, eventually with a dual-carrier approach (NRZ-DP-DC-16QAM). For tight subcarrier spacing, this may be complemented by partial response signaling (PRS; here, QPR).

5.3.4 Optical Noise Tolerance of Multilevel Modulation Formats

The trend toward higher data rates in optical transmission systems and the requirement of higher spectral efficiency has led to the use of optical multilevel modulation formats, where multiple information bits are encoded in a single symbol (see Section 5.3.3). While these modulation formats reduce the spectral width of the signal and therefore increase the spectral efficiency in WDM systems, the signal becomes more sensitive to noise added in optical amplifiers along the transmission line and therefore is limited in system reach. A comparison of different modulation formats

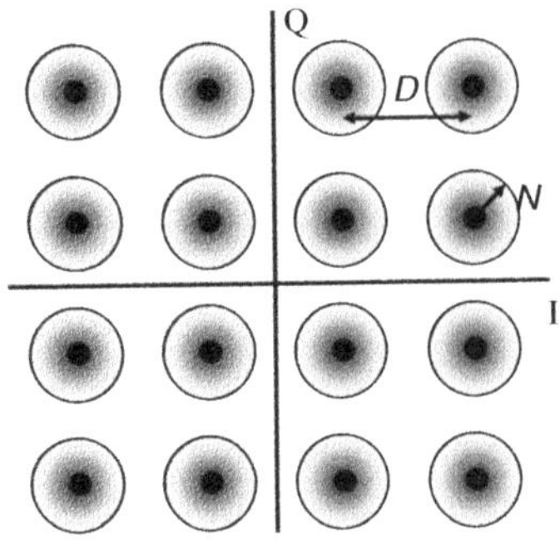

FIGURE 5.24 Representation of constellation points in the complex (I–Q) plane for the example of 16QAM. D is the distance between neighbor constellation points, and N is a measure for the noise amplitude.

for a given per-channel data rate must therefore take into account the achievable capacity of a transmission system as well as the OSNR tolerance and the respective unregenerated transmission reach. In the following, we will give a simple method to estimate the OSNR tolerances for various multilevel modulation formats.

The electrical field vector of an amplitude and/or phase-modulated optical signal can be depicted in a constellation diagram in the complex plane, as shown in Fig. 5.24 as the example of 16QAM. To encode more than one information bit into a symbol, more than two constellations points are required. In general, by using M constellation points, $b = \mathrm{ld}(M)$ information bits can be encoded into a symbol. Here, $\mathrm{ld}(x)$ is the logarithm of x with respect to base 2. The value b can be a noninteger number. In this case, multiple symbols are combined to encode an integer number of information bit.

During propagation in the transmission system, optical noise is added to the signal by optical amplifiers. Here, we assume white noise with a Gaussian amplitude distribution and uniformly distributed phase. The value N in Fig. 5.24 is a measure for the noise amplitude.

At the receiver, the decision is made, as to which constellation point is represented by the received symbol. An error occurs, when the added noise has caused the symbol to cross the threshold to a neighboring constellation point. This minimum distance to the next constellation neighbor is therefore a critical value for the performance of the modulation format. It is noted as the distance D in Fig. 5.24. We assume a linear detection, like in a coherent receiver. However, the approximate results can also be applied to direct detection receivers. The OSNR is the ratio of the average signal power P_{avg} and the optical noise power in a fixed optical bandwidth.

$$\mathrm{OSNR} = \frac{P_{\mathrm{avg}}}{N_{\mathrm{opt}}^2}. \tag{5.47}$$

Here, the average signal power P_{avg} is obtained as the average of the square of the normalized lengths of all constellation vectors. The noise present at the decision circuit is dominated by the filtered optical noise. When the symbol rate is reduced due to higher information content per symbol, a filter bandwidth, reduced by a factor

b_{rel} compared to binary signaling, can be applied and the noise impact is reduced. The SNR after filtering, responsible for the quality of detection, is approximated as

$$\mathrm{SNR} = \frac{D^2}{N^2 \cdot b_{rel}}. \tag{5.48}$$

To compare different modulation formats, the required OSNR is related to a fixed SNR, determined by the constellation distance D. If we normalize to the required OSNR for the 2PSK format, we find a penalty of an arbitrary multilevel format by the simple approximation

$$\mathrm{OSNR}_{\mathrm{Penalty}} = \frac{4 \cdot P_{avg} \cdot b_{rel}}{D^2}. \tag{5.49}$$

Please note that this is an OSNR penalty relative to the required OSNR with 2PSK format.

In Table 5.6, for various multilevel modulation formats the values of normalized average power (P_{avg}), normalized constellation point distance (D), and relative

TABLE 5.6 OSNR Penalty for Various Multilevel Modulation Formats

	Number of Levels	P_{avg}	D	b_{rel}	$\mathrm{OSNR}_{penalty}$	$\mathrm{OSNR}_{penalty}$ (dB)
Unipolar ASK	2	0.5	1	1	2	3 dB
	4	0.389	0.333	1/2	7	8.5 dB
	8	0.357	0.143	1/3	23.33	13.7 dB
	M	$\frac{(2M-1)}{6 \cdot (M-1)}$	$\frac{1}{(M-1)}$	$\frac{1}{ld(M)}$	$\frac{2(2M-1)(M-1)}{3 \cdot ld(M)}$	Fig. 5.25
Bipolar ASK	2	1	2	1	1	0 dB
	4	0.556	0.667	1/2	2.5	4 dB
	8	0.429	0.286	1/3	7	8.5 dB
	M	$\frac{(M+1)}{3 \cdot (M-1)}$	$\frac{2}{(M-1)}$	$\frac{1}{ld(M)}$	$\frac{(M^2-1)}{3 \cdot ld(M)}$	Fig. 5.25
PSK	2	1	2	1	1	0 dB
	3	1	1.732	0.631	0.841	−0.8 dB
	4	1	1.414	1/2	1	0 dB
	8	1	0.765	1/3	2.276	3.6 dB
	M	1	$2 \cdot \sin\left(\frac{\pi}{M}\right)$	$\frac{1}{ld(M)}$	$\frac{1}{\sin^2(\pi/M) \cdot ld(M)}$	Fig. 5.25
Square QAM	4	2	1.414	1/2	1	0 dB
	16	1.111	0.667	1/4	2.5	4 dB
	M	$\frac{2 \cdot (M-1)}{3 \cdot (\sqrt{M}-1)^2}$	$\frac{2}{(\sqrt{M}-1)}$	$\frac{1}{ld(M)}$	$\frac{2 \cdot (M-1)}{3 \cdot ld(M)}$	Fig. 5.25

Note: The constellations of bipolar 2ASK/2-PSK and 4-PSK/4QAM are equivalent, respectively.

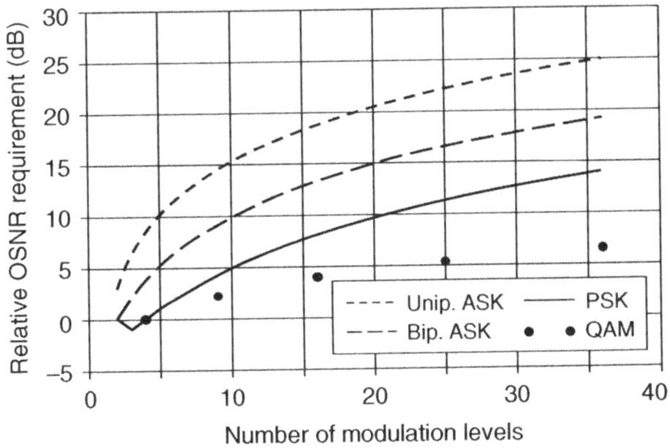

FIGURE 5.25 OSNR penalty for BER 10^{-5} for multilevel modulation formats as compared to 2PSK.

bandwidth (b_{rel}) are summarized and the resulting OSNR penalty is evaluated using Eq. (5.49). For the general case of M modulation levels, the OSNR penalty compared to 2PSK is given for each modulation type. Based on these equations, the OSNR penalties are shown in Fig. 5.25. It is found that optimum OSNR tolerance can be obtained for a 3PSK modulation, where the required OSNR is approximately 0.8 dB lower than for 2PSK or 4PSK. Using this modulation format, >1.5 bit can be encoded into one symbol or 3 bits can be encoded in two symbols.

5.3.5 Partial Response Signaling

5.3.5.1 Optical Duobinary Optical duobinary is the simplest form of *partial response* signaling in WDM transmission. Partial response (or correlative) coding intentionally produces a *known* amount of ISI in order to achieve higher spectral efficiency. Since this ISI is deterministic, it can be eliminated in the receiver. Partial response signaling has been studied extensively since the early 1960s [30–34].

For a transmitted signal $x(t) = \sum_{k=-\infty}^{+\infty} d_k q(t - kT), d_k = 0, 1$, with T being the symbol (here also bit) duration, the pulse $q(t)$ is usually chosen such that $q(kT) = 1$ for $k = 1$ and 0 otherwise in order to avoid ISI. For duobinary, the pulse $q(t)$ is chosen instead:

$$q(kT) = \begin{cases} 1, & k = 0, 1, \\ 0, & \text{otherwise.} \end{cases} \tag{5.50}$$

At sampling instances kT, the receiver recovers $d_{k-1} + d_k$, rather than d_k. By allowing this ISI, the pulse $q(t)$ can be made longer [as shown in Eq. (5.50)], hence duobinary is narrower in the frequency domain and as a consequence, less susceptible to CD. The ODB spectrum is compared to OOK and PSK in Fig. 5.7.

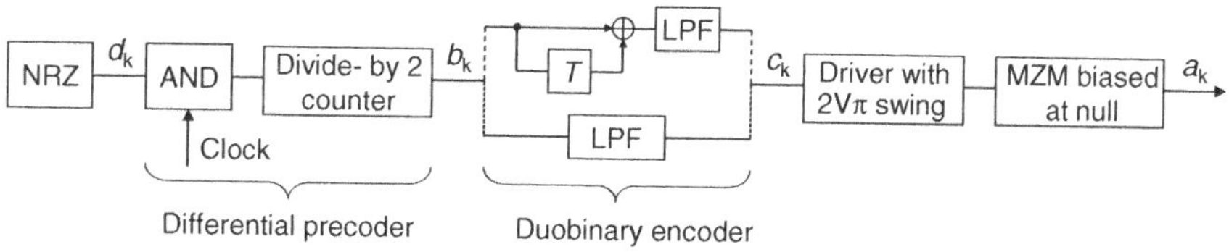

FIGURE 5.26 Optical duobinary coding.

One way to encode the duobinary signal consists of a single-stage delay-and-add filter, followed by a Nyquist low-pass filter (LPF). For $d_k = 0, 1$, this leads to three-level signals c_k that can take the values $\{-1, 0, 1\}$. The transmit signal is then given by $x(t) = \sum_{k=-\infty}^{+\infty} c_k q(t - kT)$. In fiber-optic (WDM) transmission, this encoder can be replaced by an LPF with even tighter bandwidth (so-called *filtered* optical duobinary). For 10 Gb/s signals, a Bessel LPF with 2.8 GHz can be used. If a dedicated encoder is used, the trailing LPF bandwidth is in the range of 0.5–$0.75 \cdot R_B$ (R_B being the bit rate).

The ISI introduced at the transmitter can be unraveled at the receiver by differential decoding. However, a single error might then propagate forever. Therefore, differential *precoding* is preferably performed at the transmitter, which then allows symbol-by-symbol detection. With precoding $c_k = c_{k-1} \oplus d_k$ (with $\oplus$ modulo-2 subtraction), the receiver can directly recover d_k at sampling instances kT. The differential precoder compares two successive bits.

Differential precoders similar to the ones used for DPSK (see Section 5.3.2.2) can be used for ODB. The resulting transmitter is shown in Fig. 5.26.

A precoder/LPF output example is given in Table 5.7 (note that b_k now is the encoder input).

Precoding plus (LPF) encoding leads to a triple-state electrical field (refer to Figs. 5.5 and 5.27). Since spaces ("0") of the original NRZ sequence remain spaces after pre- and encoding, ODB can be detected symbol-by-symbol with a standard IM-DD receiver. It can also be detected using heterodyne, homodyne, and intradyne techniques.

Increased spectral efficiency and CD tolerance can also be explained by the encoding, which does not allow certain sequences of c_k. In particular the sequences $\{-1\ 1\}$, $\{1\ -1\}$, $\{-1\ 0\ -1\}$, and $\{1\ 0\ 1\}$ are now forbidden. Only sequences $\{-1\ 0\ 1\}$, $\{1\ 0\ -1\}$, and those with multiple successive marks or spaces ("1"s or "0"s) can occur.

TABLE 5.7 ODB Differential Precoding and Encoding

Bit	0	1	2	3	4	5	6	7	8	9	10	11	12
d_k	–	0	1	0	1	1	1	0	0	1	0	1	0
b_k	0	1	1	0	0	0	0	1	0	0	1	1	0
c_k	–	0	1	0	−1	−1	−1	0	0	−1	0	1	0
a_k	–	0	$+E$	0	$-E$	$-E$	$-E$	0	0	$-E$	0	$+E$	0

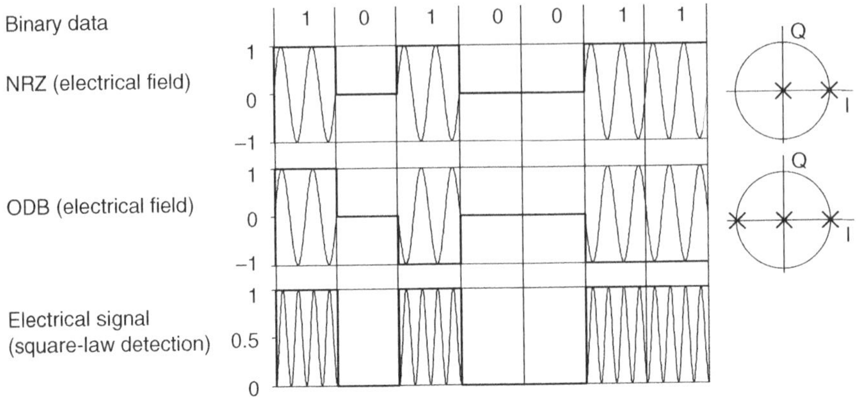

FIGURE 5.27 Duobinary coding.

To date, ODB and its complex equivalent, 9QPR, are the only correlative coding schemes relevant for WDM transmission. They are also referred to as Class-1 partial response [35]. Other correlative encoders exist, which lead to a variety of different partial response signals, for example, modified duobinary (Class-4) or AMI (alternate mark inversion).

The BER has been derived in Refs [36,37] for asynchronously detected ODB. With ideal receivers, ODB can achieve almost 1 dB advantage over OOK. The BER, however, strongly depends on the low-pass filtering. It also depends on the receive-end residual CD. The resulting dependence of the Q-factor is shown in Fig. 5.28. Here, a bit rate of 10 Gb/s has been considered. ODB can lead to a penalty of up to ~1.5 dB as compared to (NRZ) OOK, depending on implementation. Filtered ODB has better performance on longer links, that is, for higher values of residual chromatic dispersion.

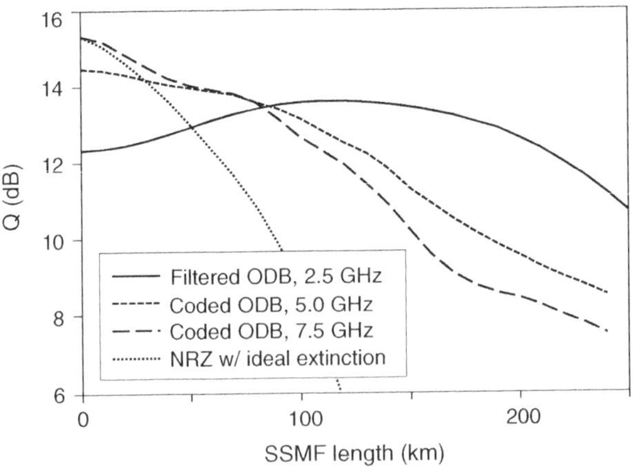

FIGURE 5.28 Comparison of filtered and precoded ODB.

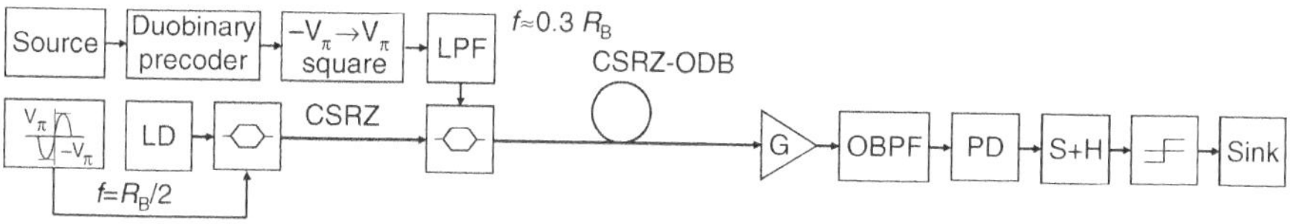

FIGURE 5.29 CSRZ-ODB with direct detection.

ODB can also be detected coherently [38]. The usual coherent detection schemes apply, that is, homodyne (intradyne) detection, and synchronous heterodyne detection. For homodyne detection, ODB achieves the second best BER performance, which can only be improved by homodyne PSK. The BER is given by

$$P_{\mathrm{B}} \approx \frac{3}{2} Q\left(\frac{\pi}{2} \sqrt{N_{\mathrm{P}}}\right). \tag{5.51}$$

$N_{\mathrm{P}} = E_{\mathrm{b}}/hf_0$ is the average per-bit photon number. With optimized threshold, homodyne ODB achieves sensitivity (at $\mathrm{BER} = 10^{-9}$) of 15 photons/bit.

The combination of ODB and RZ pulse shaping has been investigated extensively. The spectral efficiency of ODB is partly compensated by RZ coding, but the RZ tolerance against nonlinear fiber effects remains. According to Table 5.2, CSRZ-ODB has the lowest optical bandwidth of all RZ modulation schemes, measured against the symbol rate. It has been widely investigated, for example, in Refs [39,40]. The system block diagram of a CSRZ-ODB system is shown in Fig. 5.29. As compared to NRZ-OOK (Fig. 5.11), additional pulse carver including driver, LPF encoder, and differential precoder are required.

According to Ref. [36], the performance of directly detected ODB, as compared to IM-DD, depends on the chosen pulse shape. If matched filter detection is used, performance varies between gain of ~0.9 dB for si-shaped ($\mathrm{si}(x) = (\sin(x))/x$) pulses and a penalty of 3 dB for rectangular-shaped pulses, as compared to IM-DD. For cosine roll-off pulses, ODB leads to a penalty of ~1 dB.

ODB can be used for cost-effective WDM transport of 100 Gb/s signals. Then, four DWDM lanes carrying ~28 Gb/s each can be implemented (Fig. 5.30).

Both, the transmitter and receiver arrays can be integrated for low cost, power consumption, and form-factor requirements. Chirp-managed (directly modulated)

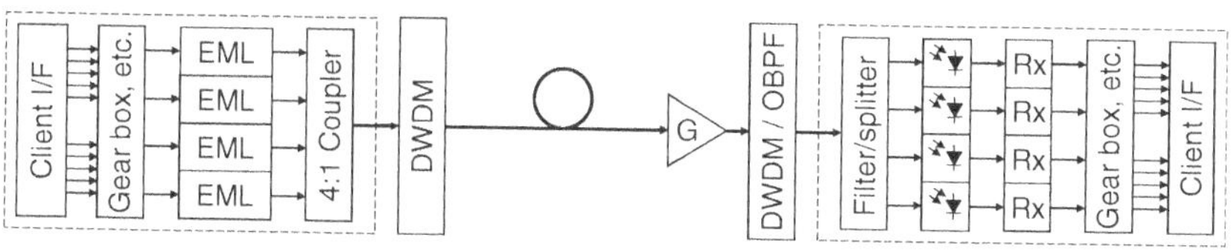

FIGURE 5.30 High-speed four-subcarrier duobinary system with externally modulated lasers (EML) and direct detection. Gear box is the synchronization circuit needed for 100 Gb/s client interfaces.

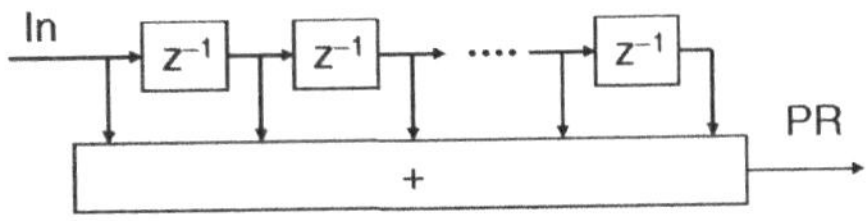

FIGURE 5.31 Generalized partial response encoding.

lasers can be used as an alternative to externally modulated lasers (EML) to further reduce cost. The approach shown in Fig. 5.30 is similar to the long-reach and extended reach CFPs, which have been standardized by IEEE for the 100GbE LR-4 and ER-4 LAN interfaces.

WDM transport according to Fig. 5.30 can achieve ~500 km reach. This is limited mainly by OSNR. The subcarrier spacing of 4 × 28G ODB transport can be 50 GHz or 25 GHz [41]. As compared to 50 GHz, 25 GHz leads to a reach penalty that is caused by subcarrier cross talk and increasing filtering effects. CD tolerance with direct detection, but without added receive-end MLSE, is in the range of ±250 ps/nm [42].

5.3.5.2 Higher Order and Quadrature Partial Response Partial response coding is not restricted to ODB. Higher order PR can be achieved by providing ISI between more than two successive symbols. This leads to multilevel transmission and increased spectral efficiency, however, in general, a maximum-likelihood sequence estimator (see Section 3.4.3) will be required at the receiver in order to eliminate the ISI. The general higher order PR encoder is shown in Fig. 5.31.

The concept of (higher order) PR can further be extended into the complex symbol space, that is, into the quadrature component. This is referred to as QPR. It has been used for military and free-space satellite communications (together with higher order PR). More recently, QPR has been studied extensively for spectrally efficient WDM transmission [43–46].

For WDM transmission, only the ODB equivalent in the complex space, 9QPR, has gained relevance so far. As shown in Fig. 5.5, 9QPR has a symmetric 3 × 3 constellation.

The BER for coherent homodyne M-ary QPR ($M = 9, 25, 49, \ldots$) can be expressed as [35]

$$P_{\mathrm{B},M\text{-QPR}} = \frac{4}{\log_2 L}\left(1 - \frac{1}{L^2}\right) Q\left[\frac{\pi}{2}\sqrt{\frac{3E_\mathrm{b}(\log_2 L)}{(L^2-1)N_0}}\right]. \tag{5.52}$$

Here, $L = \sqrt{M}$, that is, $L = 3$ for $M = 9$, and so on. Eq. (5.33) yields a penalty of ~2 dB as compared to coherent homodyne QPSK, or a gain of ~2 dB as compared to homodyne 16QAM, for E_b/N_0 kept constant. This makes 9QPR (like homodyne ODB for noncomplex binary transmission) the second most powerful complex modulation scheme. This result does not yet take an optimum receiver into consideration. For QPR, this is an MLSE that ideally can eliminate all penalty caused by partial response coding, compared to the respective nonpartial response scheme (e.g., ODB vs. BPSK, 9QPR vs. 4QAM/QPSK, and 49QPR vs. 16QAM) [47]. If this MLSE is only used for partial response decoding [i.e., *not* for additional (CD and PMD) equalization], it can be implemented at reasonable effort even for ultrahigh bit rates. Moreover, simplified MLSE variants have been demonstrated for both duobinary and 9QPR. They are referred to as ambiguity zone detectors (AZD) [32]. AZD can improve the penalty for 9QPR against (coherent) QPSK to ~1 dB at little implementation effort.

Partial response coding forbids certain sequences of symbols. If M is the number of baseband levels from partial response coding, the coded signal cannot transverse more than N levels between consecutive symbols, where N is $(M+1)/2$. The only way such forbidden sequences can occur is when a symbol has been received in error. The assumption here is that errors are caused by AWGN, and are displaced from the correct level into directly adjacent levels only. This assumption is valid under normal received S/N ratios of interest. When such a forbidden sequence occurs, it is called a partial response violation (PRV).

In an AZD, the eye at the receiver is divided vertically between the pinpoints (ideal locations of the eye centers) into two types of regions. The symbols sampled in the regions closest to the pinpoints are given ambiguity weight of 0. Other regions, or the ambiguity zones, are further from the pinpoints and given an ambiguity weight of 1. Their relative ambiguity positions above or below pinpoints are also marked for later error correction. Symbols measured in the ambiguity zones are far more likely to be in error than symbols closer to the pinpoints, and are considered to be suspect. The decoded symbols and their ambiguity weights and position markers are fed into the PRV detection and correction circuitry. When a PRV is detected, the decoder looks back D symbols to see if any decision has been made in the ambiguity zone (ambiguity weight 1). If so, a correction is made by pushing the ambiguity decision one level up or down using the position marker.

5.3.6 Frequency-Shift Keying

FSK is a nonlinear modulation format. Any linear change of the modulating signal leads to nonlinear changes of the resulting transmission signal. This can be seen from the real-value representation of the transmission signal $E(t)$:

$$E(t) = A \cdot \cos \omega(t)t = A \cdot \cos\left(\omega_0 t + \alpha \sum_k d_k \int_{-\infty}^{t} s(x - kT_S)dx\right). \qquad (5.53)$$

Here, ω_0 is the unmodulated carrier frequency, α is the modulation constant, T_S is the symbol duration, d_k contains the information (e.g., $d_k = \pm 1$), and $s(t)$ is the pulse shape that is to be applied. In FSK, $s(t)$ is rectangular shaped, $s(t) = 1\text{REC}$.

For BFSK, the constellation diagram is shown in Fig. 5.32.

The derivation of P_B of BFSK is similar to the one of IM-DD/2ASK and (D)PSK. With the definitions given in Fig. 5.32, the BER for coherent heterodyne BFSK can

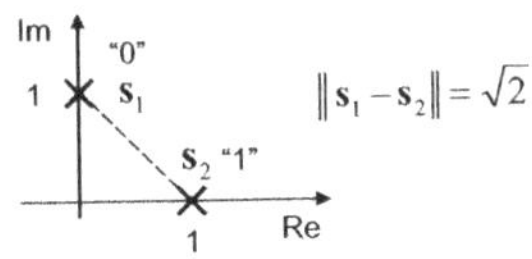

FIGURE 5.32 Constellation diagram of binary FSK.

be stated as

$$P_{\mathrm{B}} = Q\left(\sqrt{\frac{E_{\mathrm{b}}}{2N_0}}\right). \tag{5.54}$$

Coherent heterodyne FSK detection is possible with the receiver according to Fig. 5.4. An alternative coherent FSK demodulator consists of match-filter receivers for every signal state, that is, two correlators for coherent BFSP, adapted to the two signal states.

For noncoherent BFSK demodulation according to Fig. 5.2 or Fig. 5.3, the BER is

$$P_{\mathrm{B}} = \frac{1}{2}\exp\left(-\frac{E_{\mathrm{b}}}{4N_0}\right). \tag{5.55}$$

A demodulator option is available for the case of the modulation index $\eta = 0.5$. For $d_k = \pm 1$, the modulation index η can be derived from the modulation constant α through $\eta = \alpha/\pi$. This modulation index leads to *minimum-shift keying* (MSK). It can be shown that MSK is a special case of QPSK (so-called Offset-QPSK) and hence can also be demodulated linearly. Due to the equivalence with (Offset) QPSK, better BER results as well.

5.3.7 Polarization Modulation and Interleaving

Besides orthogonal multiplexing, the polarization domain can be used for modulation, which is then referred to as Polarization-Shift Keying, PolSK. PolSK has been reported with heterodyne coherent and differentially precoded direct detection [48–50]. It has also been combined with (differentially precoded) noncoherent duobinary transmission, for example, Ref. [51,52]. The respective transmission scheme is also referred to as duobinary polarization-shift keying (DPolSK).

Instead of a push–pull MZM as used for ODB, DPolSK uses a polarization modulator (Fig. 5.33). The polarization modulator converts the three-level duobinary signal into three polarization states (SOPs). Two SOPs are orthogonal to each other, the third state is an intermediate SOP.

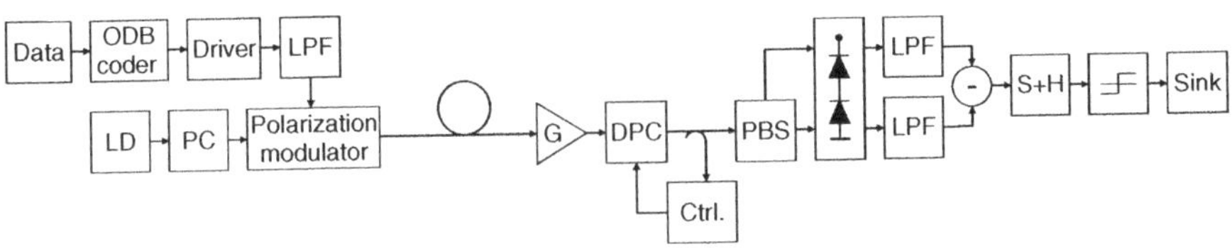

FIGURE 5.33 Duobinary polarization-shift keying (DPolSK) system.

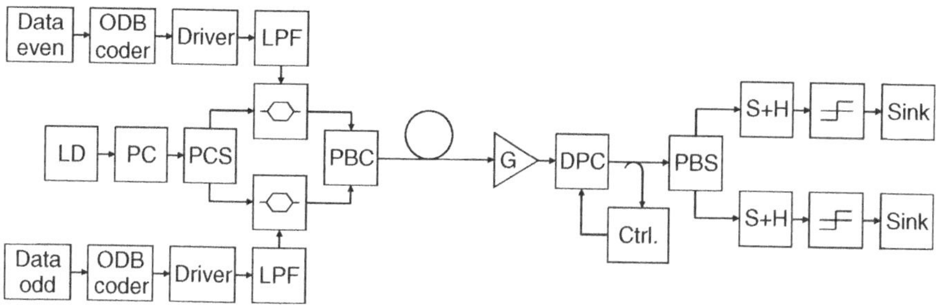

FIGURE 5.34 Polarization-interleaved optical duobinary (PI-ODB) system. R denotes an ODB receiver.

At the receiver, the signal is actively aligned to a PBS to separate the orthogonal polarization states. Dynamic polarization control (DPC) with a suitable tracking and control (Ctrl.) algorithm is necessary to compensate for the polarization drift in the fibers. Ultrafast, endless polarization tracking was demonstrated, for example, in Ref. [53]. The separated signals are fed into a balanced receiver that is followed by nondifferential direct detection.

Performance of DPolSK is similar to direct detection ODB. It has, however, better resistance against nonlinear transmission effects due to its constant-envelope property. Depending on receiver implementation, directly detected PolSK (without partial response coding) can have 3 dB sensitivity advantage over IM-DD, leading to similar results than DPSK [50].

ODB has further been used in combination with polarization interleaving (PI-ODB) [54]. The purpose here was to improve spectral efficiency of ultralong-haul WDM transmission to beyond 1 (b/s)/Hz. Each pair of WDM neighbor channels is orthogonally polarized with respect to each other. PI-ODB can thus be regarded as a special variant of PDM. The system block diagram is shown in Fig. 5.34. Even WDM channels are modulated in one polarization plane, while odd channels are modulated in the orthogonal plane.

The receiver consists of a PBS that is followed by two standard ODB receivers. The PBS is trailed by a DPC with suitable tracking algorithm. As compared to PDM, polarization interleaving allows denser spacing of the WDM carriers, thus achieving high spectral efficiency. In Ref. [54], ODB with 1 (b/s)/Hz (20 Gb/s WDM channels in 20 GHz grid) has been demonstrated without penalty.

5.4 COHERENT INTRADYNE DUAL-POLARIZATION QAM DETECTION

Coherent receivers provide superior sensitivity (or OSNR performance) through the use of a local oscillator (laser). Direct detection as performed by a slow (as compared to the optical carrier) photodiode leads to envelope detection, where all phase information is lost. In addition, it can only make use of the responsivity of the (PIN) photodiode, or the added gain and sensitivity of an avalanche mechanism [in case of avalanche photodiode (APDs)] or an optical preamplifier. Coherent detection, on the other hand, leads to a beat term between received input (in) and LO signals, which can be detected by the photodiode and which also preserves the phase information. The beat term is centered around the so-called intermediate frequency (IF) ω_{IF} (sometimes also referred to as RF, which adopts the traditional radio frequency view). Regarding ω_{IF}, we must divide three cases:

- Homodyne, $\omega_{IF} = 0$
- Intradyne, $\omega_{IF} \approx 0$ (typically <0.5 GHz)
- Heterodyne, ω_{IF} > receive signal bandwidth

Homodyne fiber-optic receivers were investigated in the late 1980s in order to increase transmission span distances. They require an optical phase-locked loop (OPLL) that can track the receive signal carrier's frequency and phase. An OPLL requires substantial effort, and the approach was made obsolete with the invention of the EDFA (invented in 1987, see Section 1.2). Today, no major work is done with regard to the traditional homodyne detection scheme.

Heterodyne detection has been considered in fiber-optic transmission with regard to several applications. Since it only requires AFC of the local laser—phase tracking is performed in the electrical IF domain—it is less complex. This advantage comes at the cost of a 3 dB penalty over the respective homodyne (de-) modulation scheme.

Intradyne detection is the most recent implementation of coherent receivers. The principle was first described in Ref. [55]. It is based on tuning the LO such that the intermediate frequency ω_{IF} becomes close to zero. Then, analog-to-digital conversion is performed, and exact phase tracking is done in the digital domain. Advantages include the avoidance of analog OPLL and also the possibility to implement polarization-diverse receivers and dispersion compensators (digital filters), which lead to almost zero penalty with regard to polarization and (chromatic) dispersion effects. Digital intradyne detection is the quasi-standard solution for 100 Gb/s transport and all future higher speed transmission. For early ultrahigh-speed examples, see for example, Refs [56–58].

In coherent detection, the detected power consists of the contributions from the input and LO signals, plus the beat term between the two, which is caused by the square-law detection:

$$P(t) = P_{in} + P_{LO} + 2\sqrt{P_{in} \cdot P_{LO}} \cdot \cos\left(\omega_{RF} t + \phi_{in} - \phi_{LO}\right). \tag{5.56}$$

Only the beat term is used for demodulation. It allows coherent phase demodulation (because the cosine argument contains the input signal's phase), and high sensitivity (because the term is weighted with the square root of the LO power). After down-conversion, the (intradyne, IF) signal is low-pass filtered and demodulated according to the modulation scheme used. A schematic diagram (not considering polarization diversity) is shown in Fig. 5.35.

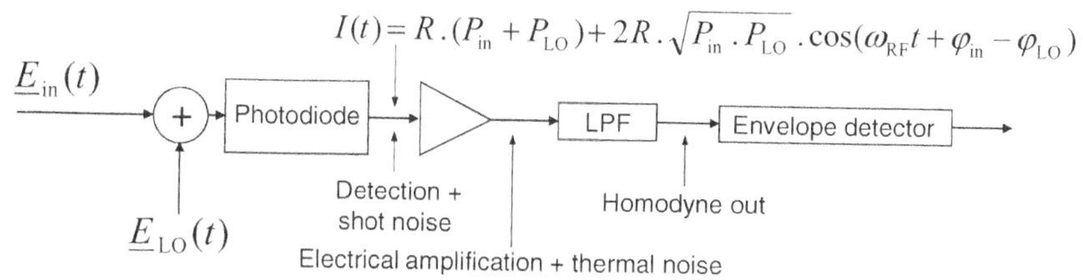

FIGURE 5.35 Coherent detection (basic scheme).

A coherent intradyne system is more complex than indicated in Fig. 5.35. This is due to the necessity of a polarization-diverse receiver that takes into account that the polarization planes of the input (receive) signal usually are not known. Hence, input and LO signal must be split into orthogonal polarizations that are then mixed and detected in independent receivers. Therefore, these systems also make use of polarization multiplexing in the transmitter in order to further reduce the Baud rate.

The resulting system configuration can be found in the literature and is shown in Fig. 5.36 for DP-QPSK [59–66].

Similar schematic system diagrams result for other QAM modulation schemes. For polarization multiplexing, the transmit laser (CW LD) is split into two orthogonal polarization planes by means of a combined polarization controller (PC)/PBS. Both polarization signals are independently modulated and recombined by a polarization beam combiner (PBC). At the receiver, the input signal is first split into orthogonally polarized signals by another PBS. The LO signal is also split and then combined with the input signal by means of two 90° hybrids. Each 90° hybrid has dual output ports for the respective inphase and quadrature components. The output signals are detected by four balanced receivers. The photodiodes are followed by fast analog-to-digital converters (ADC), which are then followed by the digital receiver. The digital receiver consists of CDR, equalizer (EQ) for dispersion compensation (and possibly nonlinear impairment mitigation), polarization demultiplexing, carrier recovery, and finally decoding.

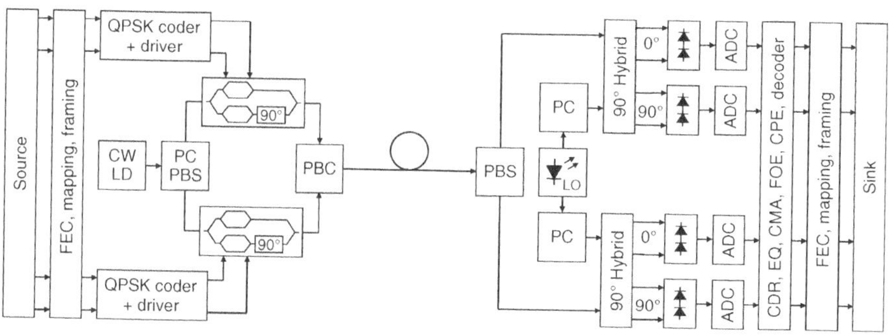

FIGURE 5.36 Coherent intradyne NRZ-DP-QPSK (NRZ-DP-4QAM) transmission.

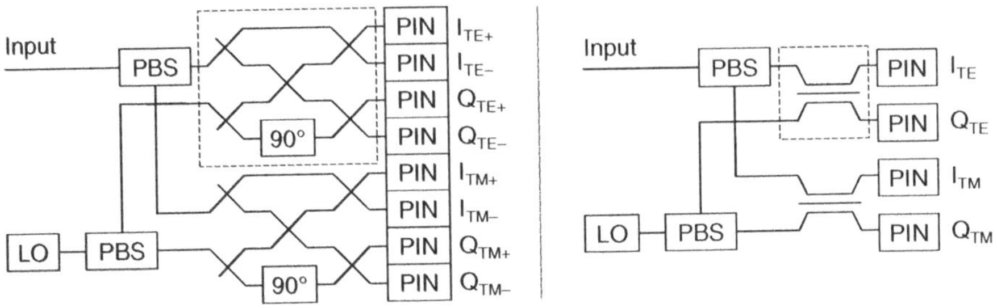

FIGURE 5.37 Two realizations of 90° hybrids in polarization-diverse coherent receivers.

This is followed by means for framing, monitoring, and error correction (FEC). The interface toward the application is provided through the client interface (I/F), which, in the case of ultrahigh-speed transport, can be implemented as a CFP pluggable. For lower speed services, the client interfaces could consist of SFP+ instead.

90° hybrids are passive combinations of 3 dB couplers and 90° phase shifters. Combined with polarization control, they provide access to the polarization and I and Q components, respectively. Several configurations exist, two of which are shown in Fig. 5.37. The first configuration consists of 3 dB couplers and dedicated 90° phase shifters (which are necessary because the 90° shifts caused by the 3 dB couplers cancel each other). The second configuration consists of 3 × 3 couplers. $I_{TE/TM}$ and $Q_{TE/TM}$ are the respective inphase and quadrature components of the orthogonally polarized TE (transverse electric) and TM (transverse magnetic) outputs of the PBS.

Figure 5.38 shows the digital receiver of the transmission system of Fig. 5.36 in detail. It consists of several stages where different functionalities are implemented [59,60]. Somewhat different structure of the digital filter stages is sometimes described in the literature, for example, refer to Refs. [56,58,67].

After detection and sampling at approximate Nyquist rate, the I and Q components have orthogonal but arbitrary planes of polarization each. The four components are now digitally processed in several succeeding filter stages.

In a first stage, bulk CD compensation is performed. This stage can be also complemented by nonlinear compensation (NLC). CD compensation is based on an

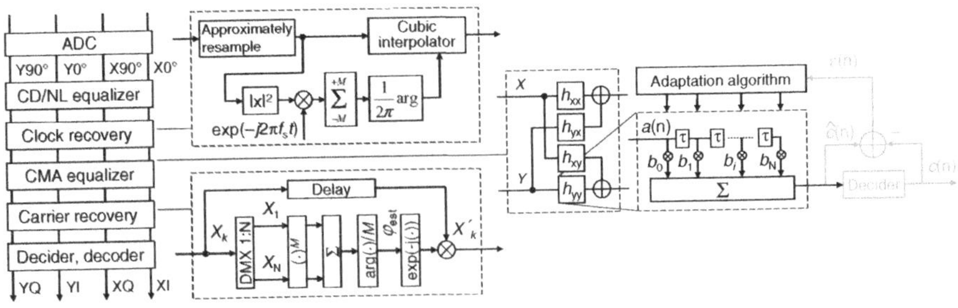

FIGURE 5.38 Coherent intradyne receiver: digital realization.

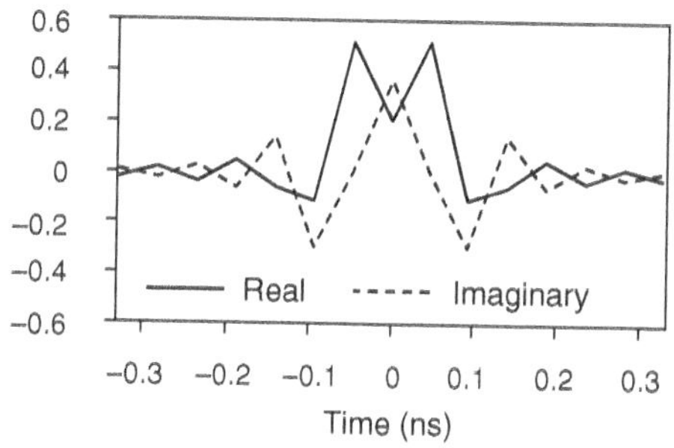

FIGURE 5.39 Example of simple FIR filter (impulse response of 90 km SSMF, truncated to seven elements).

n-tap finite impulse response (FIR) filter. Here, the sampled input data is convoluted with a vector that represents the inverse of the CD transfer function:

$$b_n = \frac{1}{N}\sum_{k=0}^{N-1} H_f^*(\omega_k)e^{j\omega_k n}, \quad H_f(j\omega) = \exp\left(j\lambda_0^2 Dz\omega^2/4\pi c\right). \tag{5.57}$$

Dz is the total CD. An example of a simple seven-tap FIR filter is shown in Fig. 5.39.

According to Ref. [64], the upper bound for the tap number at a Baud rate of B GBd is given by $0.032 \cdot B^2$ per 1000 ps/nm of chromatic dispersion. At 10.7 GBd (43G DP-QPSK), this translates to 3.7 taps per 1000 ps/nm, which can be reduced to 2.2 taps per 1000 ps/nm with minimal penalty. For 28 GBd (112G DP-QPSK), the number of taps increases by a factor of almost seven as compared to 42.8 GBd (CD increases with the square of the Baud rate).

In the next filter stage, CDR and resampling to two samples/symbol are performed. CDR is based on a digital filter-and-square timing recovery. It also performs resampling to two samples/symbol. Ideally, the ADC would already sample at an exact multiple of the Baud rate. More details on digital timing recovery can be found in Ref. [61].

PMD and residual CD equalization, and polarization recovery (demultiplexing) is performed in a multiple-input multiple-output (MIMO) filter that follows CDR or retiming. Four complex-valued FIR filters [also referred to as feed-forward equalizers (FFE), transversal filters, or tapped-delay-line filters] are used in a MIMO configuration. These filters are implemented in a butterfly structure, as adaptive equalizer between the polarization planes (Fig. 5.38). The filter taps are spaced $T/2$ in time, where T is the symbol duration. Their tap number correlates with the target dispersion-compensation (PMD and residual CD) capability.

The tap weights have to be optimized, usually using blind equalization techniques in order to avoid the necessity of training sequences. Blind adaptation can be done, for example, with a least mean square (LMS) algorithm. The earliest

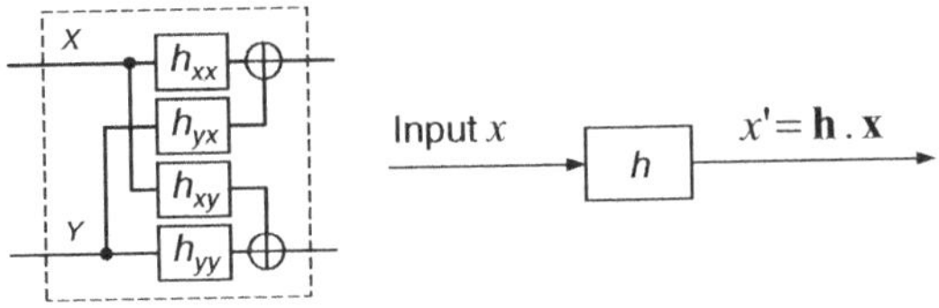

FIGURE 5.40 Linear MIMO equalizer system.

work on such adaptive filters can be traced back to the late 1950s. From this early work, the LMS algorithm emerged as a simple, yet effective, algorithm for the design of adaptive transversal filters. The LMS algorithm was devised by Widrow and Hoff in 1959 [68].

An alternative to LMS is the so-called constant-modulus algorithm (CMA) [65,69,70]. The CMA can equalize constant-modulus signals (i.e., signals with constant magnitude, e.g., M-ary PSK) and, with modification, nonconstant-modulus signals (QAM and QPR). Under ideal conditions, this results in zero-forcing equalization of the received signal.

The MIMO filter block and a single filter system are shown in Fig. 5.40.

The CMA is a stochastic gradient algorithm (SGA). Its criterion penalizes deviations in the modulus (magnitude) of the equalized signal away from a fixed value. This is done by minimizing the error ε as defined in Fig. 5.40:

$$\varepsilon^2 = \left(1 - |x'|^2\right)^2 \rightarrow \text{Minimum} \;\Rightarrow\; \frac{d\varepsilon^2}{d\mathbf{h}} = 0. \tag{5.58}$$

Here, $\mathbf{h} = \{h_1, h_2, \ldots, h_n\}$, with n being the filter tap number. Similar definitions hold for $\mathbf{x}$ and $\mathbf{y}$. The corrective CMA action is visualized in Fig. 5.41.

In the CMA, the filters are extended, MIMO-like, to both polarizations:

$$x' = \mathbf{h}_{xx} \cdot \mathbf{x} + \mathbf{h}_{xy} \cdot \mathbf{y}, \quad y' = \mathbf{h}_{yx} \cdot \mathbf{x} + \mathbf{h}_{yy} \cdot \mathbf{y}. \tag{5.59}$$

The filter coefficients $\mathbf{h}$ are modified according to Eq. (5.58) with weight μ:

$$\mathbf{h} \rightarrow \mathbf{h} - \frac{\mu}{4}\frac{d}{d\mathbf{h}}\varepsilon^2 = \mathbf{h} + \mu\varepsilon x' \bar{x}. \tag{5.60}$$

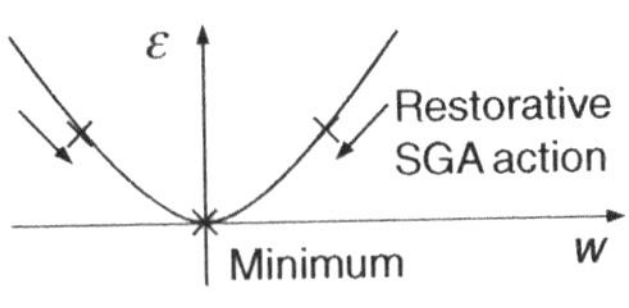

FIGURE 5.41 Corrective CMA action.

The complete set of the filter coefficients is then given by

$$\begin{aligned}
\mathbf{h}_{xx} &= \mathbf{h}_{xx} + \mu\varepsilon_x x' \cdot \overline{x}, \\
\mathbf{h}_{xy} &= \mathbf{h}_{xy} + \mu\varepsilon_x x' \cdot \overline{y}, \\
\mathbf{h}_{yx} &= \mathbf{h}_{yx} + \mu\varepsilon_x y' \cdot \overline{x}, \\
\mathbf{h}_{yy} &= \mathbf{h}_{yy} + \mu\varepsilon_x y' \cdot \overline{y}.
\end{aligned} \tag{5.61}$$

The CMA employs a cost function that does not discriminate between the two equalized signals. Hence, it is possible that the algorithm converges to a tap-weight setup that produces the same transmitted signal at both equalizer outputs, usually the one that arrived with higher power at the receiver. This problem can be circumvented by frequently monitoring the equalizer's matrix determinant and reinitializing the tap weights if necessary. This may cause discontinuity at the equalizer output. Several solutions to this problem have been proposed. An example is multiuser CMA that can avoid singularity. Multiuser CMA employs an enhanced cost function, penalizing the correlation between the two signals at the equalizer's outputs [70].

Without modification or complements, the standard CMA cannot be used for constellations other than QPSK and higher order M-ary PSK. Derivatives for blind equalization are described for BPSK [71], QAM, [72] and QPR [73], respectively.

The last filter stage performs carrier phase estimation (CPE), that is, the digital phase tracking [74]. CPE for QPSK can be done with fourth-power Viterbi and Viterbi recovery [75]. For BPSK, second-power estimation can be used, whereas for (square) QAM, modifications are required. These can consist of symbols of equal magnitude considered separately.

A QPSK signal is given by

$$E(t) = A \cdot \exp\{j[\theta_S(t) + \theta_C(t)]\}, \quad \theta_S = 0, \pm\pi/2, \pi. \tag{5.62}$$

θ_S is the symbol phase, and θ_C is the carrier phase, respectively. The symbol phase can be eliminated and the carrier phase can be estimated by fourth-power operation:

$$A^4 \exp\{j[4\theta_S(t) + 4\theta_C(t)]\} = A^4 \exp\{j4\theta_C(t)\}. \tag{5.63}$$

Due to phase noise, averaging (low-pass filtering) over N symbols, X_k is required. For QPSK and square QAM, one gets, respectively,

$$\phi_{\text{est,PSK}} = \frac{1}{M}\arg\left(\sum_{k=1}^{N} X_k^4\right), \quad \phi_{\text{est,Square-QAM}} = \frac{1}{4}\arg\left(\sum_{k=1}^{N_{\pm\pi/4}} \frac{X_{k,\pm\pi/4}^M}{\left|X_{k,\pm\pi/4}^M\right|}\right). \tag{5.64}$$

For QPSK, $M = 4$. For square QAM, only symbols at $\pm\pi/4$ are considered. A similar expression can be used for 9QPR.

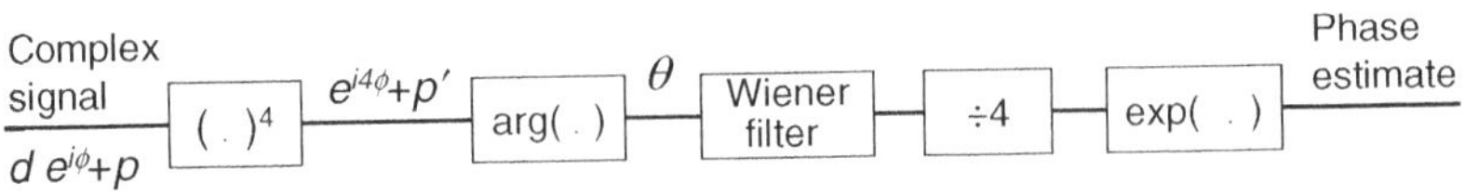

FIGURE 5.42 Phase estimation using Wiener filter.

Simple averaging over N symbols may not be sufficient to suppress phase errors. The best-possible phase estimate is derived through the maximum a posteriori (MAP) estimate, where phase φ(n) and data d(n) are jointly estimated [76]. The MAP estimate can be calculated by applying a per-survivor method to a group of symbols, and calculating the phase by successive approximation for each symbol group instance.

Alternatively, the averaging filter can be optimized. From estimation theory it is known that the best linear estimate of φ is derived if a *Wiener* filter is applied to θ. The Wiener filter solves the estimation problem for stationary or cyclostationary signals. It was introduced by Norbert Wiener in the 1940s [77]. The filter is optimal in the sense of the minimum mean square error (MMSE).

Neglecting higher order noise terms, small-angle approximation is applied, which yields a phase angle $\theta = 4\varphi + \text{AWGN}$. Wiener filtering of θ is shown in Fig. 5.42.

Errors in the phase estimate lead to an increase in the number of bit errors. They can lead to so-called cycle-slip errors, that is, data inversion in case of BPSK or QPSK. This is the reason why even in coherent intradyne systems differential precoding (i.e., DQPSK) is applied, although it is not necessary from the viewpoint of differential-delay demodulation. Figure 5.43 shows the effects of phase noise and phase plus added amplitude noise on BPSK, respectively.

Phase estimation such as the Mth-power algorithm requires that the frequency offset between transmitter and LO must be smaller than 10% of the symbol rate [78]. For 43 Gb/s (~11 GBd) DP-QPSK transmission, this leads to ~1 GHz allowable frequency offset. The frequency offset between transmitter and LO, however, can be as large as ±5 GHz. As a result, an additional digital signal processor (DSP)-based frequency offset estimator (FOE) is required to ensure that subsequent CPE algorithms can accurately recover the phase of the received signal. In Ref. [79], an ultrawide-range FOE has been described. It is based on a double-stage FOE architecture as shown in Fig. 5.44.

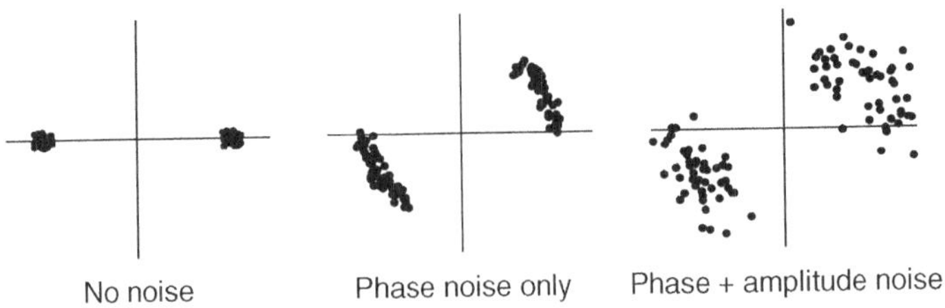

FIGURE 5.43 Phase estimation in presence of phase noise and combined phase plus amplitude noise.

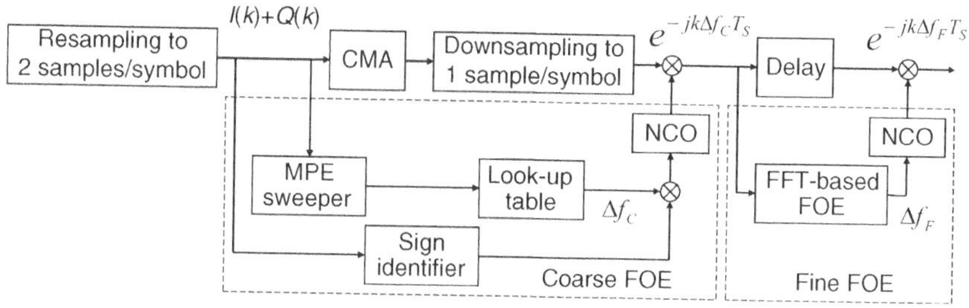

FIGURE 5.44 Cascaded FOE.

The dual-stage, cascaded FOE consists of a coarse FOE and a fine FOE. In 43 Gb/s DP-QPSK systems, the estimation range of the dual-stage FOE has been shown to be four times of what can be achieved with the *M*th-power algorithm alone.

Laser phase noise is an important impairment in coherent systems as it impacts carrier synchronization. Narrow-linewidth lasers are required to implement a phase-locked loop for carrier recovery [55].

Feed-forward carrier recovery can relax the sum linewidth to 0.001–0.0001 of the symbol rate. This is in the reach of low-cost DBR lasers. Linewidth requirements for feed-forward carrier recovery of several modulation formats are summarized in Table 5.8 [80]. When comparing them with the linewidths from Table 3.1, it becomes clear that care must be taken for constellations above 16QAM.

Comparable linewidths have been reported in Ref. [81] for QPSK and 16QAM.

The optimum block length that is used for phase-noise averaging depends on the type of noise. Under linear conditions, longer block lengths are advantageous. For nonlinear phase noise, for example, caused by 10 Gb/s IM-DD WDM channels and cascaded EDFAs, shorter block lengths are better [82]. This is demonstrated in Fig. 5.45.

DSP-based intradyne detection allows NLC for the mitigation of certain nonlinear effects. This primarily relates to intrachannel effects, that is, SPM. It is possible because SPM is deterministic and can hence be compensated. Compensation of interchannel effects (XPM and FWM) requires more complex MIMO equalizers.

NLC can be implemented, together with chromatic dispersion compensation, as a fixed equalizer that leads the adaptive CMA equalizer [67]. In order to reduce computational effort, it is desirable to compensate CD with a frequency-domain filter, followed by time-domain instantaneous NLC that compensates the nonlinear phase shift. In multispan long-haul transmission, linear and nonlinear intrachannel

TABLE 5.8 Linewidth Requirements for a Maximal Receiver Sensitivity Penalty of 2 dB @ BER = 10^{-4}

Format	QPSK	8PSK	(Square) 16QAM	64QAM
Max. linewidth ($\Delta\nu T_b$)	2.4×10^{-4}	3×10^{-5}	3×10^{-6}	3×10^{-8}
Max. linewidth @ 43 Gb/s	10 MHz	1.6 MHz	120 kHz	1.2 kHz

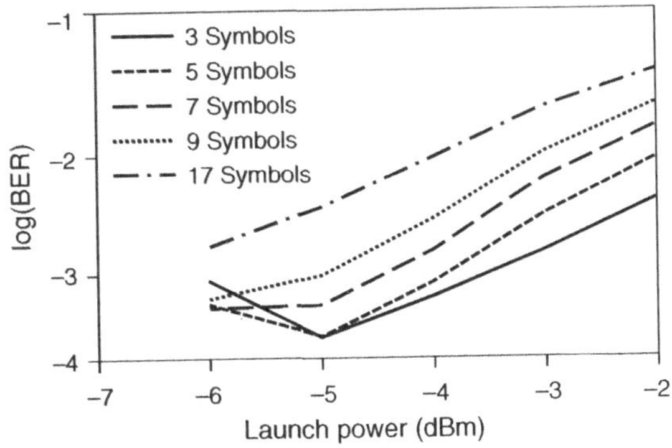

FIGURE 5.45 Effect of CPE block length.

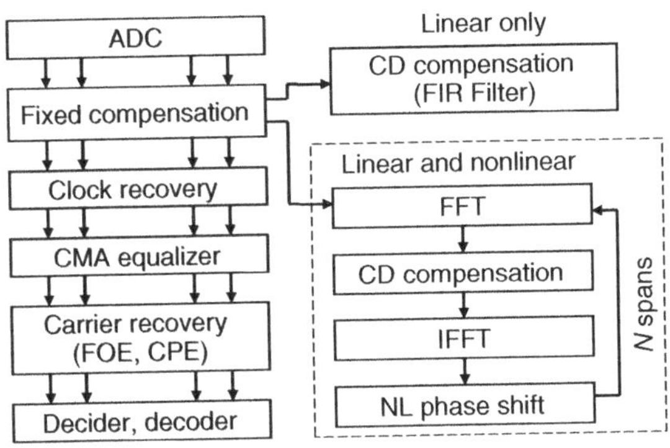

FIGURE 5.46 Linear-only or linear plus nonlinear compensation (backward propagation).

compensation can be done iteratively, switching between time and frequency domain by means of fast Fourier transform (FFT) and inverse fast Fourier transform (IFFT), respectively (Fig. 5.46). This approach is called backward propagation; it resembles the calculations that are performed with the split-step Fourier algorithm in order to solve the nonlinear Schrödinger equation [Eq. (2.32)].

Using intrachannel NLC, the optimum long-haul launch power can be increased by ~1 dB, and maximum reach for 42.7 and 85.4 Gb/s can be increased to more than 11,000 km and 8000 km, respectively [67].

Work on compute-effort reduction is ongoing.

5.5 OPTICAL OFDM

Optical orthogonal frequency-domain multiplexing (O-OFDM) was much investigated for WDM high-speed transmission since about 2005. The technique can efficiently tackle linear channel distortions. It was originally developed for 4G digital mobile (wireless) communications. So far, it does not yet have commercial relevance for WDM systems.

The basic idea behind OFDM is to reduce linear effects caused, for example, by dispersion or multipath propagation, by *increasing* the symbol duration T_S. This increase is enabled by parallelization and simultaneous transmission of the signal in subcarriers, which are multiplexed in the frequency domain. The subcarriers are orthogonally spaced according to

$$f_{SC} = \frac{1}{T_S}. \tag{5.65}$$

Orthogonality can be seen when calculating the scalar product of two subcarrier signals at frequencies f_x and f_y over one symbol duration [cf. Eq. (1.1)]:

$$\frac{1}{T_S} \cdot \int_0^{T_S} e^{j2\pi f_x t} \cdot e^{-j2\pi f_y t} dt = \begin{cases} 1, & x = y \\ 0, & x \neq y \end{cases}. \tag{5.66}$$

Due to orthogonality, subcarriers can individually be detected without interference from other subcarriers. Orthogonality can also be demonstrated in the OFDM spectrum, where the peak of one subcarrier coincides with zeros from all other subcarriers (Fig. 5.47).

In Fig. 5.47, the Fourier transforms of rectangular-shaped (1REC) OFDM pulses are shown. These REC pulses are demodulated at the receiver. For transmission, a guard interval (GI) is attached to the OFDM pulses. The duration of this GI has to be adapted to the length of the pulse response of the transmission channel (i.e., CD and PMD). In order to avoid distortion when detaching the GI from the receiver, a cyclic

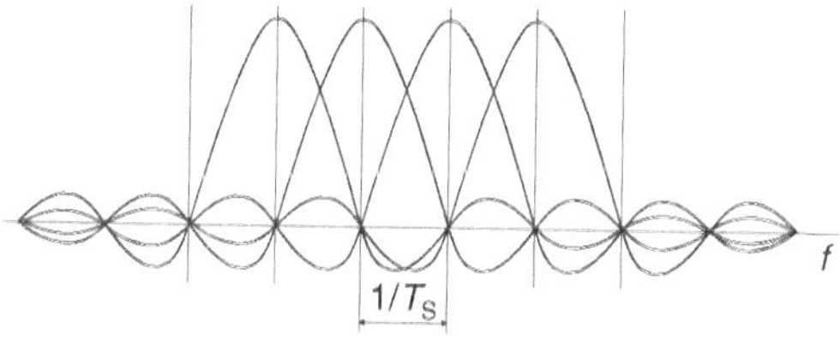

FIGURE 5.47 Orthogonality of OFDM subcarriers.

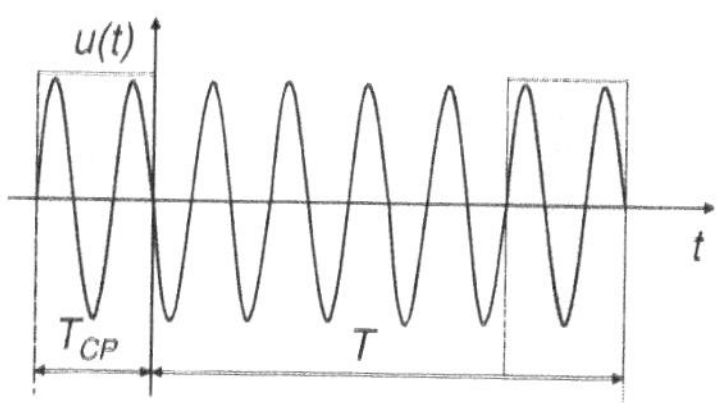

FIGURE 5.48 Attaching the cyclic prefix (CP).

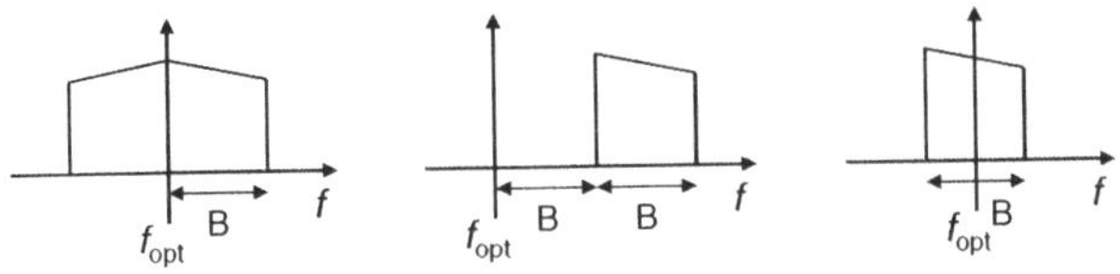

FIGURE 5.49 Dual-sideband (left), SSB (middle), and complex (I/Q) OFDM spectra.

extension of the respective subcarrier signal into the GI is required (Fig. 5.48). The cyclic extension is also referred to as cyclic prefix (CP).

Dispersion effects can at least partly be avoided via the GI. Additional compensation can be provided in coherent detection. A disadvantage of the GI is the related decrease in bandwidth efficiency.

OFDM modulators and demodulators can efficiently be implemented using IFFT and FFT, respectively. If more subcarriers N are used and symbols are hence made longer, proportionally shorter cyclic prefixes are required. Thus, relative overhead can be decreased. If overhead is to be limited to 10% for 4000 km transmission of a 10 Gb/s signal, $N = 256$ is required. For 40 Gb/s transmission, this increases to $N = 4096$ due to higher bandwidth of the optical signal and the higher symbol rate [83].

Optical OFDM can be generated and detected in different ways. Variants include real-valued, dual-sideband OFDM with amplitude modulation and direct detection, SSB generation with RF up-conversion and direct or heterodyne detection, and complex I/Q modulation with coherent (intradyne) detection. These variants are shown in Fig. 5.49.

The different optical OFDM implementations lead to different characteristics with regard to dispersion tolerance and required components effort. Important characteristics are summarized in Table 5.9.

From Table 5.9, it becomes clear that coherent I/Q variants are most useful for impairment-tolerant high-speed WDM transmission. The digital–analog-converted signals are fed to I/Q modulators. This is combined with polarization multiplexing, as shown in Figure 5.50.

Comparing Fig. 5.50 with Fig. 5.36, it becomes clear that direct I/Q-modulated coherent OFDM is one possibility of generalized polarization-multiplexed I/Q-modulated transmission. When comparing coherent DP-QPSK or DP-M-QAM (single carrier or few carrier with Nyquist WDM spacing) with coherent DP-OFDM and

TABLE 5.9 Characteristics of O-OFDM Implementations

	Dual-Sideband	SSB	Complex/Coherent
Dispersion tolerance	Like OOK	High	High, can be compensated
ADC/DAC bandwidth	B, analog	2B	B/2
Transmitter	AM	AM or I/Q	I/Q
Receiver	DD	DD or heterodyne	Coherent

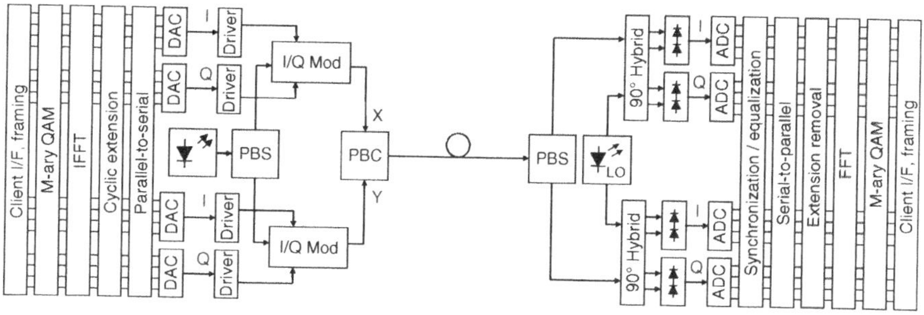

FIGURE 5.50 Direct I/Q-modulated polarization-multiplexed OFDM with intradyne detection.

M-ary QAM of the subcarriers, the trade-off of the linear and nonlinear transmission impairments must be considered. O-OFDM typically has advantages with respect to dispersion, but suffers from fiber nonlinearity more strongly due to the dense subcarrier spacing.

An important parameter in O-OFDM is laser phase noise, which can destroy the orthogonality of the subcarriers and cause interchannel interference (ICI) [81]. ICI grows linearly with N_C, with N_C being the number of subcarriers. Phase noise considerations favor a smaller number of subcarriers at increased spacing. At higher phase noise, the carrier phase can no longer be assumed as constant over an O-OFDM symbol.

5.6 COMPARISON OF MODULATION FORMATS

The modulation formats discussed so far differ in many aspects like transmission performance and implementation effort. Following, important characteristics are summarized.

The modulation schemes can be combined with NRZ and RZ pulse shaping and various duty cycles. They can be configured for binary or M-ary transmission. Depending on the modulation scheme, different modulators [electroabsorption modulator (EAM) and MZM] or directly modulated lasers and different detection schemes can be used. Table 5.10 summarizes the pulse shape, modulator, and detection options. It also lists the resulting BW relative to the bit rate R_b.

In Table 5.11, the comparison of BER performance of the basic modulation schemes OOK, PSK, and FSK is given. All schemes are normalized to the same mean per-bit energy E_b. Optimal reception (in a single polarization) is considered. For the equivalence between preamplified direct and heterodyne asynchronous detection, see Ref. [4].

For the same E_b, QPSK performs like BPSK. For same symbol energy, it performs 3 dB worse. As mentioned earlier, ODB can perform ~1.5 dB better than OOK (Section 5.3.5.1). Likewise, PolSK can perform up to 3 dB better than OOK

TABLE 5.10 Comparison of WDM Modulation Formats

			Pulse Shape/Duty Cycle				Modulator				Detection			
Format	M	BW R_b	NRZ	RZ67	RZ50	RZ33	DM	EA	MZ	PM	DD	Differential	Coherent	MLSE
NRZ-OOK	2	2	X				0	0	X		X		0	
RZ-OOK	2	4			0	0			X		X		0	
CSRZ-OOK	2	3		X					X		X		0	
NRZ-ODB	2	1	X				0^1	0^2	X		X		0	0
RZ-ODB	2	2			0	0			X		X		0	0
CSRZ-ODB	2	1.5		X					X		X		0	0
NRZ-DPSK	2	2	X					0^2	X	0		X	0	
RZ-DPSK	2	4			0	0			X			X	0	
CSRZ-DPSK	2	3		X					X			X	0	
NRZ-DQPSK	4	1	X					0^2	X	0		X	0	
RZ-DQPSK	4	2			0	0			X			X	0	
CSRZ-DQPSK	4	1.5		X					X			X	0	
CPFSK	2	>1	X				X					X	0	
PolSK	2	2	X						0	X	X		0	
DPolSK	2	1	X						0	X	X		0	
DP-BPSK	4	1	X						X	0			X	
DP-QPSK	8	0.5	X						X	0			X	
DP-9QPR	8	0.25	X						X				X	X
DP-16QAM	16	0.25	X						X				X	

PM: phase modulator (0) or polarization modulator (X), DM: direct modulation, BW: bandwidth. X refers to preferred solution, and 0 refers to optional solution. 0^1 refers to chirp-managed lasers. 0^2 assumes EAM combinations for push–pull modulation.

TABLE 5.11 Comparison of BER Performance of Basic Optical Modulation Schemes

	Homodyne	Heterodyne Synchronous	Noncoherent (DD or Heterodyne Asynchronous)
OOK	$P_B = Q\left(\sqrt{\frac{E_b}{N_0}}\right)$	$P_B = Q\left(\sqrt{\frac{E_b}{2N_0}}\right)$	$P_B = \frac{1}{2}\exp\left(-\frac{E_b}{4N_0}\right)$
PSK	$P_B = Q\left(\sqrt{\frac{2E_b}{N_0}}\right)$	$P_B = Q\left(\sqrt{\frac{E_b}{N_0}}\right)$	$P_B = \frac{1}{2}\exp\left(-\frac{E_b}{2N_0}\right)$
FSK		$P_B = Q\left(\sqrt{\frac{E_b}{2N_0}}\right)$	$P_B = \frac{1}{2}\exp\left(-\frac{E_b}{4N_0}\right)$

(Section 5.3.7). The performance for M-QAM heavily depends on M. For $M = 4$, performance is identical to QPSK.

Table 5.12 compares the resulting sensitivities (at $P_B = 10^{-9}$) for 10 Gb/s WDM transmission for the different modulation and (optimum matched filter) detection schemes. For coherent detection, no preamplifiers are assumed. With preamplification, the respective sensitivities get worse by 3 dB [23].

The implementation options listed in Tables 5.10 and 5.12 are not equally relevant. For example, NRZ-OOK is the primary solution for all WDM transport up to 10 Gb/s per channel. This includes backhaul and older backbone applications. It is also used for WDM-PON access, where it may be complemented later by heterodyne PSK/QPSK. For ultrahigh-speed WDM transmission, OOK is insufficient. Here, homodyne (intradyne) PSK and QAM are the most relevant, possibly complemented by partial response. Figure 5.51 compares the BER curves, depending on the E_B/N_0 ratio, for some high-performing modulation schemes.

In Fig. 5.51, the BER curves are plotted over E_b/N_0. Consequently, BPSK and QPSK perform equally good (curve "PSK"). In most WDM systems, the

TABLE 5.12 Sensitivity for BER = 10^{-9} for 10 Gb/s Signals

Format	Sensitivity (dBm)
OOK w/optical preamplifier	−42
DPSK asynchronous	−45.5
ASK homodyne	−46
FSK heterodyne synchronous	−46
PolSK heterodyne synchronous	−46
ODB homodyne	−48 (AZD)
BPSK homodyne	−49
QPSK homodyne	−46
9QPR homodyne	−45 (AZD)
8QAM homodyne	−42.5
8PSK homodyne	−41.5
16QAM homodyne	−39

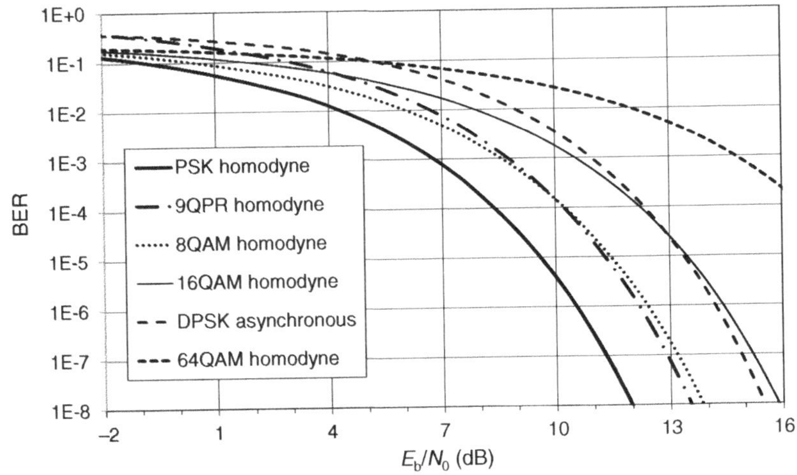

FIGURE 5.51 BER over E_b/N_0 of relevant WDM modulation schemes.

per-channel power will be limited, either by nonlinearity or by laser safety. Then, a comparison based on the same symbol energy E_s/N_0 is more meaningful. Now, BPSK achieves 3 dB advantage over QPSK. Also, 9QPR and 8QAM separate more clearly, showing the advantage of the partial response scheme. This is demonstrated in Fig. 5.52.

The curve for 64QAM is not shown here anymore in order not to compress the scale too much. Also, the DPSK curve has been omitted. It is located very near to (~0.5 dB on the right-hand side of) the QPSK curve.

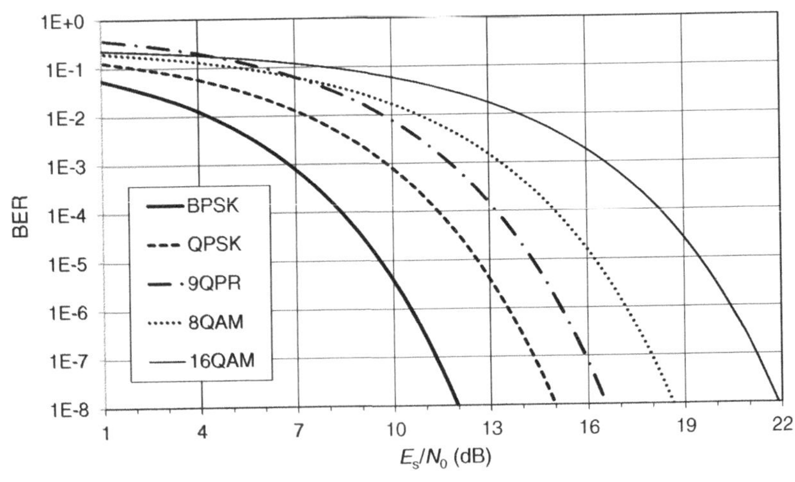

FIGURE 5.52 BER of relevant homodyne modulation schemes for same symbol energy.

Note again that all results discussed here imply ideal reception. Real transmission systems (specially the receivers) lead to clear penalties over the numbers summarized in this chapter. These penalties are in the range of ≥ 3 dB, depending on implementation details.

REFERENCES

1. K.V. Cartwright et al., The performance of dually polarized M-QAM and L-QPRS systems with crosstalk and differential phase shift, Southeastcon'91, Vol. 2, April 1991, pp. 1026–1029.
2. E. Panayirci et al., Error performance analysis of quadrature partial response trellis coded modulation (QPR-TCM) in fading mobile satellite channels, IEEE Trans. Commun., Vol. 43, No. 2/3/4, 1995, pp. 1653–1662.
3. A.F. Elrefaie and R.E. Wagner, Chromatic dispersion limitations for FSK and DPSK systems with direct detection receivers, IEEE Photonics Technol. Lett., Vol. 3, No. 1, 1991, pp. 71–73.
4. O.K. Tonguz and R.E. Wagner, Equivalence between preamplified direct detection and heterodyne receivers, IEEE Photonics Technol. Lett., Vol. 3, No. 9, 1991, pp. 835–837.
5. R.A. Linke and A.H. Gnauck, High-capacity coherent lightwave systems, IEEE J. Lightwave Technol., Vol. 6, No. 11, 1988, pp. 1750–1769.
6. R. Noé et al., 38 krad/s 3.8 Grad broadband endless optical polarization tracking using $LiNbO_3$ device, IEEE Photonics Technol. Lett., Vol. 21, No. 17, 2009, pp. 1220–1222.
7. B. Koch et al., 100 krad/s endless polarisation tracking with miniaturised module card, Electron. Lett., Vol. 47, No. 14, July 2011, pp. 813–814.
8. J. Franz, Optische Übertragungssysteme mit Überlagerungsempfang, Springer-Verlag, Berlin, 1988.
9. L.F. Mollenauer et al., Experimental observation of picosecond pulse narrowing and solitons in optical fibers, Phys. Rev. Lett., Vol. 45, No. 13, 1980, pp. 1095ff.
10. J.P. Gordon and H.A. Haus, Random walk of coherently amplified solitons in optical fiber transmission, Opt. Lett., Vol. 11, No. 10, 1986, pp. 665ff.
11. L.F. Mollenauer et al., Resistance of solitons to the effects of polarization dispersion in optical fibers, Opt. Lett., Vol. 14, No. 21, 1989, pp. 1219ff.
12. L.F. Mollenauer et al., Polarization multiplexing with solitons. IEEE J. Lightwave Technol., Vol. 10, No. 1, 1992, pp. 28ff.
13. A. Hasegawa, Optical solitons in fibers for communication systems, Opt. Photonics News, Vol. 13 February 2002, pp. 1610–1618.
14. L.F. Mollenauer et al., Polarization multiplexing with solitons, IEEE J. Lightwave Technol., Vol. 10, No. 1, 1992, pp. 28ff.
15. E. Ip and J.M. Kahn, Power spectra of return-to-zero optical signals, IEEE J. Lightwave Technol., Vol. 24, No. 3, March 2006, pp. 1610–1618.
16. Optical system design and engineering considerations, ITU-T Series G, Supplement 39 (02/2006), 2006.
17. G.P. Agrawal, Fiber-Optic Communication Systems, John Wiley & Sons, New York, 1992.

18. D. Marcuse, Calculation of bit-error probability for a lightwave system with optical amplifiers and post-detection Gaussian noise, IEEE J. Lightwave Technol., Vol. 9, No. 4, 1991, pp. 505–513.
19. T.S. Rappaport, Wireless Communications Principles and Practice, 2nd ed., Prentice Hall, 2001.
20. W.C. van Etten, Introduction to Random Signals and Noise, John Wiley & Sons, Ltd, Chichester, 2005.
21. B. Glance, Performance of homodyne detection of binary PSK optical signals. IEEE J. Lightwave Technol., Vol. LT-4, No. 2, 1986, pp. 228–235.
22. S.R. Chinn et al., Sensitivity of optically preamplified DPSK receivers with Fabry–Perot filters, IEEE J. Lightwave Technol., Vol. 14, No. 3, 1996, pp. 370–377.
23. R. Noé, Essentials of Modern Optical Fiber Communication, Springer, Berlin, 2010.
24. D. Marcuse, Derivation of analytical expressions for the bit-error probability in lightwave systems with optical amplifiers, IEEE J. Lightwave Technol., Vol. 8, No. 12, 1990, pp. 1816ff.
25. J.P. Gordon and L.F. Mollenauer, Phase noise in photonic communications systems using linear amplifiers, Opt. Lett., Vol. 15, No. 23, 1990, pp. 1351–1353.
26. H. Kim et al., Experimental investigation of the performance limitation of DPSK systems due to nonlinear phase noise, Photonics Technol. Lett., Vol. 15, February 2003, pp. 320–322.
27. X. Liu et al., Digital self-coherent detection, Opt. Express, Vol. 16, No. 2, January 2008, pp. 792–803.
28. M. Serbay et al., Comparison of six different RZ-DQPSK transmitter set-ups regarding their tolerance towards fibre impairments in 8×40 Gb/s WDM-systems, IEEE/LEOS Workshop on Advanced Modulation Formats, July 2004, Paper ThB3.
29. W. Jia et al., Generation of 20 Gb/s RZ-DQPSK signal using a directly modulated chirp managed laser, OFC/NFOEC 2011, Los Angeles, March 2011, Paper OThE4.pdf.
30. A. Lender, The duobinary technique for high speed data transmission, IEEE Trans. Commun. Electron., Vol. 82, May 1963, pp. 214–218.
31. A. Lender, Correlative digital communication techniques, IEEE Trans. Commun. Technol., Vol. 12, No. 4, 1964, pp. 128–135.
32. H. Kobayashi, Correlative level coding and maximum-likelihood decoding, IEEE Trans. Inf. Theory, Vol. 17, No. 5, September 1971, pp. 586–594.
33. G. D. Forney, Maximum-likelihood sequence estimation of digital sequences in the presence of intersymbol iterference, IEEE Trans. Inf. Theory, Vol. 18, No. 3, May 1972, pp. 363–378.
34. H. Kobayashi and D.T. Tang, On decoding of correlative level coding systems with ambiguity zone detection, IEEE Trans. Commun. Technol., Vol. 19, No. 4, August 1971, pp. 467–477.
35. D.R. Smith, Digital Transmission Systems, 3rd ed., Kluwer Academic Publishers, Norwell, MA, 2003.
36. G. Bosco et al., ASE-noise limit of direct-detection receivers: duobinary vs. IMDD, LEOS2002, 15th Annual Meeting, Nov. 2002, pp. 776–777, Paper ThQ3.
37. G. Bosco et al., Quantum limit of direct-detection receivers using duobinary transmission, IEEE Photonics Technol. Lett., Vol. 15, No. 1, 2003, pp. 102–104.

38. I. Lyubomirsky, Coherent detection for optical duobinary communication systems, IEEE Photonics Technol. Lett., Vol. 18, No. 7, April 2006, pp. 868–870.
39. Y. Miyamoto et al., Novel modulation and detection for bandwidth-reduced RZ formats using duobinary-mode splitting in wideband PSK/ASK conversion, IEEE J. Lightwave Technol., Vol. 20, No. 12, December 2002, pp. 2067–2078.
40. D.-S. Lee, et al., A simple configuration to generate 40 Gb/s duobinary CSRZ and CSRZ DPSK signals with enlarged dispersion tolerance, LEOS2003, 16th Annual Meeting, October 2003, pp. 144–145, Paper MN4.
41. K. Yonenaga et al., 100 Gb/s all-optical OFDM transmission using 4×25 Gb/s optical duobinary signals with phase-controlled optical sub-carriers, OFC/NFOEC2008, San Diego, March 2008, Paper JThA48.
42. Specification of 100 Gb/s metro card, ADVA AG Optical Networking, 2013.
43. I. Lyubomirsky, Quadrature duobinary modulation for 100G transmission beyond the Nyquist limit, OFC/NFOEC2011, Los Angeles, March 2011. Paper OThM4.
44. I. Lyubomirsky, Quadrature duobinary for high-spectral efficiency 100G transmission, IEEE J. Lightwave Technol., Vol. 28, No. 1, January 2010, pp. 91–96.
45. J. Li et al., Spectrally efficient quadrature duobinary coherent systems with symbol-rate digital signal processing, IEEE J. Lightwave Technol., Vol. 29, No. 8, April 2011, pp. 1098–1104.
46. H. Grießer et al., Quaternary duobinary modulation for superchannels with sub-Nyquist subcarrier spacing, OFC/NFOEC2012, Los Angeles, March 2012, Paper OTh1B.4.
47. J.G. Proakis, Digital Communications, 2nd ed., McGraw-Hill Book Company, New York, 1989.
48. S. Benedetto and P. Poggiolini, Theory of polarization shift keying modulation, IEEE Trans. Commun., Vol. 40, No. 4, April 1992, pp. 708–721.
49. Yan Han and G. Li, Direct detection differential polarization-phase-shift keying based on Jones vector, Opt. Express, Vol. 12, No. 24, November 2004, pp. 5821–5826.
50. S. Benedetto et al., Direct detection of optical digital transmission based on polarization shift keying modulation, IEEE J. Sel. Areas Commun., Vol. 13, No. 3, 1995, pp. 531–542.
51. A.S. Siddiqui et al., Dispersion-tolerant transmission using a duobinary polarization-shift keying transmission scheme, IEEE Photonics Technol. Lett., Vol. 14, No. 2, 2002, pp. 158–160.
52. J.M. Gené et al., Modified duobinary polarization-shift keying transmission scheme, IEEE Photonics Technol. Lett., Vol. 16, No. 1, January 2004, pp. 311–313.
53. B. Koch et al., Endless optical polarization control at 56 krad/s, over 50 Gigaradian, and demultiplex of 112 Gb/s PDM-RZ-DQPSK signals at 3.5 krad/s, IEEE J. Sel. Top. Quantum Electron., Vol. 16, No. 5, 2010, pp. 1158–1163.
54. T. Ono and Y. Yano, Key technologies for Tera-bit/second WDM systems with high spectral efficiency of over 1 b/s/Hz, IEEE J. Quantum Electron., Vol. 34, No. 11, November 1998, Invited Paper, pp. 2080–2088.
55. F. Derr, Coherent optical QPSK intradyne system: concept and digital receiver realization, IEEE J. Lightwave Technol., Vol. 10, No. 9, September 1992, pp. 1290–1296.
56. P.J. Winzer et al., 56 Gbaud PDM-QPSK: coherent detection and 2,500 km transmission, ECOC2009, Vienna, September 2009, PD2.7.

57. A. Sano et al., 240 Gb/s polarization-multiplexed 64QAM modulation and blind detection using PLC-LN hybrid integrated modulator and digital coherent receiver, ECOC2009, Vienna, September 2009, PD2.2.

58. J.E. Simsarian et al., Fast-tuning 224 Gb/s intradyne receiver for optical packet networks, OFC2010, San Diego, March 2010, PDPB5.

59. C.R.S. Fludger et al., 10×111 Gb/s, 50 GHz spaced, POLMUX-RZ-DQPSK transmission over 2375 km employing coherent equalisation, OFC2007, Anaheim, March 2007, PDP22.

60. C.R.S. Fludger et al., Coherent equalization and POLMUX-RZ-DQPSK for robust 100-GE transmission, IEEE J. Lightwave Technol., Vol. 26, No. 1, January 2008, pp. 64–72.

61. F. Gini and G.B. Giannakis, Frequency offset and symbol timing recovery in flat-fading channels: a cyclostationary approach, IEEE Trans. Commun., Vol. 46, No. 3, March 1998, pp. 400–411.

62. F.M. Gardner, A BPSK/QPSK timing-error detector for sampled receivers, IEEE Trans. Commun., Vol. COM-34, May 1986, pp. 423–429.

63. M. Oerder and H. Meyr, Digital filter and square timing recovery, IEEE Trans. Commun., Vol. COM-36, May 1988, pp. 605–612.

64. S.J. Savory et al., Digital coherent receivers for uncompensated 42.8 Gb/s transmission over high PMD fibre, ECOC2007, Berlin, September 2007.

65. S.J. Savory et al., Electronic compensation of chromatic dispersion using a digital coherent receiver, Opt. Express, Vol. 15, 2007, pp. 2120–2126.

66. G. Li, Recent advances in coherent optical communication, Adv. Opt. Photonics, Vol. 1, February 2009, pp. 279–307.

67. D.S. Millar et al., Experimental comparison of nonlinear compensation in long-haul PDM-QPSK transmission at 42.7 and 85.4 Gb/s, ECOC2009, Vienna, September, 2009, Paper 9.4.4.

68. B. Widrow and M.E. HoffJr., Adaptive Switching Circuits, IRE WESCON Conv. Rec., Part 4, 1960, pp. 96–104.

69. C.R. Johnson et al., Blind equalization using the constant modulus criterion: a review (Invited Paper), Proc. IEEE, Vol. 86, No. 10, October 1998, pp. 1927–1950.

70. A. Vgenis et al., Nonsingular constant modulus equalizer for PDM-QPSK coherent optical receivers, IEEE Photonics Technol. Lett., Vol. 22, No. 1, January 2010, pp. 45–47.

71. M. Yan et al., Adaptive blind equalization for coherent optical BPSK system, ECOC2010, Torino, September 2010, Paper Th.9.A.4.

72. X. Zhou et al., Cascaded two-modulus algorithm for blind polarization demultiplexing of 114 Gb/s PDM-8QAM optical signals, OFC/NFOEC2009, San Diego, March 2009, Paper OWG3.

73. J. Li et al., Spectrally efficient quadrature duobinary coherent systems with symbol-rate digital signal processing, IEEE J. Lightwave Technol., Vol. 29, No. 8, 2011, pp. 1098–1104.

74. M.G. Taylor, Phase estimation methods for optical coherent detection using digital signal processing, IEEE J. Lightwave Technol., Vol. 27, No. 7, April, 2009, pp. 901–914.

75. A.J. Viterbi and A.N. Viterbi, Nonlinear estimation of PSK-modulated carrier phase with application to burst digital transmission, IEEE Trans. Inf. Theory, Vol. IT-29, No. 4, July 1983, pp. 543–551.

76. M.G. Taylor, Coherent detection for optical communications using digital signal processing, OFC2007, Anaheim, March 2007.
77. N. Wiener, Extrapolation, Interpolation, and Smoothing of Stationary Time Series, John Wiley & Sons, Inc, New York, 1949.
78. Yan Han and G. Li, Coherent optical communication using polarization multiple-input-multiple-output, Opt. Express, Vol. 13, No. 19, September 2005, pp. 7527–7534.
79. S. Zhang et al., Novel ultra wide-range frequency offset estimation for digital coherent optical receiver, OFC2010, San Diego, March 2010.
80. M. Seimetz, Laser linewidth limitations for optical systems with high-order modulation employing feed forward digital carrier phase estimation, OFC/NFOEC2008, San Diego, March 2008, Paper OTuM2.
81. E. Ip et al., Coherent detection in optical fiber systems, Opt. Express, Vol. 16, No. 2, January 2008, pp. 753ff.
82. M. Alfiad et al., Effect of carrier phase estimation for 111 Gb/s POLMUX-RZ-DQPSK equalization in presence of 10.7 Gb/s OOK neighbours, IEEE/LEOS Summer Topical Meetings, Acapulco, July 2008, pp. 175–176.
83. A. Lowery and J. Armstrong, Orthogonal frequency division multiplexing for optical dispersion compensation, OFC2007, San Diego, March 2007, Paper OTuA4.

6

SYSTEM REALIZATION

6.1 ACCESS NETWORKS

Telecommunications networks are hierarchically organized. Residential and business access is based on wireless (2G/3G/4G and WiFi) and wireline [fiber point-to-point (P2P), point-to-multipoint passive optical networks (PONs), copper twisted pair, HFC (hybrid fiber coaxial)] technologies. Access concentrators (base stations, PON optical line terminations (OLTs), MSANs, etc.) are backhauled to aggregation sites of a first level. These sites are often referred to as local exchanges (LX) or central offices (COs). This backhaul can be based on coarse wavelength division multiplexing (CWDM), DWDM, or passive WDM (pWDM) (see Section 6.1.2) technology. In addition to access and backhaul, there is a strong trend toward mobile *fronthaul* [or Cloud Radio Access Networks (C-RANs)] in fourth generation LTE-Advanced (LTE-A) networks. Here, several baseband units (BBUs) are concentrated in BBU hotels (BBUHs) and connected to their respective antennas via digital CPRI (Common Public Radio Interface [1]) high-speed links. Since these links must not be statistically multiplexed, and have tight latency and jitter requirements, they must run via point-to-point fiber or wavelength division multiplexing (WDM) channels.

Several LX can be concentrated, via CWDM or DWDM, in metropolitan-area core PoPs (points-of-presence). PoPs can accommodate aggregation switches of a secondary level, and they are connected via protected DWDM rings or mesh networks. Typically, the metro core network is connected to national and international backbones via two redundant PoPs [which may accommodate core routers,

Wavelength Division Multiplexing: A Practical Engineering Guide, First Edition.
Klaus Grobe and Michael Eiselt.

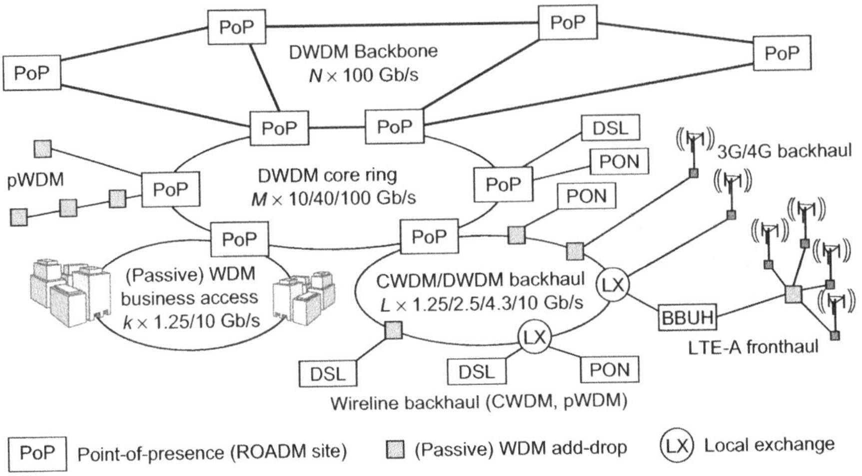

FIGURE 6.1 Hierarchical fiber-optic telecommunications network. BBUH: baseband unit hotel.

broadband remote access servers (BRASs), etc.]. The backbones use high-capacity, long-haul DWDM.

An example network is shown in Fig. 6.1

Real networks can deviate from the example shown in Fig. 6.1 in several ways. They may have less, or even more, aggregation levels. They can be based on differing combinations of ring, mesh, or point-to-point structures, and they can also make use of various (legacy) technologies like Synchronous Optical Network (SONET)/ Synchronous Digital Hierarchy (SDH), which are not mentioned in Fig. 6.1.

On the other hand, what most networks have in common is that they use some sort of MPLS-over-WDM in the core, and some sort of Ethernet (over fiber, SONET/ SDH, WDM, and even MPLS) in access and backhaul.

6.1.1 CWDM Access and Backhaul

Coarse WDM systems have been in wide use for metropolitan area backhaul and business access since the very beginning of the twenty-first century [2,3]. They are based on up to 18 CWDM channels spaced 20 nm in the wavelength region 1270–1610 nm, as defined in ITU-T Recommendation G.694.2. In the beginning, CWDM had the advantage of being cheaper than DWDM since CWDM components do not require temperature control or stabilization. This advantage is expected to decrease since CWDM transceivers are not wideband tunable. DWDM transmitters, on the other hand, are full-band (e.g., in the C-band) tunable. This allows cost reduction in manufacturing since only a single type of transmitter needs to be produced. It also allows operational cost reductions, since, for example, sparing and also network planning are greatly simplified.

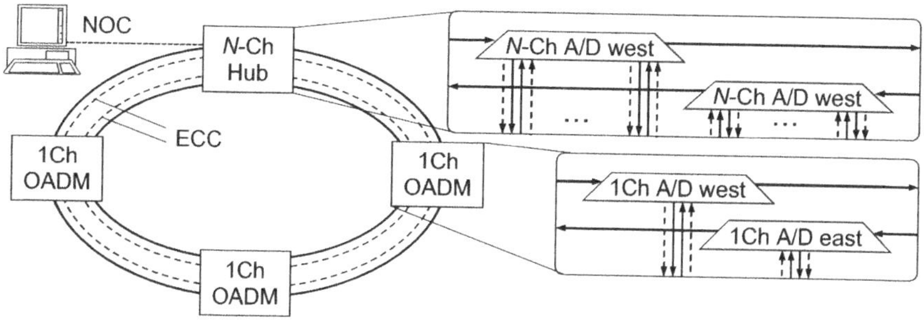

FIGURE 6.2 CWDM backhaul ring for hubbed traffic pattern. A/D: add–drop.

In many cases, CWDM was used for (DSLAM, GPON, wireless 2G/3G) backhaul, running one or multiple GbE services per wavelength. Up to 4GbE signals can be multiplexed at wire speed (after 10B/8B decoding) onto muxponders running at ~4.3 Gb/s [4]. In CWDM backhaul, this bit rate is relevant because it has similar cost than 2.5 Gb/s. This low cost was originally driven by transmission of 4 Gb/s Fibre Channel (FC) signals. In general DWDM transmission, 4 Gb/s per channel is almost irrelevant due to its lack of spectral efficiency. Capacity increase up to 16 × 10 Gb/s has also been demonstrated with CWDM [5].

Backhaul topologies are often rings, which can be protected as described in Section 8.4.2. A four-node CWDM ring with a hub node and three optical add–drop multiplexer (OADM) sites is shown in Fig. 6.2. For monitoring purposes, the OADMs can be connected to a network operations center (NOC) via an embedded communications channel (ECC). The ECC can transport narrowband management information (see Section 7.4).

CWDM systems often use CWDM-colored small form-factor pluggables (SFPs) as remote interfaces. These can be accommodated directly in the client systems, or on transponders/muxponders. Per-channel bit rates of 1.25–4.3 Gb/s are covered (where 4.3 Gb/s share the transceiver technology with 4G Fibre Channel, which is one of the drivers behind the advantageous economics). Ten gigabit per second CWDM-fixed wavelength extended form-factor pluggables (XFPs) also exist, but are typically restricted to the wavelengths 1470–1610 nm. In effect, these pluggables are versions of the respective DWDM transceivers, and, hence, unlike their lower bit rate counterparts, do not have significant cost advantages. Consequently, their relevant specifications are similar to the DWDM XFP specifications. So far, dedicated CWDM 40 or 100 Gb/s transceivers have not been built, and it is unlikely this will ever happen. Table 3.14 lists relevant parameters for simple calculations of 2.5 Gb/s CWDM systems. For 4 and 10 Gb/s, reduced chromatic dispersion (CD) allowance and also power budget must be considered. This may limit maximum reach, without added means like CD compensation (not commonly used in CWDM systems), or forward error correction (FEC) to <60 km.

Typically, in CWDM systems, effects of polarization-mode dispersion (PMD), polarization-dependent loss (PDL), and nonlinearity are covered within the

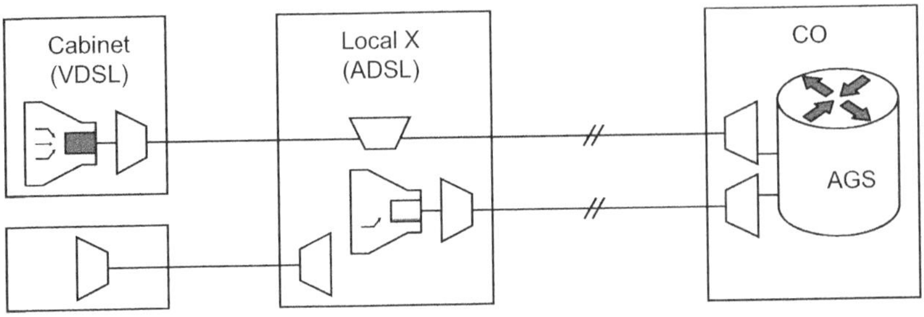

FIGURE 6.3 Concept of passive WDM as an extension of the black link approach.

transceiver specifications, due to distance limitations. If link lengths approach the respective specified CD limits, power budget penalties in the 1 dB range may have to be applied.

6.1.2 Passive WDM

A simple form of passive WDM-based networking is called *passive WDM*; it has been used in large carriers' networks for (DSL) backhaul [6,7]. Basically, it is a form of the ITU-T G.695 black link approach. Passive CWDM filters are used to allow CWDM pluggables (SFPs) to be plugged directly into client devices (DSLAMs and MSANs), and then to be multiplexed. This approach saves O/E/O conversion (transponders) or the use of multiple fibers. pWDM connections between a central office that accommodates an aggregation switch of level 1 (AGS1) and local exchanges and cabinets are shown in Fig. 6.3.

The pWDM concept allows the use of standard CWDM SFPs, which is cost-efficient for certain applications like DSLAM backhaul. Together with the low insertion loss of CWDM filters, these SFPs allow maximum reach in the range of 40–80 km (refer to and Table 6.1) . In addition, the coarse channel spacing makes most CWDM filters outdoor hardened in terms of the required temperature range (full industrial, −40 to +85°C). In fact, the 20 nm CWDM channel spacing was defined following maximum temperature-induced laser wavelength drift of approximately ±6.5 nm over this temperature range.

More recent pWDM variants also cover DWDM with up to 80 channels. For DWDM filters and transceivers, more effort is required for athermalization.

TABLE 6.1 Calculation Parameters for CWDM Systems

Add–drop loss per CWDM filter pair 1/4/8/16 channels (dB)	2/3/5/8
Environmental CWDM filter specifications	Can be outdoor hardened
CD-limited reach of 2.5 Gb/s (1 Gb/s) transceivers (km)	40–120 (depends on SFP class)
Power budget of 2.5 Gb/s (1 Gb/s) transceivers (dB)	22–32 (depends on SFP class)

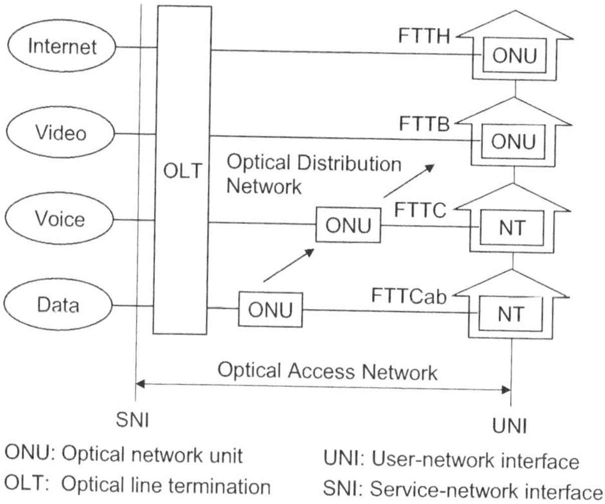

FIGURE 6.4 FTTX access: fiber-to-the-home/building/curb/cabinet. NT: network termination.

6.1.3 PONs and Optical Access Networks

Passive optical networks are the basis of Optical Access Networks (OANs) as defined in ITU-T Recommendation G.902, and also of hybrid fiber coaxial networks. PONs are often configured in tree or bus structures. Feeder and distribution fibers, together with the distribution elements in the outside plant, are referred to as Optical Distribution Network (ODN). The different degrees of optical versus electrical access down to the customers are summarized in Fig. 6.4 for the different fiber-to-the-X (FTTX) access scenarios, where X stands for cabinet (Cab), curb (C), building (B), or home (H). The reference points for the OAN are the service network interface (SNI) and the user network interface (UNI). They are defined in ITU-T Recommendation G.983.

In PONs, several customers are connected to a central office or local exchange via a passive fiber-optic infrastructure. This infrastructure splits into single-mode fibers and passive splitting components (power splitters/combiners and/or WDM filters). PONs work bidirectionally on single fibers, in almost all cases, by using different wavelengths for upstream (US) and downstream (DS).

The main PON components are as follows:

- OLT: Optical line termination (service provider head end in CO)
- RN: Remote node (mostly passive, containing splitter/combiner or filter, for example, in cabinets)
- ONU: optical network unit [at customer premises (CP), or in cabinet, etc.]

The OLT is the interface between the access network and the backbone. It is responsible for the optical transmission and reception into/from the PON. Optionally, switching or cross-connection is provided to relieve the transport network of switching responsibilities.

Remote nodes contain optical splitters/combiners or, in many WDM-PON approaches, WDM filters. RNs can be cascaded, and they may optionally accommodate active equipment, for example, PON reach extenders (amplifiers).

In the ONU, the optical signals are converted back into electrical signals (e.g., POTS (plain old telephone service), 100/1000bT Ethernet, and video). Like the OLT, the ONU is an active component that requires external power supply. The ONU acts as the residential gateway, coupling the distribution network with the in-home network. The ONU may or may not contain functions like WLAN access and the residential access router.

The ODN comprised of single-mode optical fibers and the passive optical components. Optionally, it may provide resilient paths between one or two OLTs and the RNs or the ONUs.

PONs allow sharing the cost of running fibers from the CO to the premises among many users [up to 128 in power splitter-based time-domain multiple access (TDMA)-PONs, up to several hundreds in WDM-PON and hybrid WDM/TDMA and similar PONs]. As shown in Fig. 6.5, a so-called *feeder fiber* runs from the CO to a centralized distribution point (the RN), then *distribution fibers* extend from this point to each customer location.

PONs do not require power in the outside plant for operating passive filters or splitters, thereby lowering the overall operational cost and complexity. However, active equipment, for example, for reach extension, may be deployed in PON ODNs as well. Because single fibers are shared in tree, star, bus, or ring topologies, the fiber cost is lower than in optical access with point-to-point topology. In addition, in a PON fewer fibers are terminated in the CO, which reduces floor space, energy consumption, and cabling effort.

As compared to copper-based access technologies like xDSL, higher bandwidths (total up to several 100 Gb/s) and higher distances (up to 100+ km) are possible. PONs do not suffer from electromagnetic interference (EMI), which is the case in xDSL on twisted-pair cables due to (near-end/far-end) cross talk between different users. In addition, especially WDM-based PONs are somewhat more difficult to tap for eavesdropping.

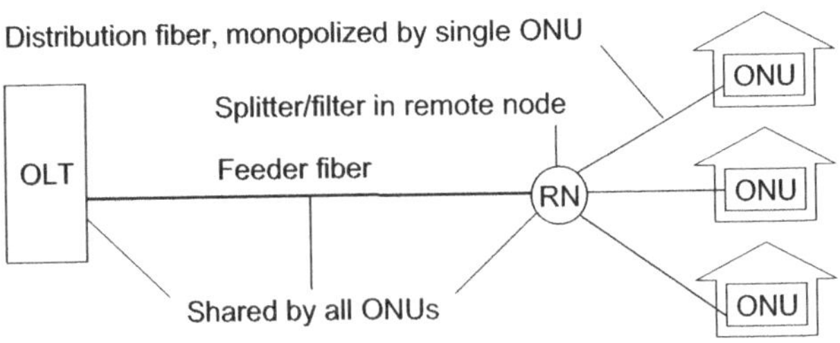

FIGURE 6.5 Principle of shared PON access.

The first fiber-to-the-home (FTTH) field trials were conducted in 1977 and the first PON was described in 1987 [8–10]. Since then, FTTX deployment differed significantly in the three main regions: North America, Europe, and Asia-Pacific. In North America and Asia-Pacific, PON deployment clearly exceeded deployments of Active Optical Networks (AON), that is, active, dedicated P2P Ethernet systems. In Europe, the situation was vice versa up to 2012. Today, global PON deployments clearly exceed AON deployments [11].

6.1.4 WDM-PON

A basic question in PONs is how the access in the *upstream* (i.e., the direction from the ONU to the OLT) is organized. The problem is that N customers need to share one feeder fiber (between the RN and the OLT) [9]. The usual technologies for multiple access to common resources can be used, that is, TDMA, WDMA (wavelength-domain multiple access), SCMA (subcarrier multiple access), and CDMA (code-domain multiple access). The downstream (from the OLT to the ONUs) can use simpler time-domain multiplexing (TDM) or WDM.

WDMA uses different wavelengths for the upstream channels (and also different wavelengths for the DS). If these wavelengths are separated in the RN by means of a WDM filter [e.g., an arrayed waveguide grating (AWG)], this is referred to as *wavelength-routed* (WR)-WDM-PON. Since all customers have a dedicated wavelength, high security against rogue (malfunctioning) ONUs and, to a certain extent, malicious users is provided through physical separation.

If all wavelengths are broadcast via a power splitter and the wavelengths are selected at the respective ONUs, this is called *wavelength-selected* (WS)-WDM-PON. Wavelength selection must be performed by a (tunable) WDM filter or a coherent receiver.

Client wavelengths are independent of other clients, thus providing a certain degree of transparency and simple per-user capacity upgrades. All components run at the client bit rate, not at the aggregate PON bit rate (like in TDMA PONs).

There is no component sharing in the CO (WDMA requires N transceivers in the OLT) so that only the feeder fiber between OLT and RN is shared. WDMA is shown in Fig. 6.6.

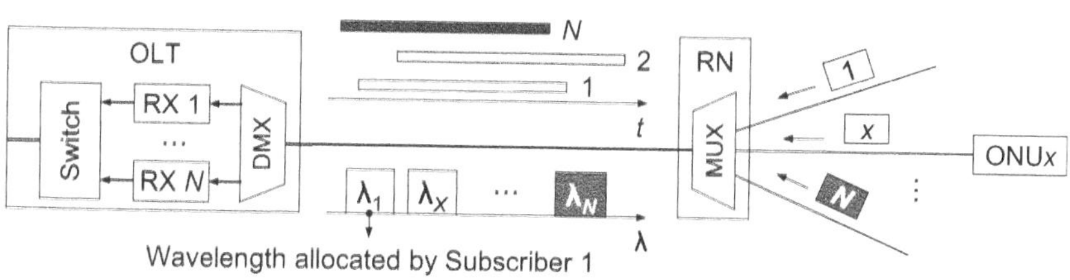

FIGURE 6.6 WDMA upstream in a WDM-PON.

TDMA-based PON-like (10G) EPON and (X)G-PON have per-user guaranteed bandwidth limitations due to the shared-medium approach, and the OLT and ONUs have to work at the aggregate bit rate. There are privacy/security issues due to the broadcast of downstream information, and network integrity can be affected by a single rogue ONU or a malicious customer that corrupts the entire upstream transmission. In addition, 1:N power splitters lead to severe power-budget penalties, thus limiting maximum link lengths (example: a 1:64 splitter imposes ~20 dB of insertion loss). Through a *ranging* protocol (i.e., detection of per-ONU distances for US collision avoidance [12]), maximum reach can be restricted to 60 km. These limitations can be solved with WDM-PONs, in particular WR-WDM-PONs where the ODN is based on WDM filters rather than power splitters. In WDM-PON, ONUs are assigned individual wavelengths (WDMA). This can provide higher guaranteed bandwidth and somewhat lower latency as compared to shared approaches.

Complete WDM-PON systems are not yet fully standardized. However, several aspects of WDM-PON have been, and are being, standardized in ITU-T Study Group 15, Questions 6 and 2 (e.g., ITU-T Recommendations G.698.3 and G.multi).

For further introductions to WDM-PON, refer to Refs [13,14].

In a WS-WDM-PON, the remote node contains a power splitter/combiner similar to the one used in TDMA PON (G-PON and EPON). Tunable WDM filters or coherent receivers with local oscillators at the ONUs select the respective wavelengths.

A basic WS-WDM-PON system with direct detection is shown in Fig. 6.7. It makes use of integrated multichannel transmit (Tx) and receive (Rx) arrays in the OLT and tunable laser diodes (T-LD) and tunable filters (TF) in the ONUs. From the OLT, all wavelengths are broadcast via the splitter. Obviously, this imposes the loss of the splitter. Hence, the power budget between OLT and ONUs can become a problem if splitting ratio is high (e.g., >1:32).

An alternative to using tunable filters is the implementation of coherent receivers. These have better selectivity and hence allow more densely spaced (i.e., more) wavelengths. They also increase the sensitivity, and hence available ODN power budget. This approach is followed in coherent ultradense WDM-PONs (Co-UDWDM-PONs) and described later.

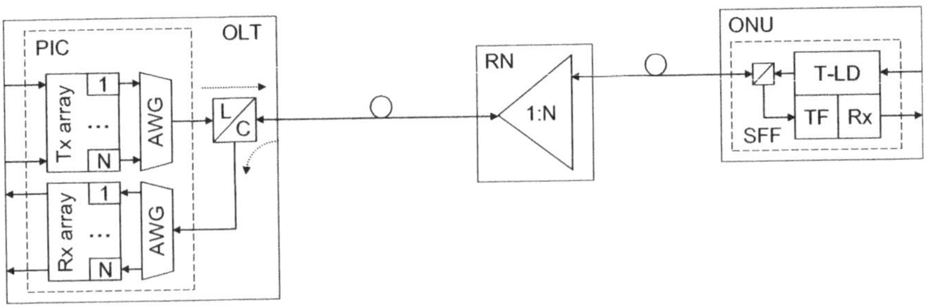

FIGURE 6.7 Basic WS-WDM-PON architecture with splitter/combiner in passive node. T-LD: tunable laser diode, TF: tunable filter, SFF: small form-factor, PIC: photonic integrated circuit.

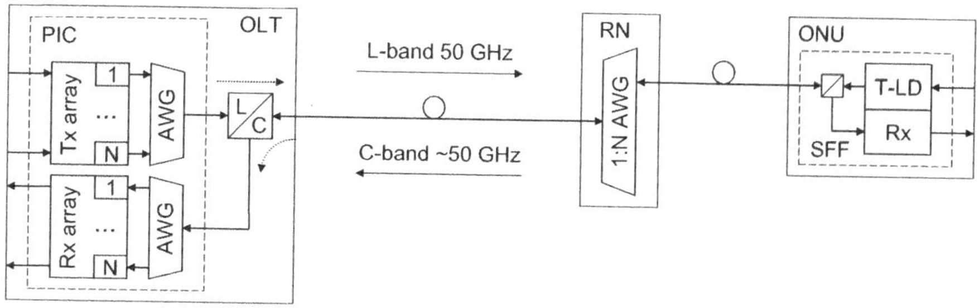

FIGURE 6.8 AWG-based wavelength-routing WDM-PON.

For comparable client count, the splitter has significantly higher insertion loss than a filter (e.g., a 1:64 power splitter has ~20 dB insertion loss and a 1:64 AWG can have <7 dB of insertion loss). In the splitter-based ODN, this prohibits the use of low-cost and low-power transceivers, for example, 1 Gb/s transceivers with PIN receivers.

In a WR-WDM-PON, a WDM filter (in most cases, an arrayed waveguide grating) replaces the power splitter/combiner in the RN. This offers lower insertion loss (50 GHz AWGs have insertion loss of 6–8 dB, nearly independentof the number of wavelengths). Then, the total number of ONUs is only limited by the number of accessible wavelengths. In Fig. 6.8, an AWG-based WR-WDM-PON is shown. In the remainder of the book, we refer to WDM-PON as the filtered (WR) WDM-PON unless otherwise specified.

The WDM-PON shown in Fig. 6.8 is based on tunable lasers in the ONUs. This is one approach of making the ONUs *colorless* in order to avoid wavelength-specific ONU variants. As compared to WS-WDM-PON, no wavelength-selective receivers are necessary, thus simplifying the ONUs.

The alternative to tunable lasers for colorless transmitters is the use of *seeded reflective* transmitters. These devices receive a seed wavelength (which has to be wavelength routed in the RN or wavelength selected in the ONU by means of a tunable filter), and reflect and modulate it with upstream information. Reflective semiconductor optical amplifiers (RSOAs), reflective electroabsorption modulators (REAMs), combinations of these devices (REAM–SOAs), or injection-locked Fabry–Pérot laser diodes (IL-FP-LDs) can be used [14].

Both (tunable) laser diodes or seeded reflective transmitters can be used in multichannel photonic integrated circuits (PICs) as OLT transceivers. A reflective PIC has to be seeded by means of a wavelength-filtered broadband source (e.g., an amplified spontaneous emission (ASE) source or a superluminiscent light-emitting diode), or by a multifrequency laser (MFL).

6.1.4.1 WDM-PON with Tunable Lasers and Laser Arrays Laser diodes can be used as transmitters for the OLT and the ONUs. The requirement for colorless or wavelength-agnostic ONUs leads to the necessity of *tunable* lasers. Up to about 2012, full-band-tunable lasers were too expensive in order to allow residential access

(FTTH) based on WDM-PON. This originally led to the seeded reflective WDM-PON approaches. Meanwhile, work is on its way, which will allow cost-effective WDM-PON based on tunable lasers.

The block diagram of a laser-based WDM-PON has been shown in Fig. 6.8. Three major elements can be identified, which are necessary for an efficient WDM-PON solution:

- A cyclic AWG in the RN to allow single fiber working (SFW) and multiband access
- An array technology (PIC) for the OLT transceivers
- An ONU transceiver based on a low-cost wavelength-agnostic (tunable) transmitter, a diplexer as directional coupler, and the related receiver

Cyclic AWGs are the most elegant way to provide SFW in WDM-PON. Other filter technologies like thin-film filters (TFF) can be used but either they lead to the requirement of additional diplexers for combining both directions of transmission (downstream and upstream) or they have higher insertion loss. In Fig. 6.8, an AWG with cyclicity according to C-band, L-band, and so on is shown. Other options exist (for example, red and blue subbands in the C-band).

OLT photonic integrated transceivers are required for reasons of cost (components' CapEx), footprint, and power consumption. The OLT PIC can consist of two arrays, one for the receivers and one for laser transmitters. If the PIC contains all channels, the lasers need not be tunable; however, a limited degree of tunability may increase the production yield of such devices and help decrease the cost.

The ONU transceiver needs to be low in cost, low in power consumption, and wavelength agnostic (colorless). The most important way to make tunable lasers low in cost is to reduce their complexity. Every subcomponent that is not absolutely necessary must be omitted. In addition, many network operators require *coolerless* transceivers to reduce power consumption. Hence, WDM-PON tunable lasers must be coolerless. Tunability should not come at significant added cost (e.g., through complicated external cavities). Hence, monolithically integrated lasers that are inherently tunable should be chosen, for example, multisection distributed Bragg reflector lasers. A dedicated per-laser wavelength locker should be omitted for cost reasons. This results in a monolithically integrated multisection laser diode without locker and cooler, which has low-cost potential. Since this device does not have its own wavelength locker and may drift in wavelength through changes of the ambient temperature, system-wide means for tuning must be established. This tuning is possible via active involvement of the OLT. This enables a *distributed, shared* wavelength locker at low cost. The principle of ONU tuning via the OLT is shown in Fig. 6.9.

The respective control commands can be sent from the OLT to the ONU via an ECC. The ECC provides a narrowband signaling channel (1–100 kb/s, depending on the implementation). It can use part of the downstream wavelength payload [e.g., an Ethernet management VLAN (virtual LAN)], or use a suitable in-band modulation

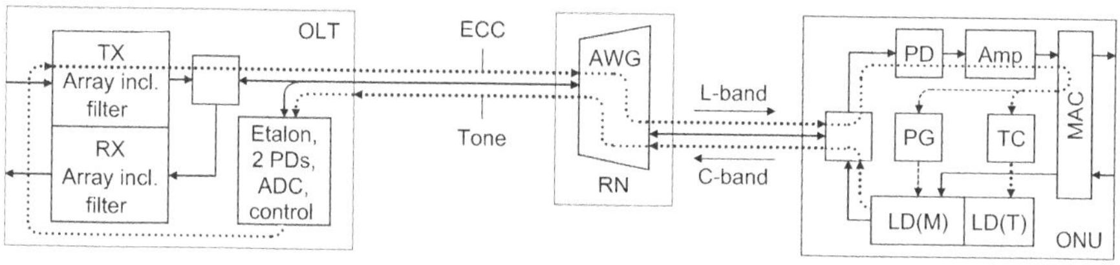

PG: pilot tone generation, TC: tuning control, LD(M): LD modulator (and gain) section, LD(T): LD tuning section

FIGURE 6.9 Tuning of tunable ONU via closed-loop control, incorporating the OLT and an ECC.

that is transparent to the payload [e.g., amplitude-modulated (AM) pilot tones]. In a WR-WDM-PON, this ECC can always be established in the downstream because the ONUs have broadband receivers. Similar tuning also works in WS-WDM-PON.

A simple approach for wavelength control at the OLT is using the received optical power at the OLT RX as a metric for wavelength tuning. Tuning is possible because two wavelength-selective elements are part of the optical path—the AWGs in the RN and in the OLT. Tuning can be based on controlling the ONU laser in order to receive maximum power at the OLT. This requires AWGs with Gaussian passband shape in order to get a well-defined tuning criterion. Furthermore, the AWG in the RN must be athermalized in order to prevent thermal drift. Tuning can be done based on gradient algorithms that maximize OLT RX power.

In case flat-top AWGs have to be used, or to further improve tuning precision, individual per-wavelength pilot tones can be used, which are simultaneously detected at a common wavelength locker at the OLT. In order to distinguish between the different ONU wavelengths, each ONU can apply a pilot tone with a distinct frequency to its optical signal. In the wavelength locker, a part of the signal is split and filtered by an etalon (i.e., a Fabry–Pérot interferometer), while the other part remains unfiltered and is detected by a reference photodiode (PD). Comparing the pilot tone amplitudes at the reference PD and the etalon PD, the OLT can calculate whether the wavelength is higher or lower than the target. Analysis for all ONU channels is done after A–D conversion of the signals of both PDs and FFTs (fast Fourier transforms). Feedback to the ONUs is provided via the ECCs.

The wavelength tuning and stabilization process consists of two phases. First, when the ONU is turned on, the laser sweeps across the upstream spectrum. Tuning over all channels does not affect neighboring channels, as the adjacent channel isolation of an AWG is typically >25 dB. The received optical power at the OLT is monitored in order to see when a signal is received. After the initial phase is completed, the pilot tone method can be used for continuous tuning.

Tone-based tuning can be based on AM or frequency-modulated (FM) pilot tones. AM/FM tones can be applied by superposition to the injection currents of the gain and the phase sections of the laser diode, respectively. When applying an AM tone to a wavelength with average power P_0, the slowly varying power $P(t)$ becomes

$$P(t) = P_0 + \Delta P_0 \sin(\omega_0 t). \tag{6.1}$$

The ratio $M_0 = \Delta P_0/P_0$ is the modulation index of the pilot tone. Using the modulation index, the resulting power penalty PEN for the (payload) data signal can be calculated:

$$\mathrm{PEN} = \frac{P_0}{P_0 - \Delta P_0} = \frac{1}{1 - M_0}. \quad (6.2)$$

A modulation index of the tone of 10% leads to a penalty for the data signal of ~0.5 dB; typical pilot tone modulation indices are ~5%. The bit error rate (BER) of the tone signal can be derived from the pilot tone power, which is given by

$$P_{\mathrm{PT}} = \frac{M_0^2}{2} R^2 P_{\mathrm{AVG}}^2 Z_{\mathrm{T}}. \quad (6.3)$$

R is the responsivity of the photodiode, Z_{T} is the transimpedance of the TIA (transimpedance amplifier), and P_{AVG} is the average data signal power. The $\mathrm{BER} = 0.5\,\mathrm{erfc}\,(Q/\sqrt{2})$ can be calculated using

$$Q = \frac{1}{2}\sqrt{\frac{P_{\mathrm{PT}}}{N' B_{\mathrm{PT}}}}. \quad (6.4)$$

B_{PT} is the bandwidth of the pilot tone detection circuit, and N' is the sum of laser RIN (relative intensity noise), electronic receiver noise, and (if applicable) signal-ASE beat noise.

When FM tones are used for tuning, the optical laser frequency can be expressed as

$$\nu(t) = \nu_0 + \Delta\nu_0 \sin(\omega_0 t + \varphi_0). \quad (6.5)$$

ν_0 is the center frequency, $\Delta\nu_0$ is the peak frequency deviation, and ϕ_0 is the phase delay between AM and FM, respectively. The time-domain signal at the etalon output can then by written as

$$s(t) = (T(\nu_0) - T'(\nu_0)\Delta\nu_0 \sin(\omega_0 t + \varphi_0))(P_0 + \Delta P_0 \sin(\omega_0 t)). \quad (6.6)$$

Here, T and T' are the etalon power transfer function and its first derivative, respectively.

For pure FM, the power spectrum $|S(j\omega)|^2$ can be approximated:

$$|S(j\omega, \Delta P_0 = 0, \varphi_0 = 0)|^2_{\omega=\omega_0} = (\pi T'(\nu_0)\Delta\nu_0 P_0)^2. \quad (6.7)$$

Since π, $\Delta\nu_0$, and P_0 are constant, spectral power of the pilot tone depends only on the first derivative of the etalon power transfer function. When the etalon transfer function is correctly matched to the desired wavelength grid, the tuning set point is the zero-crossing of the first derivative of the etalon power transfer function. As the resonance frequencies change with temperature, the etalon, like the AWGs, needs to be temperature controlled.

Digital signal processing can extract the control signal for each laser via FFT. Since each pilot tone frequency corresponds to a specific laser, the control signals for each individual laser can be extracted. This allows individually controlling the wavelengths of all lasers simultaneously in the OLT.

With a simple, bidirectional DWDM-PON, no more than 64–96 bidirectional channels are possible with SFW, depending on the free spectral range (FSR) of the cyclic AWGs and its port count. This channel count range is accessible with 50 GHz AWGs. Sometimes, more channels are required. For DWDM-PON with direct detection, this is achievable by extending the wavelength region, or decreasing the WDM grid to an extent, which is still supported by filtered direct detection.

Decreasing the WDM grid to 25 GHz is feasible because only two filters have to be traversed: one multiplexer and one demultiplexer. There are no effects of multiple cascaded filters (which have significantly narrower passbands). Therefore, passive demultiplexing is still possible. Tuning to within 25 GHz requires pilot tone-based or similar mechanism, since OLT receive power optimization may not be sufficient anymore.

Alternative increase of the channel count is possible by extending the wavelength region. This is achieved by using further wavelength bands of the cyclic AWGs. If the lowest loss region of typical single-mode fibers is considered, the S-band is the most suitable extension region. The U-band should be avoided because it is often used for PON OTDR supervision.

The wavelength bands of cyclic AWGs have been specified in ITU-T Recommendation G.698.3. They are shown, together with the spectral loss of standard single-mode fibers, in Fig. 6.10. The dashed lines indicate the lower and upper limits for the loss according to ITU-T Recommendation G.695. Throughout the S-band, maximum fiber loss is <0.4 dB/km (assuming correct fiber deployment, etc.).

The block diagram of a DWDM-PON allocating the S-, C-, and L-bands is shown in Fig. 6.11. This WDM-PON is based on band splitters [C/L, S/C, and also red/blue

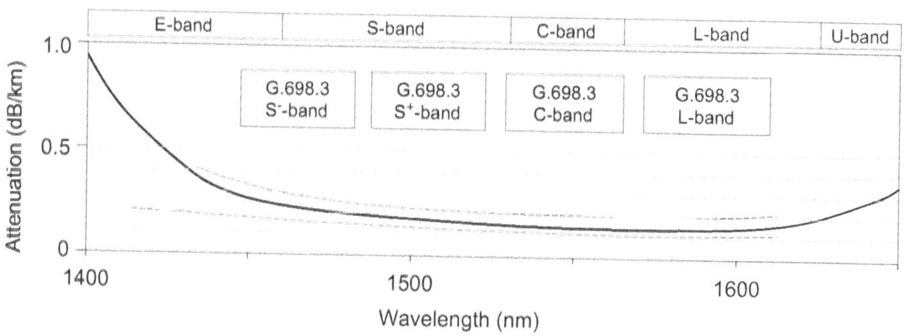

FIGURE 6.10 Accessible wavelength bands of cyclic AWGs according to G.698.3, and SSMF attenuation.

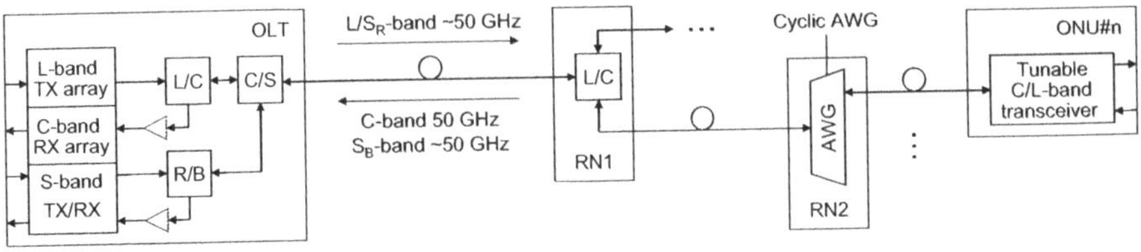

FIGURE 6.11 Laser-based WDM-PON with S-band extension for high channel count. R/B: red/blue subband splitter (separates S/minus and S/plus subbands).

(R/B) within the S-band] and AWGs. The sequence of the components in RN1 and RN2 can be changed. This allows using a single (cyclic) AWG, but requires a large number of remote diplexers.

In the PON shown in Fig. 6.11, the multiple splitter/filter stages lead to increased insertion loss. With low-cost PIN receivers, maximum reach may be limited to ~13 dB fiber power budget. Avalanche photodiode (APD) receivers are necessary for higher power-budget requirements.

One hundred ninety-two bidirectional channels (96 wavelengths each in C-band, L-band, and S/minus and S/plus bands) are possible. Increasing problems relate to laser safety and potential Raman loss/cross talk. Laser safety may pose limitations due to the high channel count. The wavelength bands used are spaced such that every second band is within spectral distance of efficient Raman coupling (the FSR of the cyclic AWGs is in the range of 5.2 THz). Power is Raman converted from the S-band to the C- and L-bands. This leads to added loss for S-band channels, and possibly added noise in the C- and L-bands, depending on the upstream/downstream assignment.

6.1.4.2 WDM-PON with Seeded Reflective Transmitters The alternative to tunable lasers in a WDM-PON consists of seeded reflective transmitters. This approach holds for both OLT PICs and ONUs (Fig. 6.12). The light sources used for seeding the reflective transmitters (transmitter arrays in the OLT) are MFLs or broadband light sources like SLEDs (superluminescent LEDs) or ASE sources.

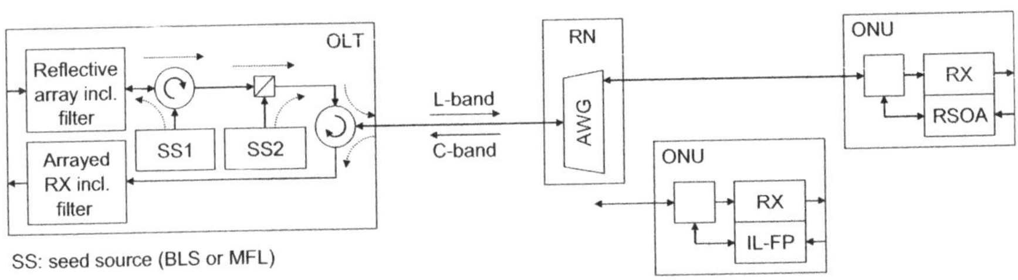

FIGURE 6.12 WDM-PON with seeded reflective (REAM) OLT transceiver array and seeded reflective ONUs (based on RSOAs, REAMs, IL-FP lasers, or combinations thereof, that is, REAM–SOA).

These seeds must be spectrally filtered (sliced) prior to entering the reflective transmitters. Due to modulation of the reflected seeds, no tuning of (ONU) transmitters is required.

The PON shown in Fig. 6.12 is based on two MFLs: one for seeding a reflective (REAM, RSOA, or combination thereof) transmitter array in the OLT, and the other for seeding the ONUs. In the first implementation, these MFLs work in two different wavelength bands, for example, C-band and L-band. Seeds and reflected modulated signals must be separated in the OLT by means of circulators. Compared to spectrally sliced broadband sources, MFLs have the advantage of higher power levels, which translates to better reach performance. However, they also lead to stronger cross talk caused by Rayleigh backscattering.

Various approaches of ONU seeding are possible. This includes different reflective devices—RSOAs, REAMs, and IL-FP lasers—as well as different schemes for providing the seed. The latter applies to both: the fibers used for seeding and the seed wavelengths.

Seeded reflective arrays can be used in the OLT as an alternative to laser arrays. In a central office that terminates several PONs, the MFL for seeding the OLT array can be shared between the different PONs, thus decreasing cost. This is shown in Fig. 6.13. Two MFLs are used for redundancy, and shared (distributed) amongst the PONs. In a similar way, seed light sources for reflective ONU transmitters can be shared.

The PON shown in Fig. 6.12 uses the C- and L-bands for upstream and downstream, respectively. This separates upstream from downstream, but Rayleigh cross talk between the upstream and its seed can limit the reach performance. Upstream and downstream can also use the same wavelengths. This increases spectral efficiency or the number of bidirectional channels within a given spectral region. Now, Rayleigh cross talk appears between DS and US. Furthermore, the seed consists of the *modulated* downstream, which must be *remodulated*. Upstream modulation must be possible irrespective of the downstream modulation. This

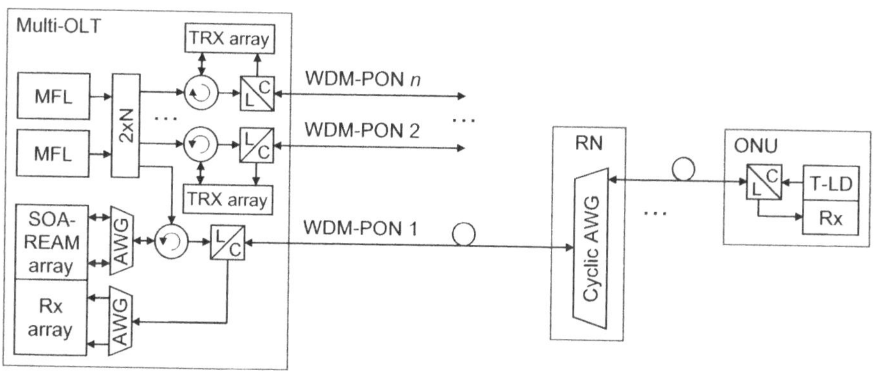

FIGURE 6.13 Multi-OLT with redundant MFLs that are shared between several PONs.

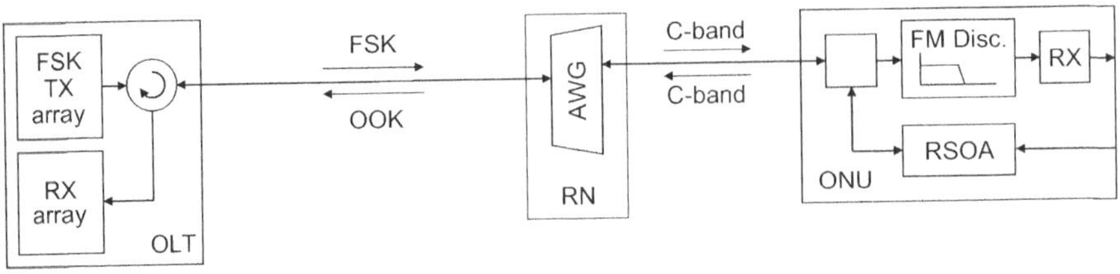

FIGURE 6.14 Wavelength reuse in seeded WDM-PON with constant-envelope FSK downstream.

can be achieved, for example, using nonreturn-to-zero (NRZ)-DPSK (differential PSK) or FSK (frequency-shift keying) modulation with constant envelope for the downstream (Fig. 6.14) [15]. This modulation is erased in a saturated RSOA in the ONU, which then performs on–off keying (OOK) onto the reflected wavelength for upstream transmission. This approach is limited by Rayleigh scattering and by cross talk due to downstream/upstream modulation (i.e., nonperfectly erased DS modulation).

Alternatives to the DPSK/FSK-OOK combination include inverse RZ (IRZ) modulation for the downstream, where the information is modulated in short dips of an otherwise constant envelope. Upstream uses return-to-zero (RZ) modulation in the remaining fraction of the bit period [16]. This system is shown in Fig. 6.15.

The total noise on the bits (0 or 1) can be separated into thermal, shot, and signal-ASE beat, plus the contributions from the Rayleigh-backscattered (RB) beat noise:

$$\sigma_x^2 = \sigma_{\text{Therm}}^2 + \sigma_{\text{Shot},x}^2 + \sigma_{\text{S-ASE},x}^2 + \sigma_{\text{RB},x}^2. \tag{6.8}$$

The subscript x indicates the dependence on the data symbol (0 or 1). σ_{Therm}, σ_{Shot}, $\sigma_{\text{S-ASE}}$ and σ_{RB} are the noise contributions due to thermal, shot, signal-ASE beat, and RB beat noise, respectively. With thc exception of thermal noise, they depend on the data symbol (subscript x).

There are two RB contributions that interfere with the upstream signal at the receiver. The first contribution, *carrier backscattering*, arises from the continuous wave seed delivered to the ONU. The second contribution, *signal backscattering*, is generated by the modulated upstream. The backscattered light reenters the reflective

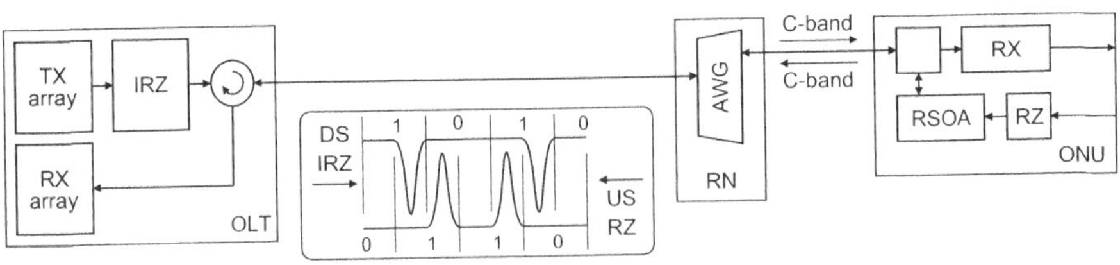

FIGURE 6.15 Wavelength reuse in seeded WDM-PON with IRZ/RZ modulation.

ONU transmitter, where it is remodulated and reflected toward the receiver. This relatively low-power light is also amplified by RSOAs or IL-FP lasers (when REAMs are used, no amplification takes place). The relative strengths of the two noise contributions depend on the reflective modulator gain, fiber lengths, and the splitting ratio of the US/DS splitters, which are used in the ONU (a 3 dB coupler is shown in Fig. 6.14 as an example, but in general the ratio is asymmetric).

The Rayleigh-backscattered beat noise consists of several contributions [17]:

$$\sigma^2_{\text{RB},x} = \sigma^2_{\text{S-RB},x} + \sigma^2_{\text{RB-RB},x} + \sigma^2_{\text{RB-ASE},x}. \tag{6.9}$$

$\sigma_{\text{S-RB}}$, $\sigma_{\text{RB-RB}}$, and $\sigma_{\text{RB-ASE}}$ are the contributions from signal-RB, RB-RB, and RB-ASE beats, respectively. They can be generated from either carrier or signal backscattering. In case of optical preamplification, the RB-ASE beat noise is given by

$$\sigma^2_{\text{RB-ASE},x} = 4D^2 P_{\text{RB},x} S_{\text{EDFA}} B_{\text{RB,ASE}}. \tag{6.10}$$

D is the photodetector responsivity, S is the EDFA output ASE PSD (power spectral density), and B is the equivalent bandwidth of the RB-ASE noise, respectively. Depending on the exact configuration, power penalties caused by Rayleigh cross talk can be as high as 3 dB.

For further analysis of seeded WDM-PON back reflection, see Refs [18–20].

In order to improve maximum reach of WDM-PONs with seeded ONUs, dedicated seed or upstream fibers can be used. These dedicated fibers can run all the way down to the ONU (end-to-end), or they can be used between the OLT and the RN only (Fig. 6.16). In particular duplicated end-to-end fibers contradict the PON SFW requirement.

Since feeder-fiber length clearly exceeds distribution-fiber length in almost all cases, dedicated seed fibers (or upstream fibers, which has a similar effect) between OLT and RN eliminate most of the Rayleigh cross talk. This is the common implementation used today for long-reach experiments with seeded ONUs, for example, Ref. [21].

An alternative to reducing Rayleigh losses is to distribute the seed sources more geographically, closer to the ONUs. Active RNs (ARNs) are used to accommodate

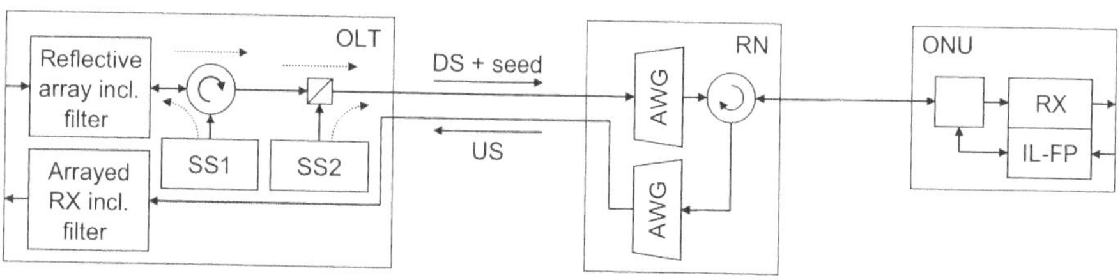

FIGURE 6.16 Decrease of Rayleigh cross talk through use of dedicated upstream feeder fiber.

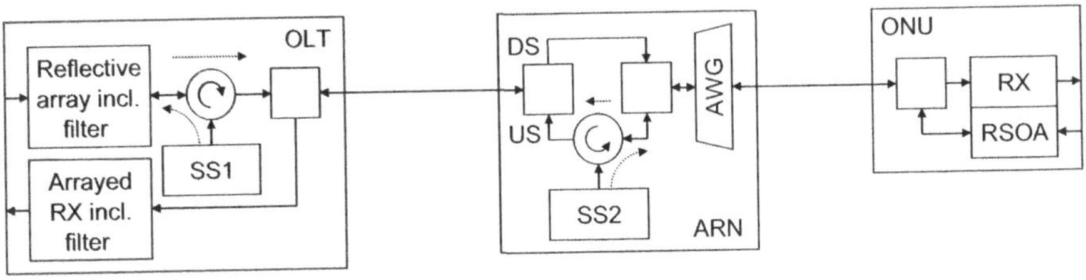

FIGURE 6.17 Decrease of Rayleigh cross talk through use of geographically distributed seeds.

the seed source (e.g., an MFL) (Fig. 6.17). Such active RN can still be low in complexity compared to fully functional local exchanges; the disadvantage is that the ODN is not fully passive anymore [13].

Seeded WDM-PONs have been used, together with per-wavelength TDMA for long-reach and high-user-count demonstrators, for example, in Ref. [21] (Fig. 6.18). This PON was based on 32 wavelength pairs, each running at symmetric 10 Gb/s with downstream TDM and upstream burst mode (BM) TDMA. Seeded REAM–SOA combinations were used for the ONUs, and the ODN consisted of WDM filters and cascaded 2:64 and 1:4 power splitters, respectively.

Total customer count of the hybrid WDM/TDMA-PON [21] was 8192, and each ONU had sustainable symmetric bandwidth of ~37 Mb/s. Downstream transmit and seed signals are multiplexed and demultiplexed via cyclic AWGs. In the ARN, one DS channel and one seed each are combined with band combiners (BC). Due to ARN-based amplification and Rayleigh cross talk suppression through dedicated upstream fibers (between the RNs accommodating the 2:64 splitter and the OLT), reach was 135 km.

For seeded reflective WDM-PON running 10 Gb/s per channel, the use of maximum-likelihood sequence estimation (MLSE) for intersymbol interference (ISI) mitigation has also been proposed [22].

A variant of the seeded reflective WDM-PON makes use of self-seeding, where a portion of the upstream light is reflected in a partial reflector at the RN [23]. The reflector is located next to the AWG in the RN, at the feeder-fiber side. It can be complemented toward a Faraday rotator mirror (FRM) in order to avoid polarization

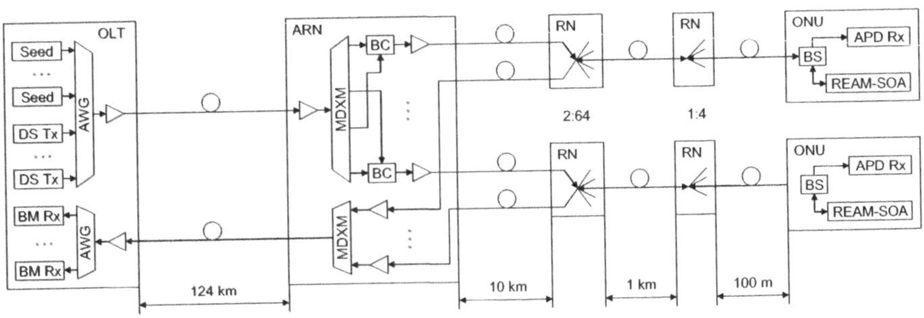

FIGURE 6.18 DWDM-TDMA-PON as an example for dedicated upstream feeder fiber.

effects. The structure ONU plus distribution fiber plus RN creates a distributed ECL (external-cavity laser).

6.1.4.3 Coherent UDWDM-PON If high client count and reach are required together with high per-ONU bandwidth, Co-UDWDM-PON can be considered. In Co-UDWDM-PON, channels are very tightly spaced, that is, with <12.5 GHz spacing. This requires coherent detection in order to provide the necessary receiver selectivity. Coherent detection also provides the high sensitivity that is required for massive power split or cascades of filters and splitters. With ultradense spacing, more than 1000 wavelengths can be provided in the low-loss region of optical fibers.

With coherent detection, quaternary PSK (QPSK) modulation can be used for optimized performance. Including implementation penalty, this can lead to ~50 dB power budget for per-channel bit rates of 1 Gb/s [24,25]. Due to coherent detection with a tunable local oscillator, power splitters have to be used, since passive filters cannot demultiplex ultradense channels. Power split in the ratio of 1:1000 leads to ~35 dB of insertion loss. Therefore, UDWDM-PONs can also use the hybrid filter plus splitter ODN, similar to the one shown in Fig. 6.18. For ultradense channel spacing as small as 3 GHz, DWDM filters with spacing of 50–400 GHz can be used. They should have flat-top transmission characteristic in order to treat all UDWDM channels equally to first approximation. This is shown in Fig. 6.19 [26].

Co-UDWDM-PON requires a polarization-insensitive receiver. This can be achieved with a polarization-diverse or polarization-scrambled receiver. Fully blown polarization-multiplexed configurations [dual-polarization (DP)-QPSK] could also be used, but in the PON context pose unacceptable cost penalties. A tunable laser diode (as local oscillator) with narrow linewidth is required. For coherent QPSK with 1 Gb/s (500 MBd), line width should be in the range of <500 kHz. External I/Q modulators are necessary for QPSK (or other complex modulation).

Integrated multichannel OLT transceivers should be used for cost, form-factor, and power consumption reasons. This can be achieved with broadband transceivers and digitally implemented SCMA. At 3 GHz DS channel spacing, with US interleaved (and shifted against the respective DS wavelengths by 1 GHz), up to 10 UDWDM channels can be provided through one broadband coherent SCMA transceiver with <30 GHz bandwidth. Such an implementation is shown in Fig. 6.20, which also shows a hybrid (filter plus splitter) ODN.

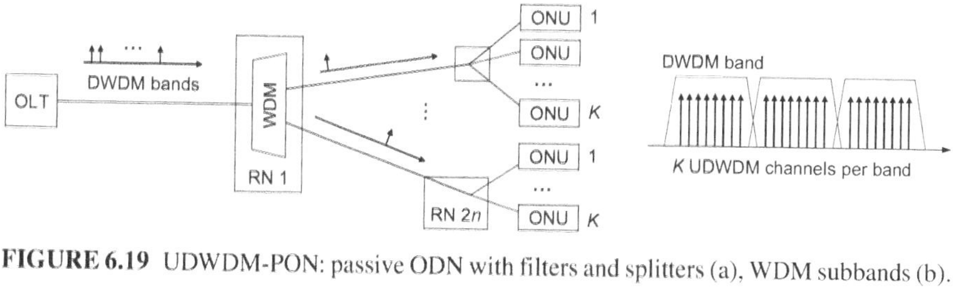

FIGURE 6.19 UDWDM-PON: passive ODN with filters and splitters (a), WDM subbands (b).

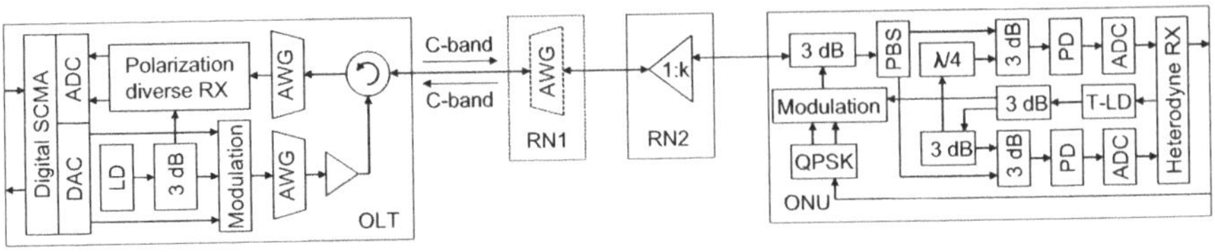

FIGURE 6.20 Heterodyne UDWDM-PON using filter plus splitter ODN.

Dual-fiber working is avoided by *heterodyne* detection. The local laser is tuned to an offset against the receive wavelength, which is larger than the payload bandwidth. This offset (the intermediate frequency in heterodyning) allows reusing the laser wavelength for upstream on the same fiber. This scheme also does not require 90° hybrids since I/Q processing is done in the electronic (digital RF) domain.

Homodyne detection schemes that avoid polarization diversity (as shown in Fig. 6.20) and the necessity for 90° hybrids have been proposed in Ref. [25], based on polarization and phase scrambling. The idea is to have the orthogonal states for phase and polarization in the same bit period *sequentially*. For the first half of the bit, the signal relative to one orthogonal component (I or H) will appear, whereas for the second half of the bit, the signal relative to the other orthogonal component (Q or V) will be seen. The principle is shown in Fig. 6.21.

Within each symbol of duration T_b, the phase of the local laser is scrambled by means of an RZ50 (return-to-zero with 50% duty cycle) signal that is synchronized with the receiver clock recovery (CLK Rec.). The RZ50 signal generates, within each symbol interval, two time slots of duration $T_b/2$ with fixed 0° to 90° phase modulation (PM). One slot represents the I component, the second slot represents the phase-shifted Q component. This is demonstrated in the insert in Fig. 6.21 (together with the subslots resulting from polarization scrambling).

By gating the photoreceiver output with the data clock and its inverse, the I and the Q components are obtained separately in the two branches, now with RZ shape. The signal power fluctuates between the I and the Q branches randomly, due to phase noise, at a rate of the order of the laser linewidth, and the combination of both outputs can assure its recovery. This operates like a phase diversity system. The local laser does not need to be phase coherent with the incoming optical carrier. This can be

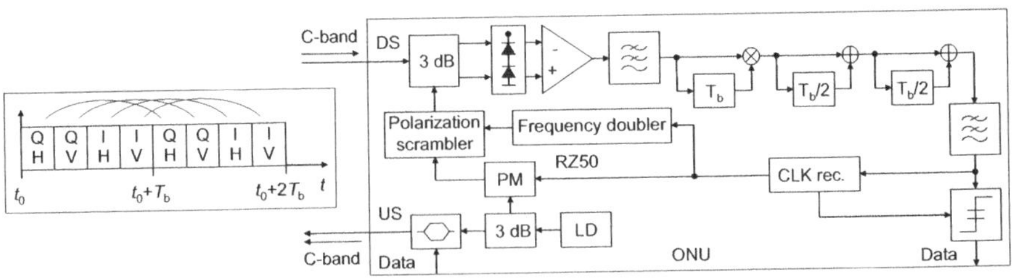

FIGURE 6.21 Homodyne detection without polarization diversity and 90° hybrids.

TABLE 6.2 Comparison of Coherent WDM-PON Schemes

Phase Handling	Polarization Handling	Sensitivity Penalty (dB)	Linewidth Tolerance (MHz)	Cost Estimation
90° Hybrid	Diversity	0	5	High
90° Hybrid	Scrambling	3	5	Medium–high
Scrambling	Scrambling	6	5.4	Medium
Heterodyning	Scrambling	6	5	Medium

regarded as an intradyne receiver with near-zero intermediate frequency. As such, dual-fiber working is required.

In Fig. 6.21, polarization is scrambled as well. The polarization scrambler is run at duplicated symbol rate in order to produce the orthogonal polarization states per I and Q time slot. This leads to four time slots of duration $T_b/4$, which contain all combinations of the orthogonal states of both phase and polarization.

The coherent receivers discussed so far reduce cost as compared to a fully blown intradyne receiver. This cost reduction comes at the cost of sensitivity (i.e., reach) penalties. Phase scrambling leads to a penalty of 3 dB, as compared to homodyning (intradyning) using 90° hybrids. Polarization scrambling leads to another 3 dB penalty, compared to polarization diversity. Heterodyning also leads to 3 dB penalty as compared to homodyning/intradyning, given the modulation scheme [here, phase-shift keying (PSK)] stays the same. This is summarized in Table 6.2.

With 6 dB penalty, a sensitivity of ~45 dB can be achieved, assuming that PSK is used at a bit rate of 1 Gb/s.

Polarization and phase scrambling requires effort. For polarization scrambling, a scrambler is required. For phase scrambling, an additional phase modulator together with synchronized I and Q delay demodulators running at duplicated bit rate are required.

6.1.5 WDM-Based PON Reach Analysis

In urban areas it is not possible to calculate with low fiber attenuation in the range of <0.25 dB/km. Reasons include an unpredictable number of splices resulting from repair work that is necessary because urban fibers are regularly affected by civil work. In addition, PONs often have to be operated on old fibers. This is particularly true for the feeder fibers, which traverse the majority of the total distance. In addition, patches, connectors, and in-house cabling for FTTH must be considered. The latter can cause significant insertion-loss penalties if fiber is not properly laid, that is, with sharp bends and so on. Table 6.3 summarizes relevant component figures, and fiber and penalty or end-of-life (EOL) margin figures as required by large network operators [27].

All components losses in Table 6.3 are typical figures, which can be found elsewhere. All power budget values assume the use of FEC. For 1 Gb/s, power

TABLE 6.3 Calculation Parameters for PON

Fiber attenuation, C/L-band/S-band	0.34 dB/km/0.41 dB/km
End-to-end penalty for intracentral-office patches and in-house cabling	5.5 dB
Insertion loss per 1:20 WDM filter/1:40 AWG/1:80 cyclic AWG	5.0 dB/5.0 dB/6.0 dB
Insertion loss per diplexer (C/L, S/C)/circulator/interleaver	1.5 dB/1.5 dB/2.0 dB
Insertion loss of $1{:}2^N$ power splitter/combiner (PS)	$N \cdot 3.5$ dB
Power budget 1 Gb/s transceiver PIN/APD/coherent heterodyne	31 dB/41 dB/51 dB
Power budget 10 Gb/s TDM/burst mode transceiver (APD, booster, and FEC)	37 dB
Booster gain adapted to maintaining laser safety class 1 M	2–6 dB

budgets as stated for PIN/APD transceivers can be achieved commercially. The figure for coherent 1 Gb/s assumes guaranteed modulated −2 dBm per-channel launch power and a sensitivity of −53 dBm.

Per-channel fiber launch power of all WDM-based PON variants depends on channel count, that is, total power limitation due to laser safety restriction, and limitations due to nonlinear effects. In many applications, total power must maintain laser safety class 1M, that is, a total fiber-coupled power of 21.34 dBm for wavelengths >1400 nm. Single-channel launch power is restricted to <7 dBm in order to avoid stimulated Brillouin scattering (SBS) countermeasures, and to keep transmitters reasonably cheap. In addition, UDWDM-PON suffers penalties from four-wave mixing (FWM) because narrow spacing allows phase matching [27]. The resulting penalties are shown in Fig. 6.22.

The resulting per-channel launch power restrictions are shown in Fig. 6.23 as a function of the channel (ONU) count.

The reach versus client-count analysis must consider the available power budgets of the respective WDM-PON transceivers, the insertion losses of all passive components in the optical end-to-end path, and penalties and fiber loss as given

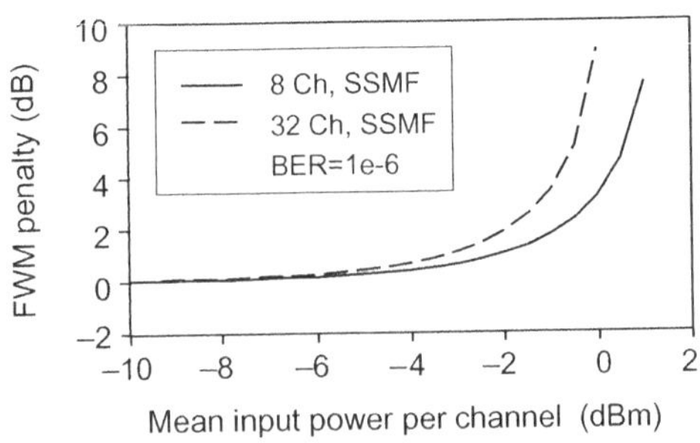

FIGURE 6.22 Penalty due to FWM for UDWDM-PON with 3 GHz grid.

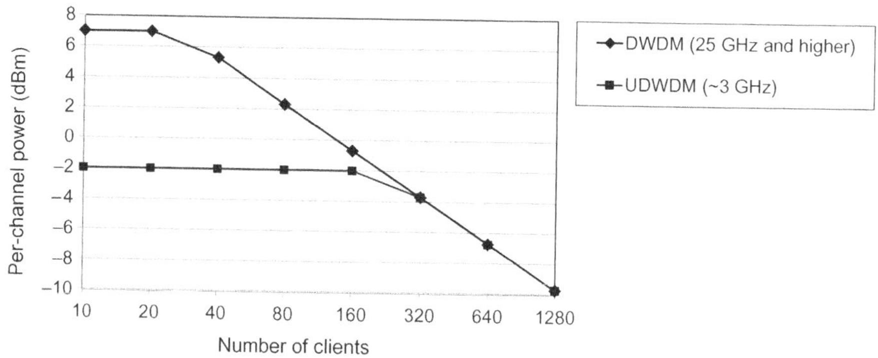

FIGURE 6.23 Per channel launch power.

in Table 6.3. Gain of optional OLT-based amplification must also be considered. Amplification in the ODN is not yet considered.

Transceiver power budget figures are available between the respective transmitter (after modulator) and the receiver. Hence, diplexers, circulators, and 3 dB couplers necessary for single fiber working must also be considered as part of the insertion loss. Resulting total insertion loss, including the cabling penalty, is shown in Fig. 6.24 for different ODN variants. It depends on client count through the scaling of $1{:}2^N$ power splitters (PS)/combiners, or the addition of WDM components (AWG, C/L splitter, S/C splitter, and interleaver).

The reach R of the WDM-based PON variants is given by

$$R = (\mathrm{PB} + G_{\mathrm{TOT}} - \mathrm{IL}_{\mathrm{TOT}})/A_{\mathrm{F}}. \tag{6.11}$$

PB is the available power budget of the respective variant, G_{TOT} is the total gain of amplifiers if applicable, $\mathrm{IL}_{\mathrm{TOT}}$ is the total insertion loss of passive components as per Fig. 6.24, and A_{F} is the per-kilometer fiber attenuation. The resulting reach is shown in Fig. 6.25.

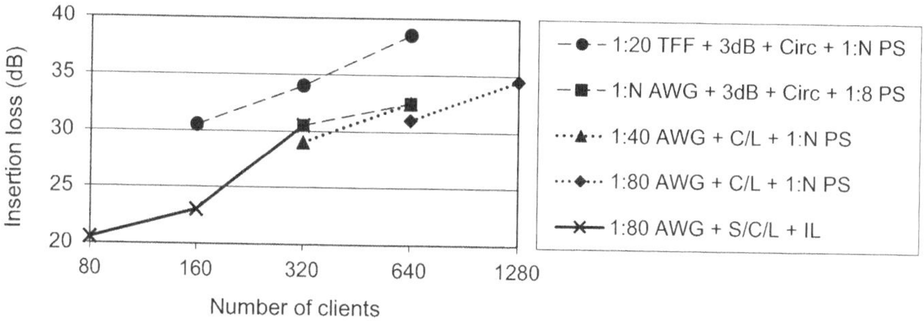

FIGURE 6.24 ODN insertion loss for WDM-based passive access.

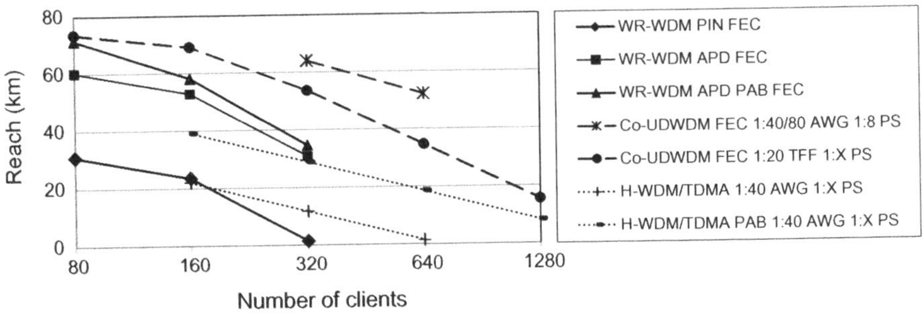

FIGURE 6.25 Reach versus client count for WDM-based passive access.

Different variants of (WR)-WDM-, hybrid (H)-WDM/TDMA-, and Co-UDWDM-PON are analyzed in Fig. 6.25, with respect to maximum reach. They differ in transceiver power budgets (WR-WDM-PON: PIN vs. APD) and/or the addition of OLT-based preamplifier/booster (PAB) combinations (applicable to WR-WDM- and H-WDM/TDMA-PON). The H-WDM/TDMA-PON uses 37 dB power-budget transceivers due to the high total ODN insertion loss. Reach for the WR-WDM-PON is stated for the C-band (after the S/C split). In the S-band, reach is somewhat lower due to higher fiber attenuation as stated in Table 6.3. For optional S-band amplification, TDFA can be used.

6.1.6 WDM-PON Deployment Reference: KT

Since 2006, Korea Telecom (KT) has deployed nationwide FTTH/FTTB networks using IEEE EPON technology, with a total of more than 4.5 million subscribers. KT is also the only large operator that has conducted commercial deployment of WDM-PON. This deployment started in 2004 with seeded/reflective DWDM-PON developed by Novera Optics. The system was capable to deliver 16 × 100 Mb/s. It was used for fiber-to-the-pole applications. A total of ~80,000 subscribers have been connected with this first WDM-PON generation.

Between 2004 and 2005, an improved version with 32 × 100 Mb/s was deployed for FTTH applications. This added ~2000 subscribers. Between 2007 and 2011, several versions of seeded WDM-PON with 16 × 1 Gb/s and 32 × 1 Gb/s, respectively, were deployed in FTTB scenarios, which added another ~2000 subscribers. This included a variant with wavelength reuse, which was developed by Corecess and ETRI.

Potential future deployments of WDM-PON include areas where sustained high bandwidths are required. This includes mobile fronthaul [so-called Cloud RAN (radio access networks)], business access, and FTTB for residential access (H. Yoon, KT, personal communication, 2012).

6.2 CORPORATE NETWORKS (STORAGE REACH EXTENSIONS)

A number of relevant WDM transport applications fall into the area of corporate or enterprise networks. These can be split into general-purpose corporate backbones and reach extensions for storage area networks (SAN) and data center connections. The latter often are point-to-point applications that require high WDM channel numbers over distances of <100 km. In addition, they often require the transport of data center specific protocols like Fibre Channel or InfiniBand (IB). Both protocols cover the OSI layers 1–4. They provide lossless transport; InfiniBand in addition provides very low latency. Both protocols started with 8B/10B coding [FC: up to 8G-FC and IB: up to quad data rate (QDR)] and then changed to 64B/66B coding for improved efficiency. Both constantly evolve toward higher bit rates.

Transport of FC (and also IB) can be done with transparent WDM transponders or, at lower FC (IB) bit rates with muxponders. Generic framing procedure (GFP) mapping into SONET/SDH/OTN (Optical Transport Network) has been defined up to 8G-FC so far in ITU-T Recommendation G.7041. In addition, 10G-FC mapping onto ODU2e has been defined in ITU-T Recommendation G.709 (Fig. 8.4).

Several data center applications do not allow the added latency, which is caused by multiplexing, framing, and the addition of FEC. Then, the respective services have to be transported by transparent WDM transponders.

Support of data center protocols is not restricted to the respective bit rates. Relevant FC vendors, for example, Brocade, use the FC interframe gaps for the insertion of added proprietary functions. In these cases, TDM multiplexing must not remove the interframe gaps for increasing efficiency. Other services, for example, IBM GDPS (Geographically Dispersed Parallel Sysplex) coupling facilities (CF), have special jitter and timing requirements. In many cases, the WDM equipment is also required to have official certifications from the FC equipment vendor.

The line rates (per physical lane, including the line coding) of relevant FC and IB variants are summarized in Table 6.4.

FC is used in most large SAN, where it carries SCSI (small computer system interface) frames. It started with FC-100; today the most important variants are FC-800 and FC-1600. There were also IBM-customized variants called FICON, FICON2, and FICON4.

To allow lossless transport, FC uses a handshaking mechanism that is based on *buffer credits*. Buffers are used on each receive side of a bidirectional FC link. Buffer credits are granted, and allow the respective remote transmitter to send one 2112 byte frame per credit. In order to avoid excessive delay caused by frequent buffer crediting, large buffer sizes (or credits) are necessary. Assuming fully utilized 2148 byte frames, the buffer credits BC in kilobyte for a given link of distance D in kilometer are [28]

- 1G-FC: $\mathrm{BC} = D/2 + 20\%$ (kB)
- 4G-FC: $\mathrm{BC} = 2D + 20\%$ (kB)
- 8G-FC: $\mathrm{BC} = 4D + 20\%$ (kB)
- 10G-FC: $\mathrm{BC} = 6D + 20\%$ (kB)

TABLE 6.4 Services/Protocols in Data Center Networks

Name	Standardization Body	Line Rate (Mb/s)	Coding
Full-speed Fibre Channel (FC-100)	ANSI INCITS T11 (see Section 9.2.6)	1062.5	8B/10B
Quadruple-speed FC, 4G-FC (FC-400)		4250	
Eight-speed FC, 8G-FC (FC-800)		8500	
10 Gb FC, 10G-FC (FC-1200)		10,520	64B/66B
16-speed FC, 16G-FC (FC-1600)		14,025	
InfiniBand SDRx1 (x4, x12)	IBTA (see Section 9.2.6) (industry standard)	2500	8B/10B
InfiniBand QDRx1 (x4, x12)		10,000	
InfiniBand FDRx1 (x4, x12)		14,026.5	64B/66B
InfiniBand EDRx1 (x4, x12)		25,781.25	

With insufficient buffer credits, throughput of an FC link decreases significantly at a certain distance called *Droop Point*. Without any additional credits, the Droop Point is at 4 km for 1G-FC. The effect of adding credits on the throughput of 1G-FC is shown in Fig. 6.26 [29,30]. Similar behavior (at higher credits) can be seen for 2G-FC and faster variants.

WDM reach extension of FC services may involve the provisioning of the necessary buffer credits in the WDM transport equipment. However, in most applications, the preferred solution is to have the buffer credits provided by the FC client equipment.

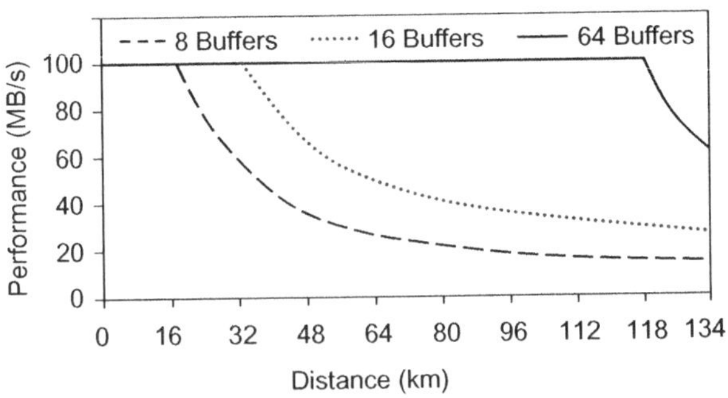

FIGURE 6.26 FC performance versus distance; parameter: buffer-to-buffer credit rate.

6.2.1 Disk Mirroring

Disk mirroring is primarily related to disaster recovery [31]. It provides redundancy for those data centers where failure/unavailability will cause significant costs. To allow real-time disk mirroring, transmission channels with bandwidths of 8–100 Gb/s are necessary. These have to be multiplied by means of WDM.

An important characteristic of disk mirroring is whether or not it can be performed *synchronously* or *asynchronously*. Synchronous disk mirroring means that the remote server can take on action immediately *without any interruption*. In asynchronous mirroring, the remote facility may not be perfectly up to date. This may cause some *limited* data loss or application interruption.

FC-based disk mirroring must provide sufficient buffer credits, depending on the link length and the FC line rate. Further limitations for synchronous coupling originate from all latencies that have to be added between the corresponding servers.

A server has a typical response time of ~2 ms. For a typical device busy rate (DBR) of 300, the server has to respond to 300 requests per second, in total being engaged for 600 ms or 60%. Thus, *further added latencies* must not exceed $L = 1.33$ ms:

$$\mathrm{DBR} \cdot (2\,\mathrm{ms} + L) < 1\,\mathrm{s}, \quad \mathrm{DBR} = 300 \Rightarrow L = 1.33\,\mathrm{ms}. \tag{6.12}$$

The resulting curve for allowed latency versus DBR is shown in Fig. 6.27.

This behavior has significant impact on the transmission latency that is allowed. This can be seen from Table 6.5, where typical latencies of different transmission systems are listed.

For servers with a high DBR, synchronous coupling is not possible in excess of ~100 km due to fiber delay. For this distance domain, latency added by the transmission equipment itself starts to count. Possibly, low-latency WDM channel cards have to be used, and FEC and added framing and multiplexing have to be avoided. For even tighter latency requirements, EDFAs have to be replaced by Raman amplification or fast regenerators, and dispersion-compensating fibers have to be replaced by lumped compensation gratings [fiber Bragg gratings (FBGs)].

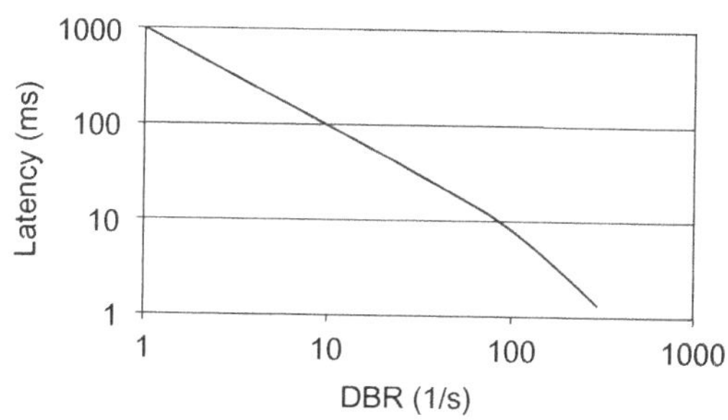

FIGURE 6.27 Maximum allowed latency as a function of DBR.

TABLE 6.5 Latencies of Different Transmission Systems

Transmission System	Latency/(Group) Delay
Transparent WDM transponder w/o framing	<10 ns
10G (OTU2) OTH transponder, FEC OFF	~3 μs
2:1 TDM muxponder with SDH framing	~5 μs
10:1 TDM muxponder with OTH framing	~25 μs
EFEC (at 10 Gb/s, per pair of cards)	~50 μs
GFEC (at 10 Gb/s, per pair of cards)	~7 μs
EDFA (single stage)	~200 ns
100 km fiber round trip	~1 ms

For distances clearly exceeding 100 km, disk mirroring is done asynchronously. Optical transmission delay is not important for asynchronous coupling. Still, sufficient buffer credits have to be provided. For ultralong distances, local "spoofing" of the FC handshaking can be done. Then, the buffer crediting is terminated by a local spoofing device. This enables almost limitless distances to be spanned. However, the spoofing devices (together with the WDM equipment) must now ensure lossless transmission.

A newer trend in FC-based SAN is Fibre-Channel-over-Ethernet (FCoE). In FCoE, FC frames are encapsulated into Ethernet, and the lowest FC layers (FC-0 and FC-1) are replaced by Ethernet. In order to allow losslessness, Ethernet was complemented by functionalities similar to FC/IB, which were defined in several IEEE standards:

- IEEE 802.1Qbb, per-priority flow control (PFC)
- IEEE 802.1Qaz, enhanced transmission selection (ETS)
- IEEE 802.1Qau, Ethernet congestion management (ECM)

Enabled by these standards, FCoE allows FC transport over an Ethernet-like infrastructure. All Ethernet equipment involved must support the standards listed earlier. The respective data center minded Ethernet variant is referred to as data center bridging (DCB). It allows the consolidation of the data center SAN and LAN onto one switching fabric, including the advantages with regard to staff training and so on.

A data center connection scenario that is based on DCB is shown in Fig. 6.28. The picture also shows additional InfiniBand links that may be used for CPU connectivity.

In long-distance DCB connections (mirroring), all intermediate Ethernet equipment must support the respective DCB standards. Otherwise, losslessness of the services cannot be guaranteed. Additional functionality of WDM transport equipment is not required.

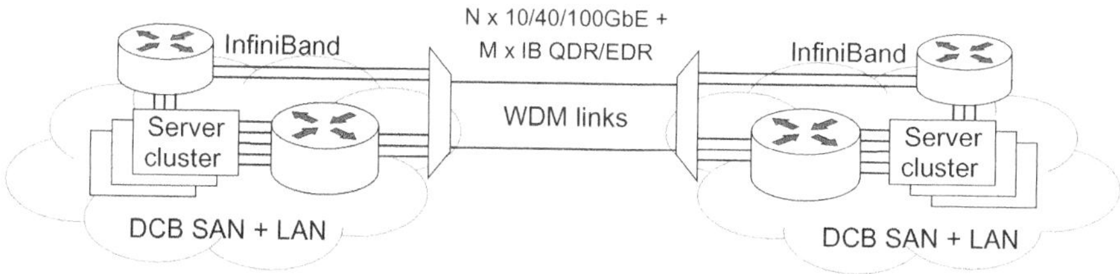

FIGURE 6.28 Data center connection via InfiniBand and DCB links.

6.2.2 Mainframe Coupling/Server Consolidation

Mainframe coupling or server consolidation are applications where IBM mainframes or Open Systems servers are coupled to create single logical servers, provide redundancy, or connect separated data centers at high availability. In some of these applications, synchronization signals are necessary, which lead to strict requirements with regard to latency.

Servers and mainframes often use Fibre Channel (8G-FC and 16G-FC) for the SAN and Ethernet (10GbE and 100GbE upcoming) for the LAN. In high-performing compute clusters, InfiniBand is often used for the CPU interconnection. Consolidation may lead to increased use of FCoE. In highly available data centers, key components are duplicated per site, and sites may be connected for even higher availability, and to create virtual servers. Data center connection then uses (at least) two separate WDM links (Fig. 6.29).

In Open Systems server consolidation, server clusters are synchronized via the Ethernet links. This is different in IBM mainframe coupling that uses special synchronization signals between the mainframe clusters. Looked at from the outside, these clusters act like a single logical server. To ensure this behavior, the synchronization signals, referred to as coupling facilities, require strict limits with regard to the total end-to-end latency. IBM mainframe coupling is shown in Fig. 6.30. The configuration is sometimes referred to as Parallel Sysplex (SYStem comPLEX), and provides redundancy of all relevant parts. The coupling facilities are based on

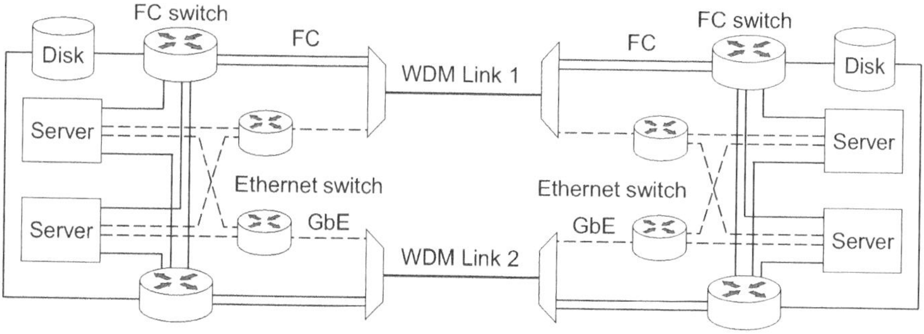

FIGURE 6.29 Server consolidation via Ethernet and Fibre Channel links.

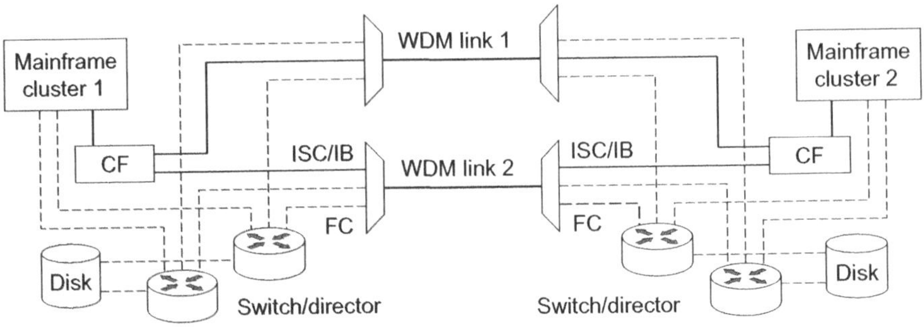

FIGURE 6.30 GDPS mainframe connection.

special versions of Fibre Channel [also referred to as intersystem connections (ISC)] or InfiniBand.

A GDPS configuration massively depends on the accuracy of its relevant signals, that is, the coupling facilities. These signals have very tight requirements with respect to jitter and latency. They must be transmitted transparently in order to avoid added latency and jitter caused by TDM, framing, and error correction. In addition, very tight requirements must be considered with regard to the differential delay between uplink and downlink of a bidirectional connection. Newer versions of GDPS can tolerate a latency deviation between both directions of a bidirectional transmission, referred to hereinafter as DDD (directional differential delay), of 500 ns. Since newer data center connections (after year 2001) require increased distances between the data center sites in the range of 100 km or more, this requires very strict consideration of all potential sources of DDD. Often, transmission distances of >100 km require optical amplifiers and other optical and electrical components like dispersion compensators or FEC. Components and transmission effects relevant for DDDs are shown in Fig. 6.31.

Not shown in Fig. 6.31 are optical connectors, splices, and all components that are not directly involved into the light path. Connectors and splices are considered delay free. The lumped components of all filter stages and the laser and photodiode are also considered free of significant DDD. All other components need to be considered in more detail.

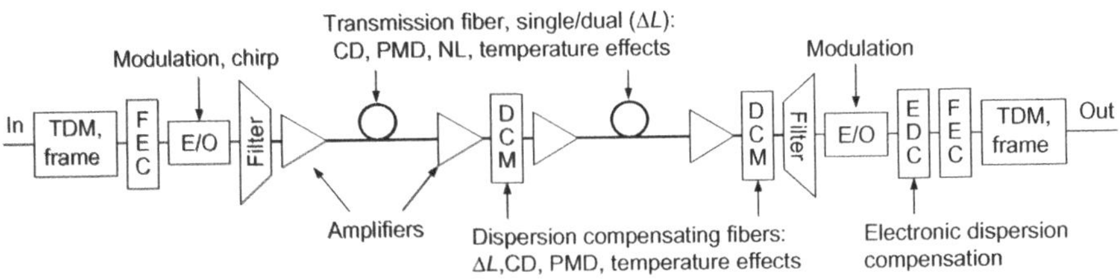

FIGURE 6.31 Transmission system for 200 km.

The transmission fiber itself is the most important component with respect to DDD. The most important effects that have to be considered are as follows:

- Length of the fibers per each direction, mismatch directly leads to DDD
- Thermal expansion of the fiber, may be different in each direction
- Temperature-induced change of the refractive index, thermo-optic coefficient (TOC)
- CD and hence group velocity as a function of wavelength

PMD, SPM (self-phase modulation), and XPM (cross-phase modulation) can be neglected here. Mechanical expansion of the fibers due to mechanical loads must be avoided by proper cabling of the fibers (e.g., in gel-filled tubes).

The length of the fiber is the most critical parameter. This holds in particular for dual-fiber working (DFW), where the length difference of both fibers—*including patch cables*—must be kept within very strict limits to allow margin for all other effects that contribute to DDD. For SFW, this effect vanishes given that equally long patch cables are used, which can be ensured more easily.

For DFW, both transmission fibers must *precisely* be measured with respect to their total length (unless it is precisely known that both fibers are part of the same cable, and no splices and patches have been inserted). Measurement must be done with a high-resolution OTDR (optical time-domain reflectometer). OTDR measurements can have a resolution of down to 1 m for a distance range of up to 250 km. Length mismatch between the two fibers has to be compensated by means of patch cables or fiber spools. The total mismatch can thus be aligned to within 1 m, or 1% of the allowed DDD. If OTDR measurements for several spans of a link are done, the measurement errors have to be added for the worst case calculation.

Optical transmission fibers show thermal expansion that can be described by the thermal expansion coefficient (TEC) α. The TEC of silica fibers is in the range of $0.5 \times 10^{-6}\,C^{-1}$ [32]. Thermal expansion is relevant for DFW only. The worst case scenario is that one fiber is heated relatively against the other. If we consider a maximum temperature difference $\Delta T = 50\,°C$ between the two fibers and a 200 km link, the resulting TEC-induced DDD can be calculated:

$$\mathrm{DDD}_{\mathrm{TEC},50°C} = 200\,\mathrm{km} \cdot 50 \times 0.5 \times 10^{-6} \cdot n/c_0 \approx 25\,\mathrm{ns}. \quad (6.13)$$

It is easy to suppress this effect by requiring both fibers to be deployed not only in the same cable, but also in the same duct. Due to the mechanical cabling and thermal conduction within the cable, fibers can then be expected to have the same temperature within strict limits, say $\Delta T \leq 1\,°C$. The resulting DDD can then be neglected ($\mathrm{DDD}_{\mathrm{TEC},\ 1°C} \approx 0.5\,\mathrm{ns} \ll 500\,\mathrm{ns}$).

The TOC causes temperature-induced changes of the refractive index and hence the group velocity. Like the TEC, the TOC is relevant for DFW where one fiber is heated relatively against the other. The TOC of silica fibers is in the range of $10 \times 10^{-6}\,C^{-1}$ [32]. If we again consider a temperature difference between the two fibers used for DFW transmission of $\Delta T = 50\,°\text{C}$, a total TOC-induced DDD for a 200 km link results:

$$\text{DDD}_{\text{TOC},50\,°\text{C}} = 200\,\text{km} \cdot n \cdot 50 \cdot 10^{-5}/c_0 \approx 500\,\text{ns}. \qquad (6.14)$$

This result is based on the assumption that the two fibers have identical length at identical temperature. Obviously, the TOC has the potential to introduce severe DDD. The TOC-induced DDD decreases to 10 ns for a temperature difference of $\Delta T = 1\,°\text{C}$, given we have identically long fibers. This again requires the fibers to be deployed in the same cable.

Chromatic dispersion of the fibers is relevant for SFW, where the signals use different wavelengths for both directions. Fibers with steep slope of the D parameter (e.g., G.652 SSMF and G.655 fibers without slope reduction like Truewave RS®) contribute to CD-induced DDD more strongly than the reduced-slope fibers. The delay can be derived by integrating D over the difference of the wavelengths used for both transmission directions. For a wavelength difference of 70 nm (which covers 1530–1600 nm) and a 200 km link, the resulting CD-driven DDD is given by $\text{DDD}_{\text{CD}} \approx 250\,\text{ns}$. This effect must be suppressed by using *similar* wavelengths for both directions of transmission. If we allow the wavelength in the counter-direction to vary by ± 1 channel (which equals ~0.4 nm for the 50 GHz grid), this leads to $\text{DDD}_{\text{CD,2Ch,200 km}} \approx 1.5\,\text{ns}$. Such a channel spacing can be achieved by interleaving the uplink and downlink channels. If this is not done, CD in SFW leads to significant DDD, although SFW eliminates all fiber length-related DDD effects.

From the components listed in Fig. 6.31, the FEC coders/decoders must be considered most seriously. In addition, (OTN) framers and TDM must be considered. The *variance of the latencies* per transponder or muxponder card is the relevant effect that contributes to DDD. Both TDM and FEC (especially for older EFEC) can produce severe DDD in the range of, and exceeding, 1 μs. For timing-critical applications like GDPS, both FEC and TDM must therefore be avoided.

EDFAs must be considered because they contain spools of Erbium-doped fibers. Here, it must be ensured that each amplifier contains the same length of Erbium fiber. It is not necessary for different amplifier types (e.g., preamplifier vs. booster amplifier) to have the same Erbium fiber length because the links are expected to be symmetrical, that is, both directions contain the same number and the same types of amplifiers.

The actual fiber length per amplifier (and per amplifier type) varies. This variance results from actual fluctuations of the gain of different Erbium fiber batches. Hence, the Erbium fiber length of actual amplifiers is adjusted on demand, during manufacturing. This variance typically does not exceed 10% of the mean per-gain-block Erbium fiber length. Both gain blocks of double-stage EDFAs need to be considered. The relevant lengths and DDDs of single-stage C-band and L-band EDFAs are summarized in Table 6.6.

Depending on the actual configuration (span losses, span lengths, etc.), a 200 km link may require up to six gain blocks. Hence, EDFA-induced DDD can be in the order of up to 120 ns, thus requiring a substantial part of the total DDD allowance.

TABLE 6.6 Lengths and DDDs of EDFAs

Amplifier Type	Erbium Fiber Length, Type (m)	Length Variance, Max. (m)	EDFA DDD (ns)
Single-stage C-band EDFA	30	±3	15
Single-stage L-band EDFA	40	±4	20

Dispersion-compensating fibers (DCF) must be considered because all effects discussed for transmission fibers apply as well. The lengths of these fibers can vary, and there may be TOC/TEC-induced effects. Typically, the DCF length is one-fifth of the nominal link length, which is to be compensated (see the parameters in Table 3.9). The lengths of DCFs can be measured by means of a high-resolution OTDR. For $<100\,\text{km}$, these OTDRs can have 0.5 m resolution or better [33,34]. Hence, up to 1.5 m length difference or 7.5 ns DDD have to be considered for the dispersion compensation, depending on the actual (multispan) configuration.

In addition to length mismatch, TEC- and TOC-induced DDD of the dispersion compensating modules DCMs has to be considered. If we require the temperature difference of the DCMs per direction $\Delta T \leq 5°\text{C}$—which is achievable by means of air condition—the TEC-induced effects can be neglected, and the TOC-induced effects (thermo-optic effect, that is, change of refractive index) may lead to a maximum DDD of 12.5 ns, given the same coefficients are valid for DCFs than for the transmission fiber:

$$\text{DDD}_{\text{TOC,DCF}} = 50\,\text{km} \cdot n \cdot 5 \cdot 10^{-5}/c_0 \approx 12.5\,\text{ns}. \tag{6.15}$$

The requirements regarding the temperature range for the DCMs can be relaxed by properly placing the DCMs. Ideally, every site contains the same amount of DCF for both directions, for example, behind a preamplifier and trailing a booster amplifier. Then, the temperature difference can be expected to be smaller. DCF add negative CD to the transmission link. Their use thus relaxes the CD-induced penalties, which result from the transmission fibers.

Electronic dispersion compensation (EDC) may also have to be considered. EDC is typically based on digital feed-forward equalizers (FFEs). FFEs are finite impulse response (FIR) filters that lead to a constant delay depending on the number of filter taps. For different wavelengths to be compensated, differential delay occurs.

Now the total DDD for an SFW reference system with 200 km reach can be estimated. SFW excludes all effects on the DDD caused by the fiber itself (length, TEC, and TOC), but different CD of the uplink/downlink wavelengths must be considered. For SFW, these effects can be suppressed by dispersion compensation based on DCF, which compensates the residual CD as close to zero as possible, for example, into ±300 ps/nm. This window must apply to all channels.

For low guaranteed DDD, all measures described earlier, that is, proper length measurements of fibers, DCF temperature control to within $\Delta T \leq 5\,°\text{C}$, same EDFA

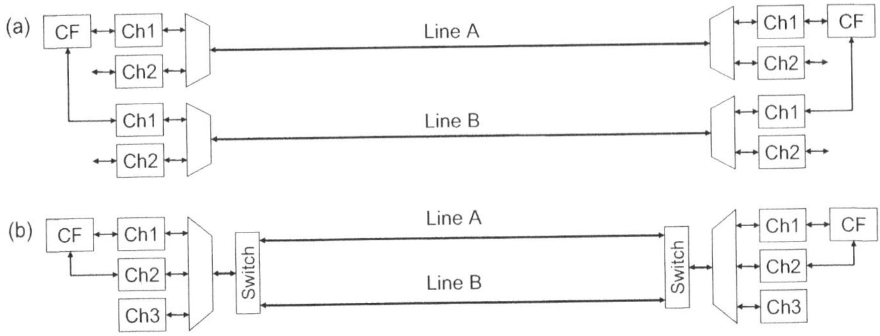

FIGURE 6.32 Redundant SFW. A: path protection, B: line protection.

and channel card types for uplink and downlink, matched patch cables, and so on must be taken.

Delay-sensitive applications also require high availability of the transmission paths and the client equipment. In mission-critical applications, all major components are duplicated. If necessary, more than two fiber connections (ducts) can be used. The options for redundant transport of the coupling facilities in a SFW GDPS system are shown in Fig. 6.32. The options shown here can also be combined for highest availability in the range of 99.9999%.

In server-consolidation applications, the protection methods can also be mixed for the different participating services. For example, FC-based SAN channels may be line-protected only. Regarding total system availability, this may be allowed since the FC switches also provide redundancy in the form of load sharing.

6.2.3 Low-Latency WDM Transport

Some applications, for example, connections between international stock exchanges, require lowest possible latency (or round-trip delay). Special WDM system and link design is required to save any 10 ns wherever possible. According to the numbers in Tables 6.5 and 6.6, purely transparent WDM transmission without any TDM multiplexing, framing, and FEC has to be used. EDFAs must be omitted, since transparent 3R regenerators are faster. Additionally, Raman amplification can be used. On long transmission links (>300 km), this leads to the necessity of adding FEC. FEC will overcompensate the advantage of Raman amplification (see the numbers in Table 6.5). Compensation of CD by means of DCF or complex (multitap) digital filters has to be avoided as well. Instead, periodic (cyclic) FBGs can be used for CD compensation. Since these are lumped devices that work on a per-channel basis (unlike DCFs that perform full-band compensation), they have delay that can be neglected in this context. The resulting WDM link design consists of a series of directly connected point-to-point systems using fully transparent transponder (no muxponders). Reach may be increased within the no-FEC limit by Raman amplification. For transparent 10 Gb/s transport, maximum length between two 3R regenerators can then exceed 300 km, depending on the fiber plant quality.

TABLE 6.7 Parameters of Low-Latency WDM and RF Links for 1000 km Total Link Distance

	WDM Link	RF Link
Span/hop length	100 km	40 km
Refractive index	1.48	1.00
3R delay	0.02 μs	140 μs
Detour factor	1.4	1.2
Number of spans/hops	14	30
Total delay	6.907 ms	8.200 ms

Some simple calculations show that an RF link, with its group velocity equaling vacuum speed of light, cannot solve the low-latency transport problem any better. First, the sustainable bandwidth is possibly too low. Typical RF systems run at a bit rate of 155 Mb/s. This bit rate may also be subject to weather influence (which is a strict no-go for mission-critical applications). Second, the advantage of higher propagation speed may be over-compensated by the necessity to apply per-hop FEC. At the relatively low bit rates, per-hop FEC can add >100 μs delay. Hop distance is 40 km at the best. A comparison of the delay for a 1000 km link between a WDM and an RF link is given in Table 6.7.

Even with larger detour factor (fiber ducts typically cannot follow direct line-of-sight), the WDM link has lower delay. Further decrease of the transport delay can be achieved by moving WDM wavelengths on SSMF to 1300 nm [because of the minimum of the fiber group index (Fig. 2.5)], or by using specialty fibers, for example, pure-silica or hollow-core photonic bandgap fibers. The latter can achieve a group index, which is significantly closer to 1.0 than in Silica SMF. However, such fibers so far have not been manufactured in long-haul chunks. The delay of RF links can be decreased by reducing the 3R delay, which is required for processing (like FEC application). This may speed up the RF link significantly, however, bandwidth remains orders of magnitude lower as compared to fiber links.

6.3 WDM TRANSPORT ENCRYPTION

WDM transport systems can offer optional encryption for services that require high security. This may become necessary, since optical fibers can be tapped. A standardized encryption system is Advanced Encryption Standard (AES), which was developed in 2000 by the National Institute of Standards and Technology (NIST) [35]. The algorithm was developed by Joan Daemen and Vincent Rijmen; it is also known as Rijndael algorithm. It was investigated cryptoanalytically in detail and is commonly regarded as relatively safe.

AES uses a block length of 128 bit, the key length is 128, 192, or 256 bit. Therefore, three variants exist, AES-128, AES-192, and AES-256. Proprietary variants can have longer key lengths (e.g., 1024 and 2048). The algorithm can be

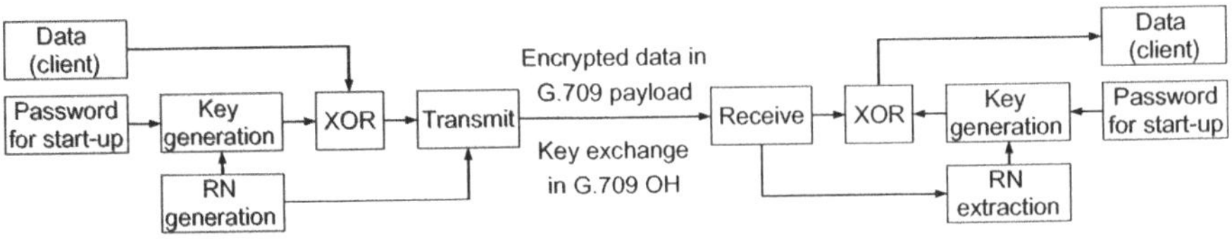

FIGURE 6.33 Application of AES on WDM transport cards. RN: random number.

implemented on WDM transport cards (transponders and muxponders) (Fig. 6.33). In this case, each pair of transport card constitutes an independent encryption system.

AES requires a key that must be exchanged securely. The Diffie–Hellman key exchange can be used even over an unsecure channel [36]. Here, both ends transmit one message each to generate a secrete key. The task of generating this key from the two messages is known as Diffie–Hellman problem. It is regarded as nonattackable. Even if an eavesdropper intercepts both messages, he or she is generally not able to generate the key. The Diffie–Hellman key exchange can be attacked by *changing* the two messages. Therefore, additional means like message authentication codes can be used.

In a WDM system, AES can be applied to the OTN optical channel payload unit (OPU) frame. Key exchange is done according to the Diffie–Hellman algorithm. A new key can be generated every few minutes, using the OTN optical channel data unit (ODU) overhead to transport the respective messages. The keys are kept only during the respective sessions, they are erased directly after the session end. No memory for session keys is provided.

AES (using Diffie–Hellman key exchange) should be complemented by secure access to the WDM transport equipment. Dedicated *Security Root Access* for security settings can be provided. The respective passwords are generated onboard the channel cards and sent to the remote systems via secure (encrypted) data communications network (DCN). Such, unauthorized access is made highly difficult.

6.4 METRO AND REGIONAL NETWORKS

Metropolitan area and regional transport networks have been relevant application domains for DWDM from the middle of the 1990s. Metro systems originate from point-to-point, single-span, unamplified DWDM systems, and the regional area was originally covered by early long-haul, multispan definitions (e.g., according to ITU-T Recommendation G.692). Today, most metro and regional installations are based on amplified multispan systems. Many of these deployments are still based on point-to-point or linear add–drop topologies. Since 1999, an increasing number of metro networks and part of the regional networks also rely on WDM *ring* architectures [37].

6.4.1 DWDM Metro/Regional Ring Networks

One reason for (WDM) ring networks is that rings require comparatively fewer fibers for redundant (i.e., protected) connections between the respective sites. As compared

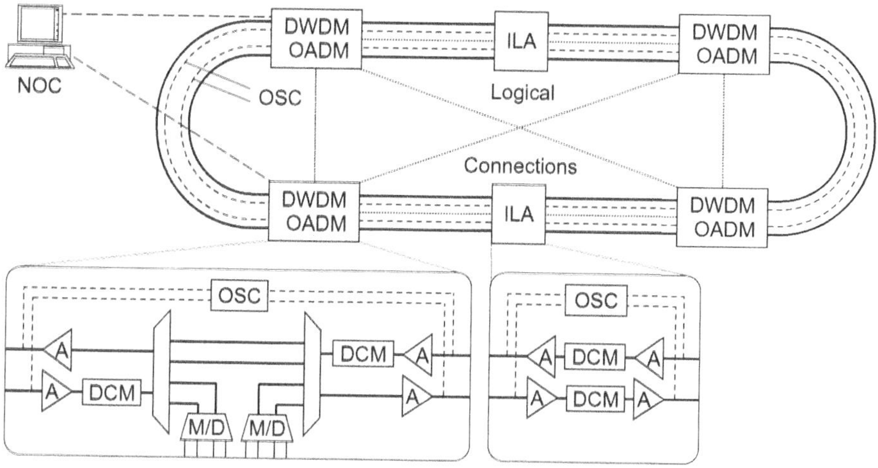

FIGURE 6.34 Metro WDM ring with banded filter structure. NOC: network operations center, A: amplifier, DCM: dispersion compensating module, OSC: optical supervisory channel, M/D: mux/demux, ILA: in-line amplifier.

to mesh structures (and also large numbers of point-to-point links), rings are also relatively easy to manage [38,39]. First-generation WDM systems also lacked the possibility of true multidegree add–drop multiplexers, which are required for meshed protection or restoration. A metro DWDM ring system is depicted in Fig. 6.34.

The static banded filter structure shown in Fig. 6.34 is increasingly being replaced by flexible ROADMs. Main advantages of ROADMs in large DWDM metro rings include the capabilities of single-channel add–drop, which increases the number of logical connections without wavelength conversions (regenerations), and remote reconfiguration.

Large metro core rings can have high numbers of logical traffic connections. Unlike backhaul rings, core rings often do not have a hubbed traffic pattern, that is, connections may have to be implemented between *any* pair of ring nodes. In non-ROADM systems (which are still in use, for example, for cost reasons), this poses planning constraints, which add to the more physics-related constraints like optical SNR (OSNR), dispersion allowance, or filtering effects.

Typical metro (and regional) system nodes consist of add–drop filters, amplifiers, means for establishing an OSC (optical supervisory channel that is part of the DCN), and optionally CD compensation modules (DCMs) (Fig. 6.34). If accumulated loss between traffic-terminating nodes becomes too high, in-line amplifiers (ILAs) can be added.

Similarly to in-line amplifiers (refer to Fig. 3.51), several optical and electrical parameters in an OADM node must be monitored, and the respective data be sent to a management system via the DCN, for example, via an OSC. This is shown in Fig. 6.35.

For metropolitan areas, it has to be considered that fiber loss can be higher than in long-haul links due to added patches or splices, which can be the result of repair work

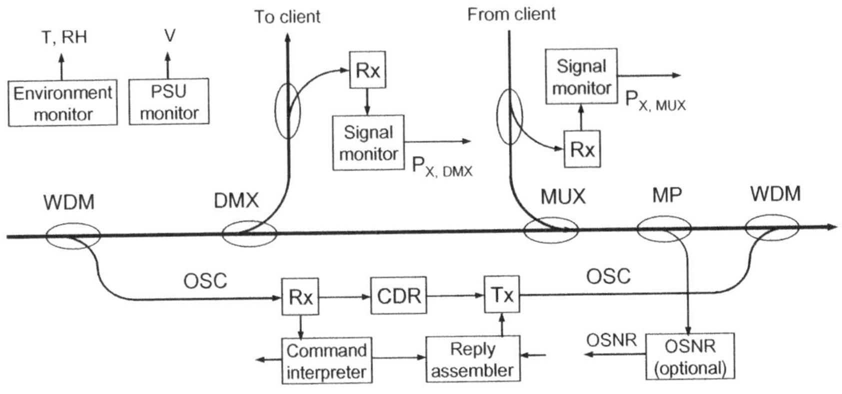

FIGURE 6.35 OADM monitoring. RH: relative humidity, WDM: add–drop filters for OSC (optical supervisory channel), MP: monitoring port, DMX/MUX: demultiplexer (drop) and multiplexer (add).

or fiber-plant reconfigurations. In certain metro networks, there is also high probability for old fiber types or vintages. Therefore, for metro WDM system design, higher EOL penalties must be used as compared to long haul.

So far, most metro WDM systems (and at least part of the regional systems) make use of CD compensation. Main reason is the mixture of services, including transparent carriers' carrier services, which does not allow the exclusive use of coherent systems with in-built equalization. Therefore, the physical-link planning rules for compensated systems apply (refer to Section 6.5.3). Due to distance (ring circumference) restrictions, effects of accumulated nonlinearity are smaller than in long haul and can be considered by a simple penalty.

Parameters for metro WDM link design are summarized in Table 6.8.

A difficulty in metro WDM link design is the potential mixture of services and bit rates, and the related transceiver technologies and parameters. On certain links, it may be impossible to align receiver sensitivities, power levels, or CD compensation requirements, which then leads to the necessity to regenerate subsets of services.

TABLE 6.8 Parameters of Metro and Regional WDM Link Design

Parameter	Value
EOL penalty (patches, etc.) (dB)	2–3
Mean fiber loss (dB/km)	0.3
Penalty for CD, PMD, and nonlinearity (dB)	1–2
Insertion loss of filter and OSC, east *or* west (dB)	8–10
Filtering penalty for serial 40G services for 50 GHz system	0.5 dB/node
Type OSNR requirement 2G5/10G/serial 40G/coherent 100G (dB)	18 (no FEC)/14/17/15
Type sensitivity 2G5/10G/serial 40G/coherent 100G (dBm)	−28/−22/−20/−20
Typ. amplifier gain and NF (dB)	20–25/~5.5

6.4.2 Capacity in WDM Ring Networks

ROADMs allow single-channel access and have a significant impact on the number of traffic connections that can be routed independently in networks. However, up to about 2010, the majority of metro WDM installations were based on bidirectional rings with $1+1$ dedicated channel protection (OCh-DPRing) and wavelength banding. It has been shown elsewhere that bidirectional rings offer significant advantages over unidirectional rings, for example, Ref. [40].

For core rings, the transport capacity almost always has to be calculated for protected services, where protection is provided through the WDM system or client-layer protection. For dedicated protection, both lead to the same result with respect to maximum capacity.

Core rings typically transport partially meshed traffic, which is a subset of fully meshed, any-to-any (A2A) traffic. Two special traffic types can often be found, hubbed traffic and site-to-adjacent-site (S2AS) traffic. The latter often occurs with SONET/SDH payloads.

If the (bidirectional) core ring has W wavelengths that are banded in B bands, and also has N nodes, then the maximum number M_{Hub} of independently routable, protected traffic connections for purely hubbed traffic is

$$M_{\mathrm{Hub}} = B \geq N. \tag{6.16}$$

Hence, the number of nodes N must not exceed the number of bands B, otherwise it is not possible to connect all nodes to the hub without expensive 3R regeneration (wavelength conversion). For completely unprotected traffic, the maximum number of routable connections is $2B$.

For pure S2AS traffic, the number of $1+1$ protected connections in the ring is also limited by Eq. (6.16). However, for unprotected S2AS traffic, the maximum number M_{S2AS} of connections is given by

$$M_{\mathrm{S2AS}} = N. \tag{6.17}$$

Now, the number of routable connections simply scales with the number of nodes. A similar result is achieved by *shared* protection, since for this traffic type the total capacity is not limited by the number of bands or wavelengths that are available (see Section 8.4.2).

For A2A traffic, the number of connections in an N-node ring is given by $N \cdot (N-1)/2$. Consequently, the number of routable protected connections in an OCh-DPRing with wavelength banding is again limited by the number of add–drop bands B:

$$M_{\mathrm{A2A}} = B \geq N \cdot (N-1)/2. \tag{6.18}$$

For unprotected A2A traffic, the number of routable connections is still limited by B, however now only $N/2 + N \cdot [(N/2) - 1]/4$ different bands are required. Therefore, for unprotected A2A traffic, the number in nodes can be increased, which is shown in Fig. 6.36.

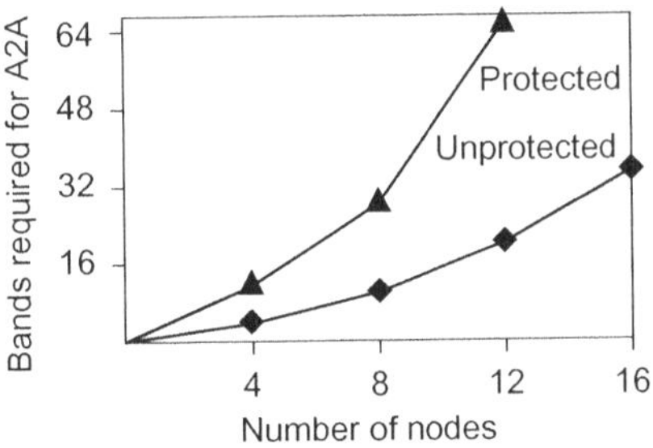

FIGURE 6.36 Number of bands required for 1 + 1 protected versus unprotected A2A traffic in a WDM ring.

So far, metro rings with a banded structure were considered. Obviously, there are very strict limitations with respect to the maximum number of nodes that are allowed in such rings. Even if we consider nonfully meshed and unprotected traffic, the maximum node number can only be little higher than eight in order to be able to route all traffic requirements without expensive wavelength conversions. Typical metro ring systems today have up to 16 bands, with 4 wavelengths per band. Since many of the ring connections only require a single wavelength, the number of nodes could significantly be increased if single-channel add–drop was possible. For fully protected A2A traffic, 11 nodes are possible for a 64-channel WDM system. Together with the requirement to flexibly assign these wavelengths, this is one of the main drivers behind ROADMs.

6.4.3 Wavelength Blocking in WDM Rings

To calculate blocking probabilities with respect to the number of wavelengths, we now consider rings with two different traffic patterns. We further assume that no wavelength blocking occurs (100% tunability or proper manual assignment of wavelengths). Fig. 6.37 shows two rings with different uplink connectivity. The numbers denote the respective number of wavelengths per each node. The two traffic patterns are partly meshed/hubbed (a) and fully hubbed (b).

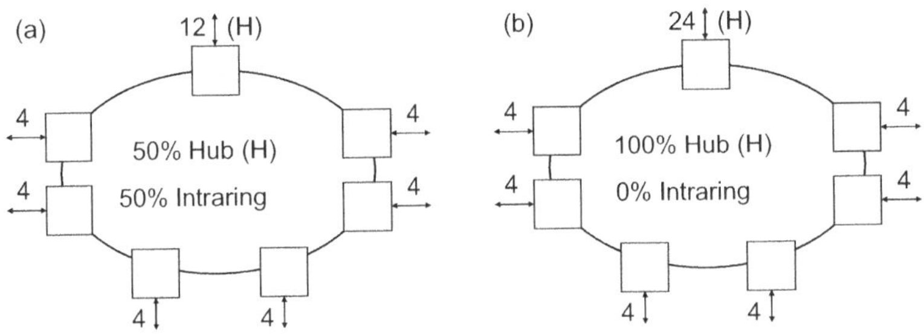

FIGURE 6.37 Ring topology with different traffic patterns.

We consider ring networks with time-dependent traffic and less wavelength channels available than would be required in the fully static case. Traffic has to be connected only for a portion of the time. A two-state, continuous-time Markov chain model is used [41]. When a station is in the idle state, it generates connection requests at a rate μ [1/s]. If a request is accepted, the station changes into active state. If a call for connection is blocked, it is lost and the station remains in the idle state and continues to generate requests at the same rate. An active station releases calls at a rate ν [1/s], with the call holding time distributed exponentially with a mean value ν^{-1} (Poisson distribution). The behavior of the model only depends on $\rho = \mu/\nu$ (Erlang).

In equilibrium, the two-state model is in active state with probability $\rho/(\rho+1)$. In the absence of blocking, the whole network of M stations offers

$$G = M\frac{\rho}{\rho + 1} \tag{6.19}$$

Erlang of traffic. Assuming that the number of wavelength channels C is smaller than that needed for full connectivity, blocking occurs and some calls would be lost. The probability of the call loss follows Engset distribution [42]:

$$P_{\mathrm{L}} = \frac{\binom{M-1}{c}\rho^C}{\sum_{i=0}^{C}\binom{M-1}{i}\rho^i}. \tag{6.20}$$

Applied to the configurations of Fig. 6.37, Fig. 6.38 shows P_{L} as a function of traffic according to Eq. (6.20), and as a function of available wavelength channels.

Probability of loss increases with traffic and decreases with the number of available wavelengths. Note that it is assumed that the wavelength channels are available, that is, nonblocking, due to full-band tunability or proper wavelength planning/assignment.

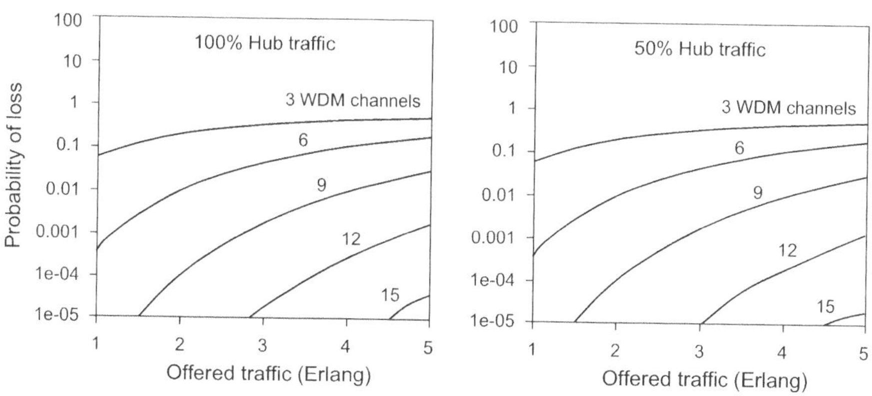

FIGURE 6.38 Probability of call loss for 100% hub traffic.

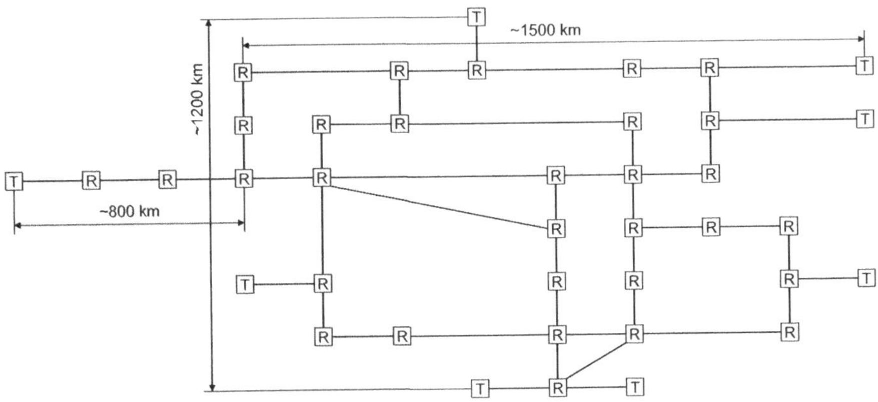

FIGURE 6.39 Meshed long-haul network with distances up to 3000 km. Not shown are the in-line amplifiers. R: ROADM nodes, T: termination.

6.5 LONG-HAUL AND ULTRALONG-HAUL SYSTEMS

With the commodity of ROADMs in meshed networks, the focus of system design moved from ultimate-reach point-to-point links toward scalability and multipath capability. Multipath restoration and fully automated control of optical connections via the Generalized Multiprotocol Label Switching (GMPLS) control plane (see Section 7.6) are enabled by a modular system design based on cascaded norm sections.

Here, we focus on meshed networks. An example network is shown in Fig. 6.39.

Key elements of this example are ROADM nodes, where traffic is added or dropped into the long-haul backbone, which is connected by multiple amplified fiber links.

6.5.1 Optical Design Challenge*

All-optical networks in conjunction with long transparent optical reach result in complex system designs [43]. Efficient modeling algorithms and tools have been developed in recent years and are a requirement for day-to-day optical system design.

Various nonlinear fiber effects have to be covered (see Section 2.2). In addition, noise introduced by optical amplifiers has to be taken into account. Further characteristics inherent to DWDM systems like spectral power ripple and tilt have to be considered as well. The influence of these effects varies with data rate, modulation format, fiber type, dispersion management, the use of FEC, and the type of optical amplification (Raman amplification and/or EDFAs). Polarization-related impairments [PDL/PDG (polarization-dependent gain) and PMD] and other limiting effects, for example, penalties induced by filter amplitudes or cross talk (XTK), or group-delay, add to the complexity.

*Sections 6.5.1 to 6.5.5 are contributed by Andreas Fäerbert.

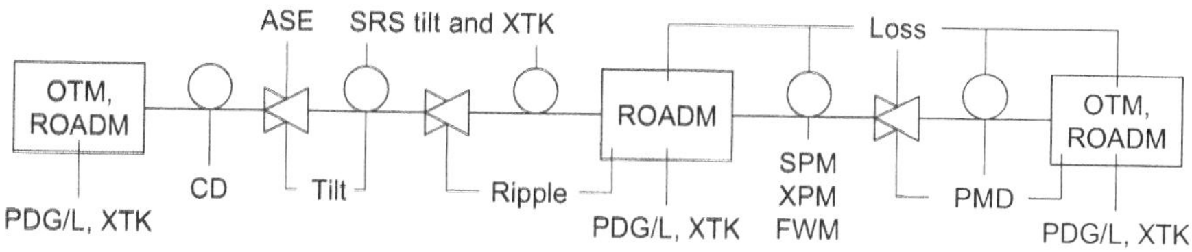

FIGURE 6.40 Signal impairments in long-haul networks. OTM: optical terminal multiplexer.

All impairments introduce certain penalties to maximum reachable distance and signal quality and need to be estimated and handled as constraints through the network design process. This is summarized in Fig. 6.40.

The simulation of nonlinear fiber transmission in conjunction with a prediction of system performance and reach is challenging. The split-step Fourier method can be used to solve the nonlinear Schrödinger equation (see Section 2.2.1), which describes optical pulse propagation [44]. However, for long distances, high optical power and high channel count, this mathematical model causes immense calculation effort. Consequently, there is a need for engineering rules enabling first-order performance estimates and identification of limiting effects even in meshed networks including multipath analysis in seconds. The optical system has to be in the optical-power range, which is sufficient to be above noise limitation but below severe accumulation of nonlinear limits. There is only a small region of valid optical system design, as explained in Fig. 6.41.

To achieve an efficient optical-system engineering tool-set, numerical simulations need to be replaced by analytical equations or even heuristic approximations to reduce calculation effort. These engineering approximations are derived and simplified from the analysis and optimization of the system configurations.

6.5.2 Engineering Models*

Key to engineering models is the separation of all limiting effects in the performance analysis. In optical amplified systems, the accumulation of noise is characterized by

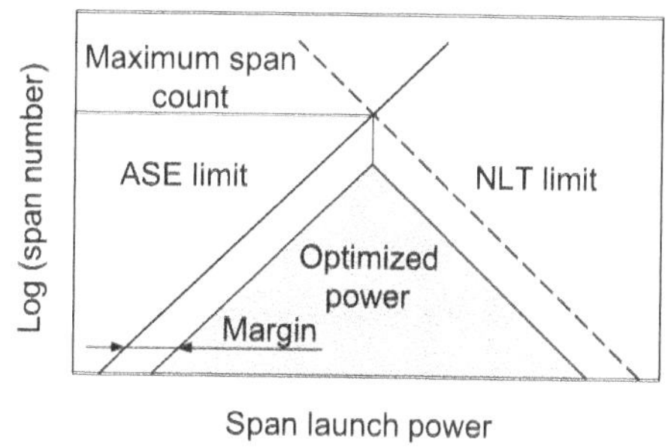

FIGURE 6.41 The "working zone" of optical-layer design. NLT: nonlinear tolerance.

the amplifiers' noise figures (NFs). For EDFAs, the NF is characterized by input power and gain. The receiver is described by the OSNR (see Section 4.2.3) necessary to guarantee a certain BER.

Due to the random and noise-like behavior of multichannel nonlinear cross talk caused by XPM, FWM, and SRS (stimulated Raman scattering), Gaussian noise models can be used. The combined effect of SPM and GVD (group velocity dispersion) is quantified by the distortion of the eye pattern in front of the electrical decision circuit.

Since the models for the limiting effects are determined independent from each other, similar conditions for all models are needed to make the influence of the different effects on the system performance comparable.

6.5.2.1 Optical Signal-to-Noise Ratio Degradation In amplified networks, the OSNR degrades due to optical noise accumulation along cascaded amplifier sites (see Sections 3.2.4 and 4.2.3). This leads to increased BER at the receiver. The main noise source is ASE generated by the optical amplifiers. The noise added by an individual amplifier is specified by the NF. It accumulates along a multispan link. The OSNR needed at the receiver imposes a lower limit on the allowable power per channel.

6.5.2.2 Spectral Ripple and Tilt Spectral tilt in DWDM systems arises from the wavelength dependence of the transmission-fiber attenuation (see Section 2.1.1.1) and from SRS (see Section 2.2.2.1). The fiber-loss tilt depends on span length and fiber type as well as the DWDM spectrum, which is actually allocated. It is independent of the launched channel power. Tilt caused by wavelength-dependent fiber loss and SRS is depicted in Fig. 6.42.

The SRS tilt depends on the total power of the DWDM channels, the DWDM spectrum used for transmission, the span length, and the fiber properties [see Eq. (2.90)].

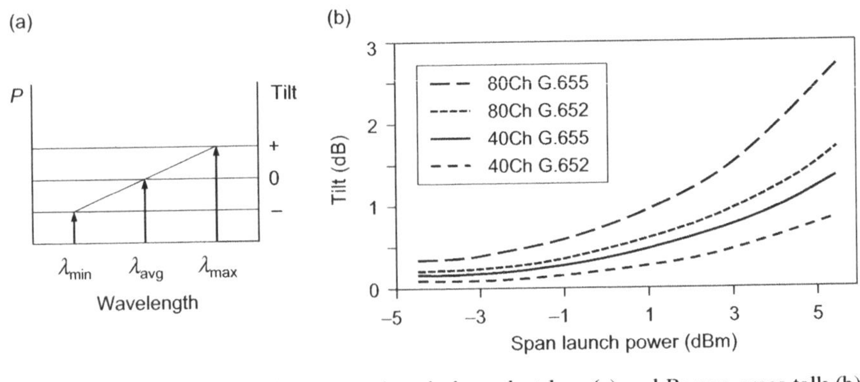

FIGURE 6.42 Spectral tilt due to wavelength-dependent loss (a) and Raman cross talk (b).

Power *ripples* are nonpredictable variances of channel power levels caused by individual components like amplifiers and filter modules. These variances can have systematic or random characteristics depending on technology (e.g., EDFA vs. Raman amplification) and individual production characteristics.

6.5.2.3 Self-Phase Modulation and Group Velocity Dispersion The Kerr effect of the transmission fiber leads to a phase modulation of an optical signal by its own intensity. The influence of this SPM must be seen in conjunction with GVD. In the anomalous dispersion regime ($D > 0$), the SPM-induced chirp has opposite sign than the GVD-induced chirp and can therefore partially compensate the CD-induced pulse broadening (see Sections 2.2.1.1 and 3.2.5.3). This compensation remains imperfect because the temporal shape of the SPM-induced chirp differs significantly from that of the GVD-induced chirp. This results in accumulation of signal distortions along the link.

For high launch powers, full CD compensation limits the maximum transmission distance, whereas an optimized undercompensation can increase the maximum transmission distance significantly. This is also referred to as distributed undercompensation (DUCS) [45]. Hence, the dispersion management is of crucial importance for optimum SPM performance.

A useful figure-of-merit for evaluating the maximum transmission distance is the product of per-span launch power P_{Launch} and the number of spans N_{Span}. For each launch level, a maximum distance can be determined: reducing the launch power by 3 dB would double the maximum distance, and so on. To achieve a certain low performance penalty, the cumulative span launch power should not be higher than a specified maximum, the nonlinear threshold (NLT). This effect is shown in Fig. 6.43.

It can be shown that the NLT law

$$\log(N_{\mathrm{Span}}) + P_{\mathrm{Launch}}(\mathrm{dBm}) = \mathrm{NLT}(\mathrm{dBm}) = \mathrm{const.} \tag{6.21}$$

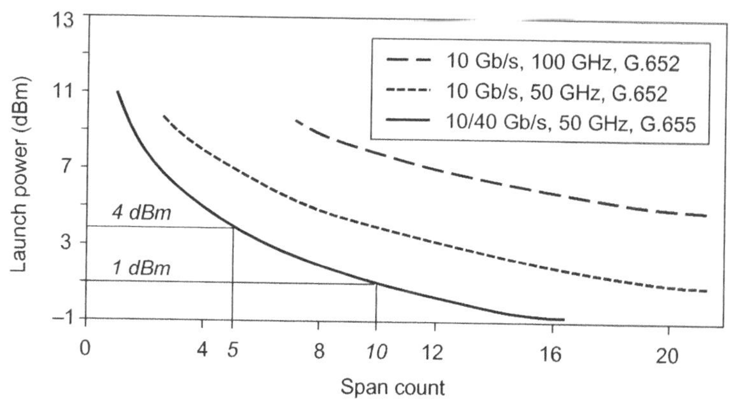

FIGURE 6.43 SPM-limited transmission distance (SSMF and DCF) versus launch power (10 Gbit/s NRZ).

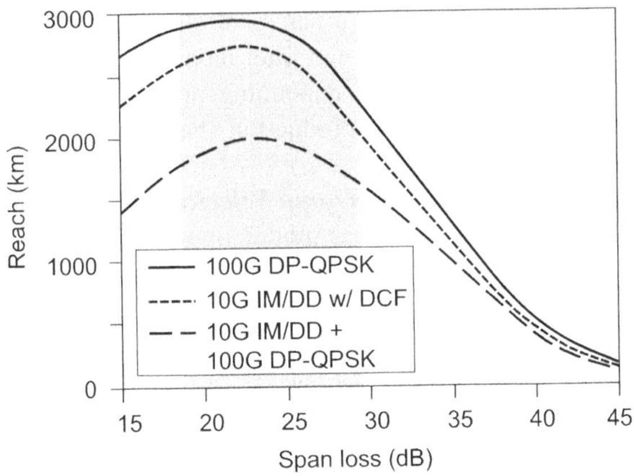

FIGURE 6.44 Transparent reach for different span losses. Shaded area marks area of interest.

holds for different fiber types, data rates, and modulation formats [46]. The NLT depends on these parameters as well as, for example, channel spacing.

As a consequence and to achieve higher transmission distances, the per-span launch power has to be reduced. This corresponds to shorter distances between amplifiers, leading to better NF and the deployment of more amplifiers. Hence, there is a trade-off between longest transparent reach and numbers of amplifiers between add–drop nodes. Typical total transparent reach versus per-span loss is shown for 10G/100G signals in Fig. 6.44.

For Fig. 6.44, fiber loss of 0.25 dB/km and a patch-panel and repair margin of 1 dB each were assumed for every span. It can be seen that lower span distances will improve the total reach at the cost of more in-line amplification to a certain amount. Beyond this point, the added noise, tilt, and ripple of each amplifier leads to a negative effect.

6.5.2.4 Cross-Phase Modulation XPM denotes the phase-modulation of a signal caused by the intensity of copropagating other signals. It leads to nonlinear coupling of channels in a DWDM system. The effect of XPM in direct detection intensity-modulated DWDM systems with NRZ-OOK signals can be described as a two-step process.

First, a signal experiences a PM caused by the intensity of copropagating signals. In the second step, the CD of the transmission fiber causes a PM-IM (intensity modulation) conversion and hence a conversion of phase fluctuations into intensity fluctuations. CD influences this process in two ways: the local dispersion of the transmission fiber leads to a walk-off between copropagating DWDM channels. The *faster* the walk-off, the *lower* the XPM-induced phase fluctuations. Hence, high local CD is beneficial for XPM suppression. The *residual* CD, measured from the point where XPM is generated in the transmission fiber up to the receiver site, determines the strength of the PM-IM conversion. Consequently, zero residual CD would almost fully suppress XPM. However, if the dispersion management of each span reassembles the bit alignment between WDM channels at the input of each span, the individual XPM contributions of all spans add up constructively, which leads to the worst-case penalty. Therefore DUCS provides a good compromise for balancing SPM and XPM influences.

For large channel numbers, XPM acts noise-like. The XPM-induced signal distortions depend on the local dispersion, the channel spacing, the bit rate, and the dispersion management. Fibers with low local dispersion, for example, nonzero dispersion-shifted fiber (NZ-DSF), cause higher XPM penalty due to the lower walk-off between adjacent WDM channels (also see Section 2.2.1.2) [47].

6.5.2.5 Four-Wave Mixing The Kerr effect also causes coherent interaction of two (degenerate case) or three wavelength (nondegenerate case) channels, transferring power into spectral regions outside these channels. This is referred to as parametric or four-wave mixing. Due to its dependence on the phase mismatch between the interacting waves, the FWM efficiency is a function of channel spacing and local dispersion of the transmission fiber (also see Eq. 2.73) [48].

With a high number of DWDM channels, modeling of FWM cross talk as a noise-like process with Gaussian amplitude characteristic is an obvious approach. With the dynamic behavior of polarization changing along the fiber, one engineering approach is to consider the worst-case polarization scenario, where all channels have the same (linear) polarization. As both the bandwidth and the required OSNR increase when going to higher data rates, the FWM-caused OSNR penalty can be shown to be independent of the data rate.

FWM is a major effect in low-dispersion fiber because in multispan systems, the contributions of the individual spans add up coherently. In many long-haul systems, CD compensation varies over the spans. As a consequence, the average OSNR degradation is only proportional to the number of spans.

6.5.2.6 Stimulated Raman Scattering Stimulated Raman scattering is based on light interaction with phonons and leads to power transfer from higher frequency channels to lower frequency channels in DWDM systems (refer to Section 2.2.2.1). Since the Raman gain coefficient increases with wavelength separation up to 100 nm, SRS strongly affects high-capacity or broadband systems. The related impairments are SRS-induced spectral tilt and noise-like cross talk. The latter is a fundamental limitation for high-capacity DWDM systems, especially if multiple transmission bands are used (C-band plus L-band) [49].

SRS-induced spectral tilt results in different power levels of DWDM channels after transmission. It can cause power differences of up to several decibel over the transmission band. Power preemphasis is typically used to lower the tilt impact as the tilt would add up linearly span by span. With many wavelengths traveling in the same fiber, the SRS-induced cross talk can be modeled as additional noise and will add up to an additional OSNR penalty.

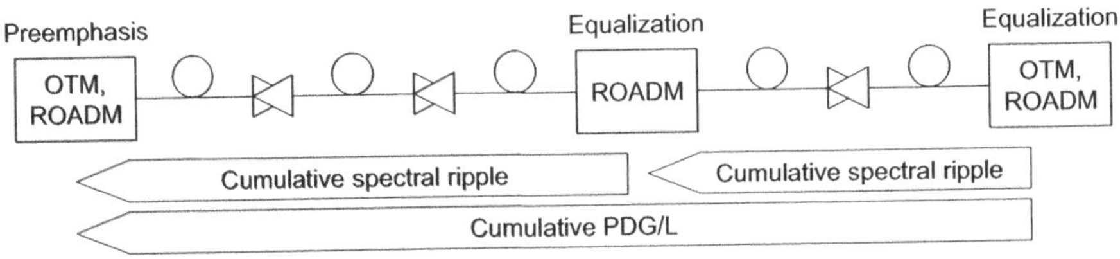

FIGURE 6.45 Accumulation of power variations and ROADM power equalization.

6.5.3 Link-Based Design Approach*

In a link-based approach, all design constraints with respect to optical power and dispersion map are tightened at the add–drop nodes, where optical signals are launched or terminated. Between these nodes, one may be able to relax the requirement of accurate CD compensation. The design is carried out by detailed modeling of deterministic and statistical power variations across the whole transmission spectrum.

6.5.3.1 Optical Power Management To calculate the total channel-power variations, the accumulation of tilt, ripple, and PDG/PDL contributions of all relevant components of the link needs to be done. Tilt and ripple accumulation starts and ends at each ROADM node. However, PDG/PDL cannot be eliminated by ROADM equalization and has to be accumulated for the end-to-end path. This is shown in Fig. 6.45.

Optical amplifiers are used to compensate the losses caused by fibers, connectors, and optical multiplex and switching components. However, the loss of these components is not constant. Spectral fluctuations induced by random insertion of loss nonuniformities cannot be compensated by optical amplification and have to be accumulated as power ripples. The ripples arise from filter loss nonuniformities combined with the amplifier gain-spectrum shape (wavelength-dependent ripple) and the transmitter output-power variations (wavelength independent). This leads to channel-power spread as shown in Fig. 6.46.

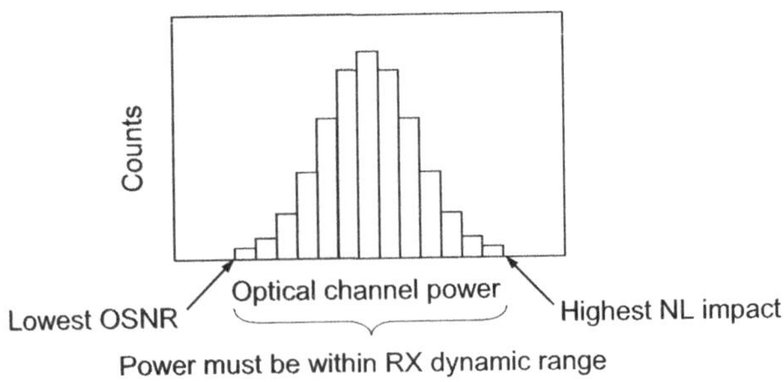

FIGURE 6.46 Histogram of WDM channel power.

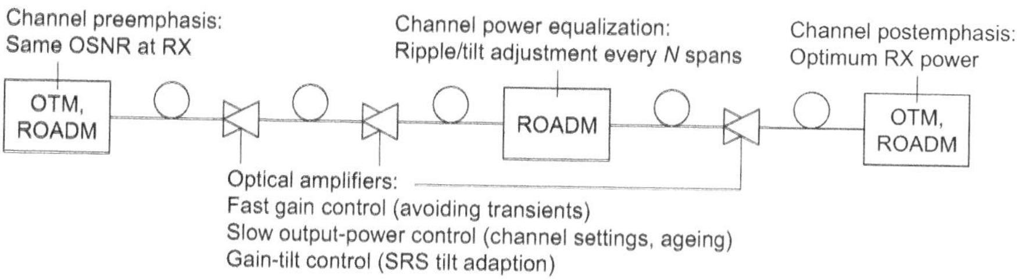

FIGURE 6.47 Optical power management by channel power adjustments.

High-power channels are subject to stronger signal distortion by nonlinear effects, whereas the BER of the lowest-power channels is impacted by the low OSNR and/or low power at the receiver. Power management is the procedure of controlling the optical-channel power levels at particular points in the networks to ensure optimum reach for all WDM channels. Optimizing OSNR performance (high launch power) has to be balanced with the impact of nonlinearities. Power management can be performed in ROADMs that are located after every N spans, as shown in Fig. 6.47. N can be in the range of 5–10 in terrestrial meshed networks. In ultralong-haul subsea point-to-point links, power equalization is rather performed by per-amplifier gain-equalizing filters (Fig. 6.56).

A major benefit of ROADM networks as compared to networks based on static passive filters is the improved transmission performance, which is enabled by (control-plane-driven) per-wavelength power adjustment.

6.5.3.2 Chromatic Dispersion Management CD design refers to the procedure of placing CD-compensating modules (DCM, based, for example, on DCF or channelized FBGs) (see Section 3.2.5) in the network to implement a desired dispersion map that allows correct amounts of net dispersion at each add–drop or in-line node. The design goal is to keep the performance penalty caused by SPM/XPM and CD for each *potential* optical connection within the network as low as possible.

The requirements for the dispersion design are derived from network parameters like maximum required per-channel data rate, line-rate mixture, maximum number of WDM channels, which must be supported end of life, and the fiber types that are deployed. If the analysis of the accumulated launch power indicates that transmission is dominated by nonlinearity, a dispersion map with DUCS in each span is preferred for 10 Gb/s (NRZ-OOK) signals. In contrast, 40 Gb/s (NRZ-DPSK) signals prefer smaller undercompensation compared to 10 Gb/s up to slight overcompensation in each span, and zero net residual dispersion at the receiver [50]. A typical long-haul dispersion map is shown in Fig. 6.48. All fiber spans are slightly and equally undercompensated.

Other dispersion maps are referred to as FOCS (full in-line and optimized postcompensation scheme) or DOCS (distributed overcompensation scheme) [45]. The main idea is to generate compensated norm sections optimized for flexible optical transparent reach (span count), suitable for meshed, dynamically switched optical networks where the optical path is unpredicted.

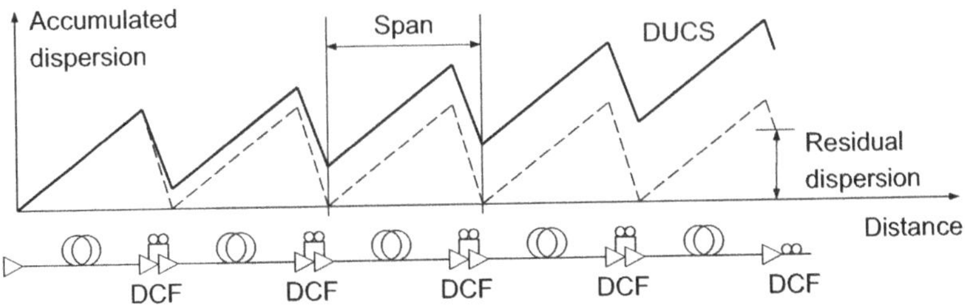

FIGURE 6.48 Compensated multispan long-haul link with distributed undercompensation (DUCS).

6.5.4 Optical-Layer Design Workflow*

A key element of the design of large meshed networks is the separation into so-called optical transparent design areas (OTDAs). Each OTDA defines an optical (sub) network where any-to-any node connectivity should be achieved without (O/E/O and 3R) regeneration.

The *reference path* of each OTDA is defined as the longest path that should work without regeneration. This reference path can be determined based on a traffic-demand matrix or by analyzing the number of spans traversed by any-to-any routes and considering potential detours caused by protection or restoration. In meshed topologies, the reference path is determined via the shortest and the second shortest path between any two nodes in the respective OTDA. The effect of protection detour on the reference path is explained in Fig. 6.49 for a simple network.

If no protection is required, the reference path is the path with the maximum span number out of all the shortest paths between any two ROADM nodes. If protection is required, the reference path is the path with the maximum span number out of the second shortest paths between any two ROADM nodes, and having no span in common with the respective shortest path.

The maximum number of supported spans N_{max} leads to the maximum per-span launch power by considering $\log(N_{max})$ for the accumulated launch power limit due to fiber nonlinearities. The choice of optical amplifiers is then guided by the target

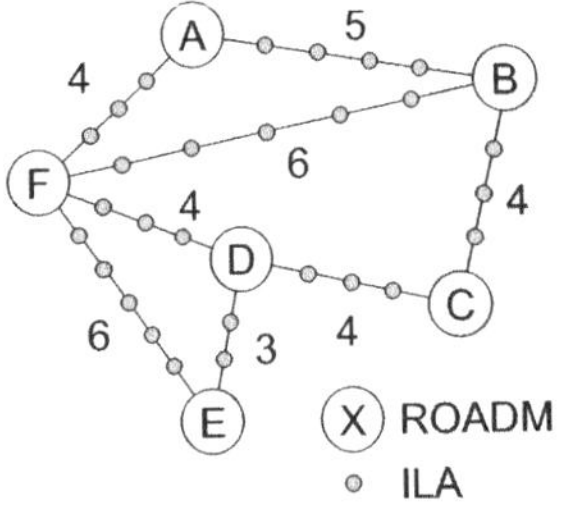

Unprotected					
	B	C	D	E	F
A	5	9	8	10	4
B		4	8	11	6
C			4	7	8
D				3	4
E					6

Protected					
	B	C	D	E	F
A	10	12	13	16	11
B		14	10	12	9
C			14	16	10
D				10	9
E					7

FIGURE 6.49 Analysis of per-link span count and reference path length in a simple meshed network.

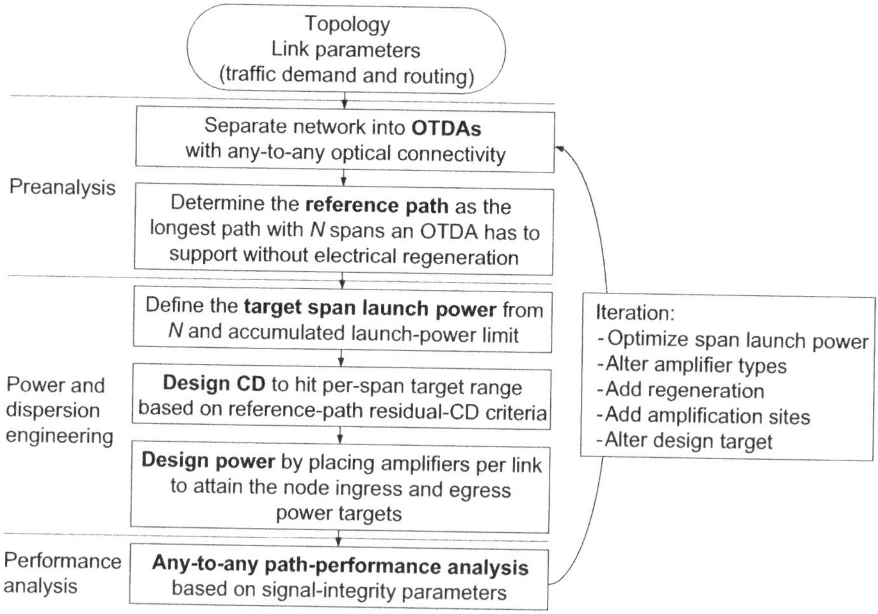

FIGURE 6.50 Design workflow for long-haul networks.

channel input power into each span. In case of a DUCS dispersion map, N_{max} of the reference path obtains the amount of undercompensation and DCM placement.

A design workflow for long-haul networks is given in Fig. 6.50.

In the process of validating the performance of each path in a network, the minimum OSNR calculated for an amplified link including all OSNR penalties to compensate for signal impairments is compared with the minimum OSNR, which is required by any of the receivers used. The OSNR penalties added must account for all impairments, for example, filter passbands, nonlinearity, PMD, or off-optimum CD compensation.

The system limit is reached if the worst-case OSNR (OSNR of the worst link at minimum power, including all penalties) is lower than the OSNR required at the receiver. In such a case, the optical signal must be electrically regenerated to ensure transmission under all operational and environmental conditions even after years of operation, when the equipment reaches its EOL (including an EOL penalty).

The OSNR system limit is explained in Fig. 6.51.

Similar to Fig. 4.9, a noise figure of $F_N = 6.5$ has been assumed here. The OSNR penalties (in decibels) are assumed to grow linearly with the span count.

6.5.5 Mixed 10 Gb/s/100 Gb/s Design*

Ten gigabit per second IM/DD (direct detection) transport requires, in most cases, a photonic layer that is compensated with respect to CD. This compensation is typically based on DCF. Alternative optical compensation can be based on FBGs

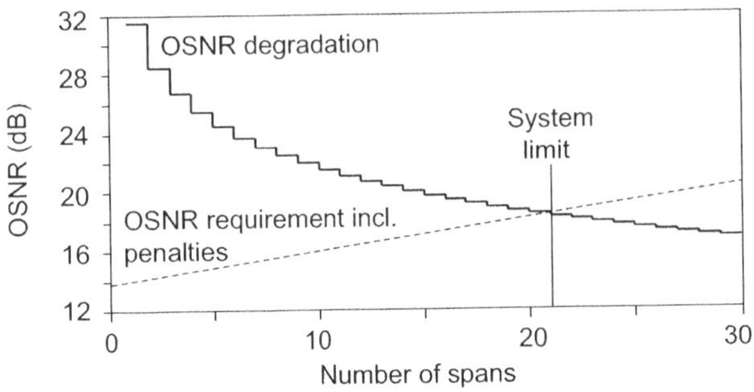

FIGURE 6.51 Transparent system limit. The OSNR penalties depend on the given link.

or similar devices (Section 3.2.5). Alternatively, EDC can be implemented at the transmitter [51]. Coherent DP-QPSK 100 Gb/s transport makes use of receive-end EDC and hence does not require CD-compensated photonic layers (see Section 5.4).

In certain transport networks [e.g., the National Research and Educational Networks (NRENs)] (see Section 6.5.7), there is a strong requirement for so-called alien-wavelength support. This refers to the capability of one network to directly and transparently support wavelengths coming from another network and, potentially, system vendor. In many cases, the capacity of alien wavelengths is 10 Gb/s. In addition, some services may not be allowed to be aggregated to the highest transport capacity. Then, combined 10 and 100 Gb/s transport is required.

Problems for combined 10 and 100 Gb/s WDM transport arise from three aspects:

- Coherent 100 Gb/s do not require optical CD compensation, whereas most 10 Gb/s systems do.
- Coherent 100 Gb/s systems can be subject to reach penalties when operated on DCF-compensated links.
- Phase-modulated signals are subject to penalties when operated together with IM/DD signals.

For combined 10 and 100 Gb/s scenarios, the performance of 100 Gb/s DP-QPSK signals is decreased by XPM-induced nonlinear phase noise (X-NLPN) (see Section 2.2.1.6) caused by the 10 Gb/s signals themselves.

The reach decrease when using DCF comes from potentially increased XPM inside the DCF, slight noise-figure increase of the amplifiers that accommodate the DCF, and the fact that DCFs also compensate the channel walk-off (leading to stronger XPM in the transmission fibers).

The resulting penalty can partly be reduced by carefully placing the DCF between the gain stages of dual-stage amplifiers and adjusting the DCF launch power. Further methods for reducing the DCF-induced XPM penalty are as follows:

- Separate 10 and 100 Gb/s transport on dedicated photonic layers (coherent overlay)

- Run a single uncompensated photonic layer
- Use CD compensators other than DCF, that is, channelized FBGs [52,53]

These methods all come at a certain cost that is summarized later.

Reduction mechanisms for X-NLPN caused by IM/DD channels include the following:

- Reducing power of 10 Gb/s channels compared to 100 Gb/s channels
- Inserting channel separation devices (such as ROADMs, interleavers, or multiplexers/demultiplexers) to separate the DWDM channels and transmit them through different paths within the same node with adequate path-length difference, to introduce differential delay between the channels
- Using channelized FBGs instead of DCFs to increase the walk-off between 10 and 100 Gb/s channels
- Increasing the spectral separation between 10 and 100 Gb/s channels (guard band)

The last three methods aim at decorrelating the 10 Gb/s from the 100 Gb/s signals along their paths. Similar effects can be achieved, for example, by inserting PMD elements.

Ten gigabit per second IM/DD and coherent, phase-modulated systems can also be completely separated on dedicated photonic layers ("coherent overlay"). This adds obvious cost, but is also regarded the only "safe" way to support upcoming ultrahigh-speed systems (running 400 Gb/s or 1 Tb/s per (super) channel) at optimum performance.

Another method of partial nonlinearity suppression is based on using IM/DD with inbuilt EDC. However, this solution only supports 10 Gb/s IM/DD channels with EDC (and not the majority of 10 Gb/s channels that do not use EDC), that is, it has *limited alien-wavelength support*. It reduces XPM, but does not avoid X-NLPN.

In addition, the carrier phase estimation (CPE) in coherent receivers can be configured to achieve optimum results in the presence of nonlinearity [54]. CPE must be operated over certain block lengths, and under nonlinear conditions, the CPE block lengths should be reduced (kept short) to achieve optimum results.

For all methods described above, certain limitations or, in a general sense, costs result. This is summarized and explained in Table 6.9.

All methods that aim at 10 and 100 Gb/s coexistence lead to certain generalized cost. These costs can consist of capital expenditures, operational complexity, reduced functionality, or reduced optical performance for 10 and/or 100 Gb/s services. Careful design and optimization of relevant parameters (e.g., combining methods 3, 4, and 5 from Table 6.9) can lead to penalties in the range of ≤ 1 dB. This penalty can apply to both bit-rate systems, or just one of them, depending on the chosen interworking parameters.

It must also be considered that photonic transport layers have a certain lifetime. Therefore, one should carefully consider additional forward-looking requirements, for

TABLE 6.9 **Cost of 10 and 100 Gb/s Coexistence**

Method		Generalized Cost	Potential Advantages
1	Coherent overlay	Almost doubles cost of photonic layers, especially fibers. Potential network (fiber) underutilization	Considered the only safe solution for 400 + Gb/s (super) channels
2	IM/DD with inbuilt EDC	No general 10 Gb/s alien-wavelength support. 10 Gb/s EDC does not reduce X-NLPN, and hence decreases 100 Gb/s reach	Can avoid compensated photonic layer and penalties associated with the use of DCF (XPM)
3	Use of FBGs instead of DCFs	Adds accumulated phase ripple to IM/DD channels and limits their reach. ~1 dB penalty after 8–10 devices. Periodic group delay restricts future deployment of wide-bandwidth or multicarrier signals (gridless approach)	Significantly reduces X-NLPN from IM/DD. Eliminates XPM from DCF on QPSK
4	Use of guard bands between 10 and 100 Gb/s channels	Decrease of system and network capacity, and added complexity in network (channel and wavelength) planning and possibly provisioning. For gradual growth to all-100 Gb/s, the band may be eliminated later	Can help to achieve optimum performance for IM/DD and QPSK
5	Power-level reduction for IM/DD	Complex (software/control plane). Reduces IM/DD performance. Still requires CD compensation for IM/DD (unless IM/DD is combined with EDC)	Can lead to less or negligible IM/DD-induced penalty for QPSK

example, added compatibility with upcoming super-channels for coherent 400 Gb/s and 1 Tb/s systems. For optimum performance, these systems require coherent-only, uncompensated photonic layers. This excludes both, coexistence scenarios based on optical CD compensation, and the general use of any IM (DD) technology.

6.5.6 Regional Differences

In today's WDM networks, there are only little regional differences regarding specific characteristics like infrastructure characteristics, service requirements, and so on. This follows from international standardization attempts and also

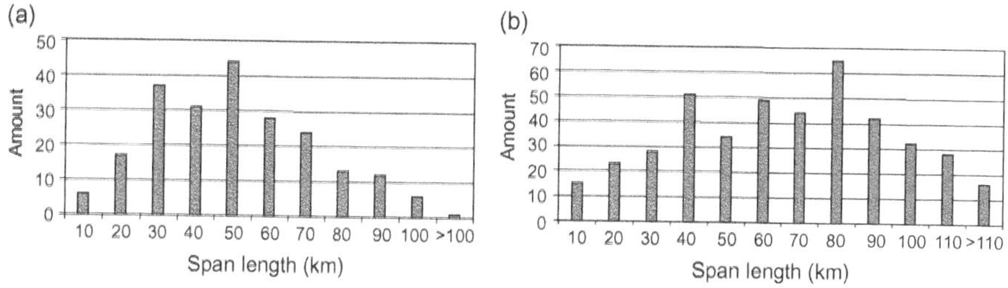

FIGURE 6.52 Span-length histograms of European (a) and US (b) long-haul networks.

from global competition and the related cost pressure. This has not always been the case. Well-known telecommunications examples of regional differences, causing noninteroperability and related added cost, include the differences in PDH (plesiochronous hierarchy) hierarchy levels and bit rates between the United States, Japan, and Europe, and also similar differences between the synchronous systems in the United States (SONET) and Europe and Japan (SDH), respectively. The latter are reflected by the different SDH [ITU-T G.707, G.783, and ETSI EN 300 417-x-1 ($x = 1, \ldots, 7, 9$)] and SONET (ANSI T1.105, T1.231, and Telcordia GR253) standards and Recommendations.

One of the few regional differences with an impact on long-haul WDM transport relates to the mean per-span length of multispan links. Differences can be identified between typical long-haul links in Europe and the United States. Fig. 6.52 shows span-length histograms that were directly derived from various long-haul installations in Europe and the United States in the first decade of the twenty-first century. In both cases, one can identify a certain concentration at 40–60 km. For US long-haul links, there is further concentration around 80 km, which also leads to a higher number of spans of $\geq$100 km length. The reason for these span-length histograms is obvious: mean distances in the denser-populated European countries are shorter, and so are the mean span lengths. It can be expected that span lengths, for example, in Japan more closely follow the European histogram (Fig. 6.52a), whereas span lengths in China, India, or Brazil may follow the US histogram (Fig. 6.52b) more closely.

The difference depicted in Fig. 6.52 has a consequence on the respective links, especially for ultrahigh-capacity, ultralong-haul installations (e.g., cross-United States, pan-European). These links typically are limited by combined nonlinear effects as well as by OSNR (refer to Sections 6.5.3 and 6.5.4). As pointed out in Section 4.2.3 and Fig. 6.44, OSNR depends on mean span length (it is also decreased by single spans with very high loss). Hence, links with spans that clearly exceed 80 km (Fig. 6.52b) may be subject to OSNR-driven span-count limitation.

6.5.7 Terrestrial Long-Haul Deployment Example: PIONIER

An example of an NREN that makes use of state-of-the-art DWDM transport is the Polish PIONIER network. It is operated by the Poznan Supercomputing and Networking Center (PSNC).

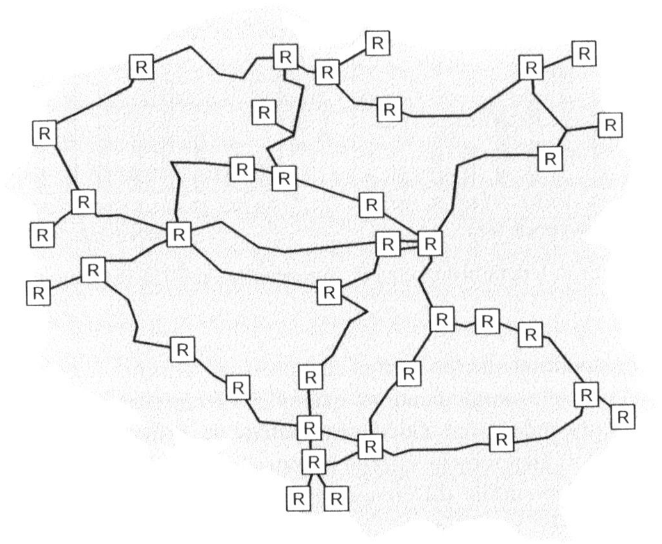

FIGURE 6.53 The Polish National Research and Educational Network, PIONIER. R: ROADM site.

PIONIER is based on WDM transport since 2003, and the actual generation was deployed in 2009–2011. Applications, within the national and the pan-European research and educational community, include 10 Gb/s any-to-any services, for example, for grid and cloud computing or virtual laboratories. The required bandwidths are provisioned flexibly on demand.

The PIONIER network connects 21 metropolitan area networks and five High-Performance Computing Centers (HPCs). In total, it consists of >100 nodes and >6600 cable kilometer (of which ~760 are in Germany, connecting PIONIER to the Scandinavian NRENs in a PoP in Hamburg). PIONIER was also the first European NREN to be based on dark fiber.

The network [ROADM sites (R) and links] is shown in Fig. 6.53.

The current PIONIER generation is based on an 80-channel DWDM system from ADVA Optical Networking. It runs on a partially meshed backbone, which consists of colorless, directionless, and contentionless (CDC) ROADMs, and which is enabled by a multilayer GMPLS control plane. The majority of traffic (as of 2012) is based on 10 Gb/s per wavelength, and the network is ready for 100 Gb/s services. The network is DCF-CD-compensated. The longest transparent paths that can be configured in the network (limited by network size rather than physics) are in the range of 2500–2800 km.

In Fig. 6.53, most of the ROADM sites located at the Polish border connect to the NRENs of the neighboring countries (e.g., to the link to Hamburg).

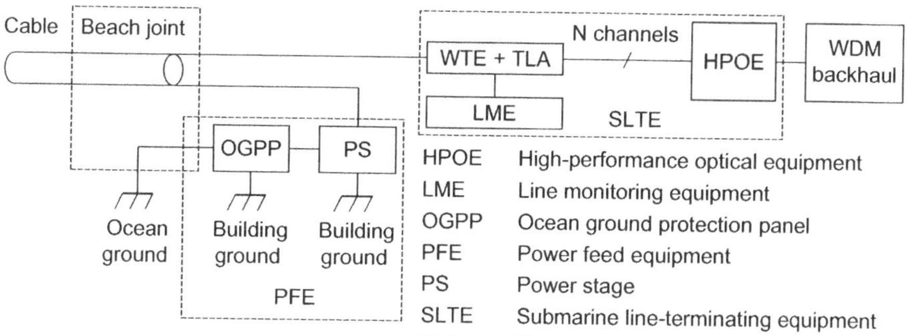

FIGURE 6.54 Termination equipment of a submarine cable system.

6.5.8 Submarine Systems

Submarine systems are a special case of (ultra) long-haul systems. In certain cases—transoceanic systems—they lead to the longest transparent link lengths, which have to be implemented globally. In addition, they pose special requirements on ruggedness, power supply, and also monitoring, which are irrelevant for most terrestrial WDM systems.

Generally, the land-based termination equipment of submarine WDM systems is split into two parts: the submarine line termination equipment (SLTE) and the power feed equipment (PFE). They are responsible for the WDM transmission including monitoring, and the power supply, respectively. A land station of a submarine system is shown in Fig. 6.54 [55,56].

Regarding WDM transmission, submarine systems have the advantage over most terrestrial systems that amplifier spacing can be made perfectly equidistant. In addition, low-noise amplifiers are used, and the spacing is kept reasonably short (50–60 km) in order to achieve low resulting NFs even in very long amplifier chains (refer Section 4.2.3). So far, dispersion-managed systems have been deployed. For further design rules, refer to Section 6.5.3.

The PFE is an area special to submarine-cable systems. The PFE has to provide power to the optical amplifiers that have to be submerged. In most cases, systems are fed from both ends. For power supply, submarine cables contain a single copper conductor (Fig. 3.40).

The PFE converts −48 V battery supply to high voltage. It then feeds the system, typically with constant current, and voltages (at the PFE) exceeding 2500 V. Generator back-up can be used to maintain the voltage, which is required to feed the battery system in case of mains power failures. The primary return current path is the seawater.

Under normal operations, dual-end fed systems feed half of the link from both sides each. However, each PFE can supply the complete cable system. If a problem occurs at one PFE, or it has to be shut down for maintenance, the other PFE

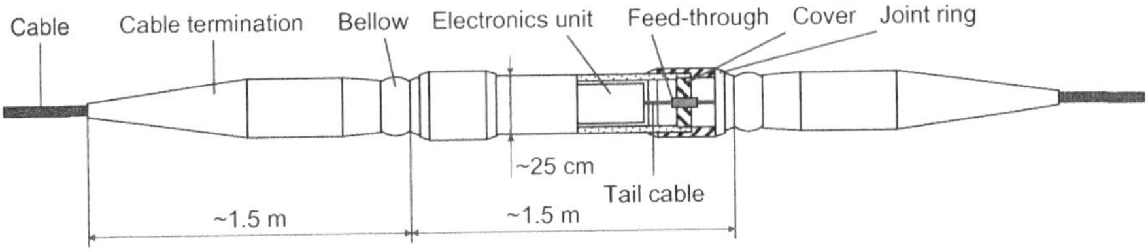

FIGURE 6.55 Submarine repeater (amplifier).

automatically takes over. Hence, dual-end fed systems switch to single-end feed if one PFE fails.

The PFE can operate on either ocean ground or building ground. To limit the risk of damage due to lightning strikes, an ocean ground protection panel (OGPP) is used. The OGPP provides isolation between ocean ground and building ground.

The submerged in-line amplifiers must be accommodated in water- and pressure-tight boxes. The complete assembly must also mechanically terminate the cable on both ends. For deep-sea deployments, it must withstand pressure up to 800 bar and tensile strength up to several 10 kN. A submarine repeater is shown in Fig. 6.55 [56,57]. Cable termination and feed-through are mechanically challenging in pressure-tight design. In addition, the inner cross-sectional area is limited, leading to highly compact amplifier design requirement.

If a submarine cable must be recovered for repair (in general, this is possible and occurs more or less regularly), the PFE can provide an additional sine-wave tone with ~5 Hz on its high-voltage output. This signal can be detected by the cable ship to relocate the cable.

In order to minimize repair action, and support fault localization, high availability of the amplifiers and means for fiber monitoring are required. The most critical part contributing to amplifier unavailability is the pump lasers. Hence, redundant pump lasers have to be implemented. In order to recover the cable system precisely, OTDR (optical time-domain reflectometry) must be supported. The resulting block diagram of a bidirectional submarine in-line amplifier is shown in Fig. 6.56 [56–58].

The figure shows 980 nm forward-pumped plus 1480 nm backward-pumped low-noise design (see Section 4.2.3) with duplicated pump lasers. It also shows cross-coupling that is required for OTDR monitoring. A dedicated OSC is not used in order to avoid the OSC-filter insertion loss. Instead, pump-power modulation is used to establish communication channels for operations and supervision [59] (also refer to the DCN Section 7.4).

For OTDR monitoring of amplified submarine systems, cross-coupling is required at the amplifiers, and coherent OTDRs (COTDRs) must be used to achieve additional measurement sensitivity in the presence of ASE [60–62]. Cross-coupling is required because each amplifier blocks reflected signals by means of inbuilt isolators (Fig. 6.56). It couples the reflected OTDR signals onto the second fiber, where it is amplified and transported together with the WDM signals. This requires that OTDR wavelengths within the EDFA gain band are used. Fig. 6.56 shows the

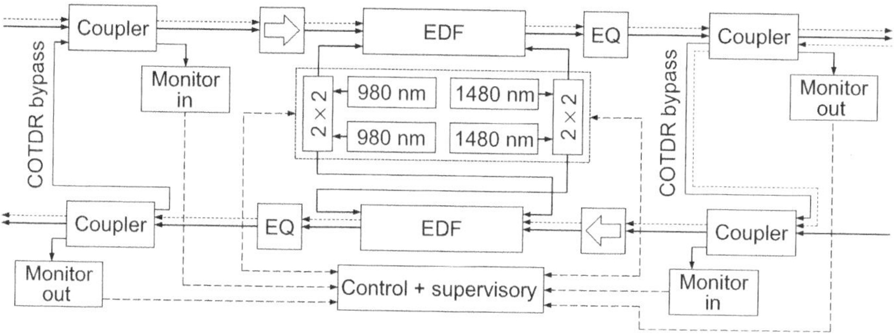

FIGURE 6.56 Submarine amplifier with COTDR cross-coupling. EQ: gain equalizer.

so-called O–I cross-coupling configuration (reflected signal coupled at amplifier output to amplifier input of reverse direction) [61]. This requires two cross-couplings. Alternatively, O–O coupling can be used (Fig. 6.58). Typically, 10%-couplers are used, leading to −20 dB for cross-coupling and ~0.5 dB loss for the main WDM path.

The block diagram of a COTDR is shown in Fig. 6.57 [60]. This set-up uses a continuous wave distributed feedback (DFB) laser and direct modulation-induced FSK and thus avoids the insertion loss of additional fast optical switches. The laser is reused for coherent self-heterodyne asynchronous detection. The band-pass filter is adopted to the FSK modulation index such that only pulses (which have shifted frequency) enter the envelope detector, but no unmodulated continuous wave (CW) light. An advantage is that this scheme applies CW power to the EDFAs.

Fig. 6.58 shows the O–O cross-coupling configuration in a long amplified system, together with the OTDR traces for both fibers in case of one fiber failing. In this case, the COTDR working on the affected fiber (Side A) can detect and localize the fault, whereas the other COTDR will measure the complete fiber span.

Care must be taken when choosing the COTDR settings [62]. With incorrect settings, either poor resolution or very long measurement time result. With state-of-the-art COTDRs, maximum distance (of amplified links) is 12,000 km, resolution can be as good as 10 m, and measurement time is 15 min over 1000 km with averaging over 2^{16} pulses [63].

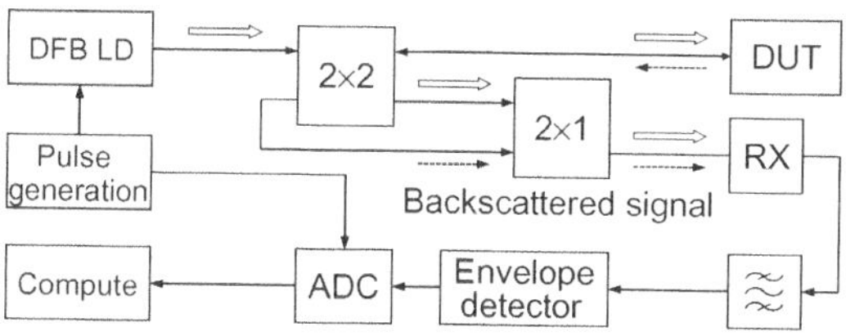

FIGURE 6.57 COTDR based on FSK. DUT: device under test.

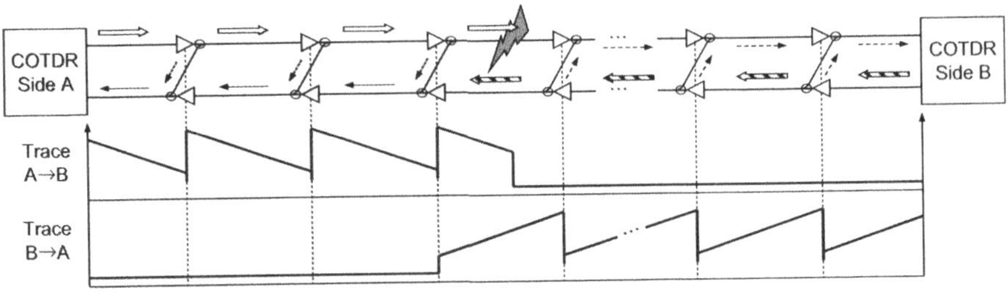

FIGURE 6.58 COTDR application in amplified links.

6.5.9 Ultralong-Haul Deployment Example: TAT-14

TAT-14 is a transatlantic cable system connecting the United States to the United Kingdom, France, The Netherlands, Germany, and Denmark. It went active on March 21, 2001, after more than two years of installation. It was deployed by KDD Submarine Cable Systems Inc. Total cost of the system was >1 billion €. TAT-14 is owned by a consortium of network operators that includes, amongst many others, AT&T Corp., Deutsche Telekom AG., France Télécom S.A., KPN B.V., TeliaSonera AB, and Sprint Nextel Corp.

For more than 10 years, TAT-14 has been the most powerful cross-Atlantic cable system. Only in 2012, two systems with higher capacity have been deployed.

TAT-14 is comprised of four active fiber pairs, configured for up to 40 DWDM channels each, in a ring topology. Landing points are Tuckerton and Manasquan (NJ, US), Bude-Haven (UK), St. Valéry (France), Katwijk (The Netherlands), Norden (Germany), and Blaabjerg (Denmark). The total length of the ring system is 15,428 km. The system, with details of the northern cross-Atlantic route, is shown in Fig. 6.59.

TAT-14 originally had a total design capacity or 1.28 Tb/s, which is calculated as [64]

$$\text{(North + South route)} \times 4\,\text{fiber pairs} \times 16\,\text{DWDM channels} \times 10\,\text{Gb/s} = 1.28\,\text{Tb/s}. \tag{6.22}$$

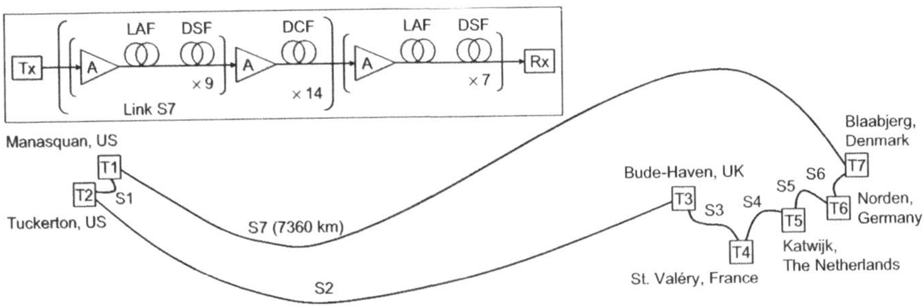

FIGURE 6.59 TAT-14 and details of the dispersion map of the northern cross-Atlantic link (S7).

According to Ref. [65], the capacity has been extended to 40 wavelengths per fiber. Out of this capacity, 640 Gb/s are ring-protected SDH capacity (64 × STM-64). This is configured as a bidirectional self-healing ring utilizing DWDM multiplexing. The system uses 75 GHz DWDM spacing.

TAT-14 uses EDFAs with a nominal NF of 4.3 dB. It uses a combination of Large-Area Fibers (LAF), DSF, and DCF. Nominal fiber loss is 0.21 dB/km, and mean amplifier spacing is 50 km.

At the northern cross-Atlantic section, the system has a cumulative residual CD slope of close to +400 ps/nm^2. This path has a distance of 7360 km over 147 spans (Fig. 6.59). At 1553.33 nm, total residual CD is approximately +1850 ps/nm without precompensation. Total PMD is in the range of 5–6 ps.

In 2008, the system was tested with RZ50-DBPSK at 40 Gb/s [66], and in 2011, general upgrade to 40 Gb/s per channel was assigned to Mitsubishi Electric.

6.6 FUTURE NETWORK CONFIGURATIONS AND CONVERGENCE

Fiber-optic networks are ever evolving. The primary goal is to achieve higher cost-efficiency per b/s transported. The main cost drivers of a network, over a certain operational period, are the Operational Expenditures (OpEx). These can take up to 80–90% of the total cost. As useful measures for the operational periods that are to be considered, one can take the lifetime of system generations. In WDM transport, this typically is 5–8 years, and for Layer-2/3 equipment, it is typically less. In addition, the return-on-invest (RoI) period that is required for certain investments can be considered. The RoI heavily depends on the type of investment, and the part of the network which is affected. Passive infrastructure build-out (trenches, ducts, cables, and fibers) can have RoI periods of several decades, whereas certain system deployments may have RoI periods of less than one year.

Relevant OpEx drivers include Operations, Administration, and Maintenance (OAM) of the network, including all active and passive systems and infrastructure, and power consumption [67]. OAM includes contributions to staff salary and real-estate cost. Power consumption cost must include the active equipment contributions as well as the overhead contribution, which are generated in PoPs, local exchanges, and other active sites (e.g., active cabinets) by the inefficiency of power supply units (PSUs), uninterruptable power supplies (UPSs), and heat, ventilation, air-conditioning (HVAC) equipment.

Consequently, major directions for future network convergence follow the efficient decrease of OAM- and power-consumption-related OpEx [68]. The related efforts are also known under the terms of network delayering or network or active-site consolidation, and Green IT, respectively. The latter further splits into the two contributions of Green-of-IT and Green-by-IT, respectively. These are the power-consumption savings that are to be achieved by or within the IT network itself, and the savings that are enabled by intelligent use of IT networks (e.g., reduction of truck rolls, etc.) [69,70].

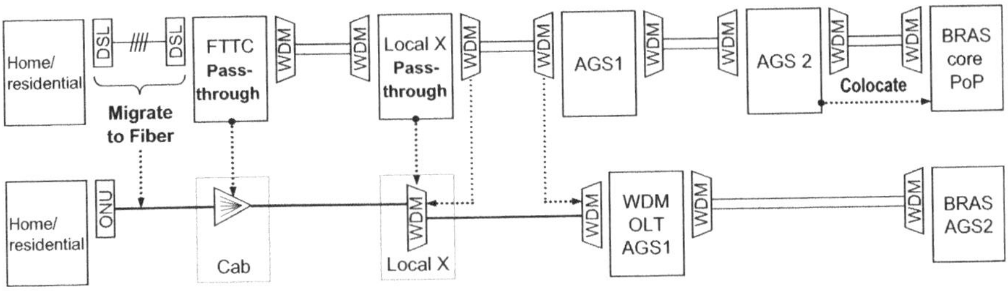

FIGURE 6.60 Metro network consolidation. BRAS is the broadband remote access server.

6.6.1 Delayering of Metro Networks

Metropolitan area networks are hierarchical networks. Up into the first decade of the twenty-first century, a mixture of different transport, aggregation, and switching technologies was used [40]. Today, many metro networks consist of at least two aggregation layers, which are mostly based on Layer-2 (Ethernet) switching. The higher aggregation layer (Aggregation Switch 2, AGS2) connects to a backbone PoP, while the lower layer (AGS1) connects to local exchanges that in turn accommodate access equipment (DSLAMs, MSANs, and PON OLTs). In order to increase cost efficiency and also to reduce power consumption, these networks over time will migrate to a converged network structure with less layers. This is demonstrated in Fig. 6.60.

Network convergence requires long-reach, high-capacity access that is based on (WDM) PON. These systems must be capable to serve $\geq$64, preferably $\geq$256, clients at maximum reach up to 90 km [71]. Accumulated access capacity of the systems is at least 128 Gb/s. Then, the AGS1 can migrate into the respective PON OLT.

Long-reach access allows shutting down a certain number of active sites (local exchanges). This in turn helps to reduce total network power consumption. First, potentially underutilized sites where HVAC is run inefficiently can be shut down. Then, the number of aggregation switches can be reduced. Power consumption of fewer but larger switches (or routers) is lower than for a higher number of smaller switches (routers) with the same total capacity [72]. In addition, site reduction helps saving real estate and other operational costs.

The trend toward active-site consolidation is contradicted by a second trend. UHDTV/3D services are one of the few services that, in the access, may require bit rates of 500 Mb/s and above. Should peer-to-peer unicast become the dominant type of service, this bit rate requirement is required for upstream and downstream. In order to prevent the backhaul and backbone networks from getting flooded with broadband UHDTV streams, it may become necessary to switch (or route) the respective traffic as locally as possible. This requires switching/routing functionality in the OLTs of the long-reach, high-capacity WDM-PON systems, and potentially also in a second layer that contradicts site consolidation.

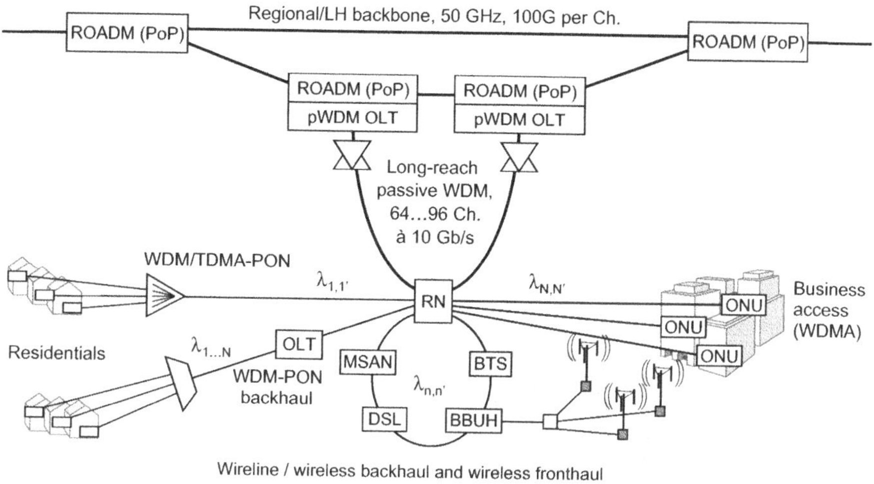

FIGURE 6.61 Metro network with consolidated WDM-PON backhaul and access.

Hence, it is not clear yet what level of site consolidation will be achieved over time. In any case, the consolidated and delayered metropolitan area network will look similar to the one shown in Fig. 6.61.

The network shown in Fig. 6.61 is based on high-capacity, long-reach WDM-PON (or pWDM), which is used for both residential and business access, and also for general purpose backhaul (including variants of Radio-over-Fiber like CPRI). Aggregation and local switching is performed in the WDM-PON OLT and also in the backbone ROADM sites.

6.6.2 Energy Efficiency of WDM

The combined areas of Information and Communications Technologies (ICT) are responsible for a significant amount of the global CO_2 footprint. In 2007, the ICT portion of global CO_2 production was in the range of 2%, with a total of ~0.83 $GtCO_2$ produced [73]. This amount is projected to significantly increase over the next years (Fig. 6.62).

From the global ICT CO_2 footprint, telecommunications infrastructure and devices are roughly in the 25% range, a majority going to PCs, printers, and peripherals. Within telecommunications, a significant part of the CO_2 footprint goes into mobile communications. According to Fig. 6.63, this part is to increase toward 2020.

A more detailed look at the per-sector and per-functionality power consumption shows that wireline telecommunications, and in particular (WDM) transport, is responsible for a minor contribution to total ICT (network) power consumption (Fig. 6.64) [74].

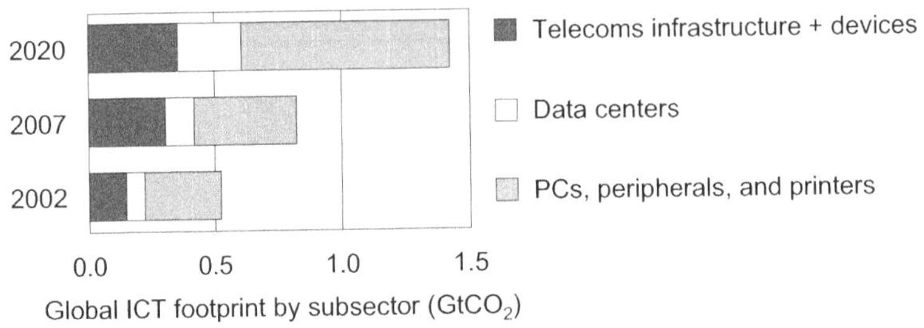

FIGURE 6.62 Global ICT carbon footprint [73].

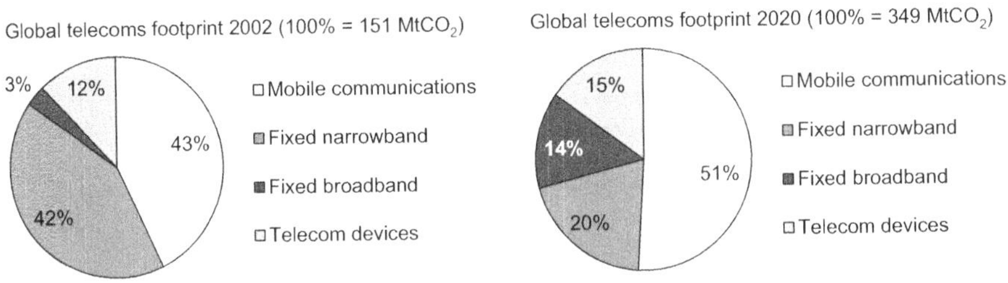

FIGURE 6.63 Global telecommunications carbon footprint [73].

Nonetheless, increased energy efficiency is relevant for (WDM) transport. Due to the exponentially increasing bit rates [75], processing (switching, routing, multiplexing, etc.) and also transport (modulation, framing, skew/deskew, regeneration, etc.) must get significantly more energy efficient per b/s. This becomes clear when looking at the power consumption as a function of the average (residential) access bandwidth (Fig. 6.65) [76].

Although WDM transport is a minor contributor to power consumption, it also must become more energy efficient. Reasons include the bandwidth increase—which does not lead to a decrease of the number of transport interfaces—and the increasing amount of digital processing, which is required with increasing bit rates.

Typical power-consumption figures of an energy-efficient WDM long-haul transport system (as of year 2012) are summarized in Fig. 6.66.

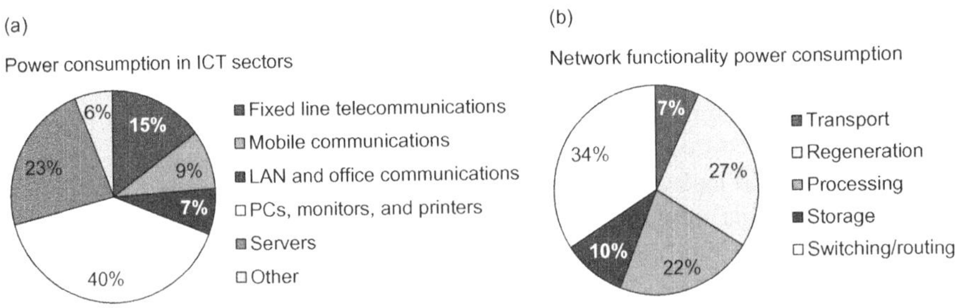

FIGURE 6.64 Power consumption in ICT sectors (a), and per network functionality (b) [74].

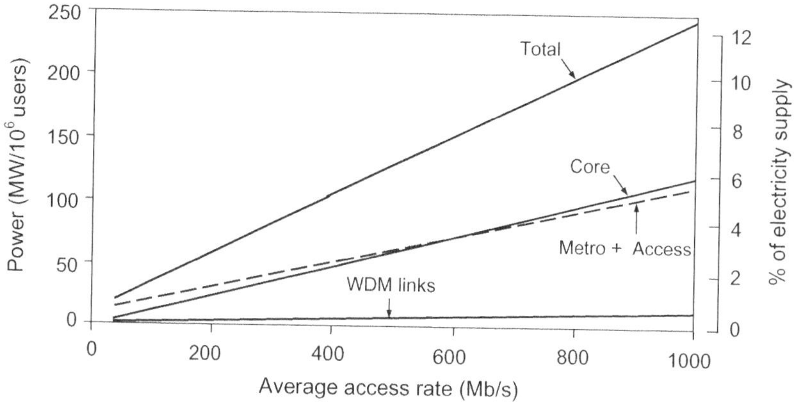

FIGURE 6.65 Relative power consumption in telecommunications networks [76].

The consumption figures in Fig. 6.66 refer to an optically compensated long-haul WDM link (i.e., supporting standard 10 Gb/s as well as coherent 100 Gb/s transmission).

Regarding *total* power consumption (or CO_2 footprint), the contribution of HVAC in the respective PoP or central office has to be added. In large sites (e.g., data centers), the HVAC contribution to total power consumption can be in the range of 25–50% [77].

Large network operators and legislation require increased energy efficiency for all ICT sectors, including WDM transport. Related improvements are sometimes tracked by means of so-called Telecommunications Equipment Energy Efficiency

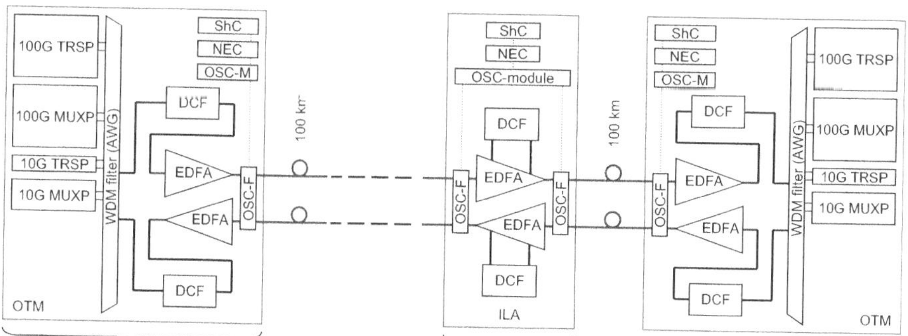

Energy consumption	
100G OTN transponder	~120 W
100G OTN muxponder (10×10G)	~140 W
10G OTN transponder (long-reach)	~30 W
10G OTN muxponder (10-Port flexible)	~50 W
EDFA (single stage)	~20 W
Shelf incl. fans, controller, and OSC	~175 W*

Energy consumption (per ILA)	
EDFA (double stage)	~30 W
Shelf incl. fans, controller, and 2×OSC	~95 W*
Total (2 EDFAs)	~155 W*

* Includes PSU loss/efficiency

FIGURE 6.66 Energy consumption figures of typical WDM long-haul components.

TABLE 6.10 WDM TEEER [78]

Equipment	FoM	N.R.	Achieve N.R. by	Notes
DWDM	Maximum throughput (Gb/s)/ AVG PC (W)	0.32	End of 2012	AVG PC = (PC @ full wavelengths + PC @ 1 wavelength)/2
CWDM		0.48		

PC: power consumption.

Ratings (TEEER) (the ANSI version is called TEER; refer to the Standard ATIS-0600015.02.2009 for transport equipment). Different TEEER is applied by large network operators like NTT in Japan or Verizon in the United States. An example of TEEER for WDM equipment, according to Ref. [78] and to be applied from the end of 2012, is given in Table 6.10. Similar calculations can be performed with ATIS-060015.02.2009, together with the Verizon Technical Purchasing Requirements VZ. TPR.9207 [79].

For relevant WDM configurations [e.g., 80-channel DWDM with fixed or flexible (ROADM) filters, and CWDM with total capacity of 40 Gb/s], Normative References (N.R.) have been defined, which have to be matched by the respective WDM systems. Systems perform the better, the clearer they exceed the N.R. Note that different N.R. have been defined in Ref. [78] for DWDM and CWDM.

Table 6.11 lists the calculated Figures of Merit (FoM) for a specific WDM transport system, the Fiber Service Platform (FSP) 3000 [80].

The configurations analyzed in Table 6.11 are a four-channel CWDM point-to-point configuration with 10 Gb/s per channel using transparent channel cards (no OTN framing, no FEC), an 80-channel point-to-point DWDM system with 10 Gb/s per channel (and also using transparent channel cards), and an 80-channel DWDM system with high-performance OTN 10 Gb/s transponders and Degree-4 (D4) colorless and directionless ROADMs. All configurations are compliant with (i.e., exceeding) the TEEER requirements. The TEEER difference between the latter two configurations can clearly be seen. In a network context, this also means that only functionalities (flexibility, performance, and monitoring capabilities) that really are required should be implemented.

TABLE 6.11 TEEER for WDM System FSP 3000 According to Ref. [78]

	CWDM $4 \times 10G$	DWDM $80 \times 10G$ AWG	DWDM $80 \times 10G$ D4 ROADM
PC @ 1 λ	45 W	80 W	606 W
PC @ full λ	69 W	1681 W	2408 W
AVG PC	57 W	881 W	1507 W
Maximum throughput	40 Gb/s	800 Gb/s	800 Gb/s
N.R.	0.48	0.32	0.32
FoM	0.70	0.91	0.53

TABLE 6.12 Energy Consumption and Further Parameters of ADC/DAC

Technology (nm)	Gates (M)	I/O Speed (Gb/s)	ADC (8 bit)	DAC (8 bit)	Availability
40	100–150	28–32	55–65 GS/s 1.2 W	55–65 GS/s 0.5 W	2011
28	150–200	>32	110–130 GS/s 1.7 W	110–130 GS/s 0.7 W	2013
20	200–250	<100	110–130 GS/s 1.2 W	110–130 GS/s 0.5 W	2015

Increased TEEER contributes to better Green-of-IT [69]. Note that much higher savings regarding CO_2 production are associated with the area of Green-by-IT, that is, savings in carbon footprint enabled by ICT, and especially telecommunications. The telecommunications-related CO_2 abatement for 2020 is estimated to be five times as high as the direct ICT contribution to CO_2 production [73].

Transmission with 100 Gb/s and above in most cases uses electronic compensation of linear and nonlinear effects. The energy consumption can then be driven by the digital processing (done in ASICs or FPGAs), together with the A/D and D/A conversion. If the processing is done in ASICs, energy consumption is in the range of the A/D and D/A converters, depending on the complexity implemented. If it is done in FPGAs, energy consumption can be somewhat higher. Table 6.12 lists consumption figures of several generations of A/D and D/A converters [81,82]. It has to be noted that a general dual-polarized I/Q transmission system requires four converters, one per lane.

According to Fig. 6.63 and Fig. 6.64, broadband WDM transport is a minor contributor to global ICT carbon footprint. Nonetheless, it must become more energy efficient. In order to increase with increasing bit rate demand, attempts to improved energy efficiency must consider every layer of (the development of) WDM transport systems and networks (the generalization to complete networks would have to also include routers, access, and data center gear):

- Power supplies
- HVAC
- TDM, switches, and cross-connects
- ADC/DAC, DSPs, FPGAs, and CPUs
- End-to-end networking aspects like network flexibility/reconfigurability

For each of these layers, energy-efficiency (shut down) modes can be developed. These modes are special operation modes of networking equipment, which allow reducing system energy consumption under specific conditions. Examples are as follows:

- Air flow reduction. Reduce effort needed for cooling, for example, in shelves that are only partly populated, or populated with low-energy cards.

- Reconfiguration, where possible, of DSPs or FPGAs. For example, in EDC, the required tap number can be adapted to the needs, and unused parts of the DSP/FPGA switched off for energy consumption reduction.
- Application of sleep and doze modes. Such modes were developed for ITU-T Recommendation G.987 (XG-PON). Similar mechanisms can be applied to termination systems that are not permanently utilized. Doze and sleep modes differ in deactivating the respective *receiver*, whereas both are periodically deactivating the *transmitters*. They can save up to ~50% energy if the main application is not always on.
- Reconfiguration on a subnetwork level. This relates to subnetworks that are at least partially meshed and that allow certain reconfigurations. In such networks, traffic can be rearranged and reassigned to transport equipment such that other parts of the transport equipment can be shut down completely. This may be allowed in periods with low traffic, for example, overnight. This functionality can be supported by enhanced control plane mechanisms ("energy based routing").

According to Fig. 6.65, *broadband access* is a large contributor to total network energy consumption. Broadband access will be based on WDM, either in the form of WDM-PON or hybrid WDM/TDMA-PON. WDM-PON can be based on direct or coherent detection. Table 6.13 summarizes consumption figures for the respective ONUs and OLTs on a per-client basis.

The figures listed in Table 6.13 will slightly reduce over time with improvements in components' energy consumption. This specially relates to the coherent UDWDM-PON with its digital coherent (heterodyne) processing. Energy consumption can further be reduced by ~50% by applying different cyclic sleeping modes as known from XG-PON.

A detailed analysis of the power consumption of WDM-based PON variants must also consider the optional use of optical amplifiers [preamplifiers and booster amplifiers, based on EDFA (C- and L-band) or TDFA (S-band)], which can be used in the PON OLT. The resulting power consumption figures for the PON variants already discussed earlier are summarized in Fig. 6.67 [83]. Note that common contribution to power consumption from (Ethernet) client interfaces, ONU PSU, and so on have been omitted here.

TABLE 6.13 Energy Consumption of WDM-Based Access Solutions [83]

Technology	Power Consumption (W)
ONU WDM-PON w/tunable LD and direct detection	<4
OLT WDM-PON w/LD array (PIC) and direct detection, per client	<2
ONU coherent UDWDM-PON	<5
OLT coherent UDWDM-PON with digital SCM, per client	<3
ONU hybrid WDM/TDMA-PON, symmetrical 10 Gb/s	<6
OLT hybrid WDM/TDMA-PON, symmetrical 10 Gb/s, per client	<1

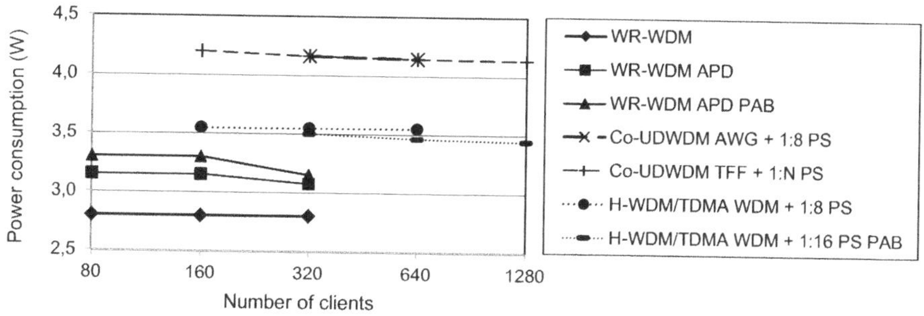

FIGURE 6.67 Per-client, end-to-end transceiver power consumption.

Power consumption of WR-WDM-PON is the lowest. This is attributed to the use of transceivers running at 1 Gb/s without further digital processing. Co-UDWDM-PON has higher power consumption driven by the use of digital coherent transceivers and the related DSP.

REFERENCES

1. CPRI Specification 2013. Available at http://www.cpri.info/spec.html.
2. K. Grobe and H. Hinderthuer, CWDM architecture and applications, Proceedings of 6th ITG Fachtagung Photonische Netze, Leipzig, May 2005, pp. 13–19.
3. J. Aldridge CWDM in metropolitan networks, in H. -J. Thiele and M. Nebeling (eds), Coarse Wavelength Division Multiplexing Technologies and Applications, CRC Press, 2007, pp. 251–267.
4. K. Grobe et al., Flexible WDM solution for DSL backhaul, Proceedings of 7th ITG Fachtagung Photonische Netze, Leipzig, May 2006, pp. 23–28.
5. H.-J. Thiele et al., 160 Gb/s CWDM capacity upgrade using 2.5 Gb/s rated uncooled directly modulated lasers, IEEE Photonics Technol. Lett., Vol. 16, No. 10, 2004, pp. 2389–2391.
6. A. Heckwolf, Employing passive WDM to cost optimise transport network operations, IIR Conference WDM & Next Generation Optical Networking, Cannes, June 2008.
7. J.-P. Elbers and K. Grobe, Passive WDM as unified access solution for the second mile, in: Proceedings of ITG 3rd Fachkonferenz Breitbandversorgung in Deutschland, Berlin, October 2008, pp. 181–186.
8. A.A. de Albuquerque et al., Field trials for fiber access in the EC, IEEE Commun. Mag., February 1994, pp. 40–48.
9. P.W. Shumate, Fiber-to-the-home: 1977–2007, Invited Paper, IEEE J. Lightwave Technol., Vol. 26, No. 9, 2008, pp. 1093–1103.
10. G. Henning, Fiber in the loop – first volume deployment within the OPAL program of the DBP telekom, 5th Conference on Optical/Hybrid Access Networks, 7–9 September 1993.
11. FTTx Market Report, IDATE Consulting & Research, July 2009.
12. C.F. Lam, Passive Optical Networks: Principles and Practice, Academic Press, San Diego, 2007.

13. A. Banerjee et al., Wavelength-division multiplexed passive optical network (WDM-PON) technologies for broadband access: a review [Invited]. J. Opt. Netw., Vol. 4, No. 11, November 2005, pp. 737–758.
14. K. Grobe and J.-P. Elbers, PON in adolescence: from TDMA to WDM-PON, IEEE Commun. Mag., Vol. 46, January 2008, pp. 26–34.
15. I. Garcês et al., Analysis of narrow-FSK downstream modulation in colourless WDM PONs, Electron. Lett., Vol. 43, No. 8, April 2007, 471–472.
16. Ericsson, RZ/IRZ WDM-PON, EU FP7 project OASE, Deliverable D4.2, 2012.
17. G. Talli et al., Modeling of modulation formats for interferometric noise mitigation, IEEE J. Lightwave Technol., Vol. 26, No. 17, 2008, pp. 3190–3198.
18. M. Fujiwara et al., Impact of back reflection on upstream transmission in WDM single fiber loopback access networks, IEEE J. Lightwave Technol., Vol. 24, No. 2, 2006, pp. 740–746.
19. S.H. Cho et al., Reducing the back reflection-induced penalty by using a receiver with an optimized decision level in a reuse single fiber bidirectional WDM-PON, ECOC2007 Berlin 2007, Paper WeP124.
20. J.H. Lee et al., Improved reflection tolerance by near-threshold operation of DFB seed light for an 80 km reach loop-back WDM-PON, OFC2009, SAN Diego, 2009, Paper JThA71.
21. C. Antony et al., Demonstration of a carrier distributed, 8192-split hybrid DWDM-TDMA PON over 124 km field-installed fibers, OFC/NFOEC2010, San Diego, March 2010. Post-Deadline Paper PDPD8.
22. A. Agata et al., Study on ISI mitigation capability of MLSE equalizers in RSOA-based 10 Gb/s WDM PON, ECOC2009, Vienna, September 2009.
23. E. Wong et al., Directly modulated self-seeding reflective semiconductor optical amplifiers as colorless transmitters in wavelength division multiplexed passive optical networks, IEEE J. Lightwave Technol., Vol. 25, No. 1, 2007, pp. 67–74.
24. H. Rohde et al., UDWDM-PON descriptions, FSAN meetings, 2010, 2011, 2012.
25. J.M. Fabrega and J. Prat, New intradyne receiver with electronic-driven phase and polarization diversity, OFC2006, Anaheim, March 2006, Paper JThB45.
26. J. Prat et al., Next generation architectures for optical access, ECOC'06, Cannes, September 2006.
27. K. Grobe et al., Combined reach, client-count, power-consumption, and cost analysis of WDM-based next-generation PON, ECOC2011, Geneva, 2011.
28. S. Guendert, Understanding the performance implications of buffer to buffer credit starvation in a FICON environment: frame pacing delay, CMG'07 International Conference, San Diego, December 2007, Paper 7080.
29. E.A. Pietschner, Datenspiegelung Über Grosse Distanzen, Presentation, EMC^2 Deutschland GmbH, 2002.
30. C. DeCusatis, Handbook of Fiber Optic Data Communication, 3rd ed., Elsevier Academic Press, Burlington, MA, 2008.
31. M. Wallace and L. Webber, The Disaster Recovery Handbook, 2nd ed., AMACOM, 2011.
32. C. Springfield et al., Heterodyne interferometric measurement of the thermo-optic coefficient in single mode fiber, Chin. J. Phys. Vol. 38, pp. 437–442.
33. Optical Time Domain Reflectometer Application, CMA, 5000, Data Sheet, NetTest, 2006. Available at http://www.otdrstore.com/nettest/info/cma5000_otdr_spec_sheet-a4-revc_002.pdf.

34. FTB-7300E—PON FTTx/MDU OTDR, EXFO specifications sheet, 2012. Available at http://documents.exfo.com/specsheets/FTB-7300E-angHR.pdf.
35. J Daemen and V Rijmen, The Design of, Rijndael: AES: The Advanced Encryption Standard. Springer, Berlin, 2002.
36. W. Diffie, and M.E. Hellman, New directions in cryptography, IEEE Trans. Inf. Theory, Vol. 22, No. 6, 1976, pp. 644–654.
37. L. Gillner et al., Field trial of a European self-healing WDM optical network, OFC2000, Baltimore, 2000, Paper TuQI-I.
38. P. Arijs et al., Design of ring and mesh based WDM transport networks, SPIE Opt. Netw. Mag., July, 2000, Vol. 1, pp. 25–40.
39. G. Maier et al., Optical network survivability: protection techniques in the WDM layer, Photonic Netw. Commun., Vol. 4, No. 3/4, 2002, pp. 251–269.
40. A.A.M. Saleh and J.M. Simmons, Architectural principles of optical regional and metropolitan access networks, IEEE J. Lightwave Technol., Vol. 17, No. 12, 1999, pp. 2431–2448.
41. T.E. Stern and K. Bala, Multiwavelength Optical Networks: A Layered Approach. Addison-Wesley Longman, Inc., 1999.
42. V.B. Iversen, Teletraffic engineering and network planning, COM Course 34340, Technical University of Denmark, September 2004. Available at http://de.scribd.com/doc/69761861/109/Engset-distribution/.
43. I. Haxell et al., 2410 km all-optical network field trial with 10 Gb/s DWDM transmission, Raman and tunable dispersion compensation, OFC2000, Baltimore, 2000, postdeadline paper, pp. 290.
44. T.R. Taha and M.J. Ablowitz, Analytical and numerical aspects of certain nonlinear evolution equations: II. numerical, nonlinear Schrödinger equation, J. Comput. Phy., Vol. 55, No. 2, 1984, pp. 203–230.
45. A. Färbert et al., Novel optimized dispersion management scheme for long-haul optical communication systems, Electron. Lett., Vol. 35, No. 20, 1999, pp. 1865–1866.
46. G. Mohs et al., Maximum link length versus data rate for SPM limited systems, ECOC2000, Munich, 2000.
47. C. Fürst et al., Influence of the dispersion map on limitations due to cross-phase modulation in WDM multispan transmission systems, OFC2001, Anaheim, 2001, Paper MF4.
48. F. Forghieri, R.W. Tkach, and A.R. Chraplyvy, Fiber nonlinearities and their impact on transmission systems, in I. P. Kaminow and T. L. Koch. (eds), Optical Fiber Telecommunications, Vol. IIIa, Academic Press, 1997.
49. C. Scheerer et al., SRS crosstalk in preemphasized WDM systems, OFC2000, Baltimore, 2000, Vol. 2, pp. 293–295.
50. G. Mohs et al., Optimized in-line dispersion compensation for self-phase modulation limited transmission, OECC/IOOC2001, Sydney, 2001.
51. R.I. Killey et al., Electronic dispersion compensation by signal predistortion using digital processing and a dual-drive Mach–Zehnder modulator, IEEE Photonics Technol. Lett., Vol. 17, No. 3, 2005, pp. 714–716.
52. Chongjin Xie, Suppression of inter-channel nonlinearities in WDM coherent PDM-QPSK systems using periodic-group-delay dispersion compensators, ECOC2009, Vienna, 2009, Paper P4.08.

53. S. Searcy and S. Tibuleac, System design tradeoffs for XPM mitigation in hybrid 100G-10G networks, OFC2013, Anaheim, March 2013.
54. M. Alfiad et al., Effect of carrier phase estimation for 111 Gbit/s POLMUX-RZ-DQPSK equalization in presence of 10.7 Gbit/s OOK neighbours, LEOS2008, Acapulco, 2008, paper WE1.3 (Invited).
55. R.J. Rapp, Submarine cables: critical infrastructure supplier perspective, 34th COLP Conference, Washington, May 2010.
56. E. Gjesteland, Technical solution and implementation of the Svalbard fibre cable, Telektronikk 3.2004, 2004, pp. 140–152.
57. Y. Shohei et al., The optical submarine repeater and its associated technologies, NEC Tech. J., Vol. 5, No. 1, 2010, pp. 13–17.
58. M. Suyama et al., WDM optical submarine network systems, Fujitsu Sci. Tech. J., Vol. 35, No. 1, July 1999, pp. 34–45.
59. M. Murakami et al., A remote supervisory system based on subcarrier overmodulation for submarine optical amplifier systems, IEEE J. Lightwave Technol., Vol. 14, No. 5, 1996, pp. 671–677.
60. Y. Horiuchi et al., Novel coherent heterodyne optical time domain reflectometry for fault localization of optical amplifier submarine cable systems, IEEE Photonics Technol. Lett., Vol. 2, No. 4, 1990, pp. 291–293.
61. O. Gautheron et al., COTDR performance optimization for amplified transmission systems, IEEE Photonics Technol. Lett., Vol. 9, No. 7, 1997, pp. 1041–1043.
62. S. Whitehead, Submarine cable testing, MW90010A coherent OTDR, Application Note, Anritsu 2012. Available at http://www.anritsu.com/en-AU/Downloads/Application-Notes/Application-Note/DWL8672.aspx.
63. Anritsu MW90010A Coherent OTDR data sheet, 2012.
64. A.A. Huurdeman, The Worldwide History of Telecommunications, John Wiley & Sons, Inc, Hoboken, NJ, 2003.
65. TAT-14 Cable System, 2012. Available at https://www.tat-14.com/tat14/index.jsp.
66. W. George et al., Serial 40G submarine deployments, OFC2009, San Diego, 2009, Paper NTuC1.pdf.
67. C.G. Gruber, CAPEX and OPEX in aggregation and core networks (Invited Talk), OFC2009, San Diego, 2009, Paper OThQ1.pdf.
68. K.von den Hoff et al., Cost reduction in the telecom industry, Arthur D. Little White Paper, 2010. Available at http://www.adlittle.com/downloads/tx_adlreports/ADL_Cost_Reduction_Telecom_Industry.pdf.
69. Y. Somemura, Key roles of green technology for access network systems, OFC2010, San Diego, March 2010, Paper OTuO6.
70. NTT DOCOMO Group CSR Report 2011, 2011. Available at http://www.nttdocomo.com/binary/about/CSR_Report_2011.pdf.
71. FSAN – NG-PON2 Operator Requirements White Paper, Section 3: NG-PON2 Requirements, Draft, April 2011.
72. M. Frey, Costs of computing, ISC'07, Dresden, June 2007.
73. Enabling the low carbon economy in the information age, SMART 2020, The Climate Group on behalf of the Global eSustainability Initiative (GeSI), 2008.

74. G. Koutitas and P. Demestichas, A review of energy efficiency in telecommunication networks, Telfor J., Vol. 2, No. 1, 2010, pp. 2–7.
75. Global mobile data traffic forecast update, 2011–2016, Cisco Visual Networking Index, White Paper, February 2012.
76. J. Baliga et al., Energy consumption of the Internet, COIN-ACOFT2007, Melbourne, 2007.
77. Best practices guide for energy-efficient data center design, U.S. Department of Energy, Revised March 2011.
78. Ecology guideline for the ICT industry (Version 2), ICT Ecology Guideline Council, March, 2011.
79. Verizon NEBS™ compliance: energy efficiency requirements for telecommunications equipment, Verizon Technical Purchasing Requirements VZ.TPR.9205, No. 5, October 2011.
80. FSP 3000, product overview and technical information, 2012. Available at http://www.advaoptical.com/~/media/Resources/Data%20Sheets/FSP_3000.ashx?pdf.
81. Fujitsu ADC/DAC preliminary data, 2011.
82. I. Dedic, 56 Gs/s ADC enabling 100GbE, OFC2010, San Diego, March 2010. OThT6.
83. K. Grobe et al., Cost and energy consumption analysis of advanced WDM-PONs, IEEE Commun. Mag., Vol. 49, No. 2, February 2011, pp. S25–S32.

7

WDM NETWORK MANAGEMENT

In order to achieve cost-efficient operation, telecommunications systems and networks are run under remote supervision and control. This management can be manually operated, and it can partly run fully autonomously. In terms of network operation and management, transport networks are often split into several layers.

7.1 LAYERED NETWORK

Most wavelength division multiplexing (WDM) networks are operated (managed) from a single, centralized Network Management Center (NMC) or Network Operations Center (NOC) [1–3]. The NMC is duplicated for high availability in many cases. Access to the management systems has to be provided to distributed sites, which accommodate network elements. In addition to the telecommunications network itself [which is referred to as the forwarding or transport plane (TP) and the centralized management system [management plane (MP)], there can be a distributed control plane (CP) that performs several discovery, signaling, and routing tasks autonomously].

Each network element of the TP is represented in the CP by its connection controller (CC). The CC typically is part of the network element, and is connected to it via the connection controller interface (CCI). In addition, the network elements and their CCs can be connected to a centralized network management system (NMS) in the MP. The resulting network architecture, including the most relevant CP interfaces, is shown in Fig. 7.1. UNI and NNI are the user-to-network and network-to-network interfaces, respectively. They are described later.

Wavelength Division Multiplexing: A Practical Engineering Guide, First Edition.
Klaus Grobe and Michael Eiselt.

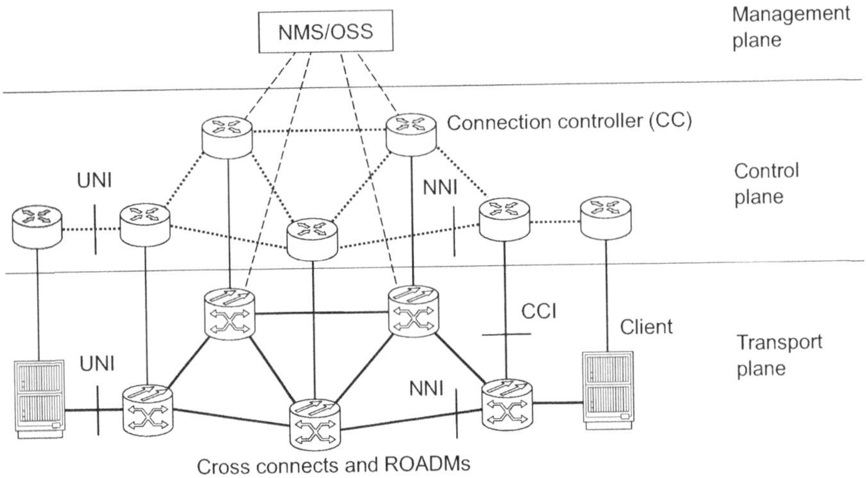

FIGURE 7.1 Forwarding, control, and management plane of a telecommunications network.

As indicated in Fig. 7.1, WDM transport networks are organized into functional groups, referred to as transport plane, management plane, and control plane. In addition, transport (WDM) networks are employed together with Data Communications Networks (DCN) that support operations within each of these groups.

The *transport plane* contains the capabilities that enable (WDM) information transmission across the transport network. The respective equipment interconnects network endpoints, providing access to the transport capabilities to/from client endpoints or other networks. The transmission media represented by the TP consist of the WDM network elements as well as the fiber plant.

The *control plane* describes a collection of capabilities that enables efficient provisioning of transport capabilities between endpoints in the TP. It refers to hardware and software components deployed along with the TP elements, which together enable connections to be provisioned across the TP. The basic CP functions include topology discovery, resource management, path determination and qualification, and connection provisioning and control. As most of these functions are dispersed throughout several network elements, the CP also employs signaling capabilities between network elements.

The *management plane* describes a collection of functionalities related to the overall operation and maintenance of a transport network. The MP comprises software and hardware components that collect, organize, and present information about the operational state of the entire network or a subnetwork (subnet view). It involves processes such as alarm collection and correlation, performance monitoring, software maintenance, and database management. It interfaces with network elements of the TP and may also interface with the CP.

Irrespective of the network planes, large transport networks are also organized in so-called domains.

An *administrative domain* refers to a collection of network resources that belong to a certain administrative entity. Examples are network operator, service provider, and end user (corporate network) domains. Interdomain functions can be implemented via sets of defined interfaces. Administrative domains do not overlap, but rather coexist.

Management domains refer to collections of entities that share certain characteristics within a given administrative domain. Examples include vendor domains, transport layer domains, geographical subnetworks, and security domains. Management domains can overlap with other management domains, but are wholly contained within a single administrative domain.

A *control domain* refers to a collection of components in the CP. These components may be distributed throughout a given network. A control domain is defined for a specific purpose, and may or may not coincide with other control domains. Certain interfaces between CP components operating within various control domains have been defined (see Section 7.6.1).

7.2 MANAGEMENT APPROACHES

Generally, two approaches for network-element (NE) and network management exist: the ITU-T Telecommunications Management Network (TMN) and the Internet Management approach [3–5].

The concept of a TMN is defined by ITU-T Recommendation M.3010. Work on this Recommendation started in 1985 in CCITT. TMN defines a number of concepts that also have relevance for Internet Management. A TMN is conceptually a separate network that interfaces a telecommunications network at several different points. The relationship between a TMN and the telecommunication network that is managed is shown in Fig. 7.2. Interface points between the TMN and the telecommunication network are formed by exchanges and transmission systems. These exchanges and transmission systems are connected via a DCN to one or more *operations systems* (OSs). The OSs perform most of the management functions. These functions may be carried out by human operators, or automatically. It is possible that a single management function will be performed by multiple OSs. In this case, the DCN is used to exchange management information between the OSs. The DCN also connects *workstations* that allow operators to interpret management

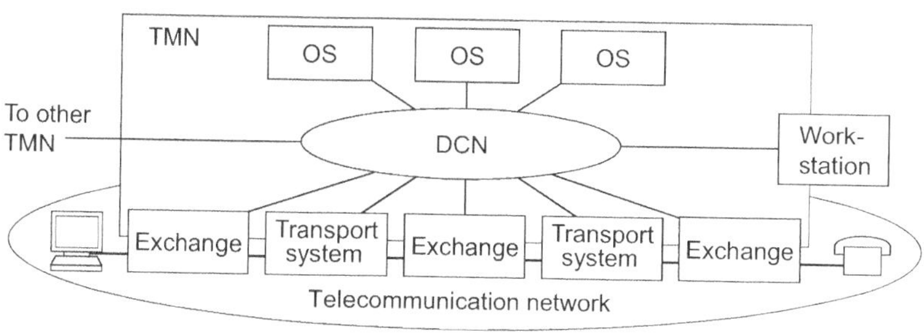

FIGURE 7.2 Concept of the telecommunications management network (TMN).

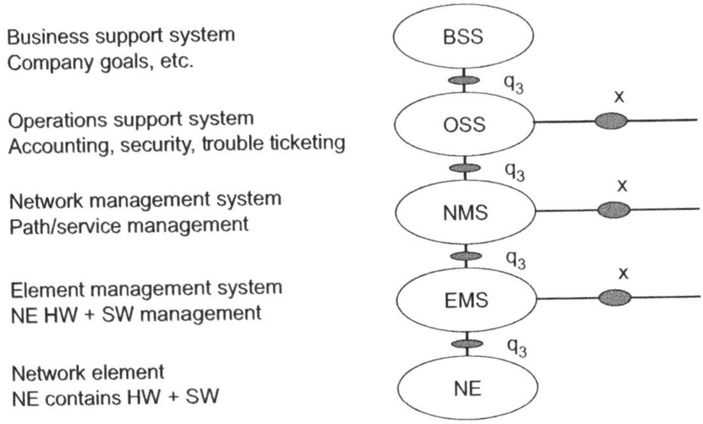

FIGURE 7.3 Layers in TMN with interfaces in-between.

information. Workstations have man–machine interfaces, the definition of these interfaces is outside the scope of TMN (workstations are therefore drawn at the border of the TMN).

An important TMN concept is the logical layered architecture (LLA). This architecture distinguishes between network element, network, service, and business management. Internet Management has traditionally focused on network element and network management only, and needs to extend in the direction of service management to allow the exchange of management information between different operators as well as between customers and operators. This extension is investigated within the Internet Engineering Task Force (IETF). The LLA can be used to identify four management system layers on top of the network elements that are managed. These are the element management system (EMS), network management system, operations support system (OSS), and business support system (BSS). The respective architecture is shown in Fig. 7.3, including the generic interfaces as defined in ITU-T Recommendation M.3010.

There is strong relationship between TMN and OSI management. The manager–agent concept that is used by TMN was originally developed by ISO/IEC. TMN function blocks are comparable to OSI protocol entities. TMN reference points are in OSI terminology comparable to underlying (management information) service providers. This relationship is shown in Fig. 7.4.

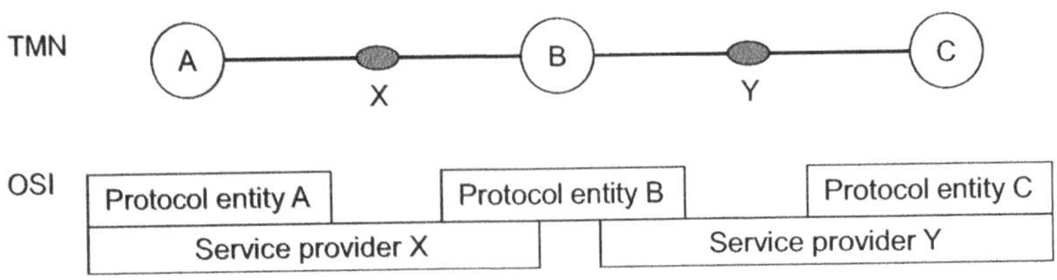

FIGURE 7.4 Comparison between TMN and OSI management.

As opposed to Internet Management, the TMN specifications suggest a conceptual *separation* between the network that is managed (the telecommunication network or transport plane) and the network that transfers the management information (i.e., the DCN). In fact, a TMN DCN can consist of in-band and out-of-band communications channels as well as a completely external DCN, for example, a backup DCN based on ISDN dial-in channels. Internet Management, on the other hand, uses the same components for the network that is managed and the network over which management information is transferred. The idea to introduce a separate network to transfer management information is comparable to the idea to introduce a separate network to exchange signaling information. In this sense, TMN resembles the CCITT Signaling System No. 7 network as defined in ITU-T Recommendation Q.700.

The most important difference between TMN and Internet Management is that the first concentrates on the *specification of management architectures* and the second concentrates on the *implementation of management protocols*. As a result, there are only a limited number of TMN products on the market, whereas there are many commercial and public-domain Internet Management products. Though SNMPv3 (the Internet Simple Network Management Protocol version 3) also has a modular architecture, it is still relatively simple compared to the architectural concepts defined by TMN.

Integration between TMN and any other protocol [Simple Network Management Protocol (SNMP), Transaction Language 1 (TL-1), vendor specific] is usually obtained via a Q Adaptor Function (QAF). The QAF translates between the q_3 reference point (Fig. 7.3), which is implemented as an OSI management protocol stack, and the so-called *m* reference point, which is implemented as an Internet Management (SNMP) protocol stack (and which is not shown in Fig. 7.3). The most critical task of the QAF is to translate between the TMN information model and the one used by the integrated protocol, for example, the Internet's Structure of Management Information (SMI).

7.3 BASIC EMS, NMS, AND OSS FUNCTIONALITY

The basic functions of the different LLA management systems are stated in Fig. 7.3. When excluding the company-specific BSS, these functions can also be summarized by the terms OAM or FCAPS. OAM stands for operations, administration, and maintenance, and it consists of the FCAPS management tasks, where FCAPS stands for fault, configuration, accounting, performance, and security management, respectively. These management tasks relate to all three planes of a telecommunications network, as shown in Fig. 7.1. An overview of the FCAPS management functions is given in Table 7.1.

More specifically, fault management can consist of tasks for continuous NE supervision, the provisioning of alarm severities, display and acknowledgement of alarms including alarm summaries (graphical, table), alarm filtering, sorting and correlation (alarm reduction), and the keeping and export of current and historic (alarm) logs.

TABLE 7.1 FCAPS Functions in Data, Control, and Management Plane

Fault	Configuration	Accounting	Performance	Security
Alarm handling	System turnup	Invoicing	Data collection	Control NE access
Fault detection	Network and service provisioning	Asset management	Report generation	Enable NE functions
Fault correction	Autodiscovery	Work order	Data analysis	Access logs

Configuration management in WDM transmission systems includes general hardware provisioning on the network element level as well as path and service provisioning on the network (or subnet) level. This includes port (WDM and TDM), cross-connect, and ROADM setups as well as necessary power level adaptations in amplifiers. These tasks can be performed by the control plane autonomously. Configuration management further includes the configuration of the DCN, forced (manual and protection) switchover, and the upload, download, and synchronization of network element management information bases (MIBs).

Accounting management includes, besides general accounting (invoicing), tasks like addition and removal of users and asset management (inventory). As such it is spread up into the OSS.

Performance management in WDM systems must enable termination point (TP) data collection with respect to countable parameters like bit rates, bit error rates (BERs), and (port, path, and service) availabilities. The respective logs must usually be available on a per 15 min and 24 h basis. This includes storage, display, and export of the performance data.

Security management includes the provisioning of access levels and access profiles (passwords), user login control, and the generation of operator activity log files. Like accounting, it is distributed up into the OSS.

The fault, configuration, and performance management tasks require continuous supervision of network elements and paths. These tasks rely on basic supervision and signaling mechanisms that are summarized in Fig. 7.5. For fault management, and also in order to maintain laser safety class levels, signaling between network elements in failure cases is required. This includes the concepts of alarm indicator signals (AISs) and remote defect identifiers (RDI) or backward defect identifiers (BDI). AIS is used for *forwarding* the presence of a failure [e.g., an interface or link (fiber and amplifier) failure] to network elements that are subsequent in a link or chain of network elements. This includes the far-end client device. RDI/BDI are used to signal the presence of failures *back* to the originating TP of an established connection. If only the outgoing direction, seen from this TP, was affected, the TP could not get knowledge of such a failure without further signaling, for example, for automated laser shutdown (ALS). The concepts of AIS and RDI/BDI are shown in the upper part of Fig. 7.5.

Relevant OAM tasks are the continuous supervision of the integrity of established paths, and the fault location. The latter is particularly relevant in multihop

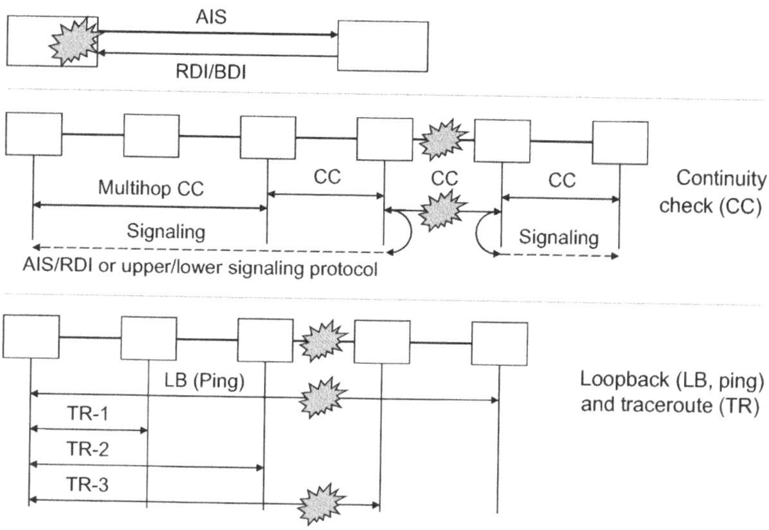

FIGURE 7.5 Basic OAM functions.

(or multispan) links. The related basic OAM functions are continuity check (CC), loopback (LB, also referred to as ping), and traceroute (TR). These basic concepts come from packet networks. In WDM transport, they can, for example, be applied to chains of amplifiers via the DCN or the control plane. They are shown in the middle and bottom parts of Fig. 7.5, respectively.

Continuity check and loopback or ping relate to each other. They aim at sending messages to the far end and receiving an acknowledgement. If a failure occurs in a multihop link, its occurrence is signaled to both ends of the connection. This signaling can be based on AIS and RDI/BDI or a signaling protocol in an upper or lower layer of the network. In failure cases, the failure location can be narrowed down or identified by means of TR. In principle, TR tries to ping all network elements in a multihop link, until it gets no response or the respective RDI/BDI answer.

The concepts of AIS, RDI/BDI, and also CC, ping, and TR are well suited for hierarchically organized multispan WDM links that are based on the Optical Transport Network (OTN) according to ITU-T Recommendation G.709 or that carry SONET/SDH payloads. An example of a SONET/SDH link (with underlying WDM) is shown in Fig. 7.6. This is similarly applicable to a WDM/OTN link, including its client equipment.

Throughout all layers of the transport hierarchy [lower order path-terminating equipment (LOPTE), higher order (HO) PTE, multiplex section terminating equipment (MSTE), and regenerator section terminating equipment (RSTE, i.e., amplifiers)], the respective AIS and RDI/BDI are generated and forwarded, including signaling toward the tributary (Trib) or client equipment. In SONET/SDH and OTN, this signaling is carried in-band in the respective overhead sections. In Fig. 7.6,

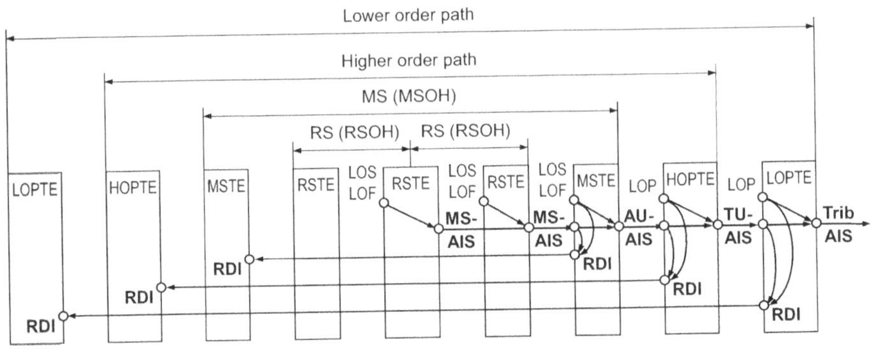

FIGURE 7.6 Principles of RDI (BDI) and AIS in WDM, OTN, and SONET/SDH systems.

MSOH and RSOH refer to the multiplex section and regenerator section overheads, respectively. LOS, LOF, and LOP refer to the possible failure scenarios, that is, loss of signal, loss of frame, and loss of (client) port, respectively.

7.4 DATA COMMUNICATIONS NETWORK

According to the TMN definition, the DCN is a dedicated network for carrying management information, which in turn can consist of any (auto) discovery, performance monitoring, routing, or signaling messages. In TMN, the DCN can be physically based on any internal or external communications channels. This is true also for Generalized Multiprotocol Label Switching (GMPLS) networks. Internal communications channels can be based on in-band or out-of-band channels, where in-band refers to management information being transported together with payload. This can be done as payload, for example, within a dedicated management Ethernet VLAN (virtual LAN), or in a dedicated overhead, for example, a SONET/SDH or OTN embedded communications channel (ECC that uses the SONET/SDH digital communications channels or OTN generalized communications channels, DCC or GCC, respectively).

Internal WDM out-of-band communication usually relies on the Optical Supervisory Channel (OSC). The OSC concept was originally developed for WDM systems in order to provide management access to in-line optical amplifiers, since these do not terminate any other part of the WDM signals they are amplifying.

As a dedicated network potentially connecting many network elements, the DCN must constitute a switched or routed network. Besides switched Ethernet approaches (e.g., for OSC domains), IS–IS and IP routing are used (refer to Section 7.6.3).

Like other Layer-3 networks, the DCN and also large OSC domains require careful administration with regard to IP addresses. This is particularly true since not all OSC systems are routed internally but rather Layer-2 switched. Network elements may require several IP addresses (for external and internal master–slave communications). In addition, OSC domains may grow over time, which may lead to the requirement to separate them into several independently switched domains. In switched OSC implementations, NEs building an OSC backbone are usually part of the same IP management subnetwork (i.e., the main IP addresses reside in the same subnetwork). Alternatively, an OSC network can also be partitioned into different domains or subnetworks. This can be used to address geographical domains or to assign one IP subnetwork address, for example, to a Layer-2 domain.

In switched OSC systems, only a limited number of NEs can be addressed internally. To allow future network growth, dedicated IP addresses can be assigned to separated OSC subnetworks or domains. Then, a Layer-3 router can be added per such domain later easily, at the cost of handling multiple IP subnetworks in the beginning.

In case of the lack of DCN routing protocol support, dynamic routing in case of line breaks and equipment malfunctioning in the DCN will need to be supported outside of the NEs. One solution is HSRP (Hot Standby Router Protocol) that will support a protected default router by using two routers connected to the OSC network, one being a backup router for the other. The switchover to the backup router is invisible to the NEs and NMS, due to a virtual IP address and MAC address taken over by the backup router in case of failure.

Many NEs that are connected to an OSC (or other DCN means) also allow the direct local connection of other equipment, which then can be accessed via the DCN. The respective NEs connected to the DCN directly act as Gateway NEs (GNEs). The NMS/OSS should be connected to such NEs in a resilient way, for example, via a self-healing ring OSC or a DCN with other means for resilience. In particular in a large DCN (e.g., one with several interconnected OSCs), the NMS/OSS should be resiliently connected to central DCN locations.

Generally, the DCN is a routed domain. It uses the IP or IS–IS discovery, signaling, and routing mechanisms. Hence, the DCN view deviates from the transport network. This is shown in Fig. 7.7. The discovery, signaling, and routing

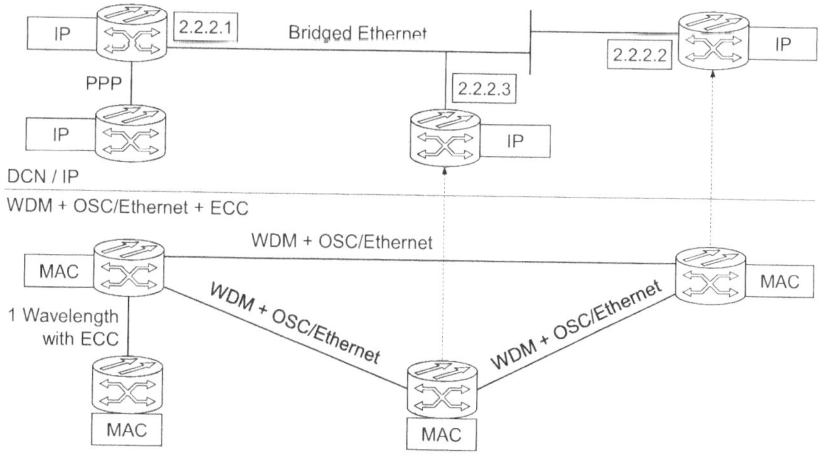

FIGURE 7.7 DCN view and TP view (including relevant addresses) in a WDM network.

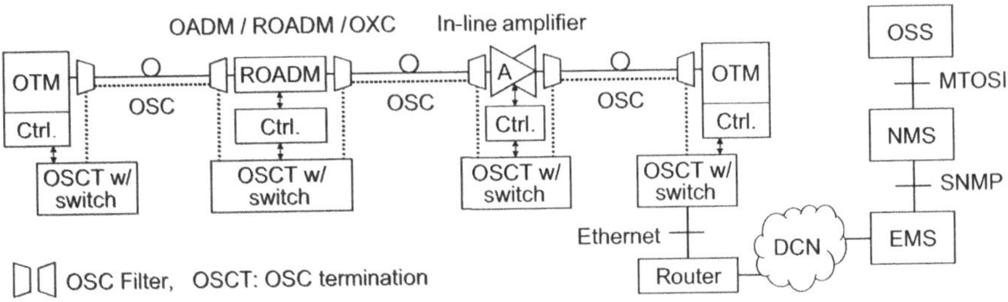

FIGURE 7.8 OSC as DCN in WDM multispan links. OTM: Optical Termination Multiplexer.

mechanisms have been complemented for use in the TP. This is described in the control plane section.

A DCN based on an OSC requires dedicated (1510 or 1625 nm) add–drop filters. It then leads to a certain transmission penalty in long-haul transmission. The OSC itself does not need to be ultralong-haul capable, since it is terminated in every in-line node. However, each OSC filter adds 1–1.5 dB of insertion loss, which then contributes to the span loss. In particular in very-high-loss spans (say, exceeding 35 dB loss), the OSC should be replaced by other means. In ultralong-haul links with high span count exceeding ~20, the accumulated OSC add–drop filter loss also becomes a problem for ultrahigh-speed transmission beyond 100 Gb/s per WDM channel. Alternatives to using an OSC exist in slight intensity modulation of EDFA pumps. This gain modulation can be detected in in-line amplifiers; the concept is known from ultralong-haul subsea (transoceanic) transmissions (Section 6.5.8) [6].

Figure 7.8 shows an example of an OSC domain that is connected via a router to the DCN and the management systems. It also shows interfaces (protocols) between the management system layers, which are relevant today.

Usually, a DCN requires bandwidths of several hundreds of kilobit per second only, unless it has to support a very high number of NEs, or very large databases have to be exchanged during software upload or download. Such bandwidths can easily be provided through the OTN and SONET/SDH ECCs. The exact bandwidths follow from the constant OTN frame sizes and SONET/SDH frame durations and number of ECC overhead bytes and are listed in Table 7.2 for the OTN OTU*x* GCC0 and SONET/SDH section and line DCC, respectively.

TABLE 7.2 Bandwidths of OTN and SONET/SDH ECCs

ECC Type	Bit Rate
OTU3 GCC0	~5.3 Mb/s
OTU2 GCC0	~1.3 Mb/s
OTU1 GCC0	327 kb/s
SONET/SDH section DCC	192 kb/s
SONET/SDH line DCC	576 kb/s

Unlike the OTN and SONET/SDH ECCs, the WDM OSC is carried on a dedicated WDM channel that falls outside the EDFA gain bandwidth. Older versions of ITU-T Recommendation G.692 defined wavelengths of 1510, 1310, and 1480 nm for the OSC. Meanwhile, the most relevant wavelengths for the OSC are 1510 and 1625 nm. The internal structure of an OSC is not standardized, but many OSCs carry full-duplex Fast Ethernet or GbE (e.g., ADVA Optical Networking FSP 3000 WDM transport system).

7.5 MANAGEMENT SYSTEM INTERFACES

As already shown in Figs. 7.3 and 7.8, standardized interfaces for network management are required. Since NEs and their EMS in almost all cases come from the same supplier, their connection can be based on a proprietary interface. All other connections (EMS–NMS, NMS–OSS, NMS–NMS, etc.) must be based on standardized interfaces in order to allow interworking between different management domains or layers, and also integration of third-party equipment into an open NMS with suitable capabilities. Nonetheless, these interworking and integration aspects are difficult at best in real management environments.

Today, the most relevant management interfaces (protocols) are SNMP, TL-1, and MTOSI (Multi-Technology Operations System Interface). The first two protocols (SNMP and TL-1) primarily target EMS–NMS connection, whereas MTOSI is a protocol for NMS–OSS connection.

SNMP is a UDP-based Internet protocol. SNMP is a component of the Internet protocol suite as defined by the IETF [7]. It consists of a set of standards for network management, including an application layer protocol, a database scheme, and a set of data objects. The layer model for SNMP is shown in Fig. 7.9.

SNMP exposes management data in the form of variables that describe the system configuration. These variables can then be queried (and sometimes set) by managing applications. In SNMP, one or more administrative computers (which are called managers) have the task of managing a group of devices or systems in a network. Each managed system executes, at all times, a software component called an agent that reports information via SNMP to the manager. The variables accessible via SNMP are organized in hierarchies. These hierarchies, and other metadata (such as type and description of the variable), are described by management information bases.

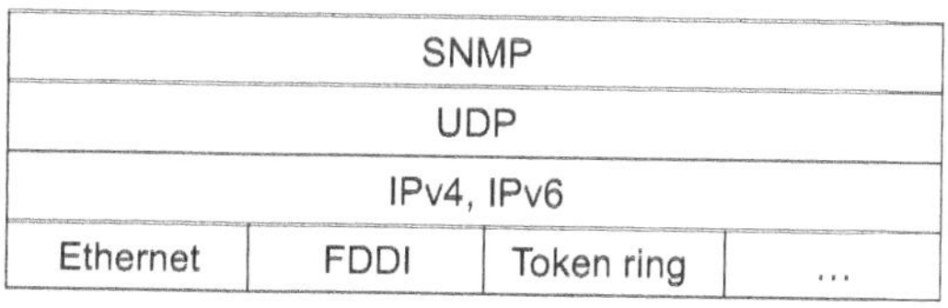

FIGURE 7.9 Layer model for SNMP.

An SNMP-managed network consists of three key components:

- Managed devices
- Agents—software that runs on managed devices
- Network management system—software that runs on the manager(s)

A managed device is a network node that implements an SNMP interface that allows unidirectional (read-only) or bidirectional access to node-specific information by the NMSs. Also referred to as network elements, managed devices can be of any type, including WDM NEs, routers, access servers, switches, hubs, and computer hosts.

An agent is a network management software module that resides on a managed device. It has local knowledge of management information and translates this information to/from SNMP.

An NMS executes applications that monitor and control managed devices. NMSs provide the bulk of the processing and memory resources required for network management. One or more NMSs may exist on any managed network (e.g., for providing resilience).

SNMP itself does not define which variables (information) a managed system should offer. Rather, SNMP uses an extensible design, where the available information is defined by MIBs that describe the structure of the management data of a device. MIBs use a hierarchical namespace containing object identifiers (OID), and the notation defined by ASN.1.

MTOSI is a standard for implementing interfaces between NMSs and/or OSSs. Large service providers and carriers use multiple NMSs and OSSs to manage complex networks. Since the various parts of the network must interact, so must the NMSs/OSSs. This interaction can utilize the MTOSI and is standardized by the TeleManagement Forum (TMF) [8].

The MTOSI standard is a unified open interface that can be used among multiple types of management systems to provide network and service management. It covers all communication technologies (from Layer-1, e.g., WDM or OTN, through higher layer technologies such as VoIP).

Current management systems employ diverse middleware technologies, a reality that is not likely to change in the immediate future. To be widely adopted, MTOSI cannot mandate specific middleware technologies for its implementation. Therefore, the MTOSI interfaces are sufficiently abstract to be middleware neutral, yet rigorous enough that vendors can map them to their middleware of choice. The CCV (common communication vehicle) is the common middleware required to implement MTOSI. It is a middleware abstraction that allows MTOSI interfaces to be bound to different middleware technologies as needed. By using the Web Services Description Language, MTOSI interfaces are composed of logical and physical definitions.

MTOSI uses XML (eXtended Markup Language)-based messaging. XML is a widely accepted technology. MTOSI also provides rules for versioning and vendor extensions to XML messages. Hence, the server and consumer application ends of the interfaces can be upgraded independently. Also, when several vendors' equipment is deployed, it is possible to manage proprietary extensions in a consistent manner.

MTOSI uses standard communication patterns to support business activities that can be implemented by a range of IT platforms and transport protocols. The underlying platforms can therefore be changed without change of the applications. This allows service providers to more easily replace existing management systems.

TL-1 is a widely used management protocol in telecommunications. It is used in most of the broadband and access networks, especially in North America. It is increasingly being used worldwide for newer applications. It is also a preferred command line interface (CLI) for telecommunications network elements. TL-1 was first defined by Bellcore (now Telcordia) in 1984 as a standard man–machine language to manage network elements for the regional Bell operating companies (RBOCs). It is based on the ITU-T Z.300 series man–machine language Recommendations. TL-1 was designed as a standard protocol readable by machines as well as humans to replace the diverse ASCII-based protocols used by different NE vendors. It is extensible to incorporate vendor-specific commands. It is defined in Generic Requirements document GR-831-CORE.

TL-1 is a cross-vendor, cross-technology man–machine language, and is widely used to manage optical (SONET/SDH and WDM) infrastructure. TL-1 is used in the input and output messages that pass between OSs and NEs. Operations domains such as surveillance, administration, and testing use TL-1 messages to accomplish specific functions between the OS and the NE.

The TL-1 language consists of four kinds of messages: input, output/response, acknowledgement, and autonomous messages. Input messages are sent by the user or the OSS. Output/response messages are the reply sent by the NE in response to an input message. Acknowledgment messages are the acknowledgment of the receipt of a TL-1 input message and are sent if the respective response message will be delayed by more than 2 s. Autonomous messages are asynchronous events or alarms messages sent by the NE.

7.6 CONTROL PLANE

WDM networks with reconfigurable and tunable components and wavelength grid enable a high degree of flexibility. This degree of network reconfigurability poses challenges regarding management, control, and monitoring, which were not seen in static WDM networks. Further, new functions in WDM networks like restoration, and also a management and control unification across several networks and technology domains are required.

Control plane technologies are one way of controlling network flexibility. Main advantages are the capability to enable restoration while guaranteeing certain QoS (quality of service) through constraint-based routing, automated hardware, topology, and service discovery, and the integration of electrical (IP/MPLS) and optical [optical channel (OCh) and Optical Multiplex Section (OMS)] technologies.

Referring to MPLS, this CP is also called Generalized Multiprotocol Label Switching. It intends to control and manage the relevant Layer-2 and transport technologies:

- IP/MPLS
- Ethernet
- OTN OCh and OMS
- WDM OCh/fibers (OMS)

The CP architectures and mechanisms have been defined by several standards development organizations, including ITU-T, IETF, and OIF (Optical Internetworking Forum). For a complete treatment of these technologies, refer to the respective standardization references and Ref. [9].

7.6.1 Control Plane Interfaces

The CP relies on standardized interfaces. These were defined by the OIF, based on the ITU-T ASON (automatically switched optical network) architectural model and by using extensions to Internet Engineering Task Force GMPLS signaling and routing. The interfaces enable end-to-end provisioning across multiple network layers and domains. The use of domains and logical reference points separates CP communication and creates an overlay model (see Section 7.6.2). This approach is consistent with the ITU-T G.8080 ASON Recommendation. It is shown in Fig. 7.10.

The UNI defines a CP relationship between a user of a network and the network itself. It defines the interface through which a client (e.g., an IP/MPLS router) can access a GMPLS network. The UNI identifies a network reference point that provides a collection of CP functionalities relevant to the creation and management of network services. This includes resource discovery, connection control, and call control. No routing functions are associated with the UNI reference point. The UNI splits into client side (UNI-C) and network side (UNI-N).

The *internal network-to-network interface* (I-NNI) defines a relationship between two CP components operating within the same control domain. The I-NNI identifies a network reference point that provides CP functionalities relevant to supporting requests for network services. This includes resource discovery, connection control, and connection routing.

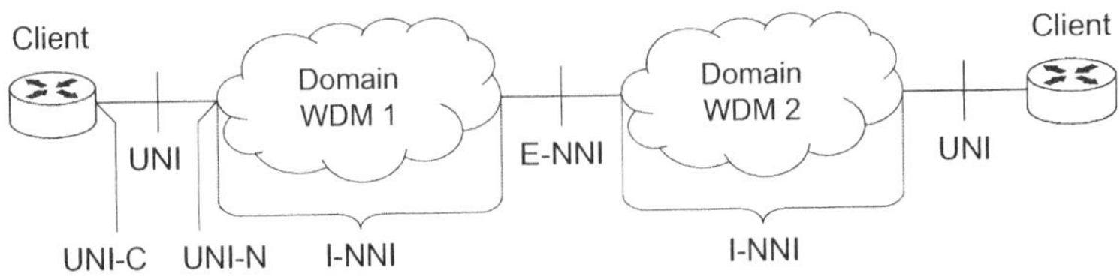

FIGURE 7.10 OIF reference model.

The *external network-to-network interface* (E-NNI) defines a relationship between two CP components operating within different control domains. It identifies a network reference point that provides a collection of CP functionality relevant to supporting network services, which cross two or more control domains. This includes resource discovery, connection control, connection routing, and call control.

7.6.2 Migration Scenarios

Two architectural models for transport networks are commonly discussed: one that favors routers and another that favors optical switches. These models are referred to as overlay model and peer-to-peer model, respectively [9]. The OIF reference points and the IETF GMPLS protocols can be assigned to both the models.

In the overlay model, the optical network provides connection services to IP routers and other client devices attached to the network through a UNI. Different layers of the network remain isolated from each other, but dynamic provisioning of bandwidth is made possible (through the E-NNI). Routers or switches request connections, and the optical network either grants or denies them. The requests can be quite detailed, for example, asking for a certain bandwidth with a particular grade of restoration. The requesting devices have no view of the network. Flexible WDM networks, together with large MPLS deployments, have a strong affinity to the overlay model. It is shown in the upper part of Fig. 7.11.

In the peer-to-peer model, optical switches and routers act as peers, using a uniform and unified control plane to establish label-switched paths (LSPs) across these devices, with complete knowledge of network resources. In this model, there is little or no distinction among UNI, NNI, and router–router (MPLS) control planes; all network elements are direct peers and fully aware of topology and resources. This

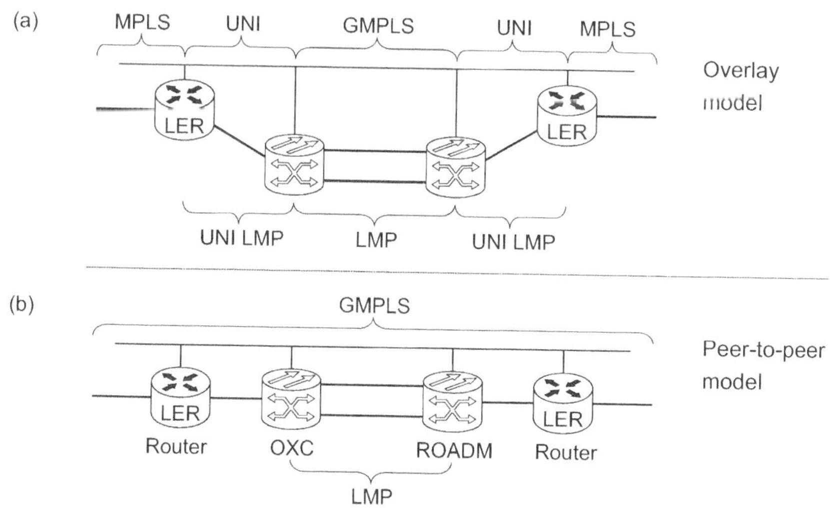

FIGURE 7.11 GMPLS migration: overlay model (a) and peer-to-peer model (b).

is a key distinction as compared to the overlay model. In the peer-to-peer model, a single CP instance can span multiple technologies/network elements, provided that the CP supports each of the technologies. This allows network operators to create a single network domain composed of different network elements, as compared to the overlay model, where the WDM network creates at least one separated domain. The peer-to-peer model is shown in Fig. 7.11b.

Both the models can be combined into a hybrid solution. From the overlay model, the hybrid takes the support for multiple administrative domains. From the peer-to-peer model, the hybrid takes the support for heterogeneous technologies within a single domain. It is likely that migration toward GMPLS will start with the overlay or hybrid model, whereas in the mid- to long-term networks will migrate toward the peer-to-peer model.

7.6.3 GMPLS Protocols

GMPLS separates the control plane and the transport plane. In addition, it also separates the CP into two parts: the signaling plane containing the signaling protocols, and the routing plane containing the routing protocols. Like MPLS, GMPLS uses modified versions of signaling and routing protocols. In addition, it makes use of the Link Management Protocol (LMP). Figure 7.12 gives an overview of the GMPLS IP protocol stack.

The GMPLS routing protocols are OSPF-TE (open shortest path first with traffic engineering) and IS–IS-TE (Intermediate-System-to-Intermediate-System with traffic engineering) routing. These are the routing protocols for both autodiscovery of the network topology and advertising resource availabilities. RSVP-TE and CR-LDP (Resource Reservation Protocol and Label Distribution Protocol, both for Traffic Engineering and Constraint-based Routing, respectively) are the signaling protocols for the establishment of traffic-engineered LSPs. The Link Management Protocol is used for control channel management, link connectivity verification, link property correlation, and fault isolation.

RSVP is a protocol that defines procedures for signaling QoS requirements and reserving the necessary resources for a router to provide a requested service to all nodes along a data path. RSVP is not a routing protocol; it rather works in conjunction with routing protocols. Routing protocols determine where packets

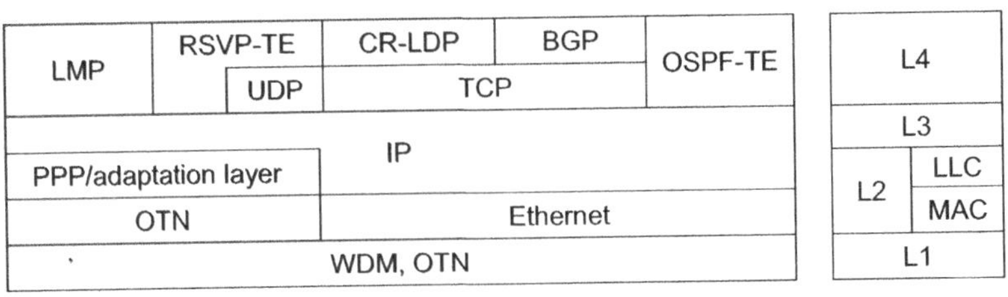

FIGURE 7.12 GMPLS IP protocol stack.

get forwarded. RSVP is concerned with the QoS of those packets that are forwarded in accordance with the routing protocol.

Reservation requests for a flow follow the same path through the network as the data comprising the flow. RSVP reservations are unidirectional in nature, and the source initiates the reservation procedure by transmitting a path message containing a traffic specification (Tspec) object. Tspec describes the source traffic in terms of peak data rate, average data rate, burst size, and minimum/maximum packet sizes.

RSVP allows the use of source routing where the ingress router determines the complete path through the network. The ingress router can use a constrained shortest path first (CSPF) calculator to determine a path to the destination, ensuring that any QoS and shared risk link group (SRLG) requirements are met. The resulting path is then used to establish the LSP.

RSVP-TE is a set of traffic engineering extensions to RSVP. RSVP-TE extensions enable RSVP to be used for traffic engineering in MPLS environments. The primary extensions add support for assigning MPLS labels and specifying explicit paths as a sequence of loose and strict routes. These extensions are supported by including label request and explicit route objects in the path message. A destination responds to a label request by including a label object in its reserve message. Labels are then subsequently assigned at each node the reserve message traverses. Thus, RSVP-TE operates in downstream-on-demand label advertisement mode with ordered LSP control.

RSVP-TE is used to establish MPLS LSPs when there are traffic engineering requirements. It is mainly used to provide QoS and load balancing across the network core, and includes the ability to control all-optical networks. The optical extensions include the ability to signal optical wavelengths and shared risk link groups, as well as bandwidth, latency, and other link characteristics.

Further extensions to RSVP-TE include the establishment of bidirectional LSPs that allow restoration schemes required in nonpacket-switched networks, for example, TDM.

OSPF is an Interior Gateway Protocol (IGP) used for routing between routers belonging to a single autonomous system or administrative domain. OSPF uses link-state technology in which routers send each other information about the direct connections and links that they have to other routers. Each OSPF router maintains an identical database describing the autonomous system's topology. From this database, a routing table is calculated by constructing a shortest path tree. OSPF recalculates routes in case of topological changes. It provides support for equal-cost multipath. Area routing is also provided, enabling routing protection and a reduction in routing protocol traffic. In addition, all OSPF routing protocol exchanges are authenticated.

IS–IS routing is a protocol used by network devices (routers) to determine the best route for datagrams through a packet-switched network. The protocol was defined in ISO/IEC 10589:2002 as an international standard within the OSI reference design. IS–IS is not an Internet standard, however IETF republished the protocol in RFC 1142 for the Internet community. It behaves similar to OSPF.

IS–IS is an Interior Gateway Protocol that is intended for use within an administrative domain. It is not intended for routing between autonomous systems, which is left to exterior gateway protocols such as Border Gateway Protocol (BGP).

IS–IS is a link-state routing protocol that operates by flooding link-state information throughout a network of routers. Each router then independently builds a database of the network's topology. Packets (datagrams) are forwarded based on the best topological path through the network to the destination. Together with Connectionless Network Protocol (CLNP), End-System-to-Intermediate-System routing (ES-IS), and Interdomain Routing Protocol (IDRP), IS–IS allows routing through entire networks.

OSPF-TE and *IS–IS-TE* with traffic engineering extensions allow accounting for optical resources, for example, fibers or wavelengths. Furthermore, hierarchical LSPs are supported, which allow nested or tunneled sets of LSPs of different media, carrying a mix of traffic. This is helpful when nesting multiple low-speed LSPs in a single high-speed LSP. For example, an optical LSP can be established with an OTU2 bandwidth. Nesting numerous STS-1/VC-4 TDM or 1000bT can be accomplished within that single optical LSP. The extensions also provide optimization of internal databases by allowing link aggregation and unnumbered links having no IP addresses (e.g., TDM links).

A relevant addition to GMPLS over MPLS is the *Link Management Protocol*. LMP is used to establish, release, and manage connections between two adjacent GMPLS-capable nodes. LMP is particularly useful in optical networks when large pools of wavelengths are present between two connected nodes.

LMP separates data and control channels, which can each be protected. Multiple control channels between two nodes allow backups and graceful switchover from one node to another, thereby maintaining the reliability and integrity of the network by protecting signaling messages. LMP helps with link fault localization that is negotiated during the initial "Hello" exchange. It provides verification of physical connectivity between two neighboring nodes, which reduces the probability of error in provisioning services.

7.6.4 Control Plane Mechanisms

7.6.4.1 Automated Network Discovery In order to be able to determine and implement paths within a given WDM network, the topology and resource availability state of the network must be known. The CP includes automated discovery processes, ensuring the respective information is correctly formed and maintained as the network grows and changes.

A common approach to automating network discovery is to extend link-state routing protocols as used in the DCN (OSPF or IS–IS, see Section 7.6.3) into the transport layer. These protocols provide basic capabilities for discovering link adjacencies and distributing connectivity information. In the DCN, this information is used to determine and distribute datagram routing information.

Network discovery within the transport layer has been achieved through traffic engineering extensions into the link-state protocols, resulting in OSPF-TE and IS–IS-TE. These extensions contain the concepts of TE links and TE routers, which correspond to links and nodes in the transport network. Standard link-state protocol mechanisms are used to exchange the respective information amongst the CP components, enabling elements to build a complete view of the transport network topology. The TE extensions also include information regarding resource availability and switching capabilities of the links and nodes.

The TE-related entities are described and managed separately from information gathered relating to the DCN, as transport network and DCNs generally are not fully congruent.

7.6.4.2 Path Computation Provisioning connections between endpoints in a transport network requires that a path be found between these endpoints, which satisfies the QoS criteria required by the client, and which fits the current resource utilization and physical constraints of the network itself. In large (meshed) transport networks, this requires automation.

Under certain network conditions, it is possible to compute paths for a connection between a set of network endpoints using graph-processing algorithms, where TE links and routers are treated as nodes and arcs in an abstract weighted graph. The optimal path through such a graph can be computed using standard shortest path first (SPF) algorithms [10,11].

WDM networks are inherently constrained in several ways. ROADMs lead to node models, which require wavelength continuity across network interfaces. In turn, networks without flexible optical switching capabilities at all nodes introduce constraints with regard to channel placement and utilization (i.e., blocking).

WDM transport is also inherently analog; its resources are subject to physical constraints including OSNR, CD, polarization-mode dispersion (PMD), cross talk, nonlinearity, and other effects. Automated path computation must take these constraints into account. In addition, WDM networks often require resilience in the optical layer (not the client layer). Then, path computation must be able to consider diversity such that, for example, working and protection paths do not share common elements like amplifiers or fibers.

As a result, automated WDM path determination requires complements beyond standard SPF. One example is constrained shortest path first (CSFP) that considers the analog link constraints [12].

End-to-end path computation is mostly performed in a distinct CP component. Individual network elements can thus offload the computation to a remote component, relieving them from the necessity to gather the end-to-end topology and constraints information. The respective component has been standardized by the IETF in the form of the *path computation element* (PCE) architecture and framework. The related RFCs are RFC4655, RFC4657, RFC4927, and RFC5440. This work defines a client–server architecture, where path computation clients (PCC) can generate requests for path computation services from PCEs. One scenario is shown in Fig. 7.13.

A PCE Protocol (PCEP) has been defined for interoperability between PCCs and PCEs. PCEs do not need to have full network view. In that case, they can act as PCCs to other PCEs.

7.6.4.3 Connection Control Once a feasible path has been determined in a WDM network, it needs to be implemented. Then, resource management operations need to

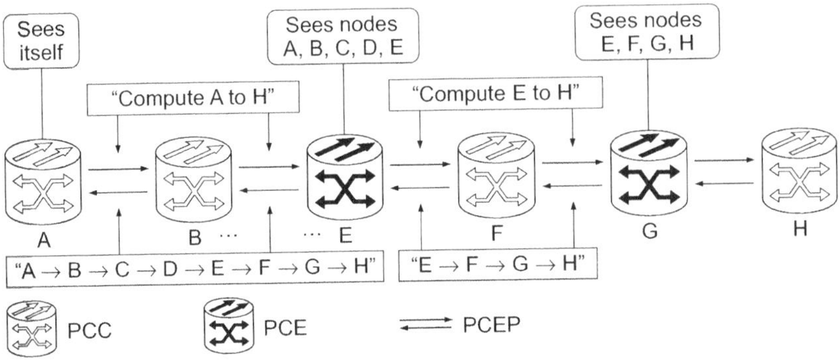

FIGURE 7.13 Path computation with several PCEs.

be performed on a set of network elements. These operations are distributed throughout the network, and affect equipment along specific paths. The resource management coordination is often delegated to an interelement signaling controller function that can be based on protocols like RSVP-TE.

RSVP-TE executes most signaling operations in a hop-by-hop manner, meaning that connection control requests are processed sequentially by each element along a given end-to-end path. If errors are encountered along the path, they are communicated back to the initiator of the request. A WDM path connection setup scenario is shown in Fig. 7.14.

When used in transport networks, signaling protocols must be able to operate out-of-band, as DCN and transport network topology may not be congruent. The protocols can also be complemented to support functions like end-to-end channel equalization along chains of amplifiers, or service restoration.

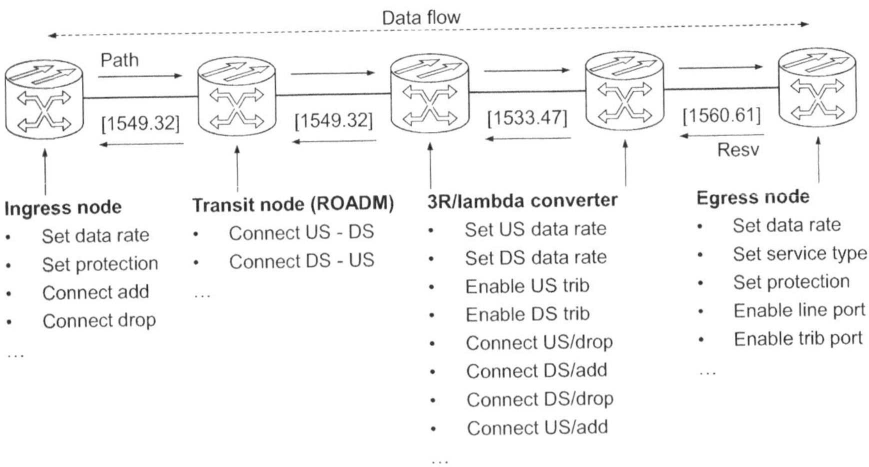

FIGURE 7.14 Path setup. US: upstream, DS: downstream.

7.6.4.4 Resource Management Resource management refers to the ability to discover, configure, and monitor the software and hardware components within a network element. Any software or hardware component of an optical element that is involved in the transmission of a wavelength must be discoverable and configurable.

Resource management converts the service request parameters associated with a given connection request into a set of local (and possibly proprietary) software and hardware configuration operations necessary to serve the request at the local node.

7.6.4.5 Entity Management Entity management refers to automation capabilities of functions that relate to transport network topology identification, parameterization, and monitoring. Entity management is sometimes referred to as link management, as one of the primary entities of interest in a given transport topology are the links between network elements.

Entity management often includes link adjacency discovery mechanisms, where endpoint identifiers of links in a transport network are identified and correlated. It often also includes a transport entity capability exchange mechanism. Here, various transport-related attributes of a given transport link are exchanged between adjacent elements. These functions are relevant to supporting higher level topology discovery functions, for example, in the context of resilience. They are also used for automatically identifying configuration errors in the TP.

Several implementation options for entity management exist. In transport networks with in-band channels, the CP may utilize these in order to discover and manage links. CP implementations that operate out-of-band may utilize specific extensions to existing protocols, or even entirely separate protocols. LMP is an example of such a separate protocol used for entity management.

7.6.4.6 Call Control Call control is an optional CP function. A call refers to the representation of a service provided by a network to the users of that network via some interface. Creation of a call may cause the creation of a certain number of connections within the network. The related separation between call control and connection control provides an abstraction between the services provided by a network and the mechanisms by which that service is actually provided.

Call control and connection control information may be processed by shared components and signaled via shared protocols. Alternatively, they can be processed by different components, but still signaled via shared protocols, or they can be processed and signaled via separate components each.

Calls are considered to progress through three basic phases during their lifetime. A call is in the establishment phase when signaling messages are being exchanged between the network and the client in order to validate and agree to the specific service characteristics of the call. Establishing a call may also trigger creation of new connections. A call is in the active phase when traffic is being exchanged between the client sites. During this phase, certain aspects of the call may be modified. Finally, a call is in the release phase when signaling messages are being exchanged between the network and the client in order to end the call. Ending the call typically releases any connections created within the network to support it, which are no longer needed.

REFERENCES

1. R.A. Weingarten, An integrated approach to centralized communications network management, IBM Syst. J., Vol. 18, No. 4, 1979, pp. 484–506.
2. I.T. Frisch, M. Malek, and S.S. Panwar, Network Management and Control, Vol. 2, Plenum Press, New York, 1994,
3. Khalil ur rehman Laghari et al., Analysis of telecommunication management technologies, Int. J. Comput. Sci. Inform. Technol., Vol. 1, No. 2, 2009, pp. 152–166.
4. R.H. Glitho and S. Hayes, Telecommunications management network: vision vs. reality, IEEE Commun. Mag., March 1995, pp. 47–52.
5. Y. Yemini, A critical survey of network management protocols and standards, in S. Aidarous and T. Plevyak (eds), Telecommunications Management into the 21st Century, IEEE Press, 1994.
6. M. Murakami et al., A remote supervisory system based on subcarrier overmodulation for submarine optical amplifier systems, IEEE J. Lightwave Technol., Vol. 14, No. 5, 1996, pp. 671–677.
7. A. Pras, SNMP: Status and Application for LAN/MAN Management, Tutorial, 1996. Available at http://wwwhome.cs.utwente.nl/~pras/presentations/ieee.pdf.
8. F. Caruso et al., Emerging industry standard for managing next generation transport networks: TMF MTOSI, 10th IEEE/IFIP Network Operations and Management Symposium 2006 (NOMS 2006), Vancouver, April 2006, pp. 1–15.
9. A. Farrel and I. Bryskin, GMPLS: Architecture and Applications, Academic Press, 2006.
10. E.W. Dijkstra, A note on two problems in connexion with graphs, Numer. Math., Vol. 1, 1959, pp. 269–271.
11. T.H. Cormen et al., Introduction to Algorithms, 2nd edn, MIT Press, 2001.
12. M. Ziegelmann, Constrained Shortest Path and Related Problems. Constrained Network Optimization, VDM Verlag Dr. Müller, 2007.

8

SELECTED NETWORK ISSUES

8.1 OVERVIEW OF THE OPTICAL TRANSPORT NETWORK

8.1.1 OTN Layers

The Optical Transport Network (OTN) is specified by ITU-T in various Recommendations (e.g., G.872 on architecture, G.709 on frames and formats, and G.798 on atomic functions and processes). It is sometimes also referred to as Optical Transport Hierarchy (OTH). It combines time-domain multiplexing (TDM) and wavelength division multiplexing (WDM) into a common transport system. The TDM part is hierarchically structured similar to Synchronous Optical Network (SONET)/Synchronous Digital Hierarchy (SDH). It is used either as an end-to-end networking entity or as a point-to-point structure between 3R regeneration points for OAM (operations, administration, and maintenance) purposes.

Optical channels (OCh) form the basis of the TDM part of the OTN. They are structured in hierarchical levels called Optical Channel Payload Unit (OPU), Optical Channel Data Unit (ODU), and Optical Channel Transport Unit (OTU).

The ODU signals are defined as the end-to-end networking entities. They are comparable to SONET/SDH virtual containers. Eight ODU signals have been defined in G.709. They address different bit rates and mapping schemes. Definitions covering bit rates according to clients ranging from GbE to 100 GbE exist. The nominal ODU*k* rates are approximately 1.25 Gb/s (ODU0), 2.50 Gb/s (ODU1), 10.04 Gb/s (ODU2), 40.32 Gb/s (ODU3), and 104.79 Gb/s (ODU4). For 10 GbE

Wavelength Division Multiplexing: A Practical Engineering Guide, First Edition.
Klaus Grobe and Michael Eiselt.

LAN PHY services (which are the clear majority of all deployed 10 GbE interfaces), an "overclocked" ODU2e with 10.40 Gb/s has also been defined. In addition, two ODUflex signals for constant bit rate (CBR) clients and clients that are mapped via the Generic Framing Procedure (GFP) (according to ITU-T Recommendation G.7041) have been specified. Their nominal bit rates are 239/238 × CBR client signal bit rate and $n \times 1.25$ Gb/s (with integer n), respectively. Jitter tolerances are ±20 ppm for most ODUs, and ±100 ppm for ODU2e (due to relaxed Ethernet jitter specifications) and ODUflex for CBR services.

The layered structure of OTN is shown in Fig. 8.1.

The OTN point-to-point single-lambda frame structure is called OTU. It is comparable to SONET/SDH section signals. The nominal OTU*k* rates are approximately 2.67 Gb/s (OTU1), 10.71 Gb/s (OTU2), 43.02 Gb/s (OTU3), and 111.81 Gb/s (OTU4). The OTU*k* frame structure contains 4 × 4080 bytes regardless of the bit rate. Therefore, the frame duration is not constant (unlike SONET/SDH, where it is constant).

Multiplexing of several wavelengths each carrying one OCh by means of WDM creates the Optical Multiplex Section (OMS) and the Optical Transport Section (OTS). The OMS and OTS can be controlled by an Optical Supervisory Channel (OSC, that is, a dedicated wavelength carrying supervision and management information only), as indicated in Fig. 8.1. The OMS refers to sections between optical multiplexer and demultiplexer, and the OTS to sections between optical amplifiers, respectively [1]. This is shown in Fig. 8.2.

Figure 8.2 also shows two OTN interfaces: IrDI (interdomain interface) and IaDI (intradomain interface). These interfaces are defined in ITU-T Recommendation G.872. They allow interworking between (IrDI) and within (IaDI) network domains. The IrDI interfaces are defined with 3R processing at each end of the interface.

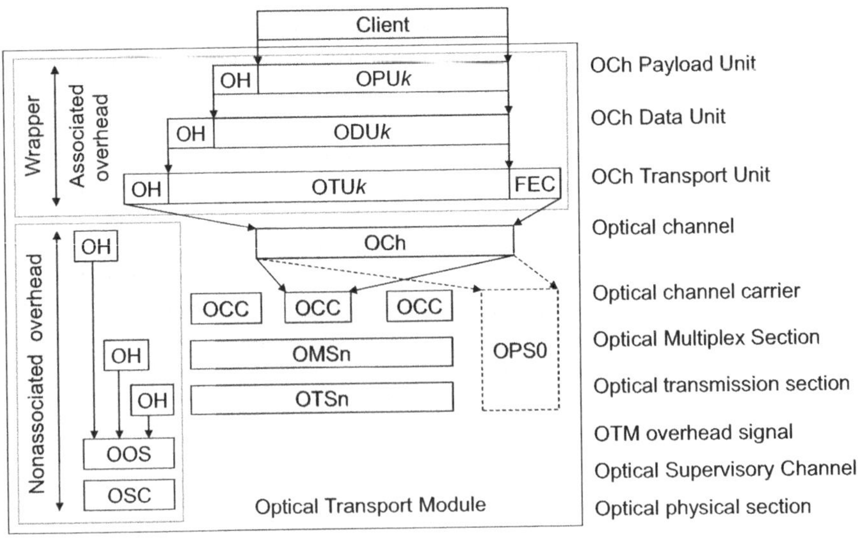

FIGURE 8.1 OTN layering and monitoring. OH: overhead.

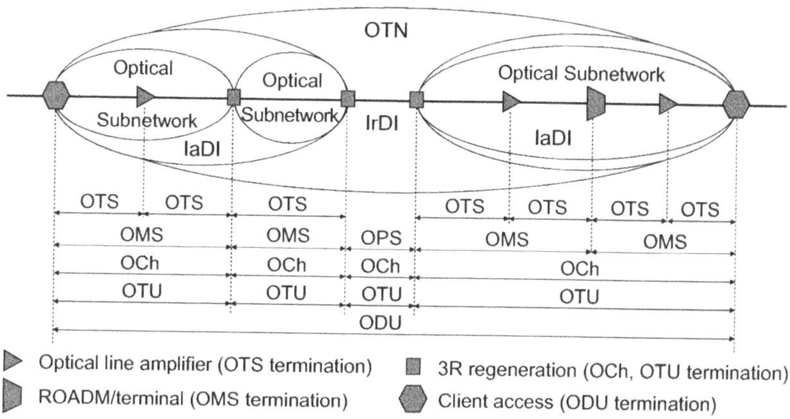

FIGURE 8.2 OTN network layers and interfaces.

So-called Optical Transport Modules (OTM) with full or reduced OTN functionality are defined in G.709. Reduction of functionality refers to the OCh that provides transparent network connections between 3R regeneration points. The reduced definitions currently apply to the IrDI.

WDM represents a dedicated (physical) layer in the network. OTN adds cross-connectivity, fault location, and protection/restoration at the optical level. This allows a managed network at wavelength granularity. Prior to full OTN standardization, the respective functionalities have been implemented in many WDM transport systems in proprietary ways.

8.1.2 OTN Mapping and Multiplexing

OTN is defined such that all relevant client signals can be mapped efficiently. Via OTN multiplexing, they can also be (time domain) multiplexed onto high bit rate wavelengths. The multiplexing is shown in Fig. 8.3. ODTU and ODTUG (Optical Data Tributary Unit (Group)) are intermediate mapping and multiplexing stages specified in G.709.

Different schemes for mapping different clients onto OTN exist. SONET/SDH client signals can directly be mapped onto OTN OPUs (which perform the client adaptation). Although OTN does not require synchronization, it can support the synchronization requirements of SONET/SDH (and also synchronous Ethernet).

For the different Layer-2 signals, different mapping mechanisms can (or must) be used. One option is to use the GFP as defined in ITU-T Recommendation G.7041. Descriptions of GFP can be found in Refs [2–4]. GFP allows standardized mapping of various Layer-2 data signals into SONET/SDH or ODU frames. The client signals can be protocol data unit (PDU)-oriented (like IP/PPP or Ethernet MAC) or block code-oriented CBR streams such as Fibre Channel (FC). Two client signal adaptation modes are defined: PDU-oriented adaptation, referred to as frame-mapped GFP (GFP-F), and block code-oriented adaptation, referred to as transparent GFP (GFP-T).

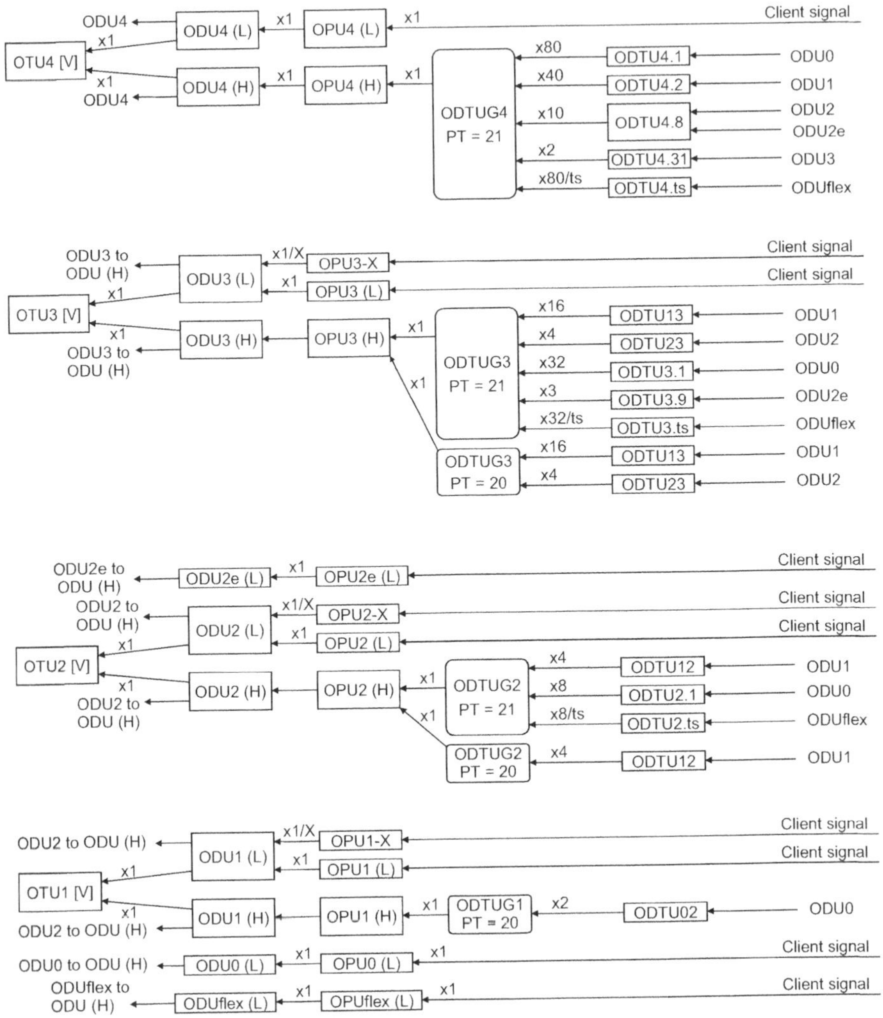

FIGURE 8.3 OTN mapping and multiplexing.

With GFP-F, a client signal frame is mapped in its entirety into one GFP frame. Client/GFP adaptation operates at the data link layer (or higher layer). GFP-F mapping is defined for Ethernet MAC and IP/PPP payloads. GFP-T mapping provides a block code-oriented adaptation in which the client/GFP adaptation function operates on the coded character stream rather than the incoming client PDUs. Block-coded client signals are decoded and mapped into fixed-length GFP frames. They can be transmitted immediately without waiting for the reception of an entire client data frame. This allows applications like LAN/SAN extension services that require low latency. GFP-T mapping is defined for FC running at various bit rates (including the IBM FICON versions) and also for GbE.

Optionally, GFP can be used with SONET/SDH in order to provide a TDM aggregation granularity of ~150 Mb/s (SONET STS-1 or SDH VC-4, respectively). Newer SONET/SDH implementations can cross-connect either contiguously or virtually concatenated containers of this size. This can be used in some cases to increase the bandwidth efficiency caused by client mapping and multiplexing over the inherent OTN capabilities. However, it requires a dedicated SONET/SDH layer (which is not always available anymore).

SONET/SDH-mapped GFP can, like native SONET/SDH clients, be mapped into the respective OTN layer. Unlike SONET/SDH networks, OTN networks do not need to be synchronized with a central clock. However, it is possible to transmit clock signals via OTN, for example, in order to provide synchronization to SONET/SDH or synchronous Ethernet network elements (also see ITU-T Recommendation G.8251).

For 10 GbE, two physical implementations—10GBASE-R (LAN PHY) and 10GBASE-W (WAN PHY)—have been defined in IEEE 802.3ae. The LAN PHY operates at 10.3125 Gb/s and the WAN PHY is reduced to 9.954 Gb/s, which makes it bit rate-compatible with SONET OC-192 and SDH STM-64. However, the different clocking concepts and jitter specifications of SONET/SDH and Ethernet still require an adaptation. For this reason, and since the LAN PHY interface has almost always been significantly cheaper than the WAN PHY interface, the vast majority of 10 GbE interfaces is of the LAN PHY type.

Direct mapping of 10GBASE-R into the overclocked ODU2 (i.e., ODU2e, running at 10.40 Gb/s) is possible. ODU2e can be transported via ODU3 and ODU4, or directly on a wavelength using an overclocked OTU2. In some older WDM transport systems, this functionality had already been implemented in proprietary ways prior to the latest G.709 Recommendation (e.g., in early releases of the ADVA Optical Networking FSP 3000 WDM transport system).

Forty gigabit Ethernet and 100 GbE can be directly mapped into OTN. For FC and other CBR services, an alternative to GFP exists. Here, the Generic Mapping Procedure (GMP) as defined in ITU-T G.709 is used for mapping into ODUflex (CBR type). The resulting mapping options are summarized in Fig. 8.4.

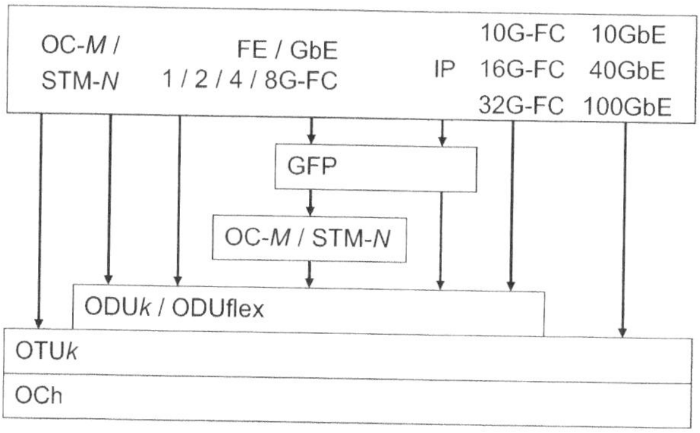

FIGURE 8.4 OTN mapping options.

TABLE 8.1 Bandwidth Efficiency of OTN Mapping

		Mapping Mechanism and Bandwidth Efficiency				
Client	Bit Rate	GFP with VCAT		ODU0/1/flex	ODU2	ODU2e
GbE	1.25 Gb/s	VC-4-7v	~96%	~80%	–	–
10 GbE LAN PHY	10.312 Gb/s	–	–	–	–	~99%
4G-FC	4.248 Gb/s	VC-4-29v	~78%	~80%	–	–
SDI Video	270 Mb/s	VC-4-2v	~90%	~99%	–	–
HDTV Video	1.485 Gb/s	VC-4-10v	~99%	~99%	–	–
OC-48/STM-16	2.488 Gb/s	–	–	~99%	–	–
OC-192/STM-64	9.953 Gb/s	–	–	–	~99%	–

All relevant client signals can be mapped into OTN with reasonable bandwidth efficiency. In certain cases, the use of a SONET/SDH grooming layer, together with virtual concatenation (VCAT) at the 150 Mb/s level, can improve bandwidth efficiency. Table 8.1 summarizes the resulting mapping efficiencies for some relevant client signals.

The numbers for GbE and 4G-FC consider the 8B/10B coding inefficiency (which is removed for GbE with GFP-F mapping). Efficiency loss caused by multiplexing into underutilized higher layer ODUs is not considered.

8.1.3 OTN Operations, Administration, and Monitoring Aspects

The OTN OCh has a frame structure for overhead (OH), payload, and forward error correction (FEC). FEC adds special redundancy and allows the detection and correction of transmission errors (refer to Section 3.4.2). The OCh is a digital signal on a dedicated wavelength, part of which (ODU*k*) is transported as an end-to-end entity. For the OCh, optical or electrical technologies can be used for switching or cross-connection. Together with the electrical intra-OCh switching, this allows cost-efficient system implementations. However, digital performance monitoring is possible only in electrical switches.

Supervision of payload traffic is mandatory in order to guarantee certain quality of service (QoS). Performance monitoring and alarm management are key elements for QoS coverage in OTN. Signals can be monitored for each layer separately. Signaling (of any monitoring or management information) can make use of a special part of the OH, the Generic Communication Channels (GCCs). Similar to the SONET/SDH DCC, the GCC forms part of the DCN (Data Communications Network) (see Section 7.4).

Beyond surveillance signals in the electrical layers, G.709 also defines surveillance signals in the optical domain to address purely optical subnetworks. The assignment of the monitoring functions to the different OTN overheads is shown in Fig. 8.5.

An important feature of the OTN, and an improvement over SONET/SDH, is *Tandem Connection Monitoring* (TCM). OTN TCM allows end-to-end monitoring

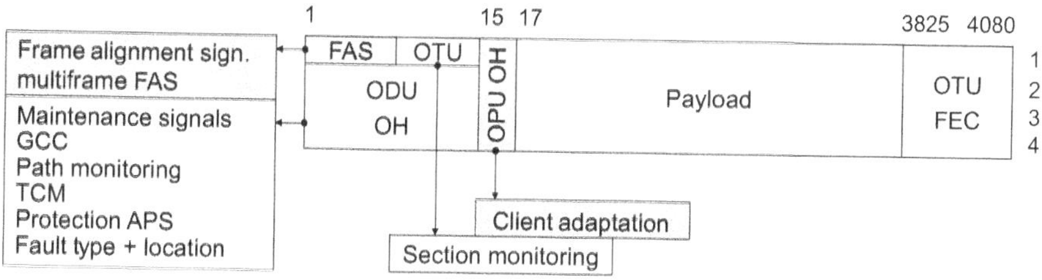

FIGURE 8.5 OTU*k* frame structure and overhead functions.

across different administrative domains (i.e., network, operator, or vendor domains). Six independent TCM levels have been defined, allowing monitoring for nested and cascaded domains.

Figure 8.6 shows an example of signal transmission over different administrative domains. Two metro carriers and one long-haul carrier with two nested vendor domains provide the end-to-end connection. Here, every network domain and also the end-to-end connection (TCM1) can be monitored simultaneously.

With OTN TCM, SONET/SDH services can be transported across several network domains transparently, without overwriting their performance-monitoring bytes. In order to supervise the quality of these client signals, additional nonintrusive monitoring (of the SONET/SDH B1 bytes) can be applied.

Several protection and restoration mechanisms have also been, or are still being, defined for OTN. The related switching events also make use of the OTU/ODU overhead, as indicated in Fig. 8.5. Linear (point-to-point) subnetwork connection (SNC) protection mechanisms are defined in ITU-T Recommendation G.873.1 for the ODU*k* level. $1+1$ and $1:N$ linear protection with different monitoring

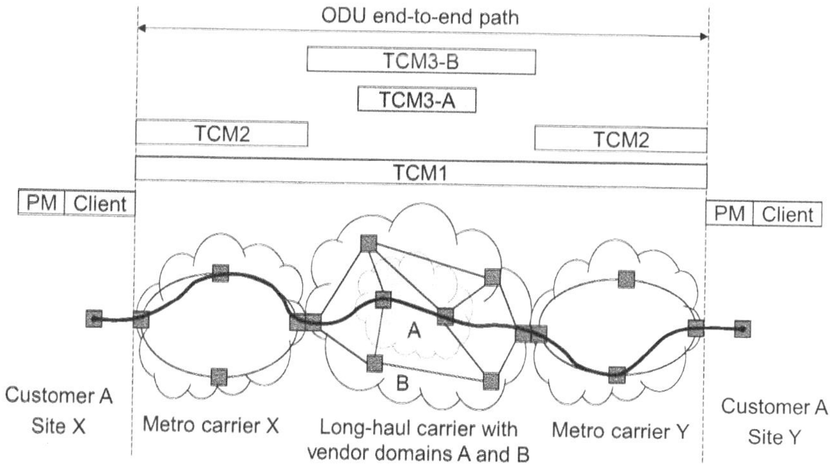

FIGURE 8.6 OTN Tandem Connection Monitoring example.

mechanisms (inherent monitoring, SNC/I, nonintrusive monitoring, SNC/N, and sublayer monitoring, SNC/S) is supported.

Further protection mechanisms include trail (OCh, OMS) $1+1$ protection as defined in ITU-T Recommendation G.872. Definition of shared ring protection for the ODU*k* level is done in Recommendation G.873.2. In addition to these mechanisms, the respective WDM systems may have further protection implementations, for example, dedicated ring protection for the OCh layer. Also refer to the protection section: Sections 8.4.1 and 8.4.2.

Restoration is under definition in ITU-T as part of the ASON (Automatically Switched Optical Networks) activity.

8.2 MONITORING IN WDM SYSTEMS

WDM technology has enabled unregenerated transmission over large distances. Multidegree reconfigurable optical add–drop multiplexers (ROADMs) and OXCs led to meshed networks with varying link lengths rather than static point-to-point links. In the resulting networks, monitoring of signal parameters and transmission impairments is necessary, for example, for the adaptation of power levels or equalizers.

8.2.1 Impairments in Transparent Optical Networks

In long-haul meshed WDM networks, WDM channels are impaired by several effects. In order to maximize fiber capacity and distances between regenerators, some of these effects need to be monitored. This holds for 10 Gb/s IM/DD channels as well as for the upcoming techniques for WDM superchannels carrying 400 GbE or 1TbE signals on multiple subcarriers with varying modulation formats over varying distances.

Impairments are generated in transmission fibers, transceivers, amplifiers, and optical cross-connects and ROADMs. They have been described in the respective preceding sections. Relevant impairments and their origin are summarized in Table 8.2.

Most of the equipment-related impairment parameters are generally known (monitored directly by the respective device or known *a priori*) and can be signaled in reconfigurable networks through either the management or the control plane. On varying-length paths, however, some fiber effects [e.g., polarization-mode dispersion (PMD)] cannot be predicted precisely and may require monitoring. The same is true for accumulated noise and filter effects (in particular, cross talk caused by faulty configurations or miss connections).

8.2.2 Measurements of the Quality of Optical Signals

Signal quality and its degradation can be described using several measures. Some of these measures are not suitable for automated (real-time) monitoring in WDM

TABLE 8.2 Transmission Impairments

Impairment	Origin
Transmitter/laser	Chirp and bandwidth, if directly modulated
	Linewidth (for use as local oscillators)
	Relative intensity noise (RIN)
	Side-mode suppression ratio (SMSR)
External modulators	Chirp (EAM), bandwidth
	(Nonlinear) Transfer function
Receiver	Noise (thermal, shot)
	Bandwidth
Fiber	Attenuation
	Chromatic dispersion (CD)
	Polarization-mode dispersion (PMD)
	Self-phase modulation (SPM)
	Cross-phase modulation (XPM)
	(Parametric) Four-wave mixing (FWM)
	Stimulated Raman scattering (SRS)
	Stimulated Brillouin scattering (SBS)
Amplifiers	Amplified spontaneous emission (ASE)
	Transients due to channel add–drop/failure
	Nonflatness of gain
	Polarization-dependent gain (PDG) (if applicable)
ROADMs, OXCs, Filters	Insertion loss (IL)
	Polarization-dependent loss (PDL)
	Linear cross talk (XT, inter- and intrachannels)

systems, and some are not suitable for identifying specific impairments. The measures are [5] as follows:

- Bit error rate (BER)
- Eye diagram
- Q-factor
- Optical spectrum

The primary goal of WDM networking is error-free transmission (where error-free relates to post-FEC BER). The most important measure therefore is the *bit error rate*. For many applications, the post-FEC BER has to be smaller than 10^{-12}, or even $<10^{-15}$ for some high-quality applications (refer to Section 3.4.2). It is difficult to observe such BERs because very long observation time is required. This is the main reason why interpolation methods such as the Q-factor method are preferred over direct BER measurements. In addition, digital bit error counters can be implemented in SONET/SDH and OTN transmission systems for long-term monitoring.

Eye diagrams are obtained by superposing the electrical time-domain signal traces at the receiver in a time slot with a duration of one or several symbols. Eye diagrams allow fast observation of distortions of the signal pulses. They are useful qualitative indicators, and experienced staff can identify several impairments from the shape of the eye opening.

The concept of the Q-factor has already been described in Section 5.3.1. It allows an estimate of the BER from parameters that can be measured at the receiver. The method relies on the Gaussian approximation of the noise and is accurate as long as total noise is amplified spontaneous emission (ASE) dominated.

Several signal parameters can be derived from the *optical spectrum*. Optical spectrum analyzers (OSAs) can be used to measure the optical signal-to-noise ratio (OSNR) and WDM channel power imbalance. However, they cannot discriminate different causes of noise or distortion. OSAs use a spectral window with typical resolution bandwidth of 0.1 nm for analysis. This window is swept across the bandwidth to be analyzed, for example, the C-band.

8.2.3 Monitoring Techniques

Monitoring can be classified into two categories: optical performance monitoring (OPM) and optical impairment monitoring (OIM) techniques. OPM techniques are *signal quality related*. OIM refers to monitoring of specific impairments. It also aims at the control of the respective impairment-induced signal degradations.

8.2.3.1 Monitoring Requirements The most relevant monitoring criteria are acquisition time and accuracy. As for every measurement, fast acquisition time contradicts high accuracy (see the remark on OTDR monitoring in Section 6.5.8). An upper bound for the required per-channel per-parameter acquisition time T_A was given [5]:

$$T_A \approx 100\,\text{ms}. \tag{8.1}$$

This upper bound considers *sequential channel-after-channel* monitoring with monitoring equipment shared between all channels and also between several fibers. A maximum of 96 WDM channels and a node degree of 9 have been assumed. Faster acquisition time may be necessary for certain parameters (e.g., PMD). This may lead to the requirement of dedicated per-channel monitoring and thus increase cost.

Regarding accuracy, work in Refs [6,7] indicates that OSNR and power monitoring should be accurate up to +/− 0.5 dB. For chromatic dispersion (CD) and PMD, accuracy targets of +/− 5% for PMD monitoring and +/− 2% for CD monitoring were stated in Ref. [5].

Further monitoring requirements result from the necessity to monitor a potentially large number of parameters (which are partly interacting) in a live network with payload traffic:

- Monitoring of more than one parameter/impairment by one method.
- Orthogonality of monitoring. The accuracy of a certain method should not depend on other parameters or impairments (e.g., CD, PMD, and nonlinearity).
- Hitless and nonintrusive monitoring. Monitoring must not affect the payload channels, and monitoring equipment should be switchable from channel to

channel or fiber to fiber for cost reasons [where applicable, that is, if no dedicated monitoring is required, for example, by related service-level agreements (SLAs)].

- Independence from bit rate or modulation scheme. For shared monitoring equipment, independence of channel bit rate and modulation scheme helps decreasing cost and improving general applicability. Otherwise, per-channel or per-bit rate, per-modulation scheme techniques are required.

At present, cost-effective monitoring techniques are available for a subset of signal parameters and impairments only. These techniques typically do not comply with the complete requirements list.

8.2.3.2 ***CD Monitoring*** In the reconfigurable WDM networks, any change in the path length leads to a different amount of accumulated CD. For adaptive dispersion compensation [e.g., tunable optical dispersion compensation (TODC), for self-coherent 40G interfaces] (see Section 3.2.5.4), CD monitoring may be required. Coherent intradyne transceivers may not require the techniques discussed here since they perform CD monitoring in the digital domain.

Pilot tone-based CD monitoring is useful for reconfigurable WDM networks since dedicated pilot tones follow their corresponding WDM signals anywhere in the network. Hence, this monitoring can be combined with channel tracing and identification (or "tagging").

AM Subcarrier and Monitoring of RF Power Fading. CD can be monitored with amplitude-modulated (AM) pilot tones [8]. The tones do not need to carry data and can be implemented with narrow linewidth. Since the tones fall within the payload data bandwidth, they experience the same CD and can be used as CD sensors.

At the transmitter, the upper and lower sidebands of the subcarrier are in phase with each other. A phase difference between both sidebands accumulates during transmission as a result of the frequency-dependent CD. The phase mismatch is related to the amount of accumulated CD, and the subcarrier photocurrent $I(f_{SC})$ fades according to

$$I(f_{SC}) = I_0 m \cos(\pi f_{SC}^2 \lambda DL / c_0). \tag{8.2}$$

Here, I_0 is the peak photocurrent, m is the subcarrier modulation depth, f_{SC} is the subcarrier tone frequency, λ is carrier wavelength, DL is the total dispersion accumulated over distance L, and c_0 is the speed of light in vacuum.

Monitor resolution and measurement range depend on the frequency of the subcarrier. If the tone frequency is too low, the magnitude of the AM tone will remain nearly constant. The resolution can be improved by increasing the tone frequency, at the cost of the measurement range. Thus, the tone frequency must be optimized to obtain reasonable resolution and measurement range. With subcarriers of 7 and 9 GHz, a range of 675 and 1200 ps/nm, respectively, has been demonstrated, with sensitivity up to 10 ps/nm related to 1 dB of power fading.

The use of high-frequency pilot tones for CD monitoring can be distorted by self-phase modulation (SPM) [9]. The SPM-induced phase modulation can generate an additional AM component due to PM–AM conversion, which in turn causes monitoring errors.

AM Subcarrier and Monitoring of Phase Delay Between the Sidebands. Here, a modulated subcarrier is added to the baseband data. During propagation, CD induces a relative time delay between the subcarrier sidebands given by [10]

$$\Delta t = f_{\mathrm{SC}}\lambda^2 DL/c_0. \tag{8.3}$$

By choosing a subcarrier frequency that is large compared to the baseband WDM bit rate, sensitive measurement of accumulated CD is possible. Measurement of the subcarrier phase delay alone limits the maximum CD that can be measured to the subcarrier period. This limit can be extended by measuring the delay experienced by low-frequency modulation applied to the subcarrier. By combining both techniques, measurement of a wide CD range with high resolution is possible. With a 16.4 GHz subcarrier, a delay of 5 ps (1/12 of the period) has been shown, leading to $DL \approx 38$ ps/nm. With a subcarrier modulation frequency of 410 MHz, the maximum DL product was ~9200 ps/nm, albeit with lower accuracy.

PM Subcarrier and Monitoring of PM–AM Conversion at the Receiver. When a phase-modulated (PM) signal propagates in a fiber, CD changes the phase difference between the optical carrier and its sidebands. As a result, the amplitude of the transmitted signal is modulated. The magnitude of the AM component generated by PM–AM conversion increases linearly with CD even when the tone operates in the low-frequency region. Thus, unlike using AM tones, it is possible to obtain a reasonable resolution even with low-frequency PM tones.

The effects of SPM and PMD on CD monitoring with PM pilot tones depend on the tone frequency. In addition, the measurement range is inversely proportional to the square of the tone frequency. By decreasing the tone frequency, it is possible to suppress the effects of SPM and PMD and extend the measurement range. However, low-frequency tones reduce the PM–AM conversion efficiency [9,11]. The use of a 2 GHz tone can increase the measurement range beyond 15,000 ps/nm [12].

Two RF Tones and Monitoring of Rx Power Ratio Between the Two Tones. Here, the influence of power loss or gain of the lightpath on the CD measurement can be eliminated. Monitoring range and sensitivity are comparatively small or low. To improve this, a CD offset can be inserted before the photodetector. Monitoring range can then be larger than 1000 ps/nm, and the sensitivity can be as high as 17 ps/nm for 1 dB radio frequency (RF) power ratio change [13].

PMD and external modulator [electroabsorption modulator (EAM)] chirp may induce CD monitoring errors. In Ref. [14], a technique is proposed to suppress these effects. The PMD effect is eliminated by optical sideband filtering and RF power ratio detection, and the error induced by chirp can be suppressed using two RF tones and a CD offset.

CD Monitoring Based on the Measurement of Phase Modulation. For intensity-modulated double-sideband signals, CD converts part of the intensity modulation into phase modulation. By measuring the phase modulation present at a specific wavelength, the dispersion of the link can be determined. Path-imbalanced Mach–Zehnder interferometers (MZIs) [15] or birefringent fiber loops [16] can be used to measure these phase modulations. The experimental setup used in Ref. [15] is shown in Fig. 8.7.

One output of the MZI is detected with a photodiode. The received RF power is measured using an RF spectrum analyzer, and is given by [15]

$$P \propto \sin^2\left(\frac{\varphi_+ + \varphi_-}{2}\right) = \sin^2\left(\frac{\pi\lambda^2 f_{\mathrm{RF}}^2}{c_0} D\right), \tag{8.4}$$

where $\phi_\pm$ are the acquired phase shifts of the carrier and its sidebands, λ is the optical wavelength, f_{RF} is the frequency of the spectral component, c_0 is the speed of light in vacuum, and D is the total accumulated dispersion of the link. The technique has an adjustable measurement range of >2300 ps/nm for 10 Gb/s NRZ data. The setup shown in Fig. 8.7 must be complemented by a (tunable) filter in order to be applicable to WDM systems.

Similar CD monitoring can be implemented using a high-birefringence fiber loop [16]. Monitoring is achieved by feeding an intensity-modulated double-sideband signal into a fiber loop that consists of high-birefringence fiber. Due to CD, a portion of the intensity modulation converts to phase modulation. The phase delay is a function of CD. By measuring the RF power at a specific frequency, determined by the length of the high-birefringence fiber, it is possible to determine the value of CD. Experimental results showed a large measurement range that is also tunable by adjusting the length of the birefringent fiber loop and independence of data rate and modulation format.

Relative Clock Phase-Shift Delay Between Upper and Lower VSB Signals. For NRZ- or RZ-modulated data, a photodetected vestigial sideband (VSB) signal contains a dedicated clock component. The clock sidebands have different wavelengths and experience a relative CD-induced delay. Thus, a clock phase shift between the lower and upper VSB signals appears [17]. Using an optical filter to select the upper and lower VSB signals and monitor the relative clock phase shift, it is possible to determine accumulated CD. This technique is sensitive and is

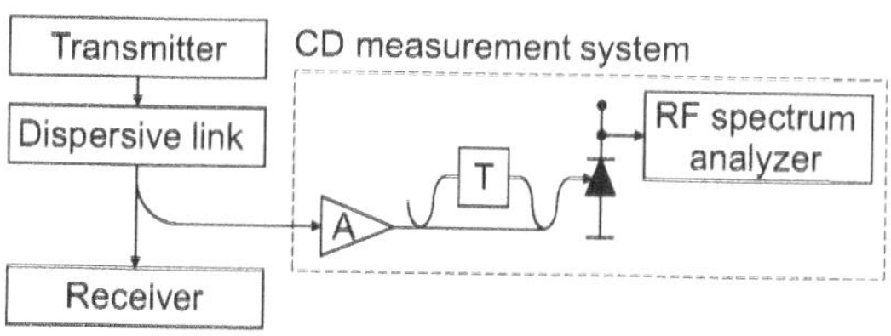

FIGURE 8.7 CD monitoring based on phase modulation measurement [15].

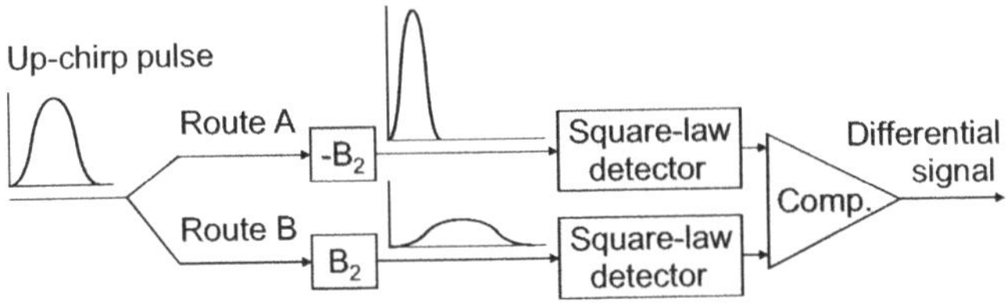

FIGURE 8.8 CD-induced chirp monitoring [18].

unaffected by first-order PMD, fiber nonlinearity, and transmitter chirp. It can be applied to WDM signals by sweeping the optical filter.

Asynchronous Dispersion-Induced Chirp Monitor. This monitor measures CD-induced chirp in optical pulses. It uses a balanced receiver with dedicated CD of $-B_2$ and B_2 in two branches A and B (see Fig. 8.8). When the input pulses have up-chirp, output pulses from CD element $-B_2$ in branch A are narrower than those in branch B. With down-chirped input pulses, the output pulses in branch A are broader than those in branch B.

Detector output signals are inversely proportional to pulse width, thus different pulse widths produce a differential signal at the comparator. The chirp monitor is bit rate flexible because no synchronization is needed [18].

Table 8.3 gives a comparison of CD monitoring methods [5].

8.2.3.3 ***PMD Monitoring*** PMD is a relevant impairment for high-speed WDM systems, in particular if legacy high-PMD fiber plant has to be used (refer to Section 2.1.3.3). With coherent intradyne detection, PMD is compensated dynamically in the digital domain on a per-channel basis. However, for 10G services with IM/DD and 40G channels with direct or self-coherent (delay demodulation) detection, PMD monitoring may be required in reconfigurable WDM networks. Several monitoring techniques have been demonstrated.

TABLE 8.3 Comparison of CD Monitoring Techniques

Monitoring Technique	RF Power Ratio [14]	MZI-Based PM Detection [15]	CD-induced Chirp Monitor [18]	Birefringent Fiber Loop [16]
Accuracy/sensitivity	High	–	High	–
Range	Moderate	High	Moderate	High
Insensitivity to other impairments	Moderate	Low	Low	Moderate
Modulation format transparency	Yes	No	Yes	Yes
Bit rate transparency	Yes	Yes	Yes	Yes
Transmitter modification	Yes	No	No	No
Multichannel	No	No	No	Yes
Multiparameter	No	No	No	No
Acquisition time	Fast	Moderate	Fast	Moderate

PMD Monitoring Based on Spectral Analysis. One class of PMD monitoring techniques is based on *spectral analysis* and *RF tones*. If launched data and pilot tone are completely polarized with DOP = 1, PMD will degrade the *degree of polarization* (DOP). The pilot tone power is a function of the total differential group delay (DGD), thus PMD can be determined by detecting the RF spectrum. By increasing the tone frequency, the monitoring range is reduced and resolution is improved [11]. The maximum DGD measurement range is bit rate independent [19]. Dual-sideband pilot tones for PMD monitoring are sensitive to CD. This problem can be solved by measuring the RF power of a single sideband (SSB) [19]. Narrow SSB tones are inherently insensitive to CD.

PMD monitoring using RF clock tones works similarly [20]. PMD causes the carrier and optical clock-tone power to split between the two PSPs and propagate at different velocities. The resulting differential delay dephases the clock tones in the two PSPs, reducing the recovered RF clock-tone power due to destructive interference.

In the absence of CD, there is no RF clock tone in NRZ electrical signals after detection since upper and lower clock sidebands beat with the carrier and cancel each other. By adding a dispersive component, for example, a chirped FBG, the optical clock sidebands can be dephased. Then, the clock signals are generated per PSP, and PMD becomes observable. The amount of added CD required to obtain maximum clock regeneration is inversely proportional to the square of the data rate of the NRZ signal under consideration. The monitoring range can be increased up to 100 ps instantaneous DGD [21].

Clock-tone RF power analysis alone does not allow simultaneous monitoring of CD and PMD [22]. Isolation of CD and first-order PMD can be achieved by monitoring the RF clock-tone power at the output ports of an unbalanced Mach–Zehnder interferometer with a quarter-bit delay in one arm. By adding and subtracting the constructive and destructive clock powers, the individual contributions of CD and first-order PMD can be derived. This method is not suitable for RZ formats since these already have dedicated clock-tone power. A DGD monitoring range up to 50 ps with an average sensitivity of 0.25 dB/ps for both OOK and DPSK signals has been demonstrated [23].

Independence of CD can also be achieved by tracking the clock-tone power of a VSB signal. For a VSB signal, the detected RF power results from beating between the carrier and only one of the optical clock tones. CD can affect the phase of the RF clock power but not the amplitude. The VSB is generated by centering a bandpass filter (BPF) at the upper part of the optical clock before the photodetector. The clock-tone power in that upper part can be used for PMD monitoring because it has a distinct peak in the RF spectrum. This method provides good monitoring sensitivity and monitoring range of half the bit time [19].

PMD Monitoring Techniques Based on DOP Measurement. The DOP of an optical signal reflects the degradation caused by PMD. DOP analysis has been proposed for use in adaptive PMD compensation [24–26]. The DOP can take values between 1, when all signal components are polarized along the same SOP, and 0 when the polarization of all components is scrambled. DOP monitoring does not require

TABLE 8.4 Comparison of PMD Monitoring Techniques

Monitoring Technique	Monitor Using DLI [23]	Degree of Polarization	RF Power of VSB [19]
Accuracy/sensitivity	–	–	High
Range	High	Moderate	Moderate
Insensitivity to other impairments	Good	Moderate	Moderate
Modulation format transparency	No	No	Yes
Bit rate transparency	No	Yes	No
Transmitter modification	No	No	No
Multichannel	No	No	No
Multiparameter	Yes	No	No
Implementation complexity	Simple	Simple	Suitable
Acquisition time	Moderate	Fast	Fast

high-speed circuits, and is insensitive to other degrading effects [25]. However, DOP-based techniques suffer from some disadvantages:

- Small DGD monitoring range for short RZ pulses
- Lack of sensitivity for NRZ signals
- Affected by higher order PMD

These limitations can potentially be overcome by centering a narrowband optical filter either at the optical center frequency or at one of the signal's sidebands. DOP measurement suffers monitoring-range reduction as pulse widths decrease. Enhanced DGD monitoring for short pulses can be achieved by broadening of the pulses in the time domain via optical filtering. This results in a near-doubling of the DOP sensitivity for PMD monitoring in 40 and 10 Gb/s NRZ systems without affecting the monitoring windows and in 25 and 50 ps increase of the monitoring range in 40 and 10 Gb/s RZ systems, respectively [25].

Eye monitoring has also been demonstrated for PMD monitoring [27]. The eye opening is correlated with the BER, and it is affected, among others, by PMD.

Table 8.4 presents a comparison of the PMD monitoring techniques [5].

8.2.3.4 OSNR Monitoring Monitoring of OSNR on a per-channel basis is required in reconfigurable WDM networks. OSNR monitoring can use *out-band* or *in-band* techniques, meaning that monitoring is achieved by measuring the ASE *between* WDM channels or *inside* the channel bandwidth, respectively. The difference is demonstrated in Fig. 8.9.

Out-Band OSNR Monitoring.

LINEAR INTERPOLATION OSNR measurements often use *interpolation*. The OSNR is obtained by measuring the total signal power in the channel passband and the noise power (ASE) in the gaps between the optical channels (normalized to 0.1 nm

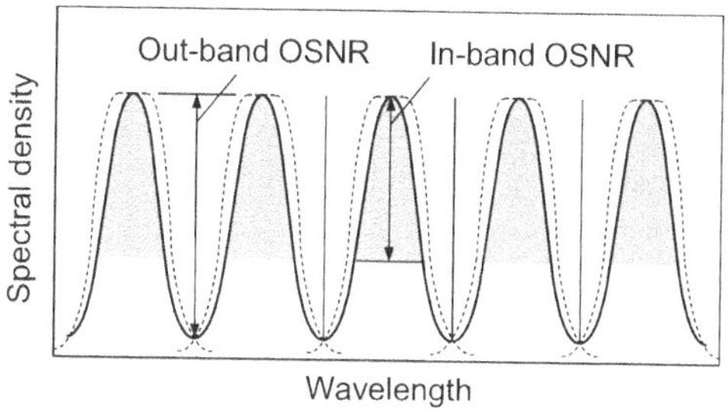

FIGURE 8.9 In-band and out-band OSNR monitoring [28].

bandwidth). This method is called linear interpolation since the noise power is averaged from ASE that is present left and right of the channel under consideration. In a reconfigurable WDM network, each channel may traverse different routes with different numbers of amplifiers and filters. Thus, adjacent channels can have different noise levels. Therefore, measurement of the true OSNR may not be possible using conventional linear interpolation. In addition, filters and ROADMs suppress the noise *between* optical channels. Thus, the noise level will be *underestimated* (see Fig. 8.9).

OSNR MONITOR USING TUNABLE OPTICAL FILTER OSNR monitoring can be achieved by using a *tunable filter*, for example, an acousto-optic filter [29]. The tunable filter has a notch with high extinction ratio (>15 dB) to remove signals in WDM channels. The ASE power is obtained by successively subtracting the power of multiple WDM channels from the total power (WDM channels plus ASE). The estimated ASE is averaged across all WDM channels.

AWG-BASED OSNR MONITOR The OSNR monitor proposed in Ref. [30] is based on an arrayed waveguide grating (AWG). It is limited by AWG cross talk for channel numbers >40, where the accumulated cross talk from nonadjacent channels becomes comparable to the noise that is to be measured. Improved OSNR monitoring is possible by combining the AWG with a periodic tunable notch filter [31].

In-Band OSNR Monitoring Techniques. To date, most OSNR measurements used the interpolation method. This method underestimates the filtered in-band noise. In order to more precisely measure the filtered noise, in-band OSNR monitoring is required.

SIGNAL DEACTIVATION SCHEMES Two principles for measuring the noise power *within* the optical channel exist [32]. The first principle is based on *signal deactivation* that requires turning off the payload signal in order to measure the noise power level correctly. The obvious drawback is the service interruption. In addition, the automatic gain control in optical amplifiers will change the ASE noise level when the signal is switched off. In *time-resolved optical gating* (TROG), high-speed optical gating can be used to overcome Erbium-doped fiber amplifier (EDFA) gain equalization. With TROG, data and probe time slots are multiplexed. At the receive-end, data and probe signals are optically demultiplexed and analyzed independently. The main disadvantage then is the added overhead, or reduced bandwidth efficiency, caused by probe multiplexing.

POLARIZATION NULLING Polarization nulling uses the different polarization attributes of noise and WDM signals. The incident WDM signal can have linear, circular, or elliptical polarization. This arbitrary SOP can be linearly polarized with a polarization controller. Linearly polarized signal and unpolarized ASE noise can then be split into two orthogonal polarization components. One polarization component consists of signal and ASE, while the other is ASE only. Signal power and polarized ASE are measured by aligning the polarizer to the signal's polarization. Only the polarized ASE (i.e., half of the total noise) can be measured in the orthogonal polarization. Using polarization nulling, it is possible to monitor true OSNR [33]. The performance of this technique can be affected by PMD, nonlinear birefringence, and (linear or nonlinear) in-band cross talk.

Table 8.5 presents a comparison of the OSNR monitoring techniques described so far.

OSNR (like other) monitoring is constantly improved. Newer, enhanced approached are briefly described hereinafter.

SPECTRAL ANALYSIS AFTER FREQUENCY DOWN-CONVERSION A method suitable for reconfigurable networks is based on spectral analysis after frequency down-conversion. Signal and noise are optically filtered to select the optical carrier and a single sideband. The SSB signal is heterodyned down to an intermediate frequency (IF) that is detected using a low-bandwidth photodetector. The electrical signal is then digitized and the IF tone is extracted with Fourier analysis. The ratio of the total average power to the amplitude of the IF tone is proportional to the OSNR [35]:

$$\mathrm{OSNR} = -10\log_{10}\left(\gamma\frac{n_0^2}{I_0}\right) = -10\log_{10}\left(\gamma\left(\frac{\alpha\cdot\beta}{8}R - 1\right)\right), \qquad (8.5)$$

TABLE 8.5 Comparison of OSNR Monitoring Techniques [34]

	Tunable Filter	AWG	Polarization Nulling
In-/out-band	Out	Out	In
Accuracy	1 dB, when ASE is uniform	1 dB, 100 GHz spacing when ASE is uniform	1 dB, 200 GHz spacing
OSNR range	10–25 dB	10–30 dB	15–30 dB
Characteristic	WDM possible, low speed, inaccurate for reconfigurable networks	WDM possible, inaccurate for high channel count and reconfigurable networks	Simple implementation only when polarization controller is used already, complex improvement of PMD tolerance

where γ depends on the effective measurement bandwidth. The noise associated with the filtered SSB optical field can be modeled as two monochromatic tones of equal amplitudes n_0. I_0 is the total signal power, α and β determine the amplitude splitting between each tone, and R is the ratio of the total power to the IF tone amplitude.

The dynamic range of the OSNR measurement is 20 dB for a single-channel 40 Gb/s NRZ system. Accuracy of better than 1 dB can be achieved for OSNR values less than 20 dB, optical input power of −30 dBm, and acquisition times as short as 10 ms. This technique is suitable for intensity-modulated formats with carrier (i.e., NRZ, RZ50, and RZ33) and can be adapted to different bit rates by changing the heterodyne local oscillator frequency.

ORTHOGONAL POLARIZATION HETERODYNE OSNR MONITORING This monitor utilizes polarization orthogonality and heterodyne mixing to measure the in-band OSNR [36]. As ASE noise is depolarized, it can be separated from the signal by heterodyne mixing and orthogonally polarizing the signal components with respect to each other. This is implemented by first splitting the signal into two branches and filtering each branch with an optical filter that is narrow relative to the signal bandwidth. The two filters have different center frequencies (f_1, f_2) and are not overlapping. The polarization of one branch is adjusted such that the SOPs of the two branches are orthogonal. The two branches are combined and the signal is detected with a photodetector. An electrical BPF centered at frequency $\Delta f = |f_1 - f_2|$ is used to select the beating terms from branches 1 and 2, and the RF power at Δf is measured with an RF power detector.

When the signals from the two branches are orthogonally polarized, the signal–signal beating vanishes and only noise terms (signal–ASE and ASE–ASE beating) remain in the RF domain. By measuring RF and signal power, the noise power can be obtained and the OSNR be calculated.

This technique has inherent immunity to PMD and robustness to nonlinear polarization scattering. It is independent of modulation format, bit length, bit rate, and spectral shape.

HIGH-BIREFRINGENCE LOOP MIRROR In polarization nulling, PMD and imperfect polarization beam splitters (PBS) lead to inaccurate measurements. A method that *exploits* these effects and simultaneously monitors PMD and OSNR in PMD-affected signals is proposed in Ref. [37]. The principle is shown in Fig. 8.10.

A demultiplexed WDM signal is fed into a polarization controller that maximizes the power at the output B of the PBS. Due to PMD and the nonideal PBS with its finite extinction ratio ER, a small fraction β of the signal power will also leak to output C. The unpolarized ASE noise power P_{ASE} is divided equally between B and C. The leakages caused by the PBS, L_{PBS}, and PMD, L_{PMD} are given by

$$L_{PBS} = ((1 - \beta)P_{signal})/\mathrm{ER}, \quad L_{PMD} = \beta P_{signal} - L_{PBS}. \tag{8.6}$$

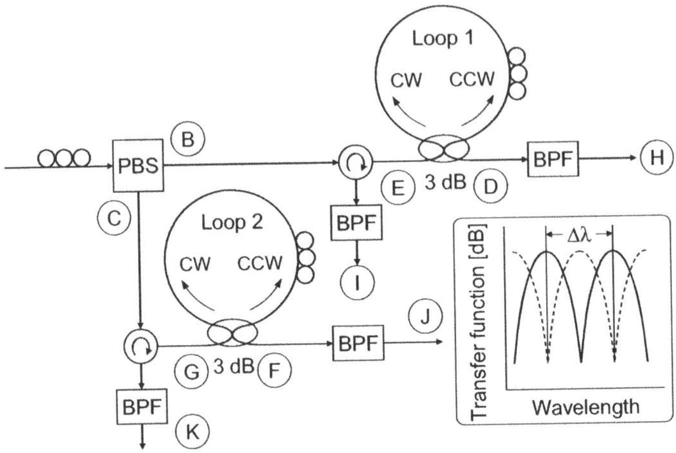

FIGURE 8.10 High-birefringent loop mirror [37]. CCW/CW: (counter) clockwise.

The fraction of the signal component at the orthogonal SOP at port C to that at B can be approximated by [38]

$$n = 0.5(1 - \sin(\pi \mathrm{DGD}\,\delta f)/\pi \mathrm{DGD}\,\delta f) = L_{\mathrm{PMD}}/(1-\beta)P_{\mathrm{Signal}}. \tag{8.7}$$

Here, δf is the signal bandwidth taken to be twice the data rate. From Eq. (8.7), the values of n and hence DGD can be calculated if β and P_{Signal} are known. Two periodic wavelength filters made of high-birefringence loop mirrors are used at the output ports of the PBS. The transfer functions at outputs D and F are complementary to those at E and G, respectively. Optical BPFs with center wavelengths identical to the input signal are placed at all loop mirror output ports. The optical power at the BPF outputs can be approximated by

$$P_{\mathrm{H}} \approx 0.5((1-\beta)P_{\mathrm{Signal}} + ((1-0.5\alpha)\alpha P_{\mathrm{Noise}})), \tag{8.8}$$

$$P_{\mathrm{I}} \approx 0.5((1-\beta)P_{\mathrm{Signal}}/\mathrm{PNR} + 0.5\alpha^2 P_{\mathrm{Noise}}), \tag{8.9}$$

$$P_{\mathrm{J}} \approx 0.5(\beta P_{\mathrm{Signal}} + (1-0.5\alpha)\alpha P_{\mathrm{Noise}}), \tag{8.10}$$

$$P_{\mathrm{K}} \approx 0.5(\beta P_{\mathrm{Signal}}/\mathrm{PNR} + 0.5\alpha^2 P_{\mathrm{Noise}}), \tag{8.11}$$

where α is the fraction of the BPF passband to $\Delta\lambda$ and PNR is the peak-to-notch ratio of the filters. The values of β, L_{PBS}, L_{PMD}, P_{Signal}, P_{ASE}, and n are obtained using the above equations, from which the OSNR (dB) $= 10\log_{10}(P_{\mathrm{Signal}}/P_{\mathrm{ASE}})$ and the DGD can be calculated.

While an overestimation occurs at low OSNR levels, measurements are quite accurate above 20 dB, irrespective of the extent of PMD.

OSNR Monitoring Using Nonlinear Optical Loop Mirror. Another loop mirror approach to in-band OSNR monitoring has been presented in Ref. [39]. The

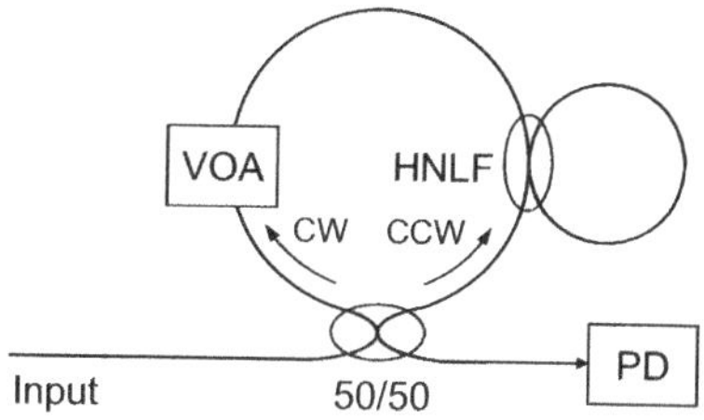

FIGURE 8.11 NOLM-based OSNR monitoring [39]. CCW/CW: (counter) clockwise.

monitoring device is a nonlinear optical loop mirror (NOLM), which is shown in Fig. 8.11. The NOLM consists of a 3 dB coupler, a variable optical attenuator (VOA), and a highly nonlinear fiber (HNLF).

With NRZ, CSRZ, and RZ formats, the probability density at the logical mark ("1") power level decreases as the OSNR is reduced. Therefore, a nonlinear device that preferentially transmits high power levels and attenuates low power levels, with a threshold in-between, can act as an OSNR monitor. A NOLM uses the nonlinear optical Kerr effect in the fiber to obtain a nonlinear power transfer function.

For any pulse shape (duty cycle), the peak power of the noise-free signal is aligned with the maximum of the power transfer function. Then, the NOLM yields the maximum output contrast, that is, the maximum output power difference between signal alone and noise alone.

The monitor can discriminate the OSNR of 40 Gb/s signals over a dynamic range of 25 dB for NRZ, 27 dB for CSRZ, and 25 dB for RZ. Due to the ultrafast response of the Kerr effect, the device is inherently scalable to Baud rates beyond 40 GBd. The device is also suitable to perform CD monitoring.

SOA-Based Optical 2R Regenerator. OSNR monitoring can also be achieved via the nonlinear response of a semiconductor optical amplifier (SOA). The technique potentially integrates CD and *Q*-factor monitoring with wavelength conversion. A single WDM channel is combined with a continuous-wave (CW) signal at the desired conversion wavelength and coupled into the SOA. The CW signal is modulated in the SOA by the data signal via cross-gain and cross-phase modulation. The wavelength-converted signal can be recovered by filtering. The total power of the wavelength-converted signal is proportional to the signal quality. Thus, a monitoring signal can be derived from the ratio of the converted signal power to the input signal power. The monitor cannot provide independent per-impairment measurements and shows weak sensitivity to OSNR. However, it correlates directly with signal quality, and can be integrated with other SOA-based functions (2R regeneration, wavelength conversion) [40].

Table 8.6 gives a comparison of the advanced OSNR monitoring techniques.

8.2.3.5 Optical Transient Monitoring (OTM) Fast power transients in optical networks occur due to the addition or removal of wavelength channels during network reconfigurations and in cases of transmitter failures and related resilience

TABLE 8.6 Comparison of Advanced OSNR Monitoring Techniques

Monitoring Technique	Analysis After Down-Conversion	Orthogonal Polarization Heterodyne	High-birefringence loop	NOLM	SOA 2R Regenerator
Accuracy/sensitivity	High	Moderate	Moderate	–	Low
Range	Moderate	Low	Moderate	High	Moderate
Insensitivity to other impairments	Low	Good	Moderate	Low	Moderate
Modulation format transparency	No	Yes	Yes	Yes	No
Bit rate transparency	No	Yes	Yes	Yes	No
Multichannel	Yes	No	No	No	No
Multiparameter	Yes	No	Yes	Yes	No
Acquisition time	Moderate	Fast	Fast	Moderate	Fast

switching events. These transients can become critical in cascaded EDFAs. EDFAs are usually operated in saturation. In the strong saturation region, the output power is nearly constant and independent of the input power. The response time of an EDFA is on the order of microsecond to millisecond, hence high bit rate signals behave like CW power to the amplifier. When WDM channels are (randomly) turned on and off, the gain in an EDFA changes with the input power variations, which results in changes of the optical channel power and in dynamic power excursions of WDM channels. This can lead to distortions in WDM networks.

A method for automatic detection and classification of transient signals that is based on the Wavelet Network (WN) has been proposed in Refs [5,41,42]. The method combines the wavelet transform (WT) with neural networks. It allows determination of the source and evaluation of any fast power transient.

Wavelet transform [43], like the Fourier transform, is a mathematical tool for signal analysis. It is suitable for power disturbance and fault signal analysis. It provides a local representation (both in time and frequency) of a given signal. WT is used to transform the original fault signal into finer wavelet scales followed by a progressive search for the largest wavelet coefficients on that scale. Large wavelet coefficients that are colocated in time across different scales provide estimates of the changes in the signal parameter.

Signals can be efficiently described by the continuous-time wavelet transform (CWT) that enables the localization of low- and high-frequency components. CWT of a given square integrable function $f(t)$ with respect to a mother wavelet $\varphi_{(\alpha,b)}(t)$ is given by

$$W_{\mathrm{T}}(a,b) = \frac{1}{\sqrt{a}} \int_{-\infty}^{\infty} \varphi^{*}{}_{(a,b)}\left(\frac{t-b}{a}\right) f(t)\mathrm{d}t. \tag{8.12}$$

The scaling parameter α and the shifting parameter b are real numbers, and the asterisk * denotes complex conjugation.

The comparison of transients in the time domain and their CWTs shows that the position of a transient is determined by the largest coefficient of its CWT. Optimum CWT scaling is therefore essential. Large CWT scaling permits a more detailed resolution, but requires a larger number of CWT coefficients for transient classification and longer compute time. Small CWT scaling corresponds to a smaller number of CWT coefficients, resulting in potential information loss. With optimized CWT scaling containing a reasonable number of CWT coefficients and conserving a large amount of information, fast transients in WDM networks can be detected and classified with an accuracy of $\sim$90%.

8.2.4 Optical Performance Monitoring

BER and Q-factor are comprehensive measures of signal quality. The respective measurement techniques relate to OPM.

8.2.4.1 BER Monitoring Several methods have been proposed for BER monitoring.

Error Monitors Using Test Sequences. The use of known test sequences and the Monte Carlo method can provide accurate BER estimations. An implementation of the multicanonical Monte Carlo simulation method, which is an efficient tool to estimate the probability of rare events, for estimating the BER in phase-modulated optical systems is proposed in Ref. [44]. This method can be regarded as iterative importance sampling. It automatically converges toward the optimum bias, so it requires less *a priori* knowledge of the system than importance sampling. It has also been applied for BER estimation in OOK systems [45].

The method shows insensitivity to BER variations during information transmission. In addition, the acquisition speed is related to the bit rate.

Error Monitors Using Optical Noise Loading. This method is used to determine the system margin in optical amplifier systems, a problem that is strongly related to the beginning- and end-of-life BERs that are too low to measure [46]. This problem can be overcome by increasing the BER in a controlled way to measurable level and thus determining this margin indirectly. Measurable BERs are produced by intentionally adding ASE. This can be achieved by attenuating the transmitter launch power or with an additional ASE source at the receiver.

Indirect BER Measurement Through Q-Factor Monitoring. A method for indirectly measuring the BER by determining the Q-factor is proposed in Ref. [47]. This method overcomes the correlation problem between Q and BER by using voltage histograms from the center of the eye that are acquired with a digital sampling oscilloscope. It uses a decision circuit to probe the rails of the eye, and it exploits the complete data stream. The Q-factor is measured by recording the BER with respect to varying decision levels. From that, an amplitude histogram is built and used for analysis.

Further methods are based on eye diagram reconstruction using asynchronous undersampling [48,49]. A digital sampling oscilloscope acquires samples at a rate that is uncorrelated with the bit sequence [48]. The samples are used to reconstruct the eye diagram. Additional digital signal processing can be used to correct the position of each sample, which then allows full eye reconstruction.

8.2.4.2 Q-Factor Monitoring Q-factor estimation is used for performance monitoring. The related Q-factor monitoring methods mostly fall into the category of histogram analysis of sampled signal values.

Sampling methods explore statistical information of an optical signal to derive the histogram and the BER of a transmission channel, in addition to measuring parameters like CD, PMD, and cross talk. They can be divided into synchronous and asynchronous sampling.

Asynchronous sampling does not require clock recovery, but loses accuracy and impairment details. The signal level is sampled randomly throughout the bit stream, without regard to the sample instant relative to the decision time. No clock recovery is required, and the method is transparent to signal bit rate and data format [47,50,51].

Averaged Q-factor can be obtained from amplitude histograms. These can be obtained by asynchronous optical sampling [52–54]. Among the sampling points that build the histogram, those whose level is higher than a predetermined threshold $\mu_{1\text{th}}$ belong to mark level ("1"), while those whose level is lower than the threshold $\mu_{0\text{th}}$ belong to space level ("0"). The averaged Q-factor Q_{avg} is defined by

$$Q_{\text{avg}} = \left|\mu_{1,\text{avg}} - \mu_{0,\text{avg}}\right| / \left(\sigma_{1,\text{avg}} + \sigma_{0,\text{avg}}\right), \tag{8.13}$$

where $\mu_{i,\text{avg}}$ and $\sigma_{i,\text{avg}}$ are the mean and standard deviation of the mark and space level distributions, respectively. The data obtained by asynchronous sampling include unwanted cross-point data in the eye diagram that decrease the measured value of the averaged Q-factor. The two thresholds are set in order to remove the cross-point data.

Asynchronous delay-tap sampling combines asynchronous sampling with a two-tap delay line, so that each sample point comprises of two measurements (x and y), separated by a fixed delay. The samples are plotted as (x, y) pairs, producing a joint map of the power and its evolution over the delay time [55]. A block diagram for asynchronous delay-tap sampling is shown in Fig. 8.12.

Figure 8.13 shows an example of the process and the resulting two-tap scatter plot. This plot contains information about the pulse shape, in particular its rise and fall times. From this, impact of several impairments can be identified.

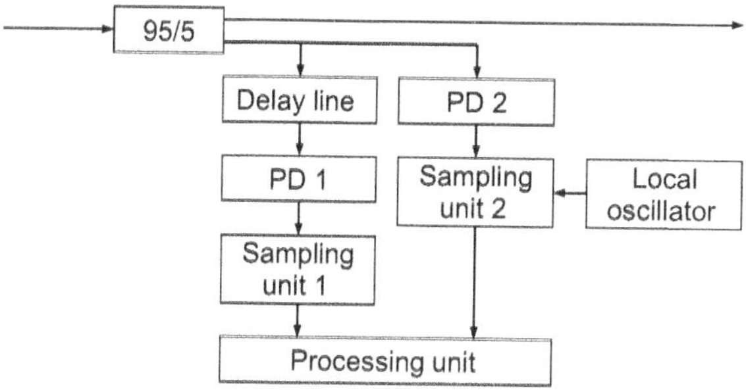

FIGURE 8.12 Block diagram of asynchronous delay-tap sampling.

In *synchronous sampling*, the signal is sampled only at the decision time, from which a histogram is constructed and analyzed [56,57]. This technique requires clock recovery and therefore is not bit rate transparent.

8.2.5 Fiber Monitoring

Fiber monitoring can be integrated into WDM systems. This holds for dual-fiber transport systems as well as for single fiber access systems. Fiber monitoring can be performed by optical time-domain reflectometers (OTDR) that can be fully integrated into the WDM system's management system [element management system (EMS) and network management system (NMS)] (see Chapter 7) and that can be remotely operated. In larger sites [central offices and points-of-presence (PoPs)], the respective OTDR can be switched onto the fiber to be tested by means of a small fiber cross-connect in order to save cost. Integration of an OTDR into a dual-fiber WDM transport system is shown in Fig. 8.14.

The OTDR is coupled to the fiber to be tested by means of a WDM filter. To allow in-service monitoring, the OTDR must use a wavelength not used by the WDM system, including its OSC wavelength. In most cases, an OTDR wavelength of 1625–1650 nm can be used. Then, the OTDR filter must allow low-loss coupling (<1.5 dB) of the OTDR wavelength, while at the same time having low pass-through

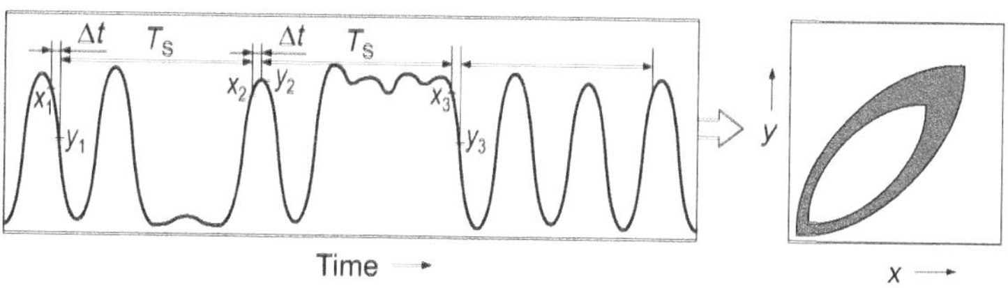

FIGURE 8.13 Asynchronous delay-tap sampling. T_S = sampling period, Δt = time offset within each sample. The two-tap scatter plot (right) shows y plotted against x for each sample pair, with $\Delta t = t_b/8$.

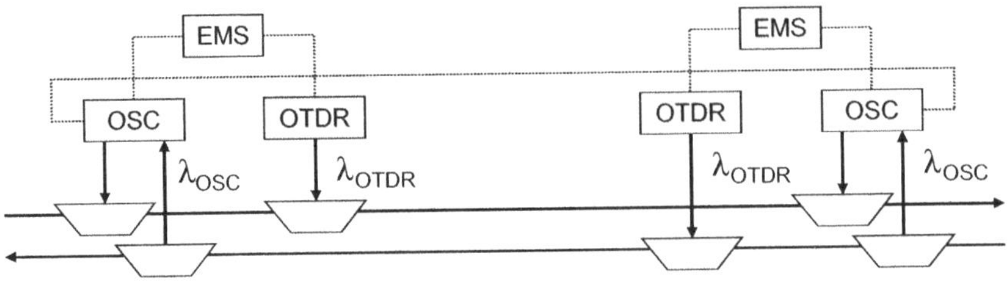

FIGURE 8.14 In-service OTDR monitoring, two-fiber WDM transport system.

loss (<1.0 dB) of the WDM wavelengths (C- or C/L-band). The OTDR connects to the DCN (OSC, as shown in Fig. 8.14), which then transmits the measurement data to the EMS or NMS (see Section 7.1). Note that according to Fig. 8.14, the OTDR signals will be blocked after one span by EDFA isolators (nonetheless, this configuration is in commercial use). To overcome this behavior, bypass techniques similar to the ones discussed in Section 6.5.8 have to be implemented. For further information on live testing of WDM systems fibers, refer to, for example, Ref. [58].

Passive optical network (PON) access fibers are another relevant area for on-demand fiber testing. OTDR measurements in conventional time-domain multiple access (TDMA)-PONs (GPON, XG-PON, EPON, and 10G-EPON) suffer from the fact that the OTDR pulses are broadcast via the PON power splitter/combiner. Should optical network units (ONUs) have similar distribution fiber lengths, their respective reflections cannot be separated anymore, leading to ambiguous results. This problem is inherently solved in a wavelength-routed WDM-PON. Now, the OTDR must be tunable across all ports of the filter (AWG) in the remote node. In-service OTDR monitoring can be performed if a wavelength band of the cyclic AWG is used, which is not in use for upstream/downstream traffic. Such integration of an OTDR into a single fiber WDM-PON is shown in Fig. 8.15 [59].

In a typical WDM-PON (downstream at L-band and upstream at C-band or vice versa), in-service monitoring can be done either in the U-band (above L-band, for example, 1650 nm like before) or in the O-band (the cyclic band around 1310 nm). As shown in Fig. 8.15, the OTDR is coupled to the respective feeder fiber in the PON optical line termination (OLT) by means of a low-loss WDM coupler. For monitoring, it must be tuned across all WDM ports sequentially. In large sites with a multitude of OLTs, the OTDR switches onto the respective WDM coupler by means of a fiber switch.

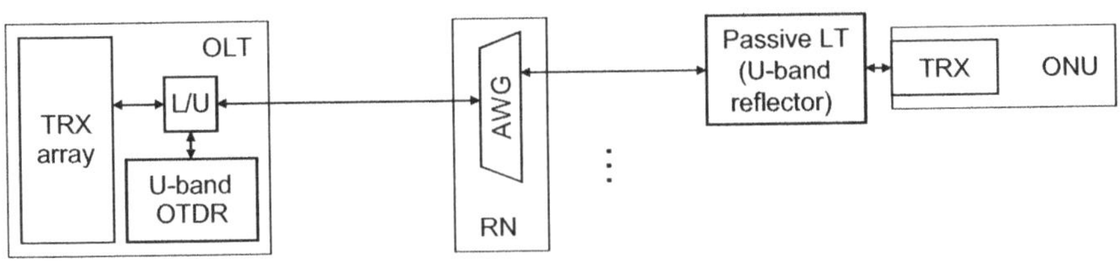

FIGURE 8.15 In-service OTDR monitoring and passive line termination (LT) in WR-WDM-PON.

8.3 FLEXIBLE WDM NETWORKS

Applications of flexibility in WDM networks differ significantly with respect to the speed that is required to perform the respective network reconfiguration. We separate slow, quasi-static, and fast reconfigurations. Quasi-static flexibility relates to (changes in the) network configurations that do not need to be performed faster than within minutes. The respective changes then may hold for hours, days, or even longer. Fast reconfigurations, within the scope of this book, relate to actions that are required in the context of resilience, that is, restoration or protection. Speed requirements range from several seconds (WDM restoration) to sub-60 ms (SONET/SDH-like WDM protection). These applications are covered in Chapter 9. We are not considering ultrafast WDM-based optical packet switching herein since this is no mature technique yet.

Some applications in flexible WDM networks require quasi-static configurations and reconfigurations, respectively. These applications include the network adaptation to (slowly) changing internal network and customers' needs. An example is reconfigurable core router bypass by means of a flexible WDM layer. The generic scenario is shown in Fig. 8.16.

Router bypass enables the offloading of routers from transit traffic. It allows higher router throughput or smaller routers in the respective sites. Proper interworking of the Multiprotocol Label Switching (MPLS) router layer and flexible WDM by means of a *multilayer control plane* must be ensured.

Another example of quasi-static flexibility that is provided by ROADMs is access to any WDM channel anywhere in large metro core rings with many ring nodes (i.e., reconfigurable single-channel add–drop). In such rings, frequent changes of the traffic pattern can occur. WDM layer flexibility automates these changes, thus reducing operational expenditures.

Flexible WDM paths not only reduce operational cost and help increase the client layer capacity but also enable higher transport capacity in WDM networks with given numbers of wavelengths and (add–drop or termination) nodes. This effect is visualized in Fig. 8.17.

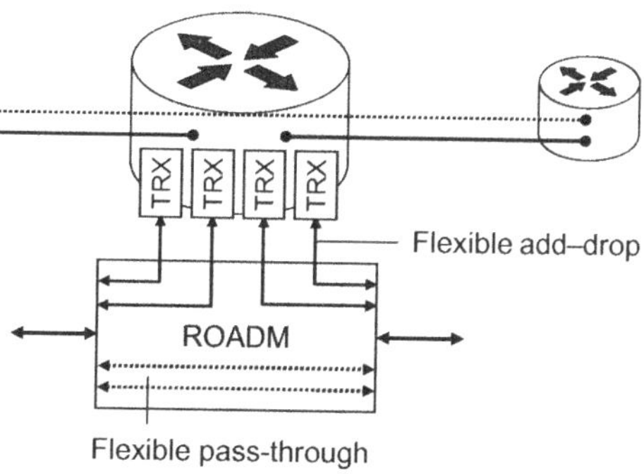

FIGURE 8.16 Flexible, on-demand router bypass.

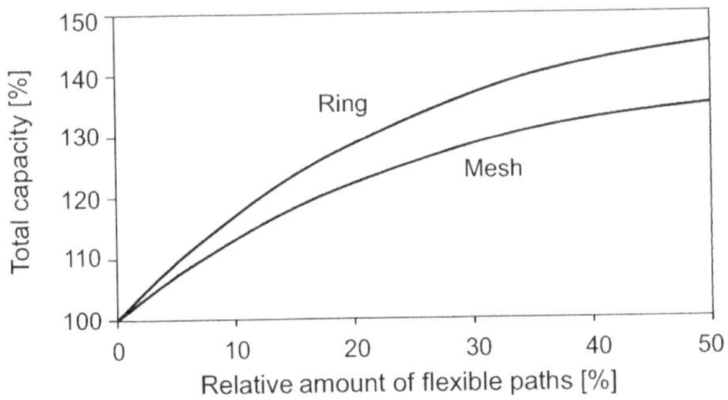

FIGURE 8.17 Capacity increase through flexible WDM paths.

Here, we assumed flexibility enabled by colorless, directionless ROADMs, supported by tunability of the transceivers within the full wavelength range of the ROADMs. With ROADMs enabling flexible single-channel add–drop, the capacity increase results from both the on-demand reconfigurations and the higher number of routable connections (equaling the number of wavelengths rather than wavelength bands). Similar effect can be seen in networks with protected services (see Chapter 9).

A further application of quasi-static WDM networks is the centralization of L3 routers or (FC) directors in regional networks in a single site (or possibly two sites for added redundancy). This obviously enables centralization of the L3 device that helps reducing the total network power consumption. It also helps reducing capacity expenditures because the WDM network part typically has a lifetime that exceeds the one of the client layer by a factor of 2–4. Hence, replacement of the client layer only requires the respective change in the central site. The remote (add–drop) sites can be left unaffected. A centralized Layer-3 scenario is shown in Fig. 8.18.

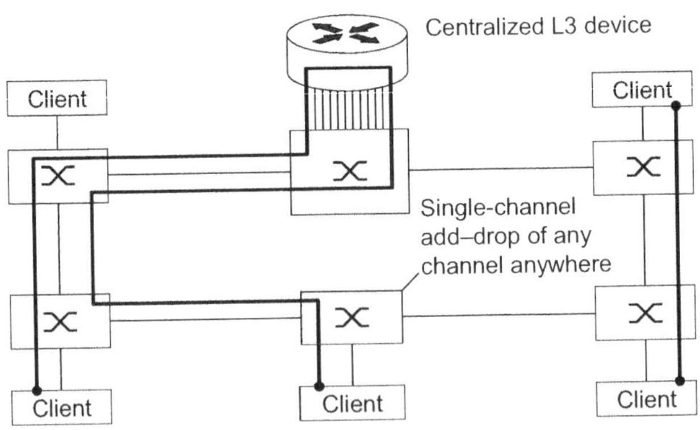

FIGURE 8.18 Centralized Layer-3, and flexible single-channel add–drop in a WDM ring.

Centralization of the client layer by means of flexible WDM helps reducing cost. One prerequisite is that the client layer lifetime is significantly shorter than that of the transport layer. This is typically the case in data center applications. Centralization is most efficient as long as the entire WDM network operates as a single transparent domain, that is, without 3R regeneration.

8.4 PROTECTION AND RESTORATION

8.4.1 Dedicated Versus Shared Protection Versus Restoration

In order to guarantee certain path availabilities and to allow related SLAs, WDM transport must be able to offer means for resilience. Availability and resilience range from unprotected links (no redundancy at all) up to client layer-protected links where the client equipment is fully duplicated and feeds to independent WDM transport links. In-between, several options for protection and restoration exist [60,61].

In *dedicated protection*, the necessary protection capacity (e.g., wavelengths) has been *calculated* and *assigned* prior to a failure. In *shared protection*, the respective path (again, for example, a wavelength) has at least been *precalculated*. It can also be preassigned, for example, to a certain group of wavelengths. In case of restoration, the path calculation can start prior to or after a failure has occurred. Assignment of the respective resource is always done after the failure. Restoration is therefore slower than dedicated protection and may perform similar to shared protection.

All dedicated protection mechanisms are able to detect failures and perform the protection switching, within 60 ms or less. Shared protection may take longer, for example, 200 ms, depending on assignment. Restoration in most cases requires >200 ms to establish the restoration path. In large networks, and without precalculation, it can take up to several (single-digit) minutes to establish the path through the network. It is sometimes used as an additional means for resilience, together with protection, to cope with multiple failures.

8.4.2 WDM Protection

WDM transport networks must be protected against fiber and equipment failures for cases where no client layer redundancy (protection and load sharing) is provided. In many cases, protection mechanisms are used that were copied from SONET/SDH. Here, the basic principles for ring (mesh) and end-to-end protection were developed. In $1+1$ protection, the respective signal is duplicated and transported via fully disjoint paths. This typically includes full $1+1$ interface (equipment) protection. In SDH, this was called SNCP. In SONET, a $1+1$-protected ring, the Unidirectional Path-Switched Ring (UPSR) was developed.

In $M:N$ protection, N services share M protection paths, thus decreasing the protection capacity ($M < N$). In SDH, the respective ring protection was named MS-SPRing (Multiplex Section Shared Protection Ring); in SONET, it was referred to as BLSR (Bidirectional Line-Switched Ring).

Overview of optical (WDM) protection is given in Refs [61–65]. In general, WDM protection must be characterized with regard to the following:

- Switching granularity: OCh versus OMS versus Optical (wavelength) Groups (OGr)
- Dedicated protection versus shared protection

Further differences (bidirectional vs. unidirectional rings and, independent of this, bidirectional vs. unidirectional switching) exist, but are less important.

Both dedicated and shared protection can make use of protection fibers or protection wavelengths. Shared protection rings can operate on OCh and OMS granularity, Dedicated protection rings in addition can work on the Optical Group level.

The simplest protection is applicable to point-to-point links. Regarding transport capacity, no shared protection can be applied to a point-to-point link. Shared protection can be applied as equipment protection to transceiver interfaces, which is rarely used in WDM transport.

On point-to-point links, four levels of dedicated protection can be implemented (refer to Fig. 8.19) . These are also applicable to ring protection and, partly, to mesh protection. In Fig. 8.19a, client layer protection is shown. Here, everything including the relevant parts of the client equipment, that is, transceivers and switch matrices, is duplicated. Client layer protection can achieve highest path availabilities, which can

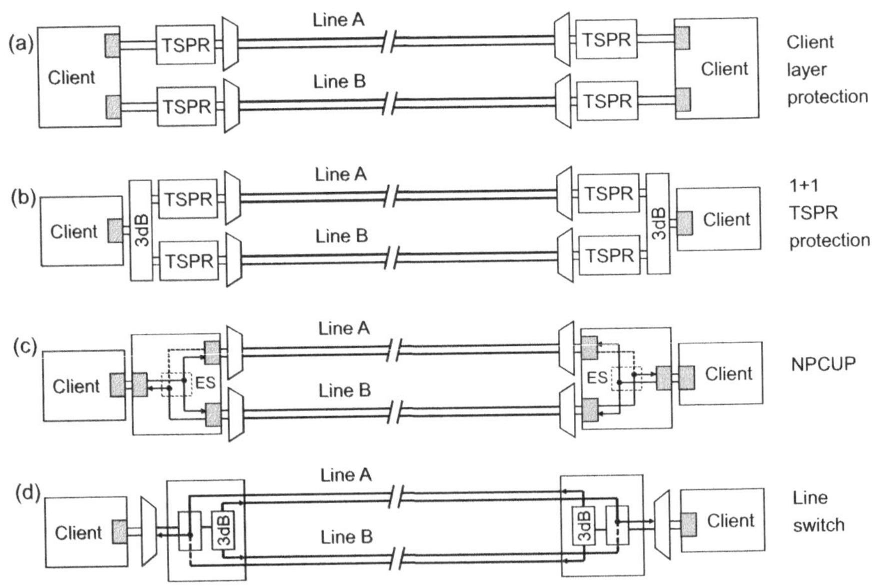

FIGURE 8.19 Point-to-point protection mechanisms. (a) Client layer protection with fully duplicated transponders. (b) 1 + 1 transponder protection. (c) Protection based on transponder with duplicated remote interfaces. (d) Line protection based on line switch and 3 dB coupler.

exceed, if combined with additional line switching, 99.9999%. Figure 8.19b shows 1 + 1 dedicated protection of the WDM transport, including transponder (TSPR, equipment) protection. It saves the duplicated client interfaces and is sufficient for the transport service provider to maintain highest SLAs. It allows to configure protection on a per-service level. Figure 8.19c shows dedicated channel protection where, on a transponder card, the WDM interfaces are duplicated [network-protected, client-unprotected (NPCUP)]. Since the client interface is not duplicated, this achieves somewhat lower availability, but offers lower cost and lower form factor. This protection scheme still offers the protection configuration on a per-channel basis. Figure 8.19d shows point-to-point protection based on fiber (line) switches. This scheme only allows switching the OMS. It can achieve significant availability improvement for cases where there is low fiber availability, for example, in urban areas. The Optical Group level typically is not supported on point-to-point links.

Early WDM protection rings have been described in Refs [66,67]. Optical protection rings are sometimes referred to as UPSR, BPSR, or BLSR. This is confusing since these names were used for SONET where they referred to time slots. For optical (WDM) rings, these names should be complemented by a trailing "O." Table 8.7 summarizes the resulting names for WDM protection rings.

This naming is identical to that in Refs [63,65], with the exception of group-switched rings. These were not yet considered since they have only been proposed later in Ref. [68].

Dedicated protection provides 100% redundant capacity for those signals that must be protected to maintain working capacity. Dedicated protection rings usually are two-fiber path-switched rings and can be further divided into unidirectional and bidirectional rings. They are the appropriate choice for hubbed ring traffic patterns. In this case, the maximum number of routable connections P between a hub and decentralized optical add–drop multiplexers (OADM) equals the number of wavelengths W:

$$P = W. \tag{8.14}$$

Dedicated ring protection is provided in most WDM systems today.

With shared protection, several working signals share a common protection capacity. In order for different signals to be able to share common protection capacities efficiently, *meshed traffic patterns* are required. In this case, shared protection can either save substantial CapEx or increase the total ring capacity. If

TABLE 8.7 Optical Protection Rings

	Dedicated	Shared (Bidirectional)
Optical Multiplex Section (line)	–	OMS-SPRing or O-BLSR
Optical wavelengths group	OGr-DPRing or O-UGSR/O-BGSR	–
Optical channel (path)	OCh-DPRing or O-UPSR/O-BPSR	OCh-SPRing or O-BPSR

only site-to-adjacent-site (S2AS) (best case) traffic is assumed, the maximum number of routable paths becomes

$$P = W \times N, \tag{8.15}$$

where N is the number of ring nodes. In Eqs. (8.14) and (8.15), W has to be replaced by the number of wavelength groups G for many older metro WDM ring systems where two-stage filters (based on TFF technology) with subsequent group splitters (GS) and per-group multiplexers and demultiplexers are used.

A 1 + 1 two-fiber OCh-DPRing (O-BPSR) is shown in Fig. 8.20. The WDM transport interfaces (transponders) are duplicated, they are fed by the client through power splitters. The duplicated interfaces feed into the east- and west-bound ring directions (clockwise and counterclockwise). In case a failure hits the working entity (fiber or transport equipment), the system switches to the protection path. The switchover itself can be managed directly between the two transponder pairs, it does not require action from a management system.

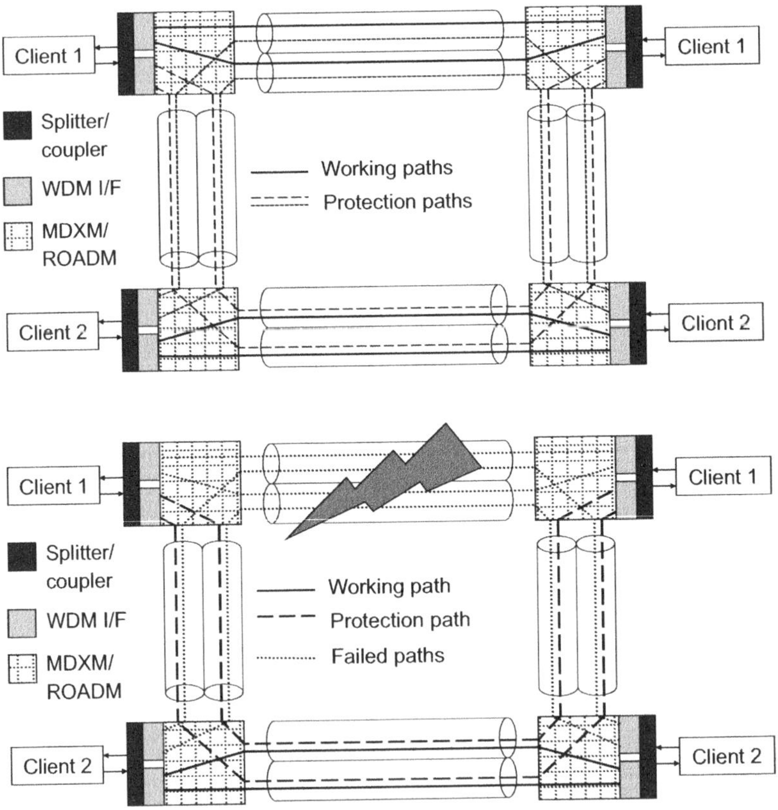

FIGURE 8.20 Protection event in a 1 + 1 OCh-DPRing.

A 1 + 1 DPRing often requires duplication of the transport interfaces, and it always blocks the protection capacity for payload traffic. Assignment of low-priority traffic to this protection capacity has been proposed; however, this traffic is lost in failure cases.

Shared protection can significantly increase ring capacity (or decrease cost associated with protection) in cases of meshed traffic patterns. In WDM SPRings, the basic idea is to provide protection wavelengths that are only used, that is, assigned to a failed working capacity, once a failure occurred. Since the respective assignments for all wavelengths and all ring spans are precalculated, switchover is still fast. A failure event in a two-fiber OCh-SPRing (O-BPSR) is shown in Fig. 8.21.

In WDM rings, OCh-SPRings have to make use of tunable interfaces to switch to the respective protection entities. Today, this is no restriction, but early proposals required switchover to dedicated protection transponders.

OCh-SPRings can be implemented as two-fiber rings. They support unprotected services and also 1 + 1 services on the same infrastructure. An important aspect is where the switchover is performed. It can be done by the transponders (end-to-end) or locally by the ROADMs that are next to the failed ring section. In the latter case,

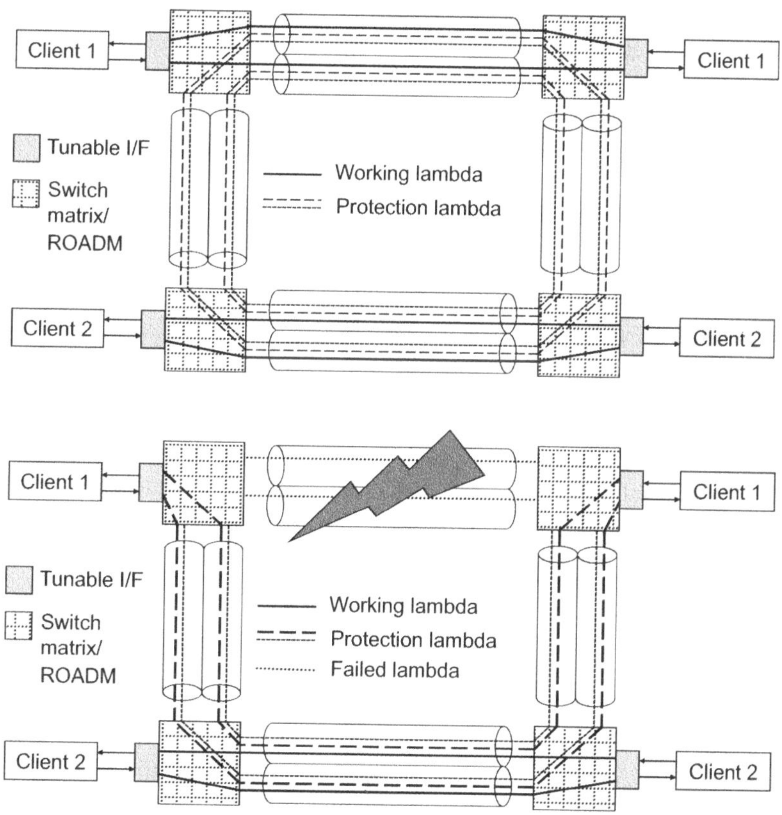

FIGURE 8.21 Protection event in an N:N OCh-SPRing.

long detours of the protection paths can result. Figure 8.22 shows one possible implementation of an OCh-SPRing with duplicated tunable lasers (WT/PT, working/protection tunables) and colorless (CL-) ROADMs.

Shared protection has also been proposed for the OMS [69]. This concept is based on using four fibers, instead of duplicating interfaces. Similar to a four-fiber SONET/SDH SPRing, ring nodes next to the failed fiber section switch the entire OMS from one fiber pair to the other. The concept is shown in Fig. 8.23.

In an OMS-SPRing, switchover is performed in the ring nodes (ROADMs) that are adjacent to the (span) failure. As shown in Fig. 8.23, this can lead to significant detour, depending on where the failure occurs. Maximum protection path length almost equals twice the ring circumference:

$$L_{\text{Prot}} = 2 \times C_{\text{Ring}} - 3 \times L_{\text{Span}}. \tag{8.16}$$

Here, L_{Prot} is the maximum protection link length, C_{Ring} is the ring circumference, and L_{Span} is the (mean) length of a single span, respectively. The same effect can be found in SDH and SONET line-switched rings. OCh-DPRings/O-BPSRs have maximum link lengths of one ring circumference minus one span.

Shared protection can lead to substantial increase of the total transport capacity of rings or meshes. For ideal site-to-adjacent-site traffic, the total ring capacity increases linearly with the number of ring nodes, independent of the number of WDM wavelengths. In a DPRing, total capacity is determined only by the number of wavelengths (see Fig. 8.24).

Metro WDM systems either use full-band ROADMs (or as the static option, AWGs) or dual-stage filters where the optical spectrum is first spilt into groups of wavelengths, followed by per-group multiplexing/demultiplexing. The wavelength group level can be used for cost-effective implementation of certain functionalities, for example, flexibility or dedicated protection [68,70,71]. The group level is the trade-off between path and line level concerning costs and granularity. It avoids couplers/switches or duplicated transponders for every wavelength channel without switching all channels (i.e., the OMS) at once. An implementation of an OGr-DPRing/O-BGSR OADM is shown in Fig. 8.25.

The group-level signals are fed into 3 dB splitters and transmitted into the clockwise and counterclockwise ring directions via GS. In the receive direction, both signals are combined and monitored by an optical switch (the monitoring points are denoted by MP in Fig. 8.25). If the working signal fails, autonomous switchover to the protection group is performed without signaling via the (ring) management system.

A comparison of the resulting path availabilities for dedicated and shared protection on the channel, group, and OMS level is given in Fig. 8.26.

Depending on implementation details, the OCh-DPRing and OCh-SPRing have cost relative to the OMS-SPRing and OGr-DPRing in the range of 150%. The main assumptions behind Fig. 8.26 regarding components availabilities are listed in Table 8.8.

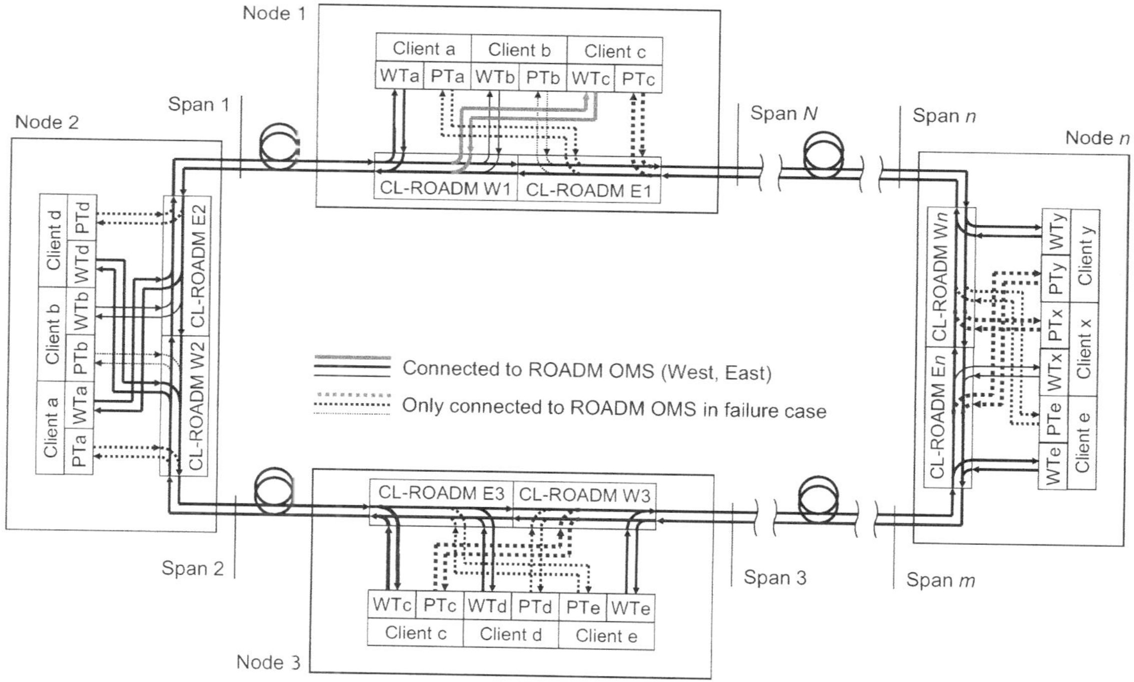

FIGURE 8.22 Implementation of an OCh-SPRing with Degree-2 ROADMs.

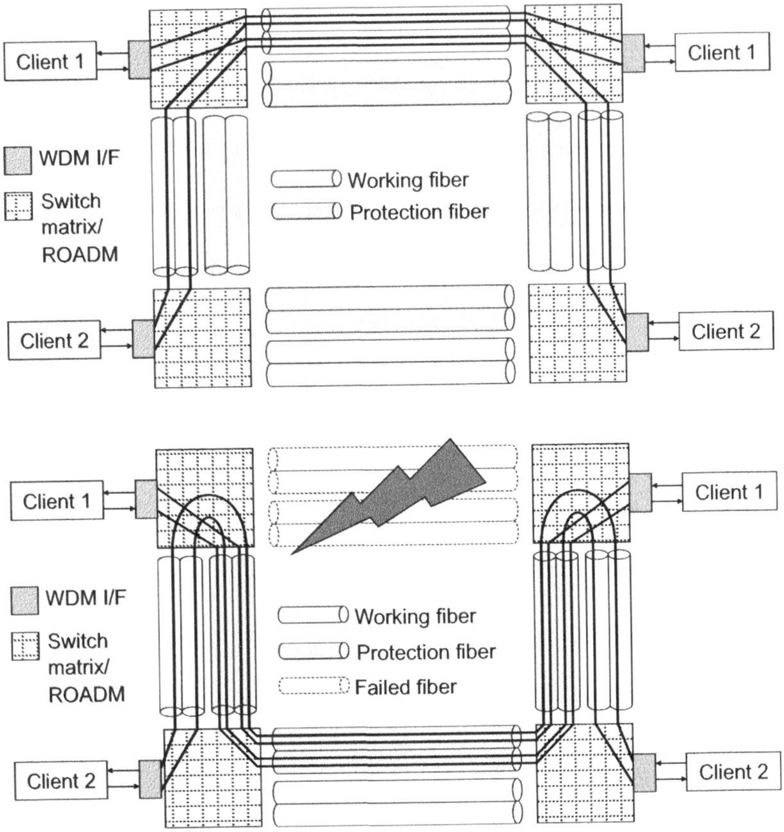

FIGURE 8.23 Protection event in a four-fiber OMS-SPRing.

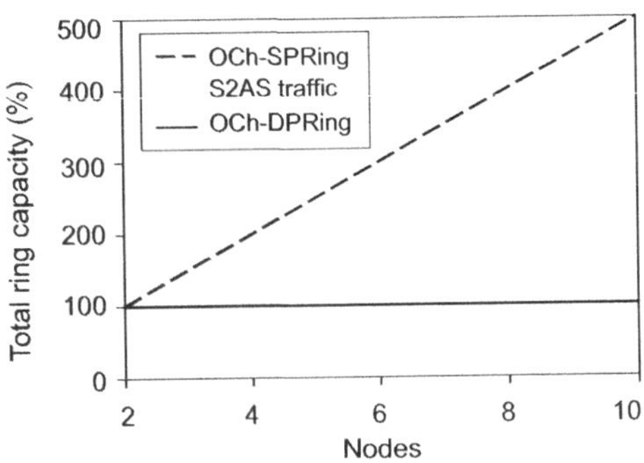

FIGURE 8.24 Total WDM ring capacity for dedicated versus shared protection.

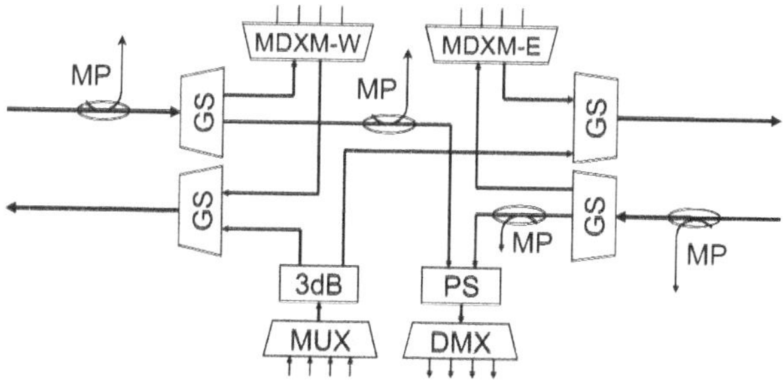

FIGURE 8.25 OADM of an optical bidirectional group-switched ring. MP: monitoring port.

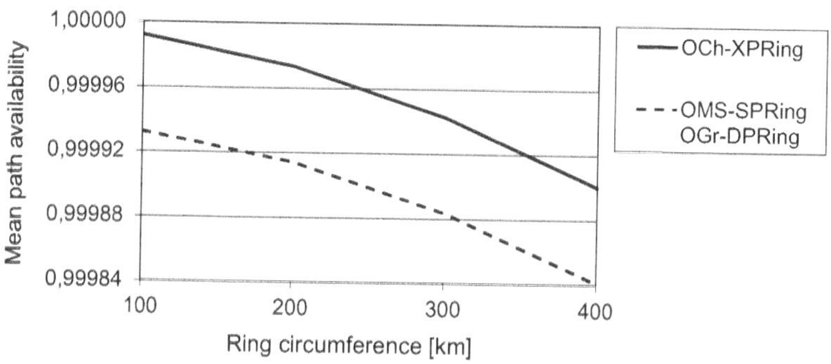

FIGURE 8.26 Path availabilities in shared and dedicated protection rings.

TABLE 8.8 Component Availabilities

Component	Availability
Chassis w/dual PSU	0.999997
EDFA	0.999993
Tunable laser card	0.999996
Passive filters	0.999999
Fiber switch	0.999987
WSS-based ROADM	0.999984
Tunable uncooled transceiver w/APD	0.999989
40-Ch transceiver array w/PINs	0.999990
Urban fiber, per kilometer	0.999943

Components that are not service-affecting have not been considered in the calculations leading to Fig. 8.26. This holds, for example, for management interface cards. For more details regarding WDM ring protection availability, refer to Refs [72,73].

8.4.3 WDM-PON Protection

WDM-PONs may also have to be protected like other WDM links. Reason is that in a WDM-PON roll-out with metro network site reduction (Section 6.6.1), many feeder fibers of different PONs may be space-multiplexed in shared ducts. Hence, civil work interrupting that duct may affect up to multiple tens of thousands of customers. Then, the respective systems need to be protected. Since (optical) residential access faces severe cost pressure (target hardware cost for an end-to-end [PoP/CO-to-client] connection is <$200), 1 + 1 protection with interface duplication is prohibited in almost all cases. Furthermore, protection of the distribution fibers is either too expensive due to the amount of required additional fibers or impossible because redundant fibers or disjoint paths are not available. Then, fiber protection between the OLT and the remote node is required, which covers the majority of link length and in particular those fibers that are multiplexed in ducts or common cables. Examples of the need for PON end-to-end protection include backhaul and business access.

OLT-RN fiber protection can be implemented in WDM-PONs using (cyclic) 2:*N* AWGs. The second AWG port carrying the OMS does not cause additional insertion loss. (This is similar to OLT-RN protection in power-split EPON/GPON using 2:*N* power splitters. Here, the second OMS port at the RN also does not cause additional loss.)

WDM-PON with seeded/reflective transmitters or tunable lasers must be considered separately. For seeded/reflective WDM-PONs, a protection scheme has been described in Ref. [74]. It is shown in Fig. 8.27a. Both, ONUs and the OLT transmitter array (TRX array) are seeded with different broadband light sources (BLS1 and BLS2). In case of a protection switch event, both BLSs are switched simultaneously at both AWGs by means of two fiber switches SW1 and SW2. Switching the BLSs at both AWGs automatically compensates the wavelength shift that is unavoidable in a 2:*N* AWG (refer to Fig. 3.69).

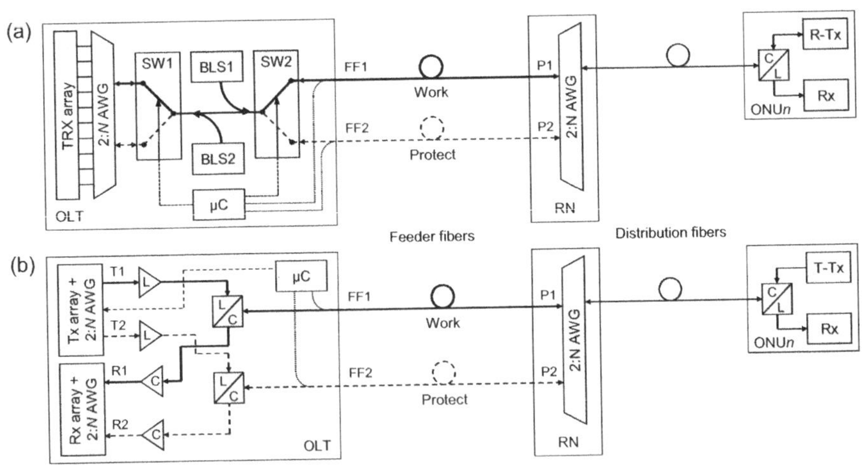

FIGURE 8.27 WDM-PON protection using 2:*N* cyclic AWGs. (a) Seeded/reflective PON. (b) Tunables.

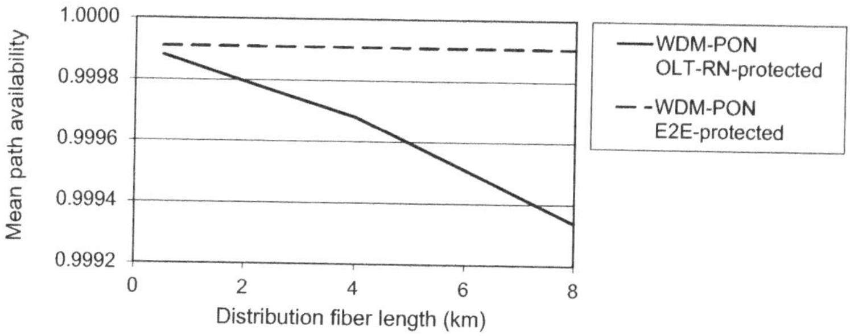

FIGURE 8.28 WDM-PON mean path availability.

A similar protection scheme is shown in Fig. 8.27b for WDM-PONs based on tunable lasers. Since no BLSs are used, the fiber switches can be omitted completely. Again, switching from one OMS fiber to the other produces a shift of the respective output ports, for example, by one wavelength channel. Hence, protection *switching is performed by tuning* all ports at the OLT and all ONUs by the respective wavelength increment. Note that no switches are required since at any of the wavelength settings (incrementally tuned or original wavelength set), one of the OMS ports always stays dark. The tuning (e.g., by ± 1 channel) is simple since the lasers have to be tuned/wavelength-controlled anyway. However, it may take >60 ms to perform the tuning for all WDM channels.

Finally, we consider the path availability that can be achieved by OLT-RN fiber switching. The unavailability obviously heavily depends on the length of the unprotected distribution fiber as the most unavailable element in the transmission link. The resulting end-to-end path availability for WDM-PONs with OLT-RN and end-to-end protection is shown in Fig. 8.28.

In Fig. 8.28, availability figures from Table 8.8 have been used.

Further WDM-PON protection schemes, in particular for end-to-end protection and covering both seeded/reflective and tunable laser approaches, are described in Refs [75–77].

8.4.4 Restoration

Restoration relates to fast (WDM) network reconfigurations in order to establish resilient paths through a network after a failure has occurred. Restoration paths can be precalculated, or path calculation only starts after occurrence of the failure. Path assignment is always done after the failure occurred. Hence, in most cases, it is not possible to establish the resilient path within 60 ms, as known from SONET/SDH and also WDM/OTN protection rings. For very sensitive, mission-critical services, restoration can be used as an addition to protection, in order to increase resilience in multiple failure scenarios.

Fast restoration can be achieved with preconfigured protection cycles or p-cycles [78]. P-cycles are efficient for CapEx minimization, and they enable (shared-) ring-like switchover speed. Precalculation must be done in nonreal time. If preassignment is also provided, the restoration scheme turns into shared mesh protection. A problem of p-cycle planning is the lack of scalability [61].

Early work on WDM restorations dates back to the late 1990s [79–81]. Switching of individual wavelengths or OMS was already considered, as well as restoration with or without wavelength conversion.

Regarding efficiency, restoration performs similar to shared protection. CapEx savings or increase of total network capacity strongly depends on the traffic pattern and also on the degree of meshing. In the real-world, partly meshed (both physically and logically) networks, restoration should always be somewhat more efficient than dedicated protection. In Ref. [80], a requirement for an additional 20–30% of wavelengths to fully restore the logical connectivity has been found, depending on the traffic load. Cost savings were derived in Ref. [79] for switching the OMS rather than single wavelengths.

In large WDM networks, restoration may lead to resilient paths that exceed the transparent field length of the respective WDM transceivers. In such networks, restoration requires pools of 3R regenerators (which can also act as wavelength converters) that are placed at certain central sites within the network. Placement of these regenerators is one optimization problem in WDM networks with restoration.

In (partly) meshed WDM networks, restoration requires directionless ROADMs. In order to reduce network blocking probability (i.e., the probability that the wavelength cannot be routed onto the resilient path because it is in use there already), the ROADMs should also be colorless and contentionless.

Consider the meshed WDM network shown in Fig. 8.29. The nodes consist of colorless directionless ROADMs, and client interfaces are full-band tunable. In addition to client WDM interfaces, regenerators may be placed at certain nodes in order to allow longer detours or midlink wavelength conversion for cases where no wavelength can be routed end-to-end anymore without conversion. Here, we consider O/E/O-type regenerators, rather than all-optical ones. Nodes can have different degrees: there are nodes of Degree-3 and Degree-5.

Now, link 1 between two end (E) nodes fails, as indicated in Fig. 8.30a. Links 3 and 5 carry the same wavelengths as link 1, indicated by the bold lines. However, these links still have resilience capacity in the form of other unused wavelengths.

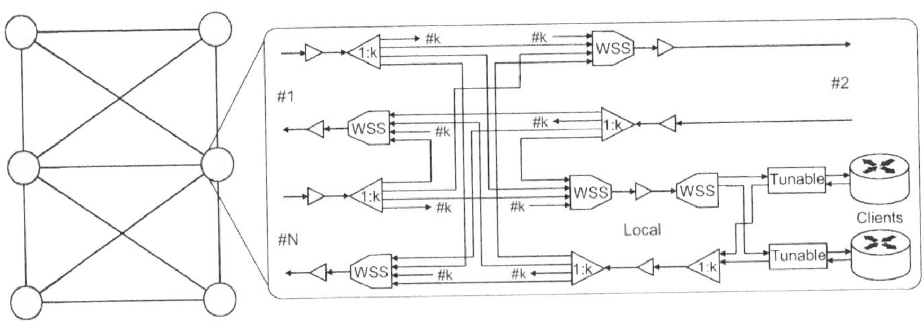

FIGURE 8.29 Colorless/directionless ROADM in meshed network suitable for restoration.

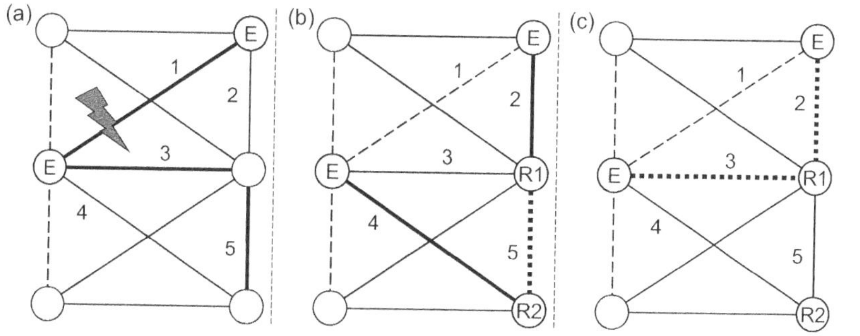

FIGURE 8.30 Restoration with directionless interfaces (a): colored (b) versus colorless (c).

Unnumbered links do not have this capacity. With colored ROADM interfaces, successful establishment of a resilient path requires regenerator nodes (R1 and R2) that perform additional wavelength conversion, as shown in Fig. 8.30b. If the end (E) nodes were also colorless, they could establish a new path by tuning to the respective wavelength. In our example, this path runs via R1, but does not require its regeneration capability.

Efficient restoration in a WDM network requires colorless, directionless, contentionless ROADMs. The WDM interfaces must be able to cope with different amounts of transmission constraints, in particular dispersion and nonlinearity. CD must be compensated per span such that the resulting total residual dispersion is within the WDM transceiver limits, or the transceivers must have adaptive CD compensation.

For intradyne transceivers with blind CD compensation, this can lead to short added latency in path establishment. Similar adaptation may be required for PMD changes (which is covered by the adaptive PMD compensation and polarization demultiplexing) and also for SPM/XPM compensation (e.g., adaptation of the backward propagation compensation). In addition, pools of regenerators may have to be placed carefully in certain central sites, depending on total network dimension and longest potential detours.

REFERENCES

1. S. Gorshe, A Tutorial on ITU-T G.709 Optical Transport Networks (OTN), Technology White Paper, PMC-Sierra, Inc., PMC-2081250, Issue 1, 2010.
2. E. Hernandez-Valencia et al., The Generic Framing Procedure (GFP): an overview, IEEE Commun. Mag., Vol. 40, No. 5, May 2002, pp. 63–71.
3. P. Bonenfant, A. Rodriguez-Moral, Generic Framing Procedure (GFP): the catalyst for efficient data over transport, IEEE Commun. Mag., Vol. 40, No. 5, May 2002, pp. 72–79.
4. P. Bonenfant, A. Rodriguez-Moral, Framing techniques for IP over fiber, IEEE Netw., Vol. 15, No. 4, July/August 2001, pp. 12–18.
5. EU, FP7 project DICONET, Deliverable D3.1, 2008. Available online.

6. F. Leplingard et al., Determination of the impact of a quality of transmission estimator margin on the dimensioning of an optical network, OFC'08, San Diego, March 2008, Paper OWA6.
7. T. Zami et al., Driving technologies addressing the future dynamic transparent core networks, International Conference on Transparent Optical Network (ICTON) 2008, Athens, June 2008, Paper Th.A2.4.
8. M.N. Petersen et al., Online chromatic dispersion monitoring and compensation using a single inband subcarrier tone, IEEE Photonics Technol. Lett., Vol. 14, No. 4, 2002, 570–572.
9. K.J. Park et al., Chromatic dispersion monitoring technique in WDM network, OFC'02, Anaheim, March 2002, Paper ThGG88.
10. T.E. Dimmick et al., Optical dispersion monitoring technique using double sideband subcarriers, IEEE Photonics Technol. Lett., Vol. 12, No. 7, 2000, 900–902.
11. C. Youn Effects of SPM and PMD on chromatic dispersion monitoring techniques using pilot tones, OFC'03, Atlanta, March 2003, Paper WP2.
12. K.J. Park et al., Performance comparisons of chromatic dispersion-monitoring techniques using pilot tones, IEEE Photonics Technol. Lett., Vol. 15, No. 6, 2003, 873–875.
13. N. Liu et al., Chromatic dispersion monitoring using the power ratio of two RF tones with a dispersion offset, OFC'04, Los Angeles, March 2004, Paper MF81.
14. N. Liu et al., PMD and chirp effects suppression in RF tone-based chromatic dispersion monitoring, IEEE Photonics Technol. Lett., Vol. 18, No. 5, 2006, 673–675.
15. A.L. Campillo, Chromatic dispersion-monitoring technique based on phase-sensitive detection, IEEE Photonics Technol. Lett., Vol. 17, No. 6, 2005, 1241–1243.
16. Y.-C. Ku et al., Chromatic dispersion monitoring technique using birefringent fiber loop, OFC'06, Anaheim, March 2006, Paper OFN2.
17. Qian Yu et al., Chromatic dispersion monitoring technique using sideband optical filtering and clock phase-shift detection, IEEE J. Lightwave Technol., Vol. 20, No. 12, 2002, 2267–2271.
18. T. Inui et al., 160-Gb/s adaptive dispersion equalization using an asynchronous dispersion-induced chirp monitor, IEEE J. Lightwave Technol., Vol. 23, No. 6, 2005, pp. 2039–2045.
19. T. Luo et al., PMD monitoring by tracking the chromatic-dispersion-insensitive RF power of the vestigial sideband, IEEE Photonics Technol. Lett., Vol. 16, September 2004, pp. 2177–2179.
20. G. Ishikawa et al., Polarization-mode dispersion sensitivity and monitoring in 40-Gb/s OTDM and 10-Gb/s NRZ transmission experiments, OFC '98, San Jose, February 1998, Paper WC5.
21. S.M.R. Motaghian Nezam et al., PMD monitoring in WDM systems for NRZ data using a chromatic-dispersion-regenerated clock, OFC'02, Anaheim, March 2002, Paper WE5.
22. Z. Pan et al., Chromatic dispersion monitoring and automated compensation for NRZ and RZ data using clock regeneration and fading without adding signaling, OFC'01, Anaheim, March 2001, Paper WH5.
23. Y.K. Lizé et al., Independent and simultaneous monitoring of chromatic and polarization-mode dispersion in OOK and DPSK transmission, IEEE Photonics Technol. Lett., Vol. 19, No. 1, 2007.

24. F. Roy et al., A simple dynamic polarization mode dispersion compensator, OFC/IOOC'99, San Diego, March 1999, Vol. 1, pp. 275–278.

25. S.M.R. Motaghian Nezam et al., Wide-dynamic-range DGD monitoring by partial optical signal spectrum DOP measurement, OFC'02, Anaheim, March 2002, p. FD8-1.

26. N. Kikuchi, Analysis of signal degree of polarization degradation used as control signal for optical polarization mode dispersion compensation, IEEE J. Lightwave Technol., Vol. 19, No. 4, 2001, 480–486.

27. F. Buchalli et al., A 40 Gb/s eye monitor and its application to adaptive PMD compensation, OFC'02, Anaheim, March 2002, Paper WE6.

28. W. Moench, J.Y. Larikova, and (Tellebas Inc.), In-Service Measurement of the OSNR in ROADM-based Networks, White Paper, JDSU, 2012. Available at www.jdsu.com/ProductLiterature/ecoc_wp_fop_tm_ae.pdf.

29. K. Asahi et al., Optical performance monitor built into EDFA repeaters for WDM networks, OFC'98, San Jose, February 1998, Paper Th02.

30. H. Suzuki, N. Takachio, Optical signal quality monitoring built into WDM linear repeaters using semiconductor arrayed waveguide grating filter monolithically integrated with eight photodiodes, Electron. Lett., Vol. 35, 1999. pp. 836–837.

31. Wenlu Chen et al., Adding OSNR monitoring functionality on AWG based power monitoring circuits, OFC'02, Anaheim, March 2002, Paper ThGG40.

32. W. Moench, J. Y. Larikova, Measuring the optical signal-to-noise ratio in agile optical networks, OFC/NFOEC'07, Anaheim, March 2007, Paper NWC1.

33. J.H. Lee et al., OSNR monitoring technique using polarization-nulling method, IEEE Photonics Technol. Lett., Vol. 13, No. 1, 2001, 176–178.

34. Wei Chen, Signal processing for optical performance monitoring and impairment mitigation, PhD thesis, Department of Electrical and Electronic Engineering, University of Melbourne, Australia, 2006. Available at www.ee.unimelb.edu.au/research/cubin/downloads/cubin_Wei_Chen_thesis.pdf.

35. L. Baker-Meflah et al., In-band OSNR monitoring using spectral analysis after frequency down-conversion, IEEE Photonics Technol. Lett., Vol. 19, No. 2, 2007, 115–117.

36. X. Chongjin et al., Orthogonal-polarization heterodyne OSNR monitoring insensitive to polarization-mode dispersion and nonlinear polarization scattering, IEEE J. Lightwave Technol., Vol. 25, No. 1, 2007, pp. 177–183.

37. E. Wong et al., Novel scheme for simultaneous polarisation mode dispersion and optical signal-to-noise ratio monitoring, ECOC 2005, Glasgow, September 2005, Vol. 3, Paper We4.P.098.

38. M. Rasztovits-Wiech et al., Optical signal-to-noise ratio measurement in WDM networks using polarization extinction, ECOC'98, Madrid, September 1998, pp. 549–550.

39. R. Adams et al., All-optical in-band OSNR monitoring at 40 Gb/s using a nonlinear optical loop mirror, IEEE Photonics Technol. Lett., Vol. 18, No. 3, 2006, pp. 469–471.

40. P. Vorreau et al., Optical noise and dispersion monitoring with SOA-based optical 2R regenerator, IEEE Photonics Technol. Lett., Vol. 17, No. 1, 2005, pp. 244–246.

41. Y. Ben-Ezra et al., Automatic detection and classification of power transients in optical communication networks, J. Opt. Netw., Vol. 5, No. 7, 2006, pp. 568–574.

42. Y. Ben-Ezra et al., Application of wavelet networks for identification of transients in optical networks, International Conference on Transparent Optical Networks, ICTON 2006, Nottingham, June 2006, Vol. 1, Paper Tu.C1.1, pp. 111–114.
43. M.B. Ruskai, Wavelets and Their Applications, Jones & Bartlett, Boston, 1992.
44. Y. Yadin and M. Orenstein, Bit error rate estimation of DPSK modulated fiber-optic systems using multicanonical Monte-Carlo simulations, OFC'05, Anaheim, March 2005, Paper OThC5.
45. R. Holzlohner, C.R. Menyuk, Use of multicanonical Monte Carlo simulations to obtain accurate bit error rates in optical communications systems, Opt. Lett., Vol. 28, No. 20, 2003.
46. V.J. Mazurczyk et al., Using optical noise loading to estimate margin in optical amplifier systems, OFC '97, Dallas, February 1997, Paper TuP5.
47. N.S. Bergano et al., Margin measurements in optical amplifier system, IEEE Photonics Technol. Lett., Vol. 5, No. 3, 1993, pp. 304–306.
48. L. Noirie et al., New transparent optical monitoring of the eye and BER using asynchronous under-sampling of the signal, ECOC'02, Copenhagen, September 2002, Paper PD2.2.
49. E. Mobilon et al., Experimental verification of an eye diagram reconstruction technique based on asynchronous undersampling, International Conference on Microwave and Optoelectronics 2005, SBMO/IEEE MTT-S, July 2005, pp. 603–606.
50. M. Fregolent et al., Adaptive optical receiver for performance monitoring and electronic mitigation of transmission impairments, ECOC'2000, Munich, September, 2000, Vol. 1, pp. 63–64.
51. C.M. Weinert, Histogram method for performance monitoring of the optical channel, ECOC'2000, Munich, September 2000, Vol. 4, pp. 121–122.
52. S.V. Kartalopoulos, Optical Bit Error Rate: An Estimation Methodology, IEEE Press, 2004.
53. N. Kikuchi et al., Performance of chromatic monitoring using statistical moments of asynchronously sampled waveform histograms, IEEE Photonics Technol. Lett., Vol. 17, No. 5, 2005, pp. 1103–1105.
54. I. Shake, H. Takara, Averaged Q-factor method using amplitude histogram evaluation for transparent monitoring of optical signal-to-noise ratio degradation in optical transmission system, IEEE J. Lightwave Technol., Vol. 20, No. 8, 2002, pp. 1367–1373.
55. S.D. Dods, T.B. Anderson, Optical performance monitoring technique using delay tap asynchronous waveform sampling, OFC'06, Anaheim, March 2006, Paper OThP5.
56. R. Wiesmann et al., Cost effective performance monitoring in WDM systems, OFC'2000, Baltimore, March 2000, Vol. 2, Paper WK2, pp. 171–173.
57. S. Ohteru, N. Takachio, Optical signal quality monitor using direct Q-factor measurement, IEEE Photonics Technol. Lett., Vol. 11, No. 10, 1999, pp. 1307–1308.
58. H. Chen et al., Live-fiber OTDR testing: traffic and measurement impairments, EXFO Technical Note 022. Available at documents.exfo.com/appnotes/tnote022-ang.pdf.
59. M. Roppelt et al., Efficient Broad- and Unicast in WDM-PON, 5. ITG-Fachkonferenz Breitbandversorgung in Deutschland, Berlin, March 2011.
60. M. Berry et al., A Classification Model for Network Survivability Mechanisms, ITG Fachtagung Photonische Netze, Leipzig, 2004.
61. G. Maier et al., Optical network survivability: protection techniques in the WDM layer, Photonics Netw. Commun., Vol. 4, No. 3–4, 2002, pp. 251–269.
62. P. Arijs et al., Architecture and design of optical channel protected ring networks. IEEE J. Lightwave Technol., Vol. 19, No. 1, 2001, pp. 11–22.

63. O. Gerstel, R. Ramaswami, Optical layer survivability: a services perspective. IEEE Commun. Mag., Vol. 38, No. 3, 2000, pp. 104–113.
64. P. Arijs et al., Design of ring and mesh based WDM transport networks, Opt. Netw. Mag., 1, No. 2, July 2000, pp. 25–40.
65. P. Bonenfant et al., The evolution of transport network survivability, IEEE Commun. Mag., Vol. 37, No. 8, 1999, pp. 44–51.
66. L. Gillner et al., Field trial of a European self-healing WDM optical network, OFC2000, 2000, Paper TuQ1-1.
67. F. Arecco et al., A transparent, all-optical, metropolitan network experiment in a field environment: the "PROMETEO" self-healing ring, IEEE J. Lightwave Technol., Vol. 15, No. 12, 1997, pp. 2206–2213.
68. M. Schlie, K. Grobe, Optical group-ring networks, 18th NFOEC, Dallas, TX, USA, September 2002.
69. X. Fang et al., All-optical four-fiber bidirectional line-switched ring, IEEE J. Lightwave Technol., Vol. 17, No. 8, 1999, pp. 1302–1308.
70. J.J. Pan et al., Scalable EDFAs simplify metro design, Lightwave Mag., Vol. 4, No. 7, July 2002, pp. 15–19.
71. J. Sauer, Flexible OADMs manage metro traffic, Lightwave Mag., Vol. 4, No. 8, August 2002, pp. 17–20.
72. I. Rados, Availability models for wavelength channel protection in WDM ring networks, Electron. Electr. Eng., Vol. 8, No. 88, 2008, pp. 81–86.
73. I. Rados, Availability improvement of WDM systems using ring structures, J. Electr. Eng., Vol. 62, No. 5, 2011, pp. 297–301.
74. K. Lee et al., A self-restorable architecture for bidirectional wavelength-division-multiplexed passive optical network with colorless ONUs, Opt. Express, Vol. 15, No. 8, April 2007, pp. 4863–4868.
75. B. Zhang et al., A survivable WDM passive optical network with colorless optical network units, OECC2006, Kaohsiung, June 2006, Paper 5E2-2-1.
76. E.S. Son et al., Survivable network architectures for WDMPON, Proceedings of Optical Fiber Communication Conference, Optical Society of America, Anaheim, CA, 2005, Paper OFI4.
77. C.C.K. Chan et al., Novel network architectures for survivable WDM passive optical networks, ECOC 2008, Brussels, September 2008, Paper Th.1.F.6.
78. W.D. Grover, D. Stamatelakis, Cycle-oriented distributed pre-configuration: ring-like speed with mesh-like capacity for self-planning network reconfiguration, Proceedings of IEEE ICC'98, Atlanta, Vol. 1, June 1998, pp. 537–543.
79. E. Karasan, E. Goldstein, Optical restoration at the wavelength-multiplex-section level in WDM mesh networks, IEEE Photonics Technol. Lett., Vol. 10, No. 9, 1998, pp. 1343–1345.
80. S. Baroni, P. Bayvel, Link failure restoration in WDM optical transport networks and the effect of wavelength conversion, OFC1997, Dallas, 1997, pp. 123–124.
81. Chuan-Ching Sue, Sy-Yen Kuo, Restoration from multiple faults in WDM networks without wavelength conversion, Lect. Notes Comput. Sci., Vol. 2093, 2001, pp. 317–325. DOI: 10.1007/3-540-47728-4_31.

9

STANDARDS RELEVANT FOR WDM

9.1 ITU-T RECOMMENDATIONS

The *ITU Telecommunication Standardization Sector* (ITU-T) is one of the three divisions of the International Telecommunication Union (ITU). ITU standardization dates back to 1865, when the International Telegraph Union was founded. It became the CCITT (Comité Consultatif International Téléphonique et Télégraphique) in 1956 and was renamed ITU-T in 1993.

The ITU-T provides standards covering all fields of telecommunications on an international basis. The resulting standards are referred to as *Recommendations* (capitalized), they become mandatory only when adopted by national law.

The technical work of ITU-T is managed by Study Groups (SGs). SGs work on different aspects such as signaling, performance monitoring, next-generation networks, and environment and climate change. The SG most relevant to wavelength division multiplexing (WDM) is SG15: Optical and Other Transport Networks. SG15 works on technologies ranging from DSL and PON (passive optical network) via SDH to WDM and Automatically Switched Optical Networks (ASONs).

ITU-T cooperates with other standardization development organizations (SDOs), for example, the Internet Engineering Task Force (IETF) or International Organization for Standardization (ISO).

Recommendations are organized in the so-called series, for example, G-series or M-series. They have names like G.600, where 600 is an identifying number. G.600 is used to refer to both the specific G.600 Recommendation and the G.6xx family of

Wavelength Division Multiplexing: A Practical Engineering Guide, First Edition.
Klaus Grobe and Michael Eiselt.

TABLE 9.1 G-Series Recommendations Families

Specification Range	Specification Type
G.600–G.699	Transmission media and optical systems characteristics
G.700–G.799	Digital terminal equipment
G.800–G.899	Digital Networks
G.900–G.999	Digital sections and digital line systems
G.7000–G.7999	Data over transport—generic aspects
G.8000–G.8999	Packet over transport aspects
G.9000–G.9999	Access Networks

Recommendations. In the WDM context, the most relevant series is the G-series. Further Recommendations can be found in the M-series (telecommunication management) and Y-series (next-generation networks, among others). An overview of the G-series families most relevant to WDM is given in Table 9.1.

Examples of the first family are the G.65x Recommendations describing (single-mode) fibers. Relevant Recommendations for OTN (and also Y-series examples) include G.709/Y.1331, G.798, G.871/Y.1301, G.872, G.873.1, G.874, G.874.1, G.875, G.959.1, and G.8251. Recommendations G.694.1 and G.694.2 describe spectral [dense wavelength division multiplexing (DWDM) and coarse wavelength division multiplexing (CWDM)] grids. Further relevant Recommendations include G.696.1, G.698.1, G.698.2, and G.698.3 (system aspects with regard to various DWDM applications) and G.973 and G.977 (submarine DWDM systems).

Further information can be found at http://www.itu.int/ITU-T/.

9.2 OTHERS

9.2.1 FSAN (NG-PON2)

Full Service Access Network (FSAN) drives applicable standards, where they already exist, into the services and products in the industry. Simultaneously, new technical specifications for PONs with the goal of better defining the FSAN are developed. Care is taken to ensure that there is no overlap or duplicative activity with other organizations. Companies that are members of FSAN can bring in technical contributions to FSAN as company contributions aligned with the current Call for Contributions for meetings following current study topics of FSAN.

FSAN is not an SDO by itself. However, designated FSAN member companies serve as editors for FSAN documents and typically offer to submit contributions to ITU-T SG 15 Q2, based on completed FSAN specification.

The single FSAN Working Group (WG) is called Optical Access Network (OAN). Within OAN, there are active Task Groups working on interoperability and next-generation PON (NG-PON). Additional Working Groups and study topics in FSAN can be initiated as required by FSAN participants.

The NG-PON Task Group deals with post-G-PON optical access. It develops carrier class solutions, with the capability of smooth migration between system generations and the reuse of the legacy fiber plant. The first output of the Task Group has been the specification of the XG-PON1 system. After finalizing the XG-PON1 basic specification, these were handed over to ITU-T for standardization as the G.987 series.

Under the term NG-PON2, the Task Group started investigations on technologies enabling further bandwidth increase. This work considered both, WDM stacking of various XG-PONs and WDM/WDMA overlay for point-to-point dedicated access.

Further information can be found at http://fsanweb.com/.

9.2.2 OIF

The *Optical Internetworking Forum* (OIF) promotes the development and deployment of interoperable networking solutions and services through the creation of so-called Implementation Agreements (IAs) for optical networking products, network processing elements, and component technologies. IAs are based on requirements developed cooperatively by end users, service providers, equipment vendors, and technology providers. They are aligned with worldwide standards, and augmented if necessary. This is accomplished through industry member participation working together to develop specifications (IAs) for the following:

- External network element interfaces
- Software interfaces internal to network elements
- Hardware component interfaces internal to network elements

A relevant recent example of OIF activities is the coherent DP-QPSK standard for remote 100 Gb/s WDM interfaces. OIF has also done relevant work on the control plane interfaces that are in use in WDM networks.

The OIF performs interoperability testing and builds market awareness and promotes education for technologies, services, and solutions. It provides feedback to other standards organizations to help achieve a set of implementable, interoperable solutions.

Further information can be found at http://www.oifforum.com.

9.2.3 IEEE-SA

IEEE stands for the *Institute of Electrical and Electronics Engineers*, and is the association's full legal name. Among others, IEEE provides a range of standards. The IEEE Standards Association (IEEE-SA) is one of the leading consensus-building bodies with regard to global technologies. Participants (both IEEE-SA members and nonmembers) from more than 160 countries work on functionalities, capabilities, and interoperability of a wide range of products and services.

Telecommunications-related areas that are covered by IEEE standardization activities include (but are not restricted to) Antennas & Propagation, Communications, Computer Technology, Electromagnetic Compatibility, Instrumentation & Measurement, and Smart Grid. Other areas include (and again are not restricted to) National Electrical Safety Code, Green & Clean Technology, Nuclear Power, and Power & Energy.

The IEEE standards that are most relevant to WDM (apart from general electrical safety, electromagnetic compatibility, and similar aspects) belong to the IEEE 802 family. These are standards dealing with local and metropolitan area networks. More specifically, they are restricted to networks carrying variable-size packets. The services and protocols specified in IEEE 802 map to the lower two layers of the OSI reference model. In fact, IEEE 802 splits the data link layer into two sublayers: logical link control (LLC) and media access control (MAC). The related work started in February 1980.

Two Working Groups are of specific interest in the WDM context: the IEEE 802.1 Working Group on Bridging and Management and the IEEE 802.3 Working Group on Ethernet.

The IEEE 802.1 Working Group develops standards in the areas of LAN/MAN architecture, internetworking among LANs, MANs, and Wide Area Networks (WANs), security, overall network management, and protocol layers above the MAC and LLC layers. One recently active area was data center bridging.

IEEE 802.3 is working on the physical layer MAC of wired Ethernet. This covers LAN and some MAN/WAN applications. It considers various types of copper or fiber cables. The respective standards describe the client most relevant to WDM networking.

Further information can be found at http://www.ieee.org/about/index.html.

9.2.4 IETF

The IETF develops and promotes Internet standards, cooperating closely with the W3C (World Wide Web Consortium) and ISO/IEC standards bodies. It deals specifically with standards of the TCP/IP and Internet protocol suite.

The IETF is an open standards organization, with no formal membership or membership requirements. It is an organized activity of the Internet Society (ISOC). ISOC is a nonprofit organization founded in 1992. It is dedicated to the open development, evolution, and use of the Internet.

IETF Working Groups are the primary mechanism for the development of IETF specifications and guidelines, many of which are intended to be standards or recommendations. WGs are created to address specific problems. Upon completion of its goals and achievement of its objectives, the Working Group is terminated.

The basic definition of the IETF standards process is in RFC 2026 (RFC stands for Request for Comments). This document has been amended several times, so other documents (RFC 5378, RFC 3979, and RFC 3932) now also describe parts of the standards process. The Internet Standards Process is intended to balance the conflicting goals of technical excellence, clear, concise, and easy understanding of the documentation, openness and fairness, and timeliness.

The IETF RFCs relevant to WDM networks relate to the Generalized Multiprotocol Label Switching (GMPLS) control plane. GMPLS grew out of Multiprotocol Label Switching (MPLS). A flavor of MPLS (MPLS-TE) provided end-to-end connections using signaling with constraint-based routing. This was extended to cover circuit-oriented optical switching technologies such as DWDM. As GMPLS appeared on the scene, the CCAMP (Common Control and Management Plane) WG was created.

Further information can be found at http://www.ietf.org/about/.

9.2.5 IEC

Founded in 1906, the IEC (*International Electrotechnical Commission*) is one of the world's leading organizations for the preparation and publication of international standards for electrical, electronic, and related technologies.

All IEC International Standards are fully consensus based and represent the needs of key stakeholders of nations participating in IEC work. Every member country has one vote and influence on what goes into an IEC International Standard. Experts from industry, commerce, government, test and research labs, academia, and consumer groups participate in IEC Standardization work.

The IEC is one of three global sister organizations (IEC, ISO, and ITU) that develop International Standards for the world. When appropriate, IEC cooperates with ISO or ITU to ensure that International Standards fit together seamlessly and complement each other.

IEC standards relevant for WDM include IEC 825-1 on laser safety.

Further information can be found at http://www.iec.ch.

9.2.6 Others (INCITS T.11 and IBTA)

The *International Committee for Information Technology Standards* (INCITS) is a forum of IT developers accredited by ANSI (American National Standards Institute). It was formerly known as the X3 and NCITS. INCITS technical standard groups and technical committees have provided various well-known standards. In particular, INCITS/T11 (X3T9.3) is working on *Fibre Channel* standardization.

INCITS coordinates technical standards activity between ANSI in the United States and joint ISO/IEC committees worldwide. This provides a mechanism to create standards that will be implemented in many nations.

The forum was established in 1960 as the "Accredited Standards Committee X3, Information Technology" and is sponsored by Information Technology Industry Council (ITI), a trade association representing providers of information technology products and services then known as the Business Equipment Manufacturers' Association (BEMA) and later renamed Computer and Business Equipment Manufacturers' Association (CBEMA). X3 was established under ANSI procedures. The forum was renamed Accredited Standards Committee NCITS, National Committee for Information Technology Standards in 1997, and the current name was approved in 2001.

Further information can be found at http://www.t11.org/ and at http://www.incits.org/.

The *InfiniBand® Trade Association* (IBTA) was founded in 1999 and is chartered with maintaining and furthering the InfiniBand™ specification. The IBTA is led by a steering committee that includes IBM, Intel, Mellanox, Oracle, and QLogic. Other members of the IBTA represent enterprise IT vendors who are actively contributing to the advancement of the InfiniBand™ specification.

The IBTA is responsible for compliance and interoperability testing of commercial products and has successfully added hundreds of products to its Integrators' List. It also promotes InfiniBand™ technology from an industry perspective through online, marketing, and public relations engagements and technical events and resources.

Further information can be found at http://www.infinibandta.org/.

10

PRACTICAL APPROXIMATIONS AND TIPS

10.1 CONVERSION BETWEEN BIT ERROR RATE AND EQUIVALENT Q-FACTOR

The exact relation between the Q-factor and the BER is given in Eq. (5.20) as

$$\text{BER} = 0.5 \cdot \text{erfc}\left(\frac{Q}{\sqrt{2}}\right).$$

As has been already discussed in previous chapters, the resulting BER is correct only if the noise distribution is Gaussian. On the other hand, the Q-factor is a more convenient parameter than the BER, as the latter varies by orders of magnitude. Therefore, a measured BER is often converted into an "equivalent Q-factor" using the equation above. This equivalent Q-factor would yield the measured BER, if the noise distribution were Gaussian. As the complementary error function is often not implemented on scientific calculators or in spreadsheet programs, the following formulas only use simple functions to approximate the relation in both directions. Please note that $Q_{\text{dB}} = 20\log_{10}(Q)$ is used in the formulas:

$$\log_{10}(\text{BER}) \approx -\left[\frac{44.2}{47.1 - Q_{\text{dB}}}\right]^{6.5}, \tag{10.1}$$

$$Q_{\text{dB}} \approx 47.1 - 44.2[\log_{10}(\text{BER})]^{-0.154}. \tag{10.2}$$

Wavelength Division Multiplexing: A Practical Engineering Guide, First Edition.
Klaus Grobe and Michael Eiselt.

The error for converting BER to Q is less than 0.1 dB for bit error ratios between 10^{-20} and 1.8×10^{-2}, and the error for the conversion from Q to BER is less than 20% of the BER for a Q-factor between 3.5 and 17.8 dB.

10.2 PROPERTIES OF A PRBS SIGNAL

For testing purposes, a pseudorandom bit sequence (PRBS) instead of "real" data is often transmitted over a link. As a main property of a PRBS data stream, all n-bit combinations of "0" and "1" appear in a PRBS-n sequence—except for all-zeros or all-ones, depending on the implementation. The number n determines the length of the sequence to $2^n - 1$. A variety of pseudorandom bit sequences are used, most prominently PRBS-7, PRBS-9, PRBS-11, PRBS-15, PRBS-23, and PRBS-31. ITU-T Recommendations o.150 [1] specifies the generation and properties of the most commonly used sequences, except for PRBS-7, while the latter one is in ITU-T V.29 [2]. The PRBS-n is generated as a shift register of length n by feedback of the XOR combination between the last bit n and one other bit, as shown in Fig. 10.1. In Table 10.1, the bits used for feedback according to the ITU-T Recommendations are shown.

Due to the nature of the PRBS data, the spectrum of the modulated data is equivalent to the spectrum of a single data pulse, sampled at the repetition rate of the PRBS word. The lowest frequency content is therefore found at $f_1 = R_B/(2^n - 1)$, where R_B is the data rate. While the lowest content of a PRBS-7-modulated 10 Gb/s

TABLE 10.1 Properties of Most Commonly Used PRBS Sequences

PRBS-n	XOR Combined Bits		Sequence Length (bits)	Sequence Length (Time) for 10 Gb/s	Lowest Frequency Content (10 Gb/s)
7	6	7	127	12.7 ns	78.74 MHz
9	5	9	511	51.1 ns	19.57 MHz
11	9	11	2,047	204.7 ns	4.885 MHz
15	14	15	32,676	3.268 μs	306.0 kHz
23	18	23	8,388,607	838.9 μs	1.192 kHz
31	28	31	2,147,483,647	214.7 ms	4.658 Hz

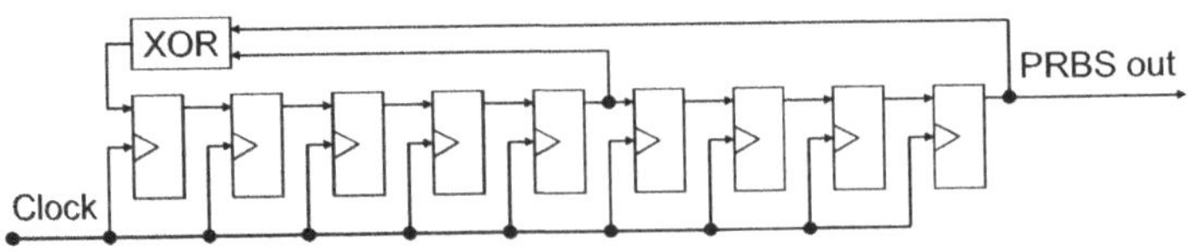

FIGURE 10.1 Schematic generation of a PRBS-9 sequence.

data stream is found at 78.7 MHz, a modulation with PRBS-31 yields a spectral line as low as 4.7 Hz. Usually, optical transmitters are AC coupled with a lower frequency cutoff around 100 kHz. The modulation with a PRBS-31 therefore results in some baseline wander over the duration of the PRBS word.

The low frequency content of the longer PRBS words is equivalent to a slow variation of the signal power over time. This can lead to certain unexpected effects. A variation on the timescale of the inversion lifetime in an Erbium-doped fiber amplifier (EDFA) (see Section 3.2.4.1) leads to a gain variation over time. A similar effect was observed in the Raman cross talk between the channels of an experimental WDM system, where all channels were modulated with the same PRBS-n bit sequences [3]. Even though chromatic dispersion leads to walk-off between the channels, the low-frequency power variations resulted in a temporal variation of the signal power at the receiver.

10.3 CHROMATIC DISPERSION VALUES AND PROPAGATION CONSTANTS

The propagation constant $\beta(\omega)$ (in 1/m) can be developed in a Taylor series around ω_0:

$$\beta(\omega) = \beta_0 + \beta_1(\omega - \omega_0) + \frac{\beta_2}{2}(\omega - \omega_0)^2 + \frac{\beta_3}{6}(\omega - \omega_0)^3 + \cdots. \qquad (10.3)$$

The expansion coefficients β_i are derivatives of the propagation constants with respect to the circular frequency ω:

$$\beta_i = \left.\frac{\mathrm{d}^i\beta(\omega)}{\mathrm{d}\omega^i}\right|_{\omega_0}. \qquad (10.4)$$

The propagation delay τ in a fiber of length L can be expressed in terms of these coefficients:

$$\tau = \beta_1 L. \qquad (10.5)$$

The derivative of the propagation delay with respect to the wavelength is referred to as chromatic dispersion D. It is related to the propagation coefficient β_2 as

$$D = L\frac{\mathrm{d}\beta_1}{\mathrm{d}\lambda} = L\cdot\frac{\mathrm{d}\beta_1}{\mathrm{d}\omega}\cdot\frac{\mathrm{d}\omega}{\mathrm{d}\lambda} = -\frac{2\pi cL}{\lambda^2}\cdot\beta_2. \qquad (10.6)$$

At a wavelength of 1550 nm, a chromatic dispersion D of 16 ps/(nm km) corresponds to an expansion coefficient β_2 of $-20.4\,\mathrm{ps^2/km}$.

REFERENCES

1. Recommendation ITU-T O.150: general requirements for instrumentation for performance measurements on digital transmission equipment, May 1996.
2. Recommendation ITU-T V.29: 9600 bits per second modem standardized for use on point-to-point 4-wire leased telephone-type circuits, November 1988.
3. M. Eiselt et al., Raman induced cross-talk with PRBS modulation, OFC2003, March 2003, Paper WE6.

INDEX

Wavelength Division Multiplexing: A Practical Engineering Guide, First Edition.
Klaus Grobe and Michael Eiselt.

WILEY SERIES IN PURE AND APPLIED OPTICS

Founded by Stanley S. Ballard, University of Florida

EDITOR: Glenn Boreman, University of North Carolina at Charlotte

ASSANTO · *Nematicons: Spatial Optical Solitons in Nematic Liquid Crystlas*
BARRETT AND MYERS · *Foundations of Image Science*
BEISER · *Holographic Scanning*
BERGER-SCHUNN · *Practical Color Measurement*
BOYD · *Radiometry and The Detection of Optical Radiation*
BUCK · *Fundamentals of Optical Fibers*, Second Edition
CATHEY · *Optical Information Processing and Holography*
CHUANG · *Physics of Optoelectronic Devices*
DELONE AND KRAINOV · *Fundamentals of Nonlinear Optics of Atomic Gases*
DERENIAK AND BOREMAN · *Infrared Detectors and Systems*
DERENIAK AND CROWE · *Optical Radiation Detectors*
DE VANY · *Master Optical Techniques*
ERSOY · *Diffraction, Fourier Optics and Imaging*
GASKILL · *Linear Systems, Fourier Transform, and Optics*
GOODMAN · *Statistical Optics*
GROBE AND EISELT · *Wavelength Division Multiplexing: A Practical Engineering Guide*
HOBBS · *Building Electro-Optical Systems: Making It All Work*, Second Edition
HUDSON · *Infrared System Engineering*
IIZUKA · *Elements of Photonics, Volume I: In Free Space and Special Media*
IIZUKA · *Elements of Photonics, Volume II: For Fiber and Integrated Optics*
JUDD AND WYSZECKI · *Color in Business, Science, and Industry*, Third Edition
KAFRI AND GLATT · *The Physics of Moire Metrology*
KAROW · *Fabrication Methods for Precision Optics*
KLEIN AND FURTAK · *Optics*, Second Edition
MA AND ARCE · *Computational Lithography*
MALACARA · *Optical Shop Testing*, Third Edition
MILONNI AND EBERLY · *Lasers*
NASSAU · *The Physics and Chemistry of Color: The Fifteen Causes of Color*, Second Edition
NIETO-VESPERINAS · *Scattering and Diffraction in Physical Optics*
OSCHE · *Optical Detection Theory for Laser Applications*
O'SHEA · *Elements of Modern Optical Design*
OZAKTAS · *The Fractional Fourier Transform*
PRATHER, SHI, SHARKAWY, MURAKOWSKI, AND SCHNEIDER · *Photonic Crystals: Theory, Applications, and Fabrication*
SALEH AND TEICH · *Fundamentals of Photonics*, Second Edition
SCHUBERT AND WILHELMI · *Nonlinear Optics and Quantum Electronics*
SHEN · *The Principles of Nonlinear Optics*
STEGEMAN AND STEGEMAN · *Nonlinear Optics: Phenomena, Materials, and Devices*
UDD · *Fiber Optic Sensors: An Introduction for Engineers and Scientists*
UDD · *Fiber Optic Smart Structures*
VANDERLUGT · *Optical Signal Processing*
VEST · *Holographic Interferometry*
VINCENT · *Fundamentals of Infrared Detector Operation and Testing*
WEINER · *Ultrafast Optics*
WILLIAMS AND BECKLUND · *Introduction to the Optical Transfer Function*
WU AND REN · *Introduction to Adaptive Lenses*
WYSZECKI AND STILES · *Color Science: Concepts and Methods, Quantitative Data and Formulae*, Second Edition

XU AND STROUD · *Acousto-Optic Devices*
YAMAMOTO · *Coherence, Amplification, and Quantum Effects in Semiconductor Lasers*
YARIV AND YEH · *Optical Waves in Crystals*
YEH · *Optical Waves in Layered Media*
YEH · *Introduction to Photorefractive Nonlinear Optics*
YEH AND GU · *Optics of Liquid Crystal Displays,* Second Edition

Printed and bound by CPI Group (UK) Ltd, Croydon, CR0 4YY
16/09/2026
14963413-0005